Radio Networks and Communication Systems

Radio Networks and Communication Systems

Edited by **Joe Myers**

WILLFORD PRESS

New York

Published by Willford Press,
118-35 Queens Blvd., Suite 400,
Forest Hills, NY 11375, USA
www.willfordpress.com

Radio Networks and Communication Systems
Edited by Joe Myers

International Standard Book Number: 978-1-68285-287-3 (Hardback)

Printed in China.

Contents

Preface

The world is advancing at a fast pace like never before. Therefore, the need is to keep up with the latest developments. This book was an idea that came to fruition when the specialists in the area realized the need to coordinate together and document essential themes in the subject. That's when I was requested to be the editor. Editing this book has been an honour as it brings together diverse authors researching on different streams of the field. The book collates essential materials contributed by veterans in the area which can be utilized by students and researchers alike.

Radio communication systems are based on the transmission and processing of radio waves and signals. This book is an assimilation of concepts such as radio networks and frequencies, wireless sensor networks, modulation of communication channels, etc. with reference to radio communication systems. There has been rapid progress in this field and its applications are finding their way across multiple industries. With state-of-the-art inputs by acclaimed experts of this field, this book targets students and professionals alike.

Each chapter is a sole-standing publication that reflects each author´s interpretation. Thus, the book displays a multi-facetted picture of our current understanding of applications and diverse aspects of the field. I would like to thank the contributors of this book and my family for their endless support.

Editor

1

Entropie analysis of floating car data systems

F. Gössel, E. Michler, and B. Wrase

Dresden University of Technology, Fakultät für Verkehrswissenschaften „Friedrich List“, Lehrstuhl Informationstechnik für Verkehrssysteme, D-01062 Dresden, Germany

Abstract. The knowledge of the actual traffic state is a basic prerequisite of modern traffic telematic systems. Floating Car Data (FCD) systems are becoming more and more important for the provision of actual and reliable traffic data. In these systems the vehicle velocity is the original variable for the evaluation of the current traffic condition. As real FCD-systems are operating under conditions of limited transmission and processing capacity the analysis of the original variable vehicle speed is of special interest. Entropy considerations are especially useful for the deduction of fundamental restrictions and limitations.

The paper analyses velocity-time profiles by means of information entropy. It emphasises in quantification of the information content of velocity-time profiles and the discussion of entropy dynamic in velocity-time profiles. Investigations are based on empirical data derived during field trials. The analysis of entropy dynamic is carried out in two different ways. On one hand velocity differences within a certain interval of time are used, on the other hand the transinformation between velocities in certain time distances was evaluated.

One important result is an optimal sample-rate for the detection of velocity data in FCD-systems. The influence of spatial segmentation and of different states of traffic was discussed.

1 Introduction

The initial point of the historical development and original area of application of the statistical information theory is communications engineering. But already Shannon suggested the application of many conclusions from information theory in other areas of science and technology (Ebeling et al., 1998; Olberg and Rákóczi, 1984). Thereby the objective is not a formal transformation of termini from one area of science into another but to discover new interrelations, to generalise and extend the applicability of known relationships and to find principles of classification and systematisation. In this way the following information-entropical approach on FCD-systems should be understood. The usage of information theory in traffic science is not new. Already Potthoff used the entropy and transinformation to analyse the structure of railway systems (Potthoff, 1972).

The original variable in FCD-systems (Michler, 2001; Gössel et al., 2002) is the vehicle velocity. The vehicle velocity can be measured directly and is suitable for the detection of the actual traffic state. Certainly vehicle velocity can be regarded as a random variable. From this point of view the information source “vehicle velocity” will provide a certain amount of information. As real FCD-systems are operating under conditions of limited transmission and processing capacity the analysis of velocity-time resp. velocity-position profiles is of special interest. Entropy considerations thereby are especially useful for the deduction of fundamental restrictions and limitations.

In the following the amount of information in velocity-time profiles will be quantitatively analysed. Thereby entropy dynamic is of special interest. Results of the analysis are consisting of an estimation of the data rate in FCD-systems, possibilities and limitations in efficient coding of velocity data and the choice of an optimal sample rate.

2 Consideration of velocity-time resp. velocity-position profiles as sources of information

The vehicle velocity is a time and value continuous variable. Entropy analysis of value continuous variables usually require the knowledge of the probability-density function of this variable. Another possibility consists in sampling and discretisation of the velocity profile and consideration of the velocity profile as a sequence of symbols. The entropy of a discrete source X with n independent symbols x_i is given by:

$$H(X) = -\sum_{i=1}^{n} p(x_i)\ ldp\ (x_i), \qquad (1)$$

Correspondence to: F. Gössel
(goessel@vini.vkw.tu-dresden.de)

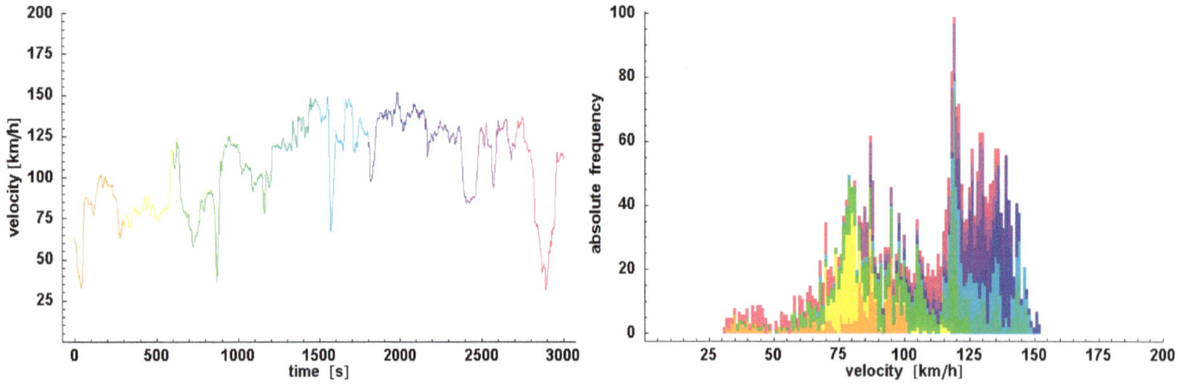

Fig. 1. Exemplary representation of a velocity-time profile (left) and the corresponding absolute frequency distribution, colours indicating segmentation (duration 300 s), quantisation 1 km/h.

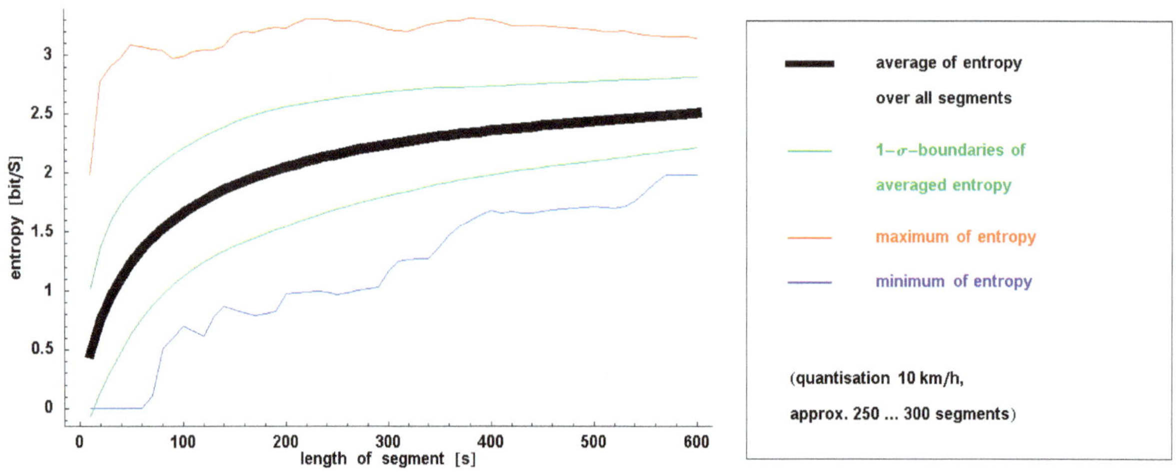

Fig. 2. Relation between entropy and length of segment.

with $p(x_i)$giving the probability of the symbol x_i. Dividing the entropy of a source X with n independent symbols by its maximum $\hat{H}(X) = ld\, n$ yields the relative entropy:

$$H_r(X) = \frac{H(X)}{\hat{H}(X)}. \tag{2}$$

In the paper the probabilities of symbols are estimated from their relative frequencies in empirical data. The basis for estimating probabilities from the corresponding relative frequencies is the stationarity of the underlying random process. In the case of velocity profiles only stepwise stationarity can be assumed. Therefore empirical investigations are restricted to certain traffic scenarios assumed to be stationary.

The further investigations are restricted to velocity-time profiles based on following two reasons. Firstly velocity-time profiles emphasise parts of the velocity profile driven at low speed which are usually more important for traffic telematic systems. Second velocity profiles are normally measured using the Global Positioning System (GPS) resulting in velocity values equally spaced in time (typical distance one second). Obtaining velocity values equally spaced in position (velocity-position profiles) needs a data conversion meaning additional effort and sources of error.

2.1 Empirical determination of (information) entropy

The empirical calculation of entropy requires the probabilities $p(x_i)$ of symbols x_i of the source "velocity-time profile". Empirical data show very different probability density functions in different parts of a velocity-time profile (see Fig. 1). Using Eq. (1) to determine the entropy shows the dependence of entropy from quantisation, length of segment (segment = space-time area of interest with stationary conditions) and traffic state.

As expected the length of the observation area (segment) has an important influence on the entropy (see Fig. 2). Considering short segments velocity usually does not change very much. This results in a small number of occupied velocity classes leading to a comparatively small entropy. But for some short segments nevertheless the entropy is quite large, especially in case of a noticeable change of speed within the segment. The consequence is a wide range of entropy. One problem in calculating the entropy from empirical data of small segments is the small amount of data not allowing a stable estimation of the frequency distribution.

On the other hand using long segments often leads to different street classes and traffic states within one segment. In these cases stationarity can not be assumed. As a trade-off the following segment length will be used: segments with duration of 10 min on motorways and segments with duration of 2 min in city traffic.

2.2 Empirical investigation in the dependence of information entropy from street class and traffic state

Street class and traffic state have considerable influence on the information entropy. Driving on motorways usually means little variation in velocity especially in case of speed limits. Assuming free traffic results in a comparatively small entropy. Substantial differences occur in Stop-and-Go situations (which are comparable with city traffic) and on motorways without speed limit (requiring a stronger adaptation of the speed on external conditions). Figure 3 gives examples of a typical velocity-time profile on a motorway and a typical velocity-time profile in city traffic. The corresponding entropies are given in Table 1.

Investigations have shown very little difference of the entropy (approximately 0,5 bit/S) on various German motorways (with different topographic characteristics and different speed limits). The entropy of velocity-time profiles in city traffic is characterised by a wide range of segment-based entropy values. Many segments achieve nearly the theoretical maximum of entropy (e.g. acceleration after stop at traffic lights), others show very little entropy (e.g. slowly approaching traffic lights). Comparing information entropy of motorways and city traffic shows significant differences in relative entropy but quite similar (absolute) entropy values.

Summarising the empirical investigations the information entropy of velocity-time profiles has proven to be quite moderate in the order of 2 bit/S. It should be mentioned that correlations between successive velocity values are yet not considered in the investigation. Consideration of these correlations will cause a further reduction of information entropy of velocity-time profiles.

3 Investigation in entropy dynamic (Ebeling et al., 1998; Fey, 1968)

Entropy dynamic provide an insight into the dependencies and correlations of symbols within a sequence. Special sequences are the independent sequence, the periodic sequence and Markov sequences of order m. Without doubt velocity-time profiles are neither independent nor periodic sequences. Due to physical reasons (inertia) the acceleration/deceleration of a vehicle is limited. This results in strong statistical dependencies between successive velocity values. Of cause these dependencies are decreasing with increasing time shift. Apparently speed-time profiles have some properties of Markov sequences.

The analysis of entropy dynamic is carried out in two different ways. On one hand velocity differences within a certain interval of time are used, on the other hand the transinformation between velocities in certain time distances was evaluated. Analysis of velocity differences is a very simple concept. Its main advantage is the effective use of the available empirical data. But this concept does not consider information of the concrete value of the initial velocity (e.g. a very high velocity is much less probable followed by an even higher velocity than by a lower velocity). This disadvantage can be avoided by the concept of transiformation. The transinformation evaluates the statistical dependencies between velocity values in a certain time distance. Compared to the analysis of velocity differences longer empirical sequences are needed.

3.1 Analysis of velocity differences

In the following all velocity differences $\Delta v = v\ (t_i + \Delta t) - v\ (t_i)$ for the run of a vehicle through a (time) segment $t_{\min} < t_i < t_{\max} - \Delta t$ are considered. Based on the resulting frequencies for certain velocity differences after a given time Δt (see Fig. 4) the associated information (entropy) can be calculated. Small values of information entropy indicate a nearly deterministic relation between the velocities at the considered times whereas a large entropy means a more random relationship.

Figure 5 shows the entropy of velocity differences over the time difference Δt calculated from empirical data of several segments on a motorway. This figure confirms the strong statistical dependencies for small time differences Δt. The given runs show an asymptotic behaviour. Close statistical dependencies can be found up to a time difference of about $\Delta t_{max} = 40\,\text{s} \ldots 100\,\text{s}$ (depending from the segment). Noticeable differences between the given runs can be found for time differences $\Delta t > 10\,\text{s}$. Thereby the single runs differ not more than about 25% from each other.

3.2 Analysis of transinformation

The transinformation $T\ (v_1, v_2)$ between two single velocities $v_1 = v\,(t_1)$ and $v_2 = v\,(t_2)$ with $t_2 = t_1 + \Delta t$ gives the amount of information which can be obtained from the knowledge of the velocity value v_1 over the velocity value v_2. The transinformation $T\ (v_1, v_2)$ is given by (Ebeling et al., 1998):

$$T\ (v_1, v_2) = H\ (v_1) + H\ (v_2) - H\ (v_1, v_2)$$
$$\text{with } H\ (v_1, v_2) \ldots \text{ compound entropy.} \qquad (3)$$

For an independent random process the transinformation between two symbols is identical zero. In case of a deterministic relationship the transinformation between two symbols is identical to the (identical) entropy of both single symbols, i.e. the transinformation represents the entire information of the following symbol.

Figure 6 gives the relationship between transinformation of two velocity values in time distance Δt over this

Table 1. Entropy calculation for Fig. 3 (sliding segments of 10 min duration (motorway) and 2 min duration (city)).

	motorway (Fig. 3)	city (Fig. 3)
Minimum of entropy of segment $\mathbf{H}_{\min}$ [bit/S]	1,36	0,25
Minimum of relative entropy of segment $\mathbf{H}_{r,\min}$	33%	9%
Maximum of entropy of segment $\mathbf{H}_{\max}$ [bit/S]	2,76	2,53
Maximum of relative entropy of segment $\mathbf{H}_{r,\max}$	66%	90%
Entropy [bit/S]	2,04	1,68
Relative entropy	49%	60%

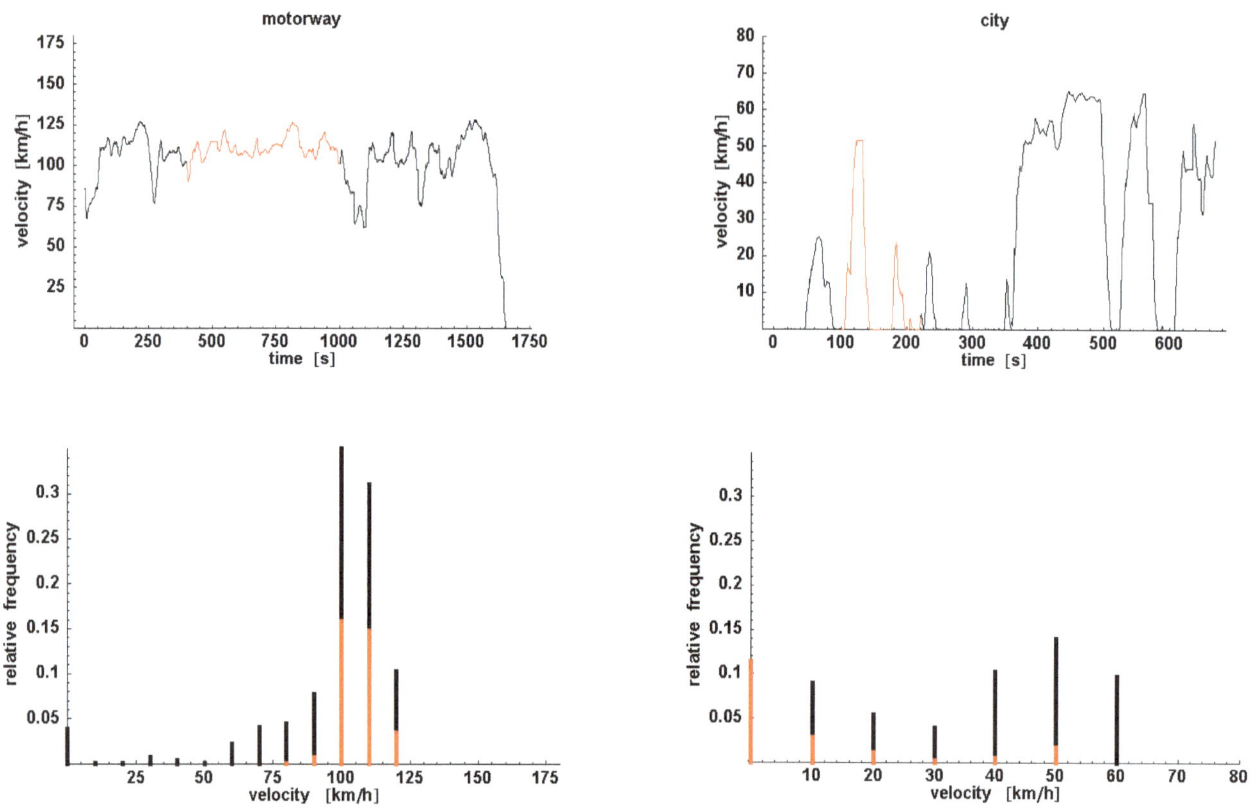

Fig. 3. Exemplary velocity-time profiles; example for single segment highlighted (red).

time distance Δt for empirical data (several segments of a test run). All runs show a nearly asymptotic decreasing behaviour. There can be distinguished between three characteristic sections. For very small time differences ($0<\Delta t<10$ s) the transinformation decreases very steep with increasing time difference. Further increasing time differences (10 s$<\Delta t<80$ s) also result in a further but slower decrease of the transinformation. For time differences $\Delta t>80$ s the transinformation is approaching a limit asymptotically.

The comparative large values of the transinformation at small time differences are an expression of inertia, i.e. the limited acceleration/deceleration ability of real vehicles. The range of medium time differences is dominated by the desire of the driver to keep a certain velocity. The not completely vanishing transinformation for larger time differences can be traced back to the choice of stationary velocity-time profiles. Stationarity means unchanging street classes and unchanging traffic condition resulting in preferring certain velocity values.

4 Optimal (acquisition) sample-rate in FCD-systems from an information entropic point of view

The acquisition sample rate is an important parameter of FCD-systems as it significantly influences the necessary effort in data acquisition, data transmission and data processing. Transmission of velocity data in very short intervals means small information content of the single values as they could distinguished with high probability from the preceding values. Contrary detection of velocity values only in very long intervals results in slow reactions of the system on a

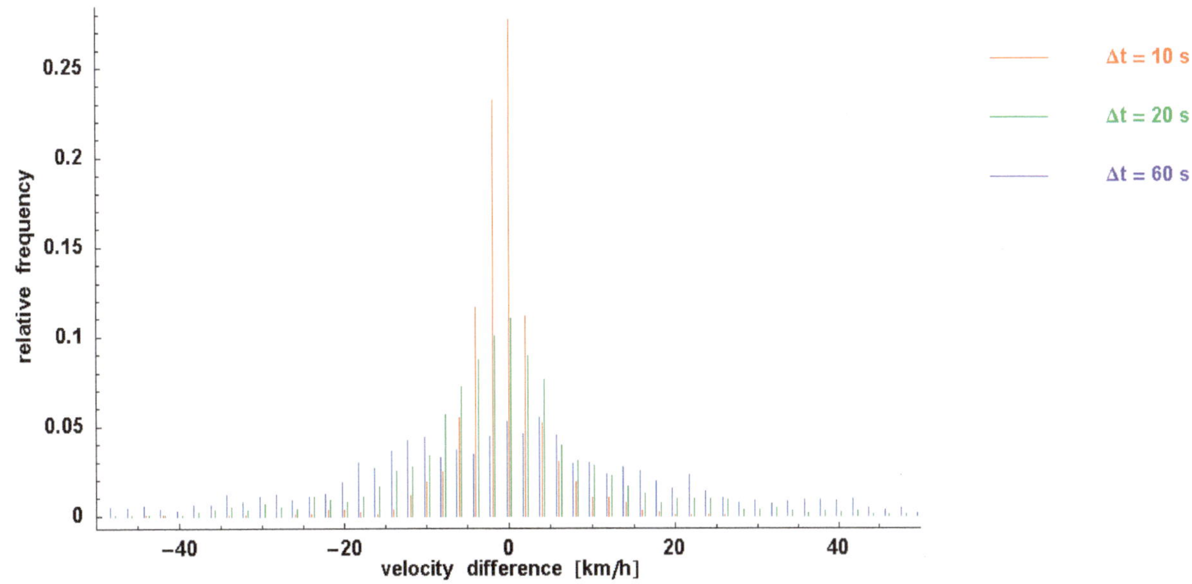

Fig. 4. Relative frequencies of velocity differences in a given time interval (parameter).

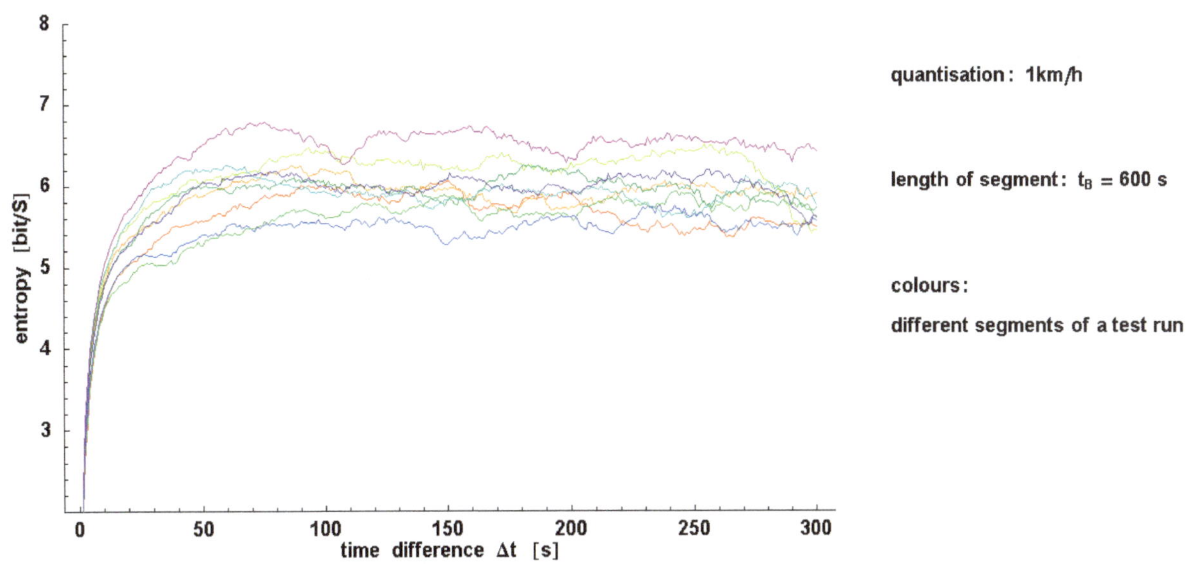

Fig. 5. Entropy of velocity differences over time difference for several segments (parameter) of a test run on a motorway.

changing environment. Considering the entropy of velocity differences (Sect. 3.1) velocity values within an interval of 10 s show very strong statistical dependencies but after about 100 ms velocity differences occur nearly randomly. Correspondingly an acquisition sample rate of $t_E < 10$ s means (considerable) redundant data whereas an $t_E > 100$ s would not completely characterise the velocity-time profile. Analysis of transinformation yields similar results. Statistical dependencies can be found up to time differences of about 80 s. Further proceeding dependencies result from the overall speed level on a certain street class and are therefore for a traffic observation system of little interest.

5 Conclusion

An acquisition sample rate of more than 0,1 Hz (i.e. ≤ 10 s time distance between samples) yields only a marginal information gain. The information gain per sample increases with the reduction of the acquisition sample rate up to approx. 0,01 Hz. At a further reduction of the acquisition sample rate the information gain per sample keeps nearly constant. Thus an acquisition sample rate between 0,01 Hz and 0,1 Hz should be taken.

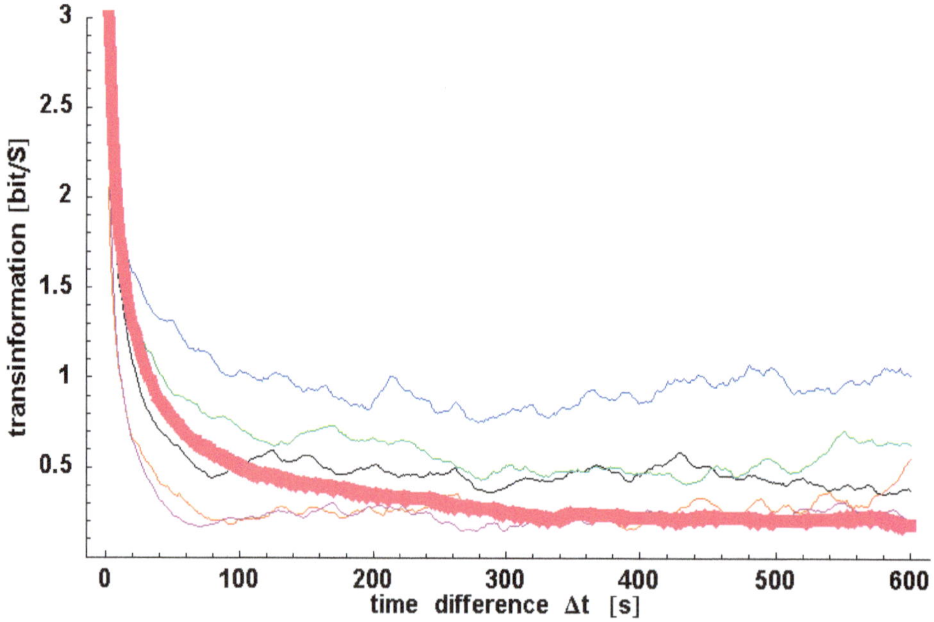

Fig. 6. Transinformation between velocity values in time distance Δt for segments (parameter) of a test run, result for whole test run highlighted (bold).

6 Summary

The paper analysis velocity-time profiles based on the (information) entropy. It could be shown that velocity-time profiles are characterised by a strong redundancy. Empirical investigations point out significant differences of the relative entropy for different street classes and traffic states. For the (absolute) entropy this significant difference could not be found. The resulting entropy of velocity-time profiles is moderate with about 2 bit/S. Investigation in entropy dynamic shows strong statistical dependencies for velocity values within short time distances. Based on entropy dynamic an optimal acquisition sample rate was distinguished.

References

Ebeling, W., Freund, J., and Schweitzer, F.: Komplexe Strukturen: Entropie und Information, Teubner Stuttgart Leipzig, 1998.

Fey, P.: Informationstheorie, Berlin, Akademie-Verlag, 1968.

Gössel, F., Michler, E., and Wrase, B.: Spectral Analysis of Floating Car Data, Kleinheubacher Berichte 2002/Advances in Radio Science, 2003.

Michler, E.: Verkehrsmonitoring/Verkehrsdatengewinnung, Kleinheubacher Berichte, 44, 296–305, 2001.

Olberg, M. and Rákóczi, F.: Informationstheorie in Meteorologie und Geophysik, Berlin, Akademie Verlag, 1984.

Potthoff, G.: Verkehrsströmungslehre, Band 4, Berlin, Transpress Verlag für Verkehrswesen, 1972.

A new high-integrated weak field sensor for automotive applications

T. Thiessen[1] **and M. Prochaska**[2]

[1]Leibniz Universität Hannover, Institut für Theoretische Elektrotechnik, Appelstr. 9A, 30167 Hannover, Germany
[2]NXP Semiconductors Germany GmbH Automotive Innovation Center Stresemannallee 101, 22529 Hamburg, Germany

Abstract. Especially in the field of automotive applications smart sensor systems for magnetic field sensing face increasing requirements concerning low cost, low power consumption and high magnetic performance. Over the past years AMR sensors play a decisive role in this application area because of their robustness and stability. In the following a high-integrated smart sensor system for magnetic field sensing is presented. A novel approach for the detection of weak magnetic fields is shown, which is based on an AC-excitation of AMR elements. In contrast to common used sensors this concept is based on a nonlinear AMR element without Barber pole construction. By means of this methodology sensitivity as well as temperature and life time stability is significantly improved, while the production costs compared to common-used sensors are reduced. Within this novel approach new signal conditioning algorithms and analog circuit topologies are presented, which are able to meet low offset and low noise requirements.

1 Introduction

For the measurement of weak magnetic fields various transducer principles are applicable like Fluxgate, Hall, Anisotropic (AMR) or Giant (GMR) Magnetoresistive sensors (Ripka, 2000). However, especially for automotive applications AMR sensors are widely used, because of their good performance concerning temperature stability, high robustness against large magnetic fields and relatively high sensitivity. In addition, usually a linear transfer characteristic, low noise and a negligible hysteresis are demanded. Most AMR sensors, which are commercially available, use a Barber pole construction for the linearization of the transfer function (Ripka, 2000; Kuijk et al., 1975). However, Barber pole sensors need a bias field applied along the easy axis to improve their stability, since the internal magnetization of the ferromagnetic sensor element has two stable positions (Dibbern, 1989). Otherwise, if for any reason a powerful magnetic field appears opposing the internal field, the magnetization flips from one into the opposite direction. Unfortunately, this stabilization field lowers significantly the sensitivity (Tumanski, 2001). So far Barber pole sensors fulfill sensitivity, low noise and hysteresis requirements. However, recent developments especially in the field of automotive applications demand weak field sensing systems which have to satisfy further demands such as low power consumption, low temperature and low life time drift. Typical examples are wheel speed sensing in automotive transmission systems, current measurement or electronic compassing. In addition, the magnetic field strength in automotive systems is often so violent that the measurement range of common-used weak field sensors is inadequately. AC-biasing techniques – such as modulation or flipping methods – are well-known to improve the performance of MR sensors (Tumanski, 2001)–(Razavi, 1997). However, usually they are not able to fulfill low power as well as low cost requirements of industrial and automotive applications.

In the following a novel approach is presented which is based on a superhet structure of the signal processing unit (Razavi, 1997). The initial point is a nonlinearized AMR element which guarantees a high sensitivity in view of the fact that a bias field along the anisotropic axis is not needed. Since the sinusoidal excitation signal along the hard axis driven by on-chip coils is small, low power requirements are fulfilled. The superhet structure leads to a linearization of the sensor output signal in frequency domain, where hystereses as well as offset effects are eliminated. Since offset effects are mainly responsible for temperature and life time drift this approach also leads to a highly stable sensor characteristic, where the natural robustness against high magnetic fields is

Correspondence to: T. Thiessen
(tina.thiessen@gmx.de)

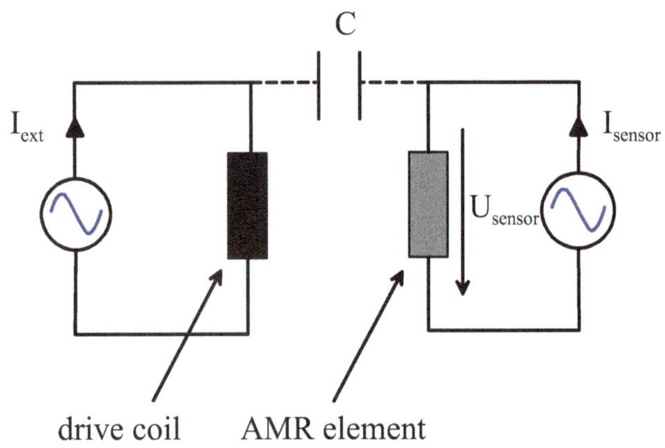

Fig. 1. Simplified circuit diagram of current fed AMR sensor with sinusoidal excitation.

preserved. Furthermore, the elimination of offsets caused by the AMR effect enables in addition with the lock-in character of the sensor signal processing an efficient amplification of output signals of the sensor elements (Meade, 1994).

However, a basic superhet structure consisting of filters and mixer circuits is not able to fulfill low cost demands. Furthermore, AC-driven systems show a capacitive coupling effect between MR layer and coil layer which requires a complex signal processing to separate the speed information from the capacitive coupled frequency. Finally, most AC-driven AMR sensors possess a small air gap range, since large magnetic inputs lead to distortions, which cannot be eliminated by a filter arrangement.

In the following a new method is presented by which the conversion of the speed signal from bandpass to baseband takes place in the AMR element itself and needs no mixer. Coincidentally the new approach separates the speed information from the capacitive coupled frequency. To eliminate the capacitive coupled frequency and reduce the requirements of the signal processing unit a difference principle is used, which leads to significant cost down. Furthermore the method features a large air gap capability, high sensitivity and a high (dynamic) air gap range. The reader should note that the presented methodology is also applicable to other xMR technologies such as GMR.

2 Methodology

As shown in Fig. 1, the invention is based on a nonlinearized AMR element driven by a sinusoidal current and an on-chip coil, which creates a sinusoidal magnetic excitation in y-direction. Between the MR layer and the coil layer occurs a capacitive coupling effect represented by C. The transfer characteristic of a nonlinearized AMR element is given by

$$R = R_0 + \Delta R \cos^2 \alpha \quad (1)$$

where R_0 represents the resistance of the AMR element, ΔR the maximum change of R_0 and α with

$$\sin \alpha = \frac{H_y}{H_x + H_0} \quad (2)$$

the angle between the internal magnetization of a ferromagnetic thin film and the direction of the current through the film (Dibbern, 1989). H_x corresponds to the external magnetic field along the anisotropic axis, H_y is the field perpendicular to the easy axis and H_0 can be regarded as a material constant compromising the so called demagnetizing and anisotropic fields (Tumanski, 2001). For small magnetic fields H_y in y-direction and a negligible field in x-direction ($H_x \rightarrow 0$) AMR sensors without Barber poles can be described by

$$\begin{aligned} R &= R_0 + \Delta R \left(1 - \left(\tfrac{H_y}{H_0}\right)^2\right) \\ R &= R_0 \quad \text{for} \quad H_y > H_0 \end{aligned} \quad (3)$$

If we generate by means of an on-chip coil a sinusoidal modulation field $H_{exc}(t)$ and assume without a loss of generality an external sinusoidal field $H_{ext}(t)$, it follows

$$H_y = \hat{H}_{ext} \sin(\omega_{ext} t) + \hat{H}_{exc} \sin(\omega_{exc} t) \quad (4)$$

where $\omega_{ext}=2\pi f_{ext}$ and $\omega_{exc}=2\pi f_{exc}$. Sinusoidal fields H_{ext} are typical in the field of wheel speed sensing, e.g. anti lock braking systems (ABS) or automotive transmissions, where the amplitude can vary caused by mechanical torsions up to the factor 8.

If we put Eq. (4) into Eq. (3) we arrive at

$$R_{sensor}(t) = R_0 + \Delta R \left(1 - \frac{\hat{H}_{ext}^2}{2H_0^2} - \frac{\hat{H}_{exc}^2}{2H_0^2}\right) + \cdots$$

$$+ \Delta R \frac{\hat{H}_{ext}^2}{2H_0^2} \cos(2\omega_{ext} t) + \Delta R \frac{\hat{H}_{exc}^2}{2H_0^2} \cos(2\omega_{exc} t) - \cdots$$

$$- \Delta R \frac{\hat{H}_{ext} \hat{H}_{exc}}{H_0^2} \left(\cos\left((\omega_{exc} - \omega_{ext})t\right) - \cdots \right. \quad (5)$$

$$\left. - \cos\left((\omega_{exc} + \omega_{ext})t\right)\right), \quad (6)$$

which represents a amplitude modulation of the external signal $H_{exc}(t)$ by $H_{ext}(t)$. First, we assume, that the output voltage of a current fed AMR sensor is given by

$$U_{sensor} = R_{sensor} \cdot I + U_C(t) \quad (7)$$

where I is the constant current through the AMR resistor and

$$U_C(t) = \hat{U}_C \sin(\omega_{exc} t) \quad (8)$$

is the capacitive coupling voltage induced by the coil layer. The spectrum of the output voltage can be seen in Fig. 2.

To isolate the needed signal a common superhet structure can be used consisting of a band-pass filter, a mixer and a

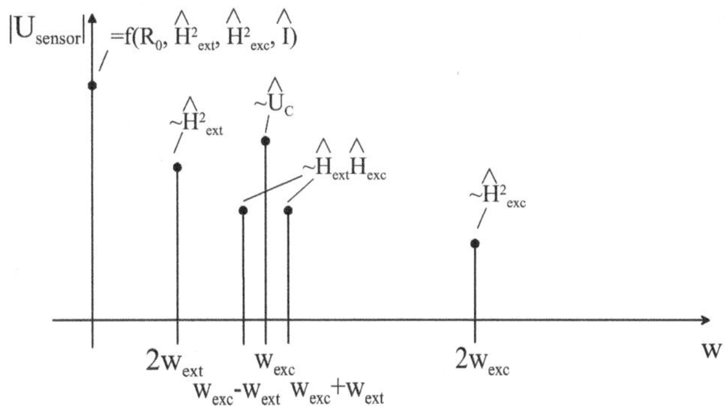

Fig. 2. Spectrum of the output voltage U_{sensor} with constant current through the AMR resistor.

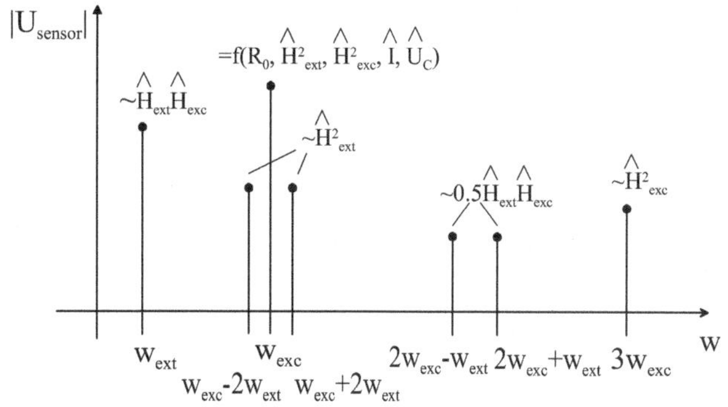

Fig. 3. Spectrum of the output voltage U_{sensor} with sinusoidal current through the AMR resistor.

low-pass filter (Meade, 1994). In comparison with conventional smart sensor systems this method needs three further components: two filters and a mixer circuit, which increase costs of the signal processing unit. To reduce the needed chip area of the filter arrangement, a higher excitation frequency seems to be useful. However, there is a capacitive coupling between the on-chip coil and the AMR layer, which depends on the modulation frequency. Especially in the case of a higher excitation frequency an additional offset appears so that the advantages of AC-biasing are eliminated. To produce relief we will show a methodology, which drastically reduces the needed filters and eliminates the must of the mixer circuit. Moreover, by means of this approach capacitive coupling is eliminated by the method itself.

If we perform the mixing by means of Ohm's law, i.e. we choose an adequate sinusoidal current through the AMR resistors (see Fig. 1) with

$$I(t) = \hat{I}\sin(\omega_{\text{exc}}t) \tag{9}$$

it follows

$$\begin{aligned}
U_{\text{sensor}} &= \left[R_0 + \Delta R\left(1 - \frac{\hat{H}_{\text{ext}}^2}{2H_0^2} - \frac{\hat{H}_{\text{exc}}^2}{2H_0^2}\right)\right]\cdots \\
&\cdot \hat{I}\sin(\omega_{\text{exc}}t) + \Delta R\frac{\hat{H}_{\text{ext}}^2}{4H_0^2}\cdot\hat{I}\left[\sin((\omega_{\text{exc}} - 2\omega_{\text{ext}})t) + \cdots\right. \\
&\cdots + \left.\sin((\omega_{\text{exc}} + 2\omega_{\text{ext}})t)\right] + \cdots \\
&\cdots + \Delta R\frac{\hat{H}_{\text{ext}}^2}{4H_0^2}\cdot\hat{I}[\sin(3\omega_{\text{exc}}t) - \sin(\omega_{\text{exc}}t)] - \cdots \\
&\cdots - \Delta R\frac{\hat{H}_{\text{ext}}^2\cdot\hat{H}_{\text{exc}}^2}{2H_0^2}\cdot\hat{I}\left[2\sin(\omega_{\text{ext}}t) + \cdots\right. \\
&\cdots + \left.\sin((2\omega_{\text{exc}} - \omega_{\text{ext}})t) - \sin((2\omega_{\text{exc}} + \omega_{\text{ext}})t)\right] + \cdots \\
&\cdots + \hat{U}_C\sin(\omega_{\text{exc}}t)\,.
\end{aligned} \tag{10}$$

The spectrum of the output signal is shown in Fig. 3. To isolate the needed signal at ω_{ext} a simple low pass filter can be used. It turns out that by means of the modulation of the AMR elements the capacitive effects are separated, offset effects are eliminated and consequently temperature and life time drift are significantly reduced. In addition, since the wanted signal is offset free it can be amplified strongly. For this reason the sensitivity can be controlled directly by the amplification which makes large sensing distances possible. On this account external fields down to sensor noise can failsafe detected. Moreover, since in contrast to Barber pole sensors a bias magnet is not needed, the sensitivity is improved, where the production costs are reduced. However, the low pass filter has high requirements to exfiltrate higher frequencies without a phase shift of the output signal.

To reduce the requirements of the filter one can use a difference principle desribed below. The initial point is Eq. (5)

$$\begin{aligned}
R_{\text{sensor}}^{+}(t) &= R_0 + \Delta R - \Delta R\frac{\hat{H}_{\text{ext}}^2}{H_0^2}\sin^2(\omega_{\text{ext}}t) - \cdots \\
&\cdots - 2\Delta R\frac{\hat{H}_{\text{ext}}\hat{H}_{\text{exc}}}{H_0^2}\sin(\omega_{\text{ext}}t)\sin(\omega_{\text{exc}}t) - \cdots \\
&\cdots - \Delta R\frac{\hat{H}_{\text{exc}}^2}{H_0^2}\sin^2(\omega_{\text{exc}}t),
\end{aligned} \tag{11}$$

where H_y is

$$H_y = \hat{H}_{\text{ext}}\sin(\omega_{\text{ext}}t) + \hat{H}_{\text{exc}}\sin(\omega_{\text{exc}}t). \tag{12}$$

If we excite a second comparable AMR element with

$$H_y = \hat{H}_{\text{ext}}\sin(\omega_{\text{ext}}t) - \hat{H}_{\text{exc}}\sin(\omega_{\text{exc}}t) \tag{13}$$

we arrive at:

$$\begin{aligned}
R_{\text{sensor}}^{-}(t) &= R_0 + \Delta R - \Delta R\frac{\hat{H}_{\text{ext}}^2}{H_0^2}\sin^2(\omega_{\text{ext}}t) + \cdots \\
&\cdots + 2\Delta R\frac{\hat{H}_{\text{ext}}\hat{H}_{\text{exc}}}{H_0^2}\sin(\omega_{\text{ext}}t)\sin(\omega_{\text{exc}}t) - \cdots
\end{aligned}$$

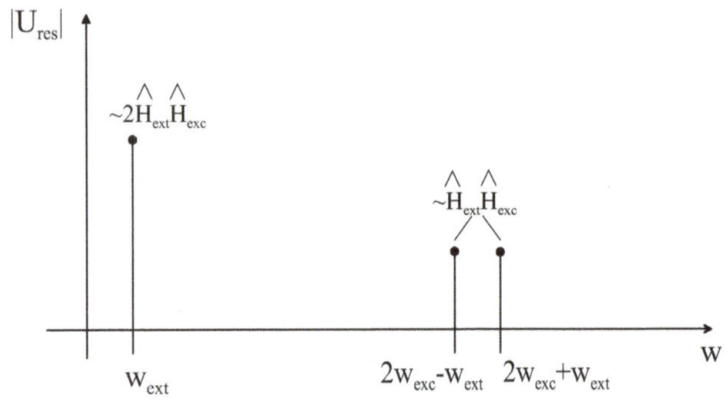

Fig. 4. Spectrum of the resulting output voltage U_{res} by using the difference principle.

$$\cdots - \Delta R \frac{\hat{H}_{\text{exc}}^2}{H_0^2} \sin^2(\omega_{\text{exc}} t) \tag{14}$$

The resulting output voltage

$$\begin{aligned} U_{\text{res}} &= U_{\text{sensor1}} + U_{\text{sensor2}} \\ &= R_{\text{sensor}}^{+}(t) \cdot \hat{I} \sin(\omega_{\text{exc}} t) + \hat{U}_C \sin(\omega_{\text{exc}} t) + \cdots \\ &\cdots + R_{\text{sensor}}^{-}(t) \cdot \hat{I} \sin(-\omega_{\text{exc}} t) + \hat{U}_C \sin(-\omega_{\text{exc}} t) \\ &= -\Delta R \frac{\hat{H}_{\text{ext}} \hat{H}_{\text{exc}}}{H_0^2} \hat{I} \cdot [2 \sin(\omega_{\text{ext}} t) + \cdots \\ &\cdots + \sin((2\omega_{\text{exc}} - \omega_{\text{ext}})t) + \sin((2\omega_{\text{exc}} + \omega_{\text{ext}})t)] \end{aligned} \tag{15}$$

represents the wanted speed signal and components around the double excitation frequency. These higher frequencies can easily be filtered by a low pass with low order (see Fig. 4). The complete sensor concept is shown in Fig. 5. By means of this invention the conversion of the speed signal from bandpass to baseband takes place in the AMR element itself and needs no mixer. Furthermore, the disturbing capacitive coupled signal is eliminated by this new approach and the requirements of the filter are reduced by the difference principle. Therefore, the invention is based on a simple low pass filter with low order and high edge frequency preventing frequency dependent phase shift between magnetic input and electric output signal. In addition, the presented structure fulfils low cost requirements concerning the signal conditioning ASIC.

If the amplitude of the magnetic input signal is large – this means that the magnetic input signal fulfils $|H_y| > H_0$ and the transfer function loses its quadratic characteristic – higher harmonics can appear. It is impractical to separate these perturbations from the wanted signal by filter operations. On this account a traditional concept which is based on a superhet structure is not able to measure magnetic signal with amplitudes larger than H_0. In contrast, as shown above the suggested sensor concept suppresses signals which are not

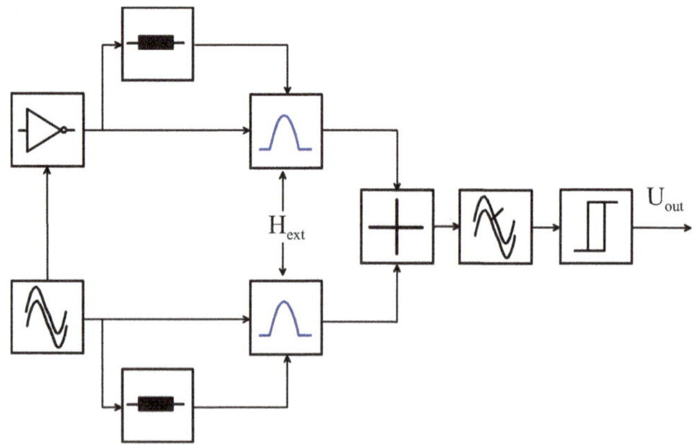

Fig. 5. Block diagram of the signal processing unit.

in phase with the wanted signal. So this structure is able to detect magnetic inputs with $|H_y| > H_0$. If we assume

$$\begin{aligned} H_y &= \hat{H}_{\text{ext}} \sin(\omega_{\text{ext}} t) + \hat{H}_{\text{exc}} \sin(\omega_{\text{exc}} t) \\ &\approx \hat{H}_{\text{ext}} \sin(\omega_{\text{ext}} t) \end{aligned} \tag{16}$$

for

$$\hat{H}_{\text{ext}} \gg \hat{H}_{\text{exc}} , \tag{17}$$

we are able to calculate by using Eqs. (3) and (16) the Fourier series of the resulting resistance $R_{sensor}(t)$ for

$$\hat{H}_{\text{ext}} > H_0 . \tag{18}$$

From this follows

$$\begin{aligned} R_{\text{sensor}}(t) &= R_0 - \frac{K}{\pi}\left(\arcsin(P) - P \cdot \sqrt{1 - P^2}\right) + \cdots \\ &+ \sum_{n=1}^{\infty} \frac{K}{\pi} \cdot \left[\frac{4}{2n} \sin(2n \arcsin(P)) - \cdots \right. \\ &- \frac{2}{2n-2} \sin((2n-2)\arcsin(P)) - \cdots \\ &\left. - \frac{2}{2n+2} \sin((2n+2)\arcsin(P))\right] \cdot \cos(2n\omega_{\text{ext}} t) \end{aligned} \tag{19}$$

where

$$K = \Delta R \frac{\hat{H}_{\text{ext}}^2}{H_0^2} \tag{20}$$

and

$$P = \frac{H_0}{\hat{H}_{\text{ext}}} . \tag{21}$$

From Eq. (19) follows that high amplitudes of the external signal do not thwart the basic properties of this principle. However, if Eq. (17) is valid, higher harmonics of H_{exc} can change the amplitude of the mixed signal so that quantitative measurements are hindered. On the other hand often qualitative field sensing is sufficient, i.e. in the field of wheel speed sensing.

Fig. 6. Layout of the AMR layer.

3 Experimental results

To verify the presented methodology integrated prototypes were implemented. The signal processing units are realized according to Fig. 5. The layout of the sensing elements is similar to the KMZ51 earth field sensor (Stork, 1997). Since the methodology is based on a nonlinear transfer characteristic of the AMR elements, in contrast to the KMZ51 Barber poles were not implemented. Furthermore, only one drive-coil along the easy axis was realized to excite the sensing elements. As a matter of principle offset effects are eliminated and so temperature and life time drift are also essentially reduced. Instead of meander patterned sensor (Ciureanu, 1992), this system consists of two strips of 36 single resistors, which are not arranged in a Whetstone bridge. This leads to a further cost down, since on chip trimming of the bridge resistors is not needed. To guarantee as low as possible tolerances a sensor layout as shown in Fig. 6 was realized. The resistance of each stripe of sensing elements is approximately 14 kΩ , while the drive coil resistance is about 200 Ω and the coil field factor is approximately $25\frac{(A/m)}{(mA)}$. So by means of relatively small excitation current through the on-chip coils a high sensitivity S can be achieved, i.e. $S>20\frac{(mV/V)}{(kA/m)}$ for $H_x \leq 50\frac{A}{m}$ and $\hat{H}_{\text{exc}}>100\frac{A}{m}$.

The spectrum of the output signal of one stripe of sensing elements near the excitation signal is shown in Fig. 7, where ω_{ext}=20 kHz and ω_{exc}=1 MHz. For the generation of

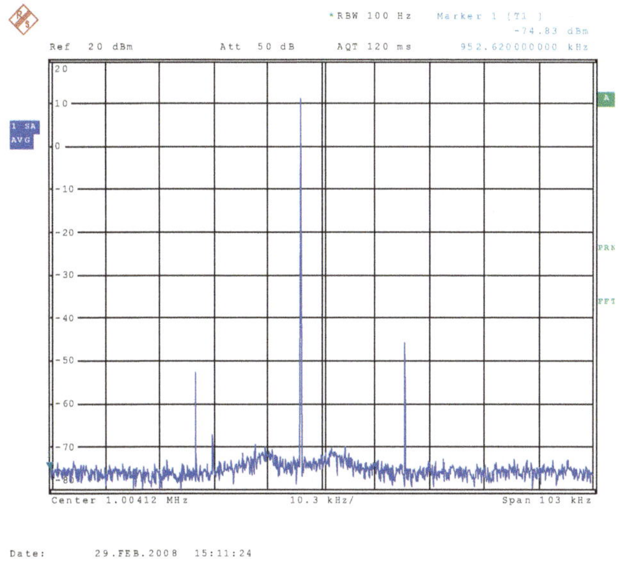

Fig. 7. The spectrum of the AMR element series R^+ near f_{exc}.

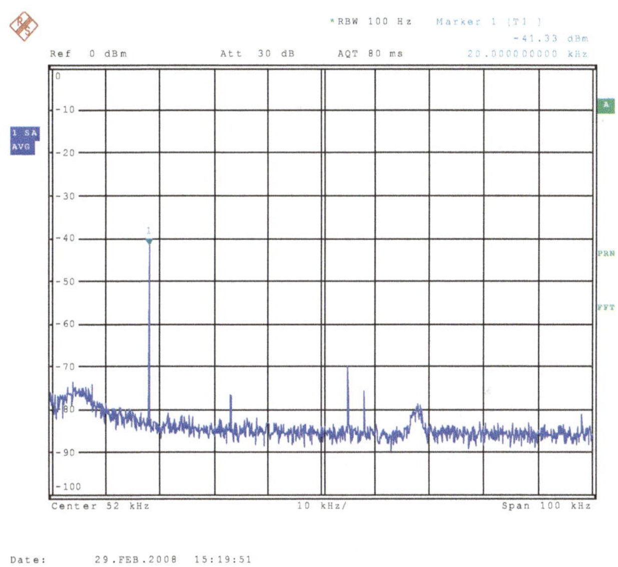

Fig. 8. The spectrum of the output signal of the smart sensor system.

the external field a Helmholz coil was used, while the AC-excitation of AMR elements was produced by current-fed on-chip drive-coils. The current through the coils is to be chosen to 5 mA· sin(2π 1 MHz·t). As expected there is a large spectral line at f_{exc} which is a result of the capacitive coupling between the AMR and coil-layer. However, if we use the presented methodology the wanted signal is directly mixed to the base-band (see Fig. 8). Following the difference principle shown in Fig. 5 the output signals of both AMR elements are added. The remaining, undesired spectral lines can easily be removed by a simple low-pass filter. The required chip area depends on the modulation frequency. For a higher modulation frequency the low-pass filter is smaller and the phase shift of the sensor output signal compared to $H_{\text{ext}}(t)$ is lower. Further investigations show that in contrast to common used weak field sensors both techniques are robust against large external fields. So in case of constant magnetic fields the measurement range in $\pm H_0$, while periodical signals with amplitudes of up to $6H_0$ can be detected.

4 Conclusions

In this paper a novel method for weak field sensing is presented. The highly integrated prototype demonstrates feasibility and outstanding performance with respect to the requirements in industrial and especially automotive applications such as low cost, low power and high robustness. The sensor system is able to detect fields in the order of the sensor noise and possess a large measurement range. As a matter of principle offset effects are eliminated and on this account temperature and life time drift are essentially reduced. Since trimming of bridge resistors as well as a bias magnet is not needed the production costs are essentially decreased. Summarizing, performance and cost efficiency are outstanding in comparison to each commercial available magnetic sensor system and justify the relevance of AMR technology for industrial and especially automotive applications.

References

Ciureanu, P.: Magnetoresistive Sensors, in: Thin film resistive sensors, edited by: Ciureanu, P. and Middelhoek, P., Institute of Physics Publishing, 253-436, 1992.

de Niet, E. and Vreeken, R.: A magnetoresistive head with magnetic feedback, IEEE T. Magn., 15(6), 1625–1628, September 1979.

Dibbern, U.: Magnetoresistive Sensors, in: Sensors – A Comprehensive Survey, edited by: Göpel, W., Hesse, J., and Zemel, J. N., Vol. 5, Magnetic Sensors, edited by: Boll, R. and Overshott, K. J., Wiley-VCH, Weinheim, 342–379, 1989.

Flynn, D.: Demodulation of a magnetoresistive sensor signal to achieve a low-cost, stable and high-resolution vector magnetometer, Elsevier Science, Sensor. Actuat. A-Phys., 50, 187–190, 1995.

Kuijk, K. E., van Gestel, W. J., and Gorter, F. W.: The Barber Pole, a linear magnetoresistive head, IEEE Trans. Magn., 11(5), 1215–1217, September 1975.

Meade, M. L.: Lock-in amplifiers: principles and applications, IEE Series in Measurement, Peregrinus, 5–22, 1994.

Petrou, J., Dimitropoulos, P. D., Hristoforou, E., and Neagu, N.: New 2D fluxgate device based on the phase modulation of magnetization rotation in AMR films, Elsevier Science, Sensor. Actuat. A-Phys., 129, 107–111, 2006.

Razavi, B.: RF Microelectronics, Prentice Hall, 54–96, 1997.

Ripka, P.: Magnetic Sensors and Magnetometers, Arttech House, 130–150, 2000.

Ripka, P., Tondra, M., Stokesa, J., and Beech, R.: AC-driven AMR and GMR magnetoresistors, Elsevier Science, Sensor. Actuat. A-Phys., 76(1), 225-230, August 1999.

Stork, T.: Electronic Compass Design using KMZ51 and KMZ52, Application Note AN00022, NXP Semiconductors, 1997.

Tumanski, S.: A new Type of Thin Film Magnetoresistive Magnetometers – an Analysis of Circuit Principles, IEEE T. Magn., 20(5), 1720–1722, September 1984.

Tumanski, S.: Thin Film Magnetoresistive Sensors, IOP Publishing, 33–145, 2001.

Ueda, M., Endoh, M., Yoda, H., and Wakatsuki, N.: AC Bias Type Magnetoresitive Sensor, IEEE T. Magn., 26(5), 1572–1574, 1999.

3

System considerations and RF front-end design for integration of satellite navigation and mobile standards

A. Miskiewicz[1,2], **A. Holm**[1], **and R. Weigel**[2]

[1]Infineon technologies AG, Neubiberg, Germany
[2]University of Erlangen-Nuremberg, Erlangen, Germany

Abstract. The paper presents the challenges involved in a system design of a robust reconfigurable RF front-end for navigation and mobile standards. Receiver architecture is chosen from the point of view of inter-system interference and 130nm CMOS process characteristics. System concept covers the implementation of GPS, Galileo, UMTS, GSM and CDMA2000 using a Zero-IF architecture with reconfigurable analog and digital path. Feasibility studies of the system cover analysis of the wireless regulations and performance criteria, such as overall gain, noise figure (NF), and 1dB compression point (P1dB) of the RF chain, phase noise requirements and VCO tuning range [1]. The presented chip was fabricated in 130 nm CMOS technology. System considerations are confirmed with the chip measurements of gain, noise figure, and linearity. Prospects for the future work are presented including technology shrink.

1 Introduction

Current market of mobile handsets with navigation capabilities is driven by two main factors: government regulations and commercial demand. With introduction of E-911 mandate by the United States, mobile operators are required to provide localisation of mobile subscribers calling an emergency service. On the other hand, there is a demand for enhancing the penetration of Location Based Services (LBS), which provide users with a wide variety of geographically-referenced information. Both of these functions require precise positioning information, which cannot be derived from classical network solutions (such as triangulation in mobile networks). The use of navigation services is inevitable if the quality of the service needs to be high, especially in urban environment, where the signals are weak.

Correspondence to: A. Miskiewicz
(anna.miskiewicz@infineon.com)

With the agreement between the government of the United States and the European Union, which leads to a full interoperability of GPS and Galileo, a new market for combined navigation receiver has been created. Integration of both systems into a Global Navigation Satellite System (GNSS) should supply users with highly precise and reliable positioning service. As market demand for integrated mobile phones with GPS (and Galileo) functionality is already a fact, it is desirable to account for that expansion now and develop a highly competitive multifunctional hand-held receiver. Such a receiver should support both navigation and mobile standards, not only those already widely used, but also those under development. On the other hand, a low-cost production, low current consumption, and small chip size require a highly-reconfigurable architecture. Implementation of both GPS and Galileo has been the core task of chip development. Mobile communications standards increase the competitiveness of the receiver in terms of operation capabilities, as long as cost, chip size and current consumption are acceptable. Therefore, a trade-off between the number and type of implemented standards and chip overall performance has been made. Considering the above requirements, the standards presented in Table I have been selected for the multi-mode receiver.

2 Satellite navigation receiver overview

Both GPS and Galileo are spread-spectrum systems, where the navigation message is multiplied with the pseudo-random noise (PRN) code specific to the satellite transmitter. The PRN codes are orthogonal, therefore all satellites can send the signals using the same centre frequency and bandwidth. For Galileo signal, the result of navigation message and PRN code multiplication is additionally multiplied with a subcarrier frequency of 1.023 MHz (Fig. 1). The last stage of the signal generation for both GPS and Galileo

Table 1. Implemented standards.

A-GPS	Galileo	UMTS	CDMA2000	GSM
L1	E1	II-III	PCS 1900	850
		V-VI	Cell 850	900
		VIII-IX		1800
				1900

Table 2. Satellite navigation standards.

	GPS L1 [3]	Galileo E1 [4]
Centre frequency	1575.42 MHz	1575.42MHz
Modulation type	DSSS-BPSK	DSSS-MBOC(1,1)
Spreading code chip rate	1.023 Mcps	1.023 Mcps
Spreading code length	1023 chip	4092 chip
Data rate (nav. message)	50 sps	250 sps
Bandwidth (RF)	2.046 MHz	4.092 MHz
Nominal S/N @GPS antenna	−19 dB	−19 dB
Max. sensitivity (indoor/outdoor)	−130/−155 dBm	−127/−152 dBm

Table 3. Out-of-band blockers.

Blocker	Freq (MHz)	P_B (dBm)
UMTS b3 UL	1710	10
GSM 900 UL	915	10

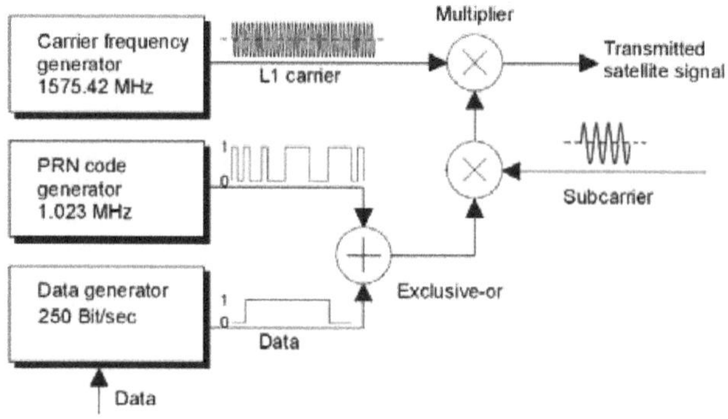

Fig. 1. Simplified diagram for Galileo signal generation.

is an up-conversion using a multiplication with a carrier of 1575.42 MHz.

Maximum power level of signals arriving from the satellite to the receiver on the Earth surface is around -130dBm for GPS and −127 dBm for Galileo signals. For the case where no direct Line-of-Sight (LoS) between transmitter and receiver is available, the signals can be further attenuated by more than 20dB due to tree foliage, multipath propagation in urban canyons or attenuation through walls and roofs for in-door navigation [2]. Therefore, the GPS and Galileo specifications define only maximum available signal power at the Earth surface and not the minimum sensitivity, as it is the case for mobile standards. For state-of-the-art low-cost receivers integrated in a mobile platform the bandwidth of the GPS and Galileo signals is limited to the first main lobe for GPS signal and two main lobes for Galileo standard. The bandwidth is mainly defined by the off-chip SAW filtering. Tab. II presents overall RF specification of GPS standard in L1 band and Galileo standard in E1 band.

Chip cost and size considerations suggest a re-use of an already existing processor to run GPS algorithms, rather than to add another processor on the positioning chip. Based on the system integration requirements, a host-based hardware solution has been suggested. Figure 2 presents the proposed navigation chip partitioning.

3 RF front-end

According to Galileo and GPS specification documents, there are no specific test cases to be considered in GNSS receiver design. However, multimode receiver application requires consideration of test cases, which are relevant for the definition of RF front-end blocks, such as amplifiers, inter-stage filters and mixers [5]. These requirements are imposed by large blocker levels and possible intermodulation products, which may fall into GPS and Galileo receive band.

Based on the specifications for mobile standards implemented in the chip, a set of GNSS-signal specific test cases has been created and simulated (Table 3–5).

The purpose of the Zero-IF RF front-end is to receive, amplify, filter, down-convert and sample the signal to enable further baseband signal processing.

As presented in Fig. 3, a configuration with two separate antennas is proposed. There are two paths – one for GNSS reception and one reconfigurable chain for proposed mobile standards. After the second amplifier, an integrated switch is implemented, which enables choosing the type of received signal. Further blocks are common for both mobile and navigation standards. This solution offers high reconfigurability and lower cost of implementation for multiple standards.

After the signal is received in the antenna, it is filtered and amplified in a low-noise amplifier, which is followed by RF filtering. These steps are performed off-chip. The main target of the navigation path is low noise figure of the chain and a RF bandwidth of at least 4MHz [6]. For mobile standards the reconfigurability was the main focus in the design process. The analog part is short, featuring only the second LNA, mixer and analog baseband filter. Due to moderate analog gain in the front-end, the requirements for DC offset are relaxed. Further, the baseband filter attenuation of the adjacent channels is moderate. Additional filtering is integrated in the digital front-end, where DC offset compensation is also implemented. For the common path, mixer of a high

Table 4. Second-order intermodulations.

Blocker 1	Freq (MHz)	P_B1 (dBm)	Blocker 2	Freq (MHz)	P_B2 (dBm)
GSM 900 DL	876	−25	802.11b, g	2452	−10
GSM 850 UL	890	10	802.11b, g	2476	−10

Table 5. Third-order intermodulations.

Blocker 1	Freq (MHz)	P_B1 (dBm)	Blocker 2	Freq (MHz)	P_B2 (dBm)
UMTS b2 UL	1850	10	UMTS b1 DL	2124	−25
UMTS b2 DL	1990	−25	802.11b, g	2412	−10
UMTS b3 DL	1805	−25	802.11a	5190	−10

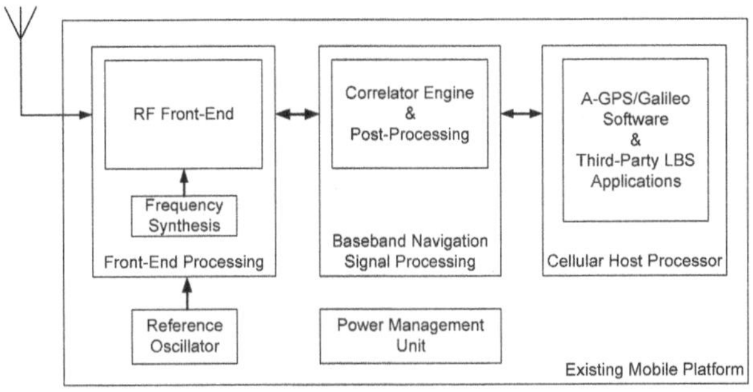

Fig. 2. GPS/Galileo chip partitioning.

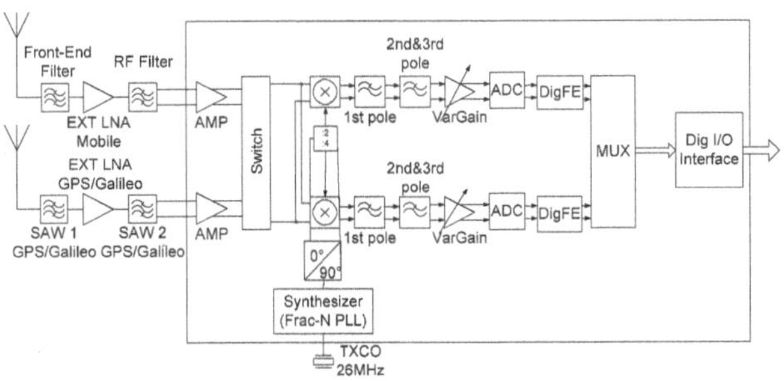

Fig. 3. Implemented receiver front-end.

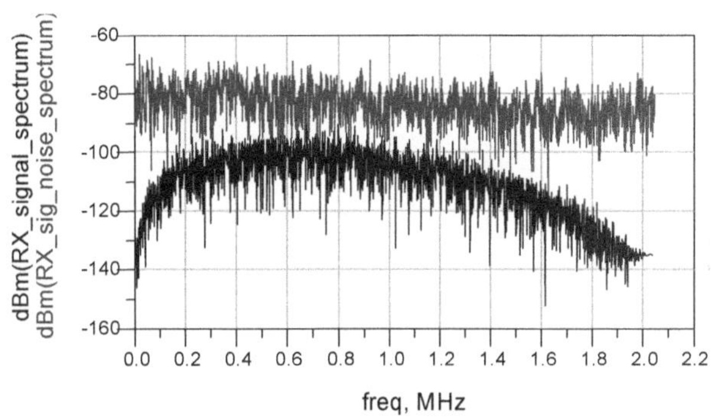

Fig. 4. Simulated Galileo signal spectrum and signal plus noise spectrum after the receiver front-end.

linearity has been implemented, which allows relaxed front-end filtering. Due to constant gain approach in the front-end no switching transients are present in the chip, however external gain steps are necessary for mobile standards. The setup is finalized with a high-resolution continuous-time Sigma-Delta ADC, which is immune to high level interferers and enables low-gain analog front-end implementation. Proposed front-end architecture implements additional baseband digital processing, including filtering, decimation, fractional sample rate conversion, and compensation of unwanted RF-FE effects.

The front-end has been simulated for GPS and Galileo signals. Figure 4 presents the typical Galileo spectrum at the output of the analog front-end part. The influence of the front-end has been determined by comparison of the input spectrum at the antenna and the spectrum after down-conversion and filtering.

The simulated signal-to-noise deterioration in the front-end is acceptable for the satellite navigation signal reception. Based on the level-plan calculation appropriate values of gain, NF, and P1dB have been selected for RF blocks in the chain. Table 6 presents an overview of typical block performance.

4 Measurements

The chip was fabricated in 130 nm CMOS technology and packaged in 5×5 mm^2 WFBGA package. The analog performance for GPS and Galileo was measured including off-chip filters and LNA. The gain results were corrected for the additional gain in digital front-end. The input match-

Table 6. Block performance for gps/galileo path.

	Gain (dB)	NF (dB)	IIP3 (dBVrms)
SAW1	−0.8	0.8	N/A
EXT LNA	20.2	2.0	−15.2
SAW2	−1.1	1.1	N/A
AMP	10.5	4.5	−7.7
Mixer+1st pole	32.0	19.9	−6.3
2nd & 3rd pole	6.0	42.4	21.0
Overall	66	3.2	−45.3

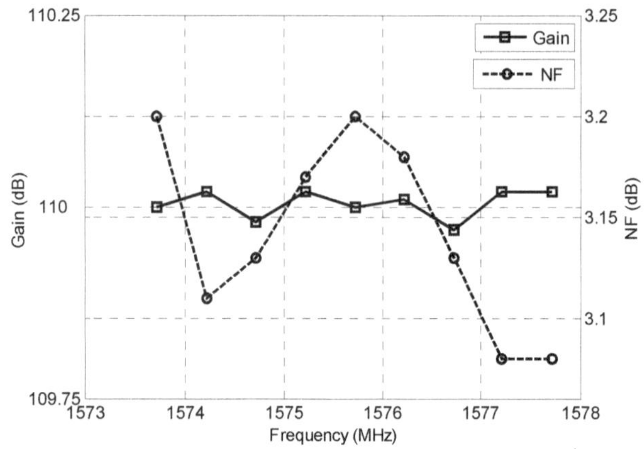

Fig. 5. Measured gain and noise figure for GPS/Galileo path.

ing, gain, noise figure, and linearity conform to the simulated results. For a setup with an external low-noise amplifier, the noise figure is 3.2 dB. Due to the fact, that the amplifier and mixer blocks are wideband, the measured out-of-band 1dB compression point is comparable to the in-band value. The measured current consumption equals 60 mA and 58 mA for 1.5 V and 2.85 V domains respectively. The current consumption may be optimized by the adjustment of the digital front-end settings and lower the linearity of the analog blocks, at the cost of additional required selectivity in the off-chip components.

5 Summary

Highly reconfigurable multimode, multiband receiver integrated in 130nm CMOS technology has been presented. The chip enables implementation of assisted-GPS and future assisted-Galileo standards using only one receiver chip. The reconfigurability is possible because of the wideband analog blocks, high dynamics ADC and flexible digital front-end. The further steps include integration of mobile transmitter paths in the chip and technology shrink to the next node of 65nm CMOS technology.

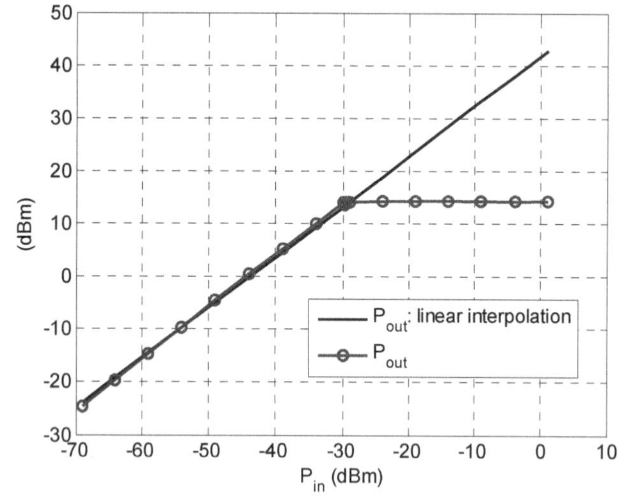

Fig. 6. Measured 1 dB compression point.

Acknowledgements. The authors wish to acknowledge the financial support of the Bavarian Ministry for Economic Affairs, Infrastructure, Transport and Technology in the scope of the innovation policy of Bavaria and the German Space Agency (DLR).

References

Janssens, J. and Steyaert, M.: CMOS Cellular Receiver Front-Ends, New York: Kluwer Academic Publishers, 2002.

Kaplan, E. D. and Hegarty, C.: Understanding GPS: Principles and Applications, Boston, Artech House, 2006.

Global Positioning System: Standard Positioning Service Signal Specification, 2nd Edition, June 1995.

OS SIS ICD, Galileo Open Service, Signal In Space Interface Control Document, Draft 1, February 2008.

Gustafsson, M., Parssinen, A., Bjorksten, P., Makitalo, M., Uusitalo, A., Kallioinen, S., Jallivuori, J., Korpi, P., Rintamaki, S., Urvas, I., Saarela, T., and Suhonen, T.: A Low Noise Figure 1.2-V CMOS GPS Receiver Integrated as a Part of a Multimode Receiver, IEEE Journal of Solid-State Circuits, 42(7), 1492–1500, 2007.

Lim, K., Lee, S.-H., Min, S., Ock, S., Hwang, M.-W., Lee, C.-H., Kim, K.-L., and Han ,S.: A Fully Integrated Direct-Conversion Receiver for CDMA and GPS Applications, IEEE Journal of Solid-State Circuits, 41(11), 2408–2416, 2006.

4

Improving MIMO-OFDM decision-directed channel estimation by utilizing error-correcting codes

P. Beinschob, M. Lieberei, and U. Zölzer

Department of Signal Processing and Communications, Helmut-Schmidt-Universität/University of the Federal Armed Forces Hamburg, Germany

Abstract. In this paper a decision-directed Multiple-Input Multiple-Output (MIMO) channel tracking algorithm is enhanced to raise the channel estimate accuracy. While DDCE is prone to error propagation the enhancement employs channel decoding in the tracking process. Therefore, a quantized block of symbols is checked on consistency via the channel decoder, possibly corrected and then used. This yields a more robust tracking of the channel in terms of bit error rate and improves the channel estimate under certain conditions.

Equalization is performed to prove the feasibility of the obtained channel estimate. Therefore a combined signal consisting of data and pilot symbols is sent. Adaptive filters are applied to exploit correlations in time, frequency and spatial domain. By using good error-correcting coding schemes like Turbo Codes or Low Density Parity Check (LDPC) codes, adequate channel estimates can be acquired even at low signal to noise ratios (SNR). The proposed algorithm among two others is applied for channel estimation and equalization and results are compared.

1 Introduction

MIMO-OFDM combines two powerful techniques for next generation wireless communication systems. For development accurate channel estimates of certain environments are required to derive realistic parameters for Monte-Carlo simulations of MIMO channels. They can be obtained by applying a MIMO system and an appropriate channel estimation algorithm in the respective environment. In Akhtman and Hanzo (2007) a channel tracking algorithm based upon a MIMO Recursive Least Square (RLS) algorithm is proposed. MIMO RLS channel estimation has obvious advantages over traditional interpolation approaches. On the one hand, time and spatial diversity can be exploited which makes it suitable for spatial multiplexing MIMO system design approaches even in fast time-varying environments. On the other hand, due to the filter structure the output depends only on the current input, therefore the channel estimate is available almost delay-free rather then after whole frame processing as in interpolation if time variance is reflected.

The paper is structured as follows. The system model is described in Sect. 2. The employed MIMO-OFDM RLS algorithm is reviewed in Sect. 3. The enhancement to this algorithm is shown in Sect. 4. Finally, the performance of Turbo Codes and LDPC Codes as channel coding schemes in this context is compared using the mean square error of the channel estimate and the MIMO system's bit error rate (BER) versus SNR in Sect. 5, before concluding in Sect. 6.

2 System model

The vector of received values $\mathbf{r}$ at the time sample m of a MIMO system is a superposition of $L \cdot N_T$ previously send samples and the current N_T samples. It is given by

$$\mathbf{r}[m] = \sum_{l=0}^{L} \mathbf{H}[l, m] \cdot \mathbf{s}[m - l] + \mathbf{w}[m], \tag{1}$$

where $\mathbf{s}[m]$ denotes the current vector of symbols of every transmit antenna, $\mathbf{w}$ an identically, independently distributed (i.i.d.) additive white Gaussian noise term and $\mathbf{H}$ the MIMO channel matrix. The past sent samples are denoted by $\mathbf{s}[m-l]$, for $l \neq 0$. For simulations the data symbols of the K subcarrier are modulated by an inverse Fast Fourier Transform (IFFT). Every value corresponding to a transmit antenna of the resulting vectors is transmitted using the formula above. If a time variant channel is assumed, for every sample m the channel matrix differs from the previous one. It is modeled such that the movement of the terminal results

Correspondence to: P. Beinschob
(patric.beinschob@hsu-hamburg.de)

in a Doppler shift of the send signal. This property is integrated in $\mathbf{H}[l, m]$ by first generating independent channel matrices from every m, l and second filtering over time m for every tap l with the Jakes spectrum, that reflects the spread of the Doppler shifts in the frequency domain. In a simulation only a finite number of Doppler frequencies can be realized, therefore the Jakes spectrum is sampled at 16 non-zero frequencies. The result is a continuously time-varying channel, even for small terminal velocities (Pätzold, 1999).

In the frequency domain the system model in Eq. (1) can be described in the time invariant case as

$$\mathbf{r}[n, k] = \mathbf{H}[n, k] \cdot \mathbf{s}[n, k] + \mathbf{w}[n, k], \quad (2)$$

where n denotes the time index of an OFDM symbol and k its subcarrier index. The MIMO channel coefficients $H_{r,t}, r{=}1, \ldots, n_R, t{=}1, \ldots, n_T$ are normal i.i.d. random variables for every tap. As described above they may vary over time and frequency n, k in a MIMO-OFDM frame. Through the whole discussion and in the simulations a frame consists of 256 OFDM symbols of which the first 128 are pilots to train the adaptive filters. The source bits are coded using modern coding schemes like Turbo Codes and LDPC Codes. The codes are constructed such that one codeword fits in an OFDM symbol. Thus for a M-QAM modulation scheme and code rate R_C the codeword length set to

$$n_C = K \cdot R_C \cdot \log_2 M. \quad (3)$$

It is important for the ability to track time-variant channels that the symbols in a transmitted OFDM symbol can be decoded at once.

For all simulations presented pilots and data symbols are send to compute a bit error rate as a confidence level for the estimated channel impulse response.

The pilot tones are organized as in code division multiplexing (CDM) manner. A single symbol is mapped to a sequence. The sequences are pairwise orthogonal thus preventing mutual interference. Each transmit antenna sends its own sequence. A set of orthogonal sequences are (Frank-Zadoff-Chu, 1972) sequences which are used for channel estimation in the training phase.

3 Adaptive MIMO RLS channel estimation

The employed MIMO-OFDM-Recursive Least Squares (RLS) algorithm, proposed in Akhtman and Hanzo (2007), is due to find a solution to the Wiener-Filter problem:

$$\min E\{(\mathbf{r}[k] - \mathbf{H}[k] \cdot \mathbf{s}[k])^2\}. \quad (4)$$

The global solution is called Wiener-Hopf equation

$$\mathbf{H}[k] = (\mathbf{\Phi}^{-1}[k]\boldsymbol{\theta}[k])^H. \quad (5)$$

with

$$\mathbf{\Phi}[k] = E\left\{\mathbf{s}[k] \cdot \mathbf{s}^H[k]\right\} \quad \forall k$$
$$\boldsymbol{\theta}[k] = E\left\{\mathbf{s}[k] \cdot \mathbf{r}^H[k]\right\} \quad \forall k. \quad (6)$$

In general the expectation values are unknown and must be estimated. To derive the RLS algorithm Eq. (4) is modified to get the following optimization problem: The channel estimate $\tilde{\mathbf{H}}[n, k]$ at time instant n on subcarrier k should map the data symbols $\mathbf{s}[n, k]$ in such a way that the distance to the actually received symbols is minimized

$$\min \varepsilon = \min \|\tilde{\mathbf{r}}[n, k] - \tilde{\mathbf{H}}[n, k] \cdot \mathbf{s}[n, k]\|_2. \quad (7)$$

The RLS implementation of this solution replaces the expectation value with an exponentially weighted sum. Therefore the matrices can be calculated recursively:

$$\begin{aligned}\mathbf{\Phi}[n, k] &= \sum_{m=1}^{n} \xi^{n-m}\mathbf{s}[m, k]\mathbf{s}^H[m, k] \\ &= \xi\mathbf{\Phi}[n-1, k] + \mathbf{s}[n, k]\mathbf{s}^H[n, k] \quad (8)\end{aligned}$$

and

$$\begin{aligned}\boldsymbol{\theta}[n, k] &= \sum_{m=1}^{n} \xi^{n-m}\mathbf{s}[m, k]\mathbf{r}^H[m, k] \\ &= \xi\boldsymbol{\theta}[n-1, k] + \mathbf{s}[n, k]\mathbf{r}^H[n, k]. \quad (9)\end{aligned}$$

Still Eq. (5) is the global minimum and used to calculate the channel matrix $\mathbf{H}$. By the ξ weighting, newer samples have greater impact on the channel estimate thus adaptivity on time variant channels can be achieved. For lower SNR a $\xi \to 1$ is in favor of averaging noise out but the algorithm loses its ability to adapt to time varying channels (Haykin, 2001). So a compromise has to be found, it is set to $\xi{:=}0.95$.

By employing such an estimator for every subcarrier, the complete channel transfer function $\mathbf{H}^k_{r,t}[n]$ for every SISO channel of the MIMO system is calculated for every OFDM symbol.

Next step is to distinct the corresponding meaningful CIR taps from noise. This is done by an eigenvalue tracking algorithm called Projection Approximation Subspace Tracking (PAST) (Yang, 1995). The meaningful CIR taps build a subspace in the signal space which is reasonably smaller than the signal space. In tracking only the subspace complexity can be reduced especially in focus on the third step, in which the CIR taps are predicted for the next OFDM symbol.

The CIR subspace $\boldsymbol{\alpha}^l_{r,t}[n]$ can be calculated using an RLS algorithm which minimizes the following cost function

$$\min_{\mathbf{W}} J(\mathbf{W}) = \mathrm{E}\{\|\mathbf{H} - \mathbf{W}\mathbf{W}^H\mathbf{H}\|^2\}, \quad (10)$$

in which $\mathbf{W}$ denotes the subspace transformation matrix. Again replacing expectation value with an exponentially

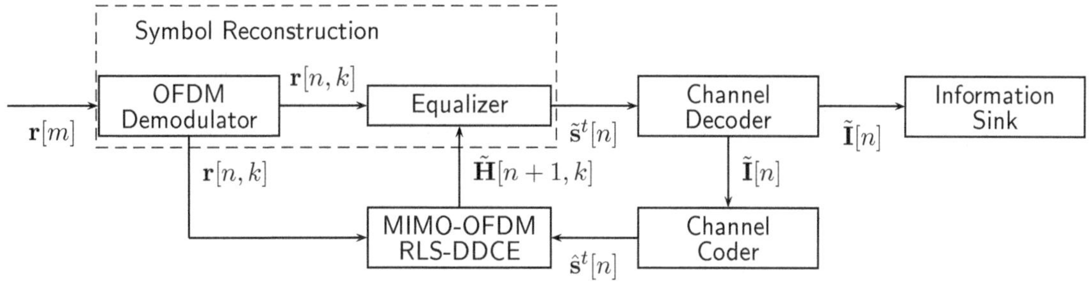

Fig. 1. Block diagram of proposed algorithm enhancement.

weighted sum yields

$$J(\mathbf{W}[n]) = \mathrm{E}\{\|\mathbf{H}[n] - \mathbf{W}[n]\mathbf{W}^H[n]\mathbf{H}[n]\|^2\}$$
$$= \sum_{i=1}^{n} \eta^{n-i} \|\mathbf{H}[i] - \mathbf{W}[i]\mathbf{W}^H[i]\mathbf{H}[m]\|^2. \quad (11)$$

For $\boldsymbol{\alpha}[n]{=}\mathbf{W}^H[n]\mathbf{H}[n]{\approx}\mathbf{W}^H[n{-}1]\mathbf{H}[n]$, again the discrete Wiener-Hopf equation is the minimum, which can be computed by

$$\mathbf{W}[n] = \mathbf{C}_{\alpha\mathbf{H}}[n]\mathbf{C}_{\mathbf{HH}}^{-1}[n] \quad (12)$$

where

$$\mathbf{C}_{\alpha\mathbf{H}}[n] = \sum_{i=1}^{n} \eta^{n-i}\mathbf{H}[i]\boldsymbol{\alpha}^H[i] \quad (13)$$
$$= \eta\mathbf{C}_{\mathbf{H}\alpha}[n-1] + \mathbf{H}[n]\boldsymbol{\alpha}^H[n] \quad (14)$$

and

$$\mathbf{C}_{\mathbf{HH}}[n] = \sum_{i=1}^{n} \eta^{n-i}\mathbf{H}[i]\mathbf{H}^H[i] \quad (15)$$
$$= \eta\mathbf{C}_{\mathbf{HH}}[n-1] + \mathbf{H}[n]\mathbf{H}^H[n]. \quad (16)$$

Again, η is a weight factor that reconsiders time variant behavior and is also set to $\eta{:=}0.95$.

The prediction of the CIR subspace is then straightforward RLS prediction by minimizing

$$J(\alpha_l) = \sum_{i=1}^{n} \beta^{n-i}|\alpha_l[i+1] - \mathbf{q}_l^H[n]\alpha_l[i]|^2. \quad (17)$$

and the recursion of the filter coefficients $\mathbf{q}_{r,t}^l[n]$ can efficiently be calculated with

$$\mathbf{q}_l[n] = \mathbf{q}_l[n-1] + \mathbf{k}_l[n-1]e_l^*[n], \quad (18)$$

where

$$e_l[n] = \hat{\alpha}_l[n] - \mathbf{q}_l^H[n-1]\hat{\boldsymbol{\alpha}}_l[n-1] \quad (19)$$

is the a-priori prediction error, which is given by $\hat{\boldsymbol{\alpha}}_l[n]{=}(\alpha_l[n], \alpha_l[n{-}1], \ldots, \alpha_l[n{-}M_P{+}1])^T$ that covers the history of the l-th CIR tap. The RLS prediction gain vector is given by

$$\mathbf{k}_l[n] = \frac{\mathbf{P}_l[n-1]\hat{\boldsymbol{\alpha}}_l[n]}{\beta + \hat{\boldsymbol{\alpha}}_l^H[n]\mathbf{P}_l[n-1]\hat{\boldsymbol{\alpha}}_l[n]} \quad (20)$$

and the recursively calculated, inverse of the covariance matrix of the relevant CIR taps by

$$\mathbf{P}_l[n] = \frac{1}{\beta}\left(\mathbf{I}_{M_P+1} - \mathbf{k}_l[n]\hat{\boldsymbol{\alpha}}_l^H[n]\right)\mathbf{P}_l[n-1]. \quad (21)$$

Likewise, β is a weight factor, but it has been stated that the impact on the algorithm's convergence is very low (Schafhuber and Matz, 2005). It is set to $\beta{:=}0.9$.

4 The DDCE enhancement

In the discussed algorithm the symbols $\mathbf{s}$ are estimated by equalizing the corresponding received symbols $\tilde{\mathbf{s}}{=}\mathbf{H}^{-1}\mathbf{r}$. A traditional DDCE scheme would then map the symbol estimates $\tilde{\mathbf{s}}$ on the nearest constellation symbols. This can be described by a quantization operation $\mathcal{Q}$: $\hat{\mathbf{s}}{=}\mathcal{Q}\{\tilde{\mathbf{s}}\}$ which maps the infinite set of equalized symbols to the finite set of constellation symbols. The output symbol depends only on the euclidean distance of the symbol estimate to the constellation symbols as commonly used in "hard decision" demodulation. The algorithm reviewed in the previous section depends highly on correctly decided symbols $\hat{\mathbf{s}}$ that are used instead of the at the receiver unknown send symbols $\mathbf{s}$. Once a symbol is decided wrong the channel estimate is changed and further symbols are less likely to be correctly equalized. Through the PAST algorithm frequency coupling the error on a subcarrier spreads upon the others and finally the predictor outputs for all subcarriers are affected. The whole process of error propagation starts and the algorithm is unlikely to recover if not re-initialized.

Identified the weakness in the quantization it is possible to make an improvement. As said, to achieve minimum delay in processing a time varying channel, the channel coding is done on every OFDM symbol separately. Therefore, the codeword length is linked to the number of subcarriers

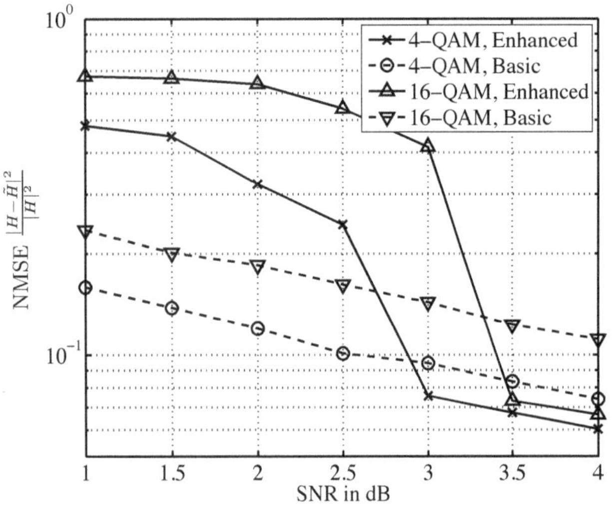

Fig. 2. Channel estimation error expressed in Normalized Mean Square Error (NMSE).

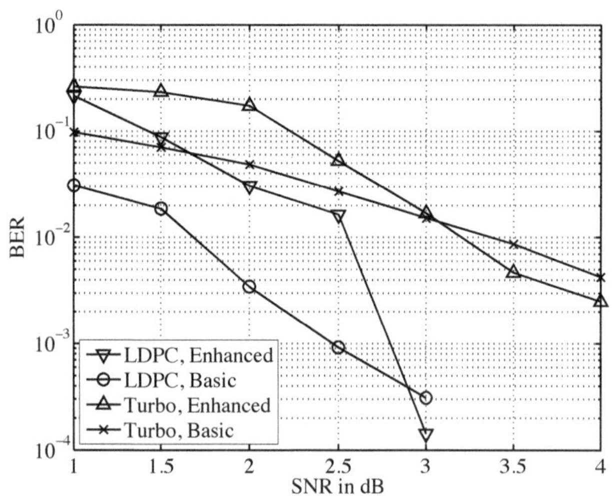

Fig. 3. BER comparison between basic DDCE and enhanced DDCE implementation for Turbo and LDPC Codes.

as stated in Eq. (3); the associated symbol vector of the t-th antenna is denoted with $\mathbf{s}^t$.

The single symbol mapping or quantization operation $\mathcal{Q}$ will be replaced by decisions based upon a codeword $\tilde{\mathbf{I}}=\mathcal{C}^{-1}\{\tilde{\mathbf{s}}^t\}$ and $\hat{\mathbf{s}}^t=\mathcal{C}\{\tilde{\mathbf{I}}\}$ we can write a new quantization operation as $\mathcal{Q}'\{\cdot\}:=\mathcal{C}\{\mathcal{C}^{-1}\{\cdot\}\}$. Through the decoding and reencoding symbol errors will be corrected and the corrected versions are used to update the channel estimate in Eqs. (8) and (9).

Therefore the estimated symbols $\tilde{\mathbf{s}}^t$ are soft demodulated. The corresponding Log-Likelihood Ratios (LLR) $\tilde{\mathbf{I}}$ are input of the used decoder. Both Turbo Codes' and LDPC Codes' decoder use LLRs. The possibly corrected decoder output must be modulated and is then equivalent to the mapped versions $\hat{\mathbf{s}}$ and can be fed into the MIMO RLS algorithm. The whole scheme is visualized in Fig. 1.

However, if the noise is too high and too many errors occur in a codeword the underlying decoding procedure might introduce more errors through unsuccessful iterative decoding than there are originally. This is in particular the case for LDPC Codes. Thus, below the threshold of a sufficiently probable correct decoding, the proposed enhancement performs worse than the unmodified algorithm. Only above that threshold a gain in channel estimation in terms of mean square error can be realized.

5 Simulation results

For the following results the system parameters are given in Table 1. The MIMO transmission scheme is chosen as spatial multiplexing, i.e. two independent data streams are transmitted over the two transmit antennas. It lies in the structure of the MIMO RLS algorithm to find an optimal combination of the four receive signals to reconstruct the transmit signals.

5.1 Turbo vs. LDPC codes

In Fig. 2 a measure for the channel estimate quality, the normalized mean squared error (NMSE), is given for LDPC and Turbo Codes of M=4 and M=16 QAM. It is calculated by

$$\mathrm{NMSE} = \frac{\sum_{n=0}^{N-1}\sum_{k=0}^{K-1}\sum_{t=1}^{n_T}\sum_{r=1}^{n_R}\left|H_{r,t}[n,k]-\tilde{H}_{r,t}[n,k]\right|^2}{\sum_{n=0}^{N-1}\sum_{k=0}^{K-1}\sum_{t=1}^{n_T}\sum_{r=1}^{n_R}\left|H_{r,t}[n,k]\right|}. \quad (22)$$

The decoder algorithm implementation for LDPC codes is based on belief propagation (Pearl, 1988), for Turbo Codes the Log-MAP algorithm is used. Both decoder approximate a-posteriori probabilities from the soft decision inputs and decide them. For the 16-QAM modulation scheme the code rate R_C was lowered to 1/4 to match the same E_b/N_0 as with 4-QAM and R_C=1/2. It is shown that the proposed enhancement improves the channel estimate quality in terms of MSE. All solid lines, which denote the enhanced scheme, finally are below the dashed ones, which denote the basic DDCE approach, even for 16-QAM modulation which requires higher accuracy in magnitude. The minimum of MSE is reached when also the BER in Fig. 3 first hits its minimum, i.e. error-free transmission of 15 frames. If the transmission is not error-free, in the case the SNR is too low, the proposed scheme performs much worse than the basic algorithm. This holds in terms of MSE and BER. In this case the decoding algorithms try to determine the possibly corrupted message bits from a large portion of corrupted check bits. The iterative result may end in a decoded bit word that has a larger

Table 1. MIMO-OFDM system parameters.

$N_T \times N_R$	2×4
number of subcarriers K	128
samples of guard interval	8
number of OFDM symbols per frame	256
number of training symbols	128
channel model	COST207 BU
channel order L	8
bandwidth	12.5 MHz
carrier frequency	2.412 GHz
normalized Doppler frequency $f_{d,n}$	0.0018921
terminal velocity v	77 km/h

Hamming distance than the unprocessed message bits to the original message. This introduces even more symbol errors in RLS algorithm than the basic algorithm would. So applying this scheme at very low SNR yields no gain, even more loss. But above the threshold of virtually error-free decoding a gain in MSE can be realized.

5.2 DDCE schemes vs. interpolation

In Fig. 4 the bit error rates for three MIMO-OFDM systems (4-QAM modulated) and a single-carrier system are compared: First a MIMO-OFDM RLS enhanced DDCE system with LDPC Codes as channel coder, second again a RLS enhanced DDCE system with Turbo Codes consisting of two recursive systematic convolutional coder with generator polynomials (7, 5) punctured to code rate 1/2. The third is a MIMO-OFDM system with a polynomial interpolation over distributed pilots in the time-frequency grid (Lieberei and Zölzer, 2008). The last system is the FS-BLAST where the interference is canceled out successively from the MIMO layers (Weikert, 2007). The number of pilots is equal to the RLS DDCE systems. Furthermore, its channel code is the same as the first system's. It is a half-rate LDPC code built from a regular parity check matrix (MacKay, 1999).

The RLS DDCE systems clearly outperform the interpolation and the FS-BLAST systems under the conditions considered here, SNR range, channel order and Doppler shifts.

5.3 Future work

For the simulations LDPC codes where generated using a regular parity check matrix. Recent results have shown that irregular parity check matrices have even better error-correcting properties (Luby et al., 1997). So additional improvements can be made if these codes are used. However, the code word length has impact on the performance. In the presented system a larger code word would need more subcarriers thus means a longer OFDM symbol duration. A larger OFDM symbol duration would decrease the ability

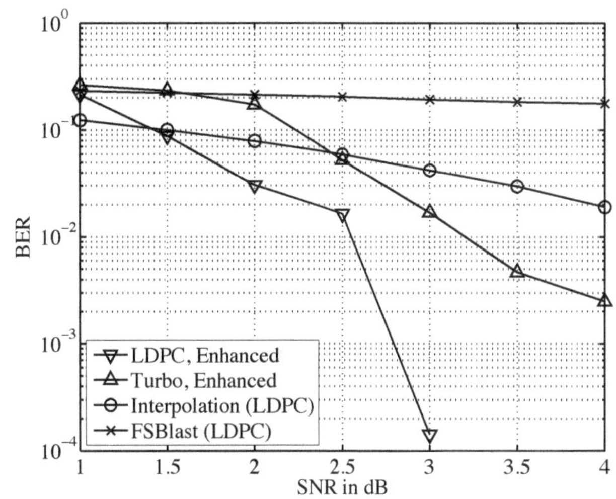

Fig. 4. BER comparison enhanced DDCE scheme with a traditional interpolation approach (LDPC coded).

to track a time-varying channel. So short and yet powerful LDPC Codes are highly desirable.

6 Conclusions

A channel tracking algorithm is extended for a DDCE that uses a channel decoder/encoder as a symbol quantizer. It is feasible for the algorithm to recover from errors through the error-correcting properties of the codes. This results in significant lower failure rate due to error propagation. A threshold in SNR can be determined which depends on the error-correcting performance of the applied code. Above this threshold, the proposed DDCE code enhancement improves the channel estimate and bit error rate that verifies the channel estimate. But it is limited due to the decoding operation a OFDM symbol must hold a complete code word. Below the threshold the system's performance is degraded by the proposed scheme. The choice of channel coder has impact on the overall performance.

Even LDPC Codes from regular parity check matrices outperforms widespread Turbo Codes in this context. Additionally the MIMO-RLS scheme is compared to an interpolation approach where the noise averaging ability of the RLS leads to a better bit error rate.

References

Akhtman, J. and Hanzo, L.: Advanced Channel Estimation for MIMO-OFDM in Realistic Channel Conditions, IEEE International Conference on Communications (ICC), 2528–2533, 24–28 June, 2007.

Pätzold, M.: Mobilfunkkanäle, Vieweg Verlag, 1st Edition, 1999.

Chu, D. C.: Polyphase Codes with Good Periodic Correlation Properties, IEEE Transactions on Information Theory, 18(4), 531–532, 1972.

Haykin, S.: Adaptive Filter Theory, Prentice Hall, 4th Edition, 2001.

Yang, B.: Projection Approximation Subspace Tracking, IEEE Transactions on Signal Processing, 43(1), 95–107, 1995.

Schafhuber, D. and Matz, G.: MMSE and adaptive prediction of time-varying channels for OFDM systems, IEEE Transactions on Wireless Communications, 4(2), 593–602, 2005.

Pearl, J.: Probabilistic reasoning in intelligent systems: Networks of plausible inference, Kaufmann, 1988.

Lieberei, M. and Zölzer, U.: MIMO-OFDM on a Flexible Laboratory Radio System, International Symposium on Communications, Control and Signal Processing (ISCCSP), 1184–1187, 2008.

Weikert, O.: Blinde Demodulation in MIMO-Übertragungssystemen, Dissertation, Helmut-Schmidt-Universität, 2007.

MacKay, D. J. C.: Good error-correcting codes based on very sparse matrices, IEEE Transactions on Information Theory, 45(2), 399–431, 1999.

Luby, M., Mitzenmachen, M., Shokrollahi, A., Spielman, D., and Stemann, V.: Practical Loss-Resilient Codes, Proceedings of the 29th annual ACM Symposium on Theory of Computing, 150–159, 1997.

Two methods for transmission line simulation model creation based on time domain measurements

D. Rinas and S. Frei

Dortmund University of Technology, Dortmund, Germany

Abstract. The emission from transmission lines plays an important role in the electromagnetic compatibility of automotive electronic systems. In a frequency range below 200 MHz radiation from cables is often the dominant emission factor. In higher frequency ranges radiation from PCBs and their housing becomes more relevant. Main sources for this emission are the conducting traces. The established field measurement methods according CISPR 25 for evaluation of emissions suffer from the need to use large anechoic chambers. Furthermore measurement data can not be used for simulation model creation in order to compute the overall fields radiated from a car. In this paper a method to determine the far-fields and a simulation model of radiating transmission lines, esp. cable bundles and conducting traces on planar structures, is proposed. The method measures the electromagnetic near-field above the test object. Measurements are done in time domain in order to get phase information and to reduce measurement time. On the basis of near-field data equivalent source identification can be done. Considering correlations between sources along each conductive structure in model creation process, the model accuracy increases and computational costs can be reduced.

1 Introduction

Electromagnetic Compatibility plays an important role in the development of automotive systems. Beside the radiating cable bundles integrated Electronic Control Units (ECUs) are sources for electromagnetic emissions.

Standardized component field measurement methods, like the ALSE antenna method provided in CISPR 25 for evaluation of electro-magnetic emissions from automotive systems, suffer from the need of large and expensive anechoic chambers. Also a single field strength value is often not sufficient to characterize the EMI behavior of a complex system. Furthermore it is not possible to use the measurement data for behavioral simulation model creation. Having simulation models, a statement about the radiating electromagnetic fields can already be made in early phases of development.

Basically the electro-magnetic emission must be distinguished in the emission of circuit boards and their housing and the emission of their connecting cable bundles. Where in lower frequency range up to 200 MHz radiation from the bundles is dominant the importance of emission from ECUs grows with the rising frequency.

In Rinas et al. (2010) a method for estimating the emissions from cable bundles is presented. Here the electromagnetic near-field at several points near a cable bundle is measured. With measured data a single transmission line model with equivalent current distribution is generated. The measurements are done in Time Domain with a standard oscilloscope to get phase information and to reduce measurement time.

To create a behavioral model of PCB the electromagnetic near-field in a plane above the radiating structure is measured. Knowing the fields in an indefinitely extended plane above the test object all information is available to calculate any field vector above this plane (Balanis et al., 1996). This can be used to solve an inverse problem and to identify the equivalent sources of the planar structure. This approach is discussed in Vives-Gilabert (2007); Baudry et al. (2008); Isernia et al. (1996); and Tong et al. (2010). Several methods dealing with the inverse problem are ill-posed due to the presence of errors in measurement data, or suffer from the problem of converging to local minima. When optimization methods are applied this leads to long computation time (Isernia et al., 1996; Pierri et al., 1999).

Correspondence to: D. Rinas
(denis.rinas@tu-dortmund.de)

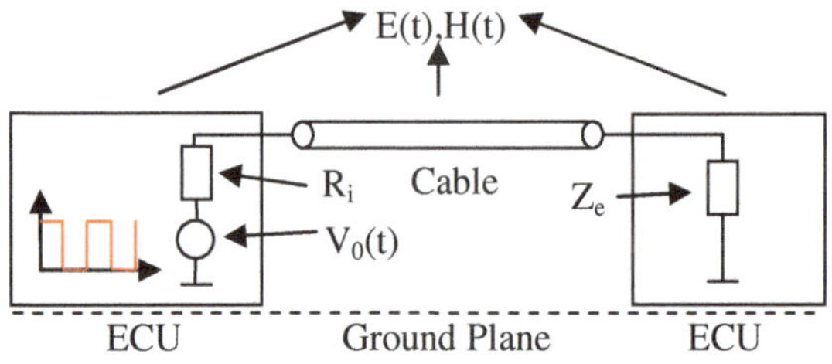

Fig. 1. Simple configuration to investigate.

As main sources for emission of PCBs are the conducting traces, the distribution of the approximating sources can be bounded to the routing geometry. Furthermore techniques for correlating the properties of the approximating sources to compute a more physically model can be integrated in the model creation process. This can reduce the number of free source parameters and computation time. Furthermore the possibility of error correction is given and accuracy of the results increases.

In this paper a method for source identification of transmission lines, in form of cables bundles and conducting paths on PCBs, with considering correlations between the sources is presented. Investigations are done in a frequency range up to 500 MHz. The generated model enables different types of postprocessing e.g. far field estimations.

2 Source identification of cable bundles

As shown in Rinas et al., 2010 the current distribution of a radiating single cable (Fig. 1) can be estimated by measuring the magnetic field in at least two field points along the cable.

When calculation is based on two measurement sets

$$I_z = \frac{1}{2Z}(V_e + ZI_e)e^{j\beta(l-z)} + \frac{1}{2Z}(V_e - ZI_e)e^{-j\beta(l-z)} \tag{1}$$

with

$$\begin{pmatrix} V_e \\ I_e \end{pmatrix} = \begin{bmatrix} a_1\, b_1 \\ a_2\, b_2 \end{bmatrix}^{-1} \cdot \begin{pmatrix} I_1(z_1) \\ I_2(z_2) \end{pmatrix} \tag{2}$$

leads to current I_z at position z. Where V_e and I_e are voltage and current at the end of transmission line, Z is the wave impedance, β the propagation constant, l the length of the line, a and b include the propagation functions and I_1 and I_2 are the current values corresponding to the magnetic field data.

The current distribution can also be estimated directly by increasing the amount of measured field points along the cable depending on the frequency with respect to

$$N = l/\Delta d \tag{3}$$

$$\Delta d > \frac{1}{10}\lambda \tag{4}$$

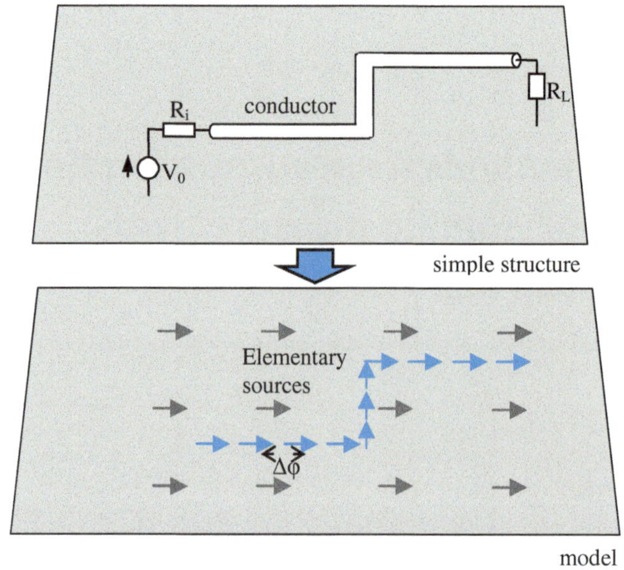

Fig. 2. Drawin g of the test board (above), model of the board (below) with equally distributed sources (grey) and sources only along the current path (blue).

where N is the amount of measurement points and Δd the spatial discretization width between the neighboring points. For achieving I_z the discrete current values are approximated to a sinusoidal function with nonlinear regression algorithms.

In order to create an equivalent emission model elementary electric dipoles are placed in a line along the cable path in a height h over ground adopted from the cable, following 3. The dipole moments $M_1 \ldots M_K$ are calculated with

$$M_k = \Delta d \frac{I_k + I_{k+1}}{2} \tag{5}$$

where I_k is the discretized current I_z.

3 Source identification of conducting traces on planar structures

According to Table 2 transmission lines on planar structures, esp. PCBs, can also be approximated with elementary dipoles. With knowledge of the current paths, e.g. based on CAD-data, the approximating dipoles can be distributed along these paths (Fig. 2). Where R_i and R_L are the terminating impedances and V_0 is the exciting input voltage. The number of dipoles is chosen following Eq. (3) This approach, in comparison with the method of equally distributed dipoles, can considerable reduce the amount of necessary elementary sources, decrease computational cost and increase the model accuracy.

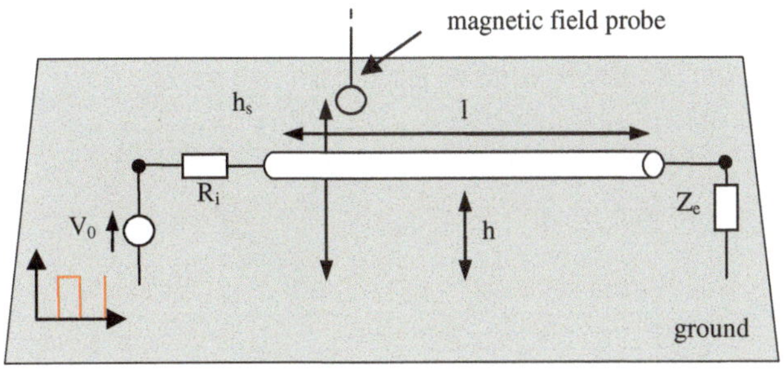

Fig. 3. Single conductor under test.

Dependencies between the neighboring elementary sources of each current path are used to correlate phases of the dipoles. For the spatial distribution Δd the phase shift between two adjacent dipoles is set to

$$\Delta\varphi = \frac{2\pi}{\lambda}\Delta d \tag{6}$$

as shown in Fig. 2. With adherence to Eq. (6) phase jumps between two sources different from $\Delta\phi$ are prevented and the model becomes more physically.

To compute the dipole moments M the magnetic near-field has to be measured in a plane above the planar structure. Afterwards the system of equations

$$I_0 = \mathrm{A}^{-1}H \tag{7}$$

where A contains the wave vector k_0 and the fixed geometric parameters for each dipole and H the complex magnetic field at K near-field points is solved.

For integrating the correlation between sources into model creation the inverse problem can be treated with the expandable minimization function F following 4.

$$F = \left(\sum_{m=1}^{M}\alpha_m(x_1,y_1,z_1)M_m e^{-j\varphi_m},\ldots,\right.$$
$$\left.\sum_{m=1}^{M}\alpha_m(x_N,y_N,z_N)M_m e^{-j\varphi_m}\right)^T$$
$$-(H_0(x_1,y_1,z_1),\ldots,H_0(x_N,y_N,z_N))^T \rightarrow \mathrm{Min} \tag{8}$$

With α_m are the fixed geometric parameters at observation point x_n, y_n, z_n, the amplitude and M_m is and the phase $e^{-j\phi m}$ of each dipole. $H_0(x_n, y_n, z_n)$ stands for the measured reference field at observation points. Optimization methods are used for solving the minimization problem.

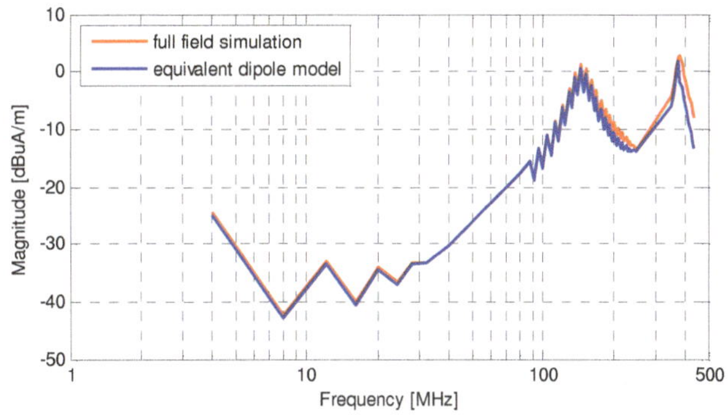

Fig. 4. Magnetic field at p_1 in comparison with full field simulation.

Table 1. Magnetic Field of Measurement based Model at p_1 in comparsion with full field simulation.

Frequency	Magnitude [dBμA/m] Sim.	Model
4 MHz	29.4963	26.3542
20 MHz	32.4243	32.9482
60 MHz	36.7945	37.3562
124 MHz	43.1243	45.2324
164 MHz	47.7698	50.1270

4 Results

4.1 Cable bundles

This chapter presents the results for model creation process of a single radiating cable and a small bundle. Some results are presently derived from an electromagnetic full field solver (EMCoS Consulting and Software, www.emcos.com).

The input signal V_0 (Figs. 1 and 3) is a pulsed signal with an amplitude $V_0 = 5\ V$, a fundamental frequency of $f_0 = 4$ MHz, a pulse/pause ratio $t_h/t_l = 1$, and a rising and falling edge of $t_{r,f} = 2.5$ ns. Source impedance R_i is 50 Ohm.

4.1.1 Single conductor – measurement based results

The cable (Fig. 3) consists of a single conductor placed in the height h = 50 mm over a ground plane. It has a length of l = 490 mm and a thickness of d = 1 mm. It is terminated with a $Z_e = 50\ \Omega$ impedance. The magnetic field is measured at two positions $y_1 = 120$ mm and $y_2 = 360$ mm along the cable at a height $h_S = 20$ mm over ground.

With the identified sources the magnetic field is calculated at a position $p_1 = (3, 3, 3)^T [m]$. Exemplary the fields of the fundamental wave and the 2th, 7th, 15th and 20th harmonics of the pulsed input signal are presented in Table 1. As a comparison a full field simulation of the cable under test given by a MoM solver (EMCoS Consulting and Software, www.emcos.com) is shown. The results agree with a maximum error of 3 dB.

Table 2. Near-Fields of different Models in Comparison with full field simulation.

Full field simulation	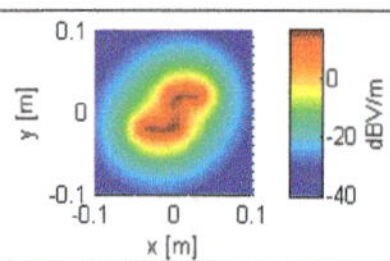	
	Equally distributed sources	Sources along current path
No phase correlation of sources	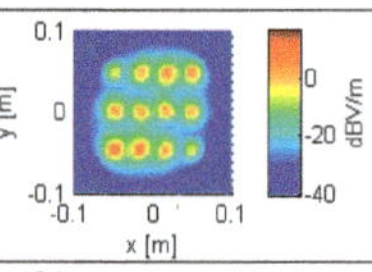	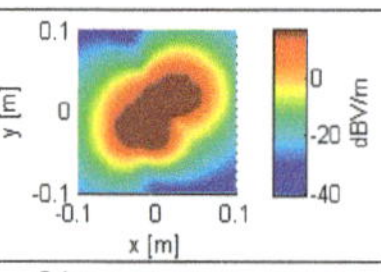
With phase correlation of sources	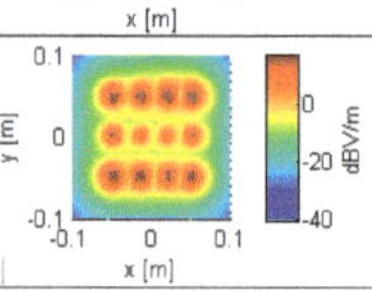	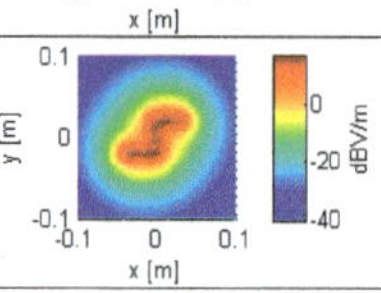

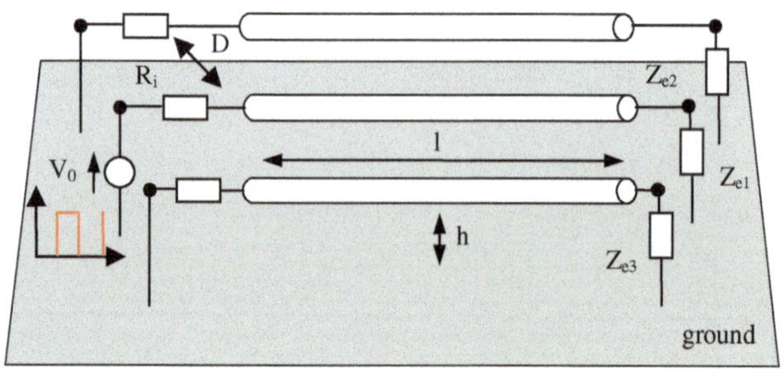

Fig. 5. Multiconductor under test.

4.1.2 Single conductor – simulation based results

The following investigations are presently based on computer simulations.

The cable (Fig. 3) consists of a single conductor placed in the height h = 10 mm over the ground plane. It has a length of l = 600 mm and a thickness of d = 0.3 mm. It is terminated with a serial circuit of resistor $R_e = 50\ \Omega$ and inductance $L_e = 1\ \mu$H. The model parameters are taken from the cable under test. The source internal resistance is set to $R_i = 50\ \Omega$. The magnetic field is measured at two positions $y_1 = 200$ mm and $y_2 = 400$ mm along the cable at a height $h_S = 20$ mm over ground.

Figure 4 shows the magnetic field at point p_1 in a frequency range of 1 MHz to 500 MHz. The results in comparison with the full field simulation agree with maximum error of about 3 dB.

4.1.3 Multiconductor – simulation based results

The following investigations are presently based on computer simulations.

The multiconductor (Fig. 5) consists of three single conductors placed in the height of h = 10 mm over ground

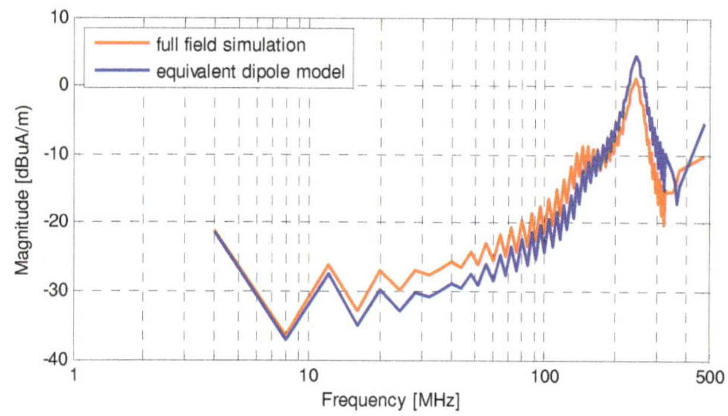

Fig. 6. Magnetic field at p_1 in comparison with full field simulation.

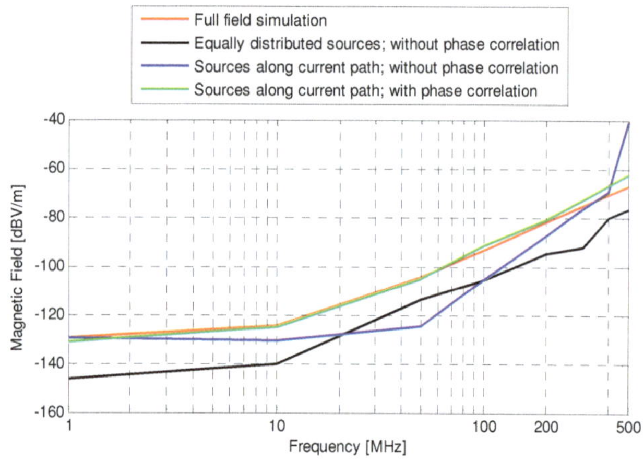

Fig. 7. Magnetic field at p_1 for different source distribution in comparison with full field simulation.

plane. The length of each conductor is l = 600 mm; the thickness is d = 0.3 mm and they are arranged in a distance of D = 5 mm. The internal source resistances are set to $R_i = 50\Omega$. The terminations are $Z_{e1} = 50\Omega || j\omega \cdot 0.1$ nF, $Z_{e2} = 50\Omega$, $Z_{e3} = 50\Omega + j\omega \cdot 1\ \mu$H. The excitation is impressed in the center conductor.

In Fig. 6 the magnetic field at point p_1 in comparison with the full field simulation is shown. The magnetic field agrees with a maximum error of 3 dB at most frequencies.

4.2 Printed circuit boards – simulation based results

A simple planar structure is analyzed. It consists of a 0.1×0.1 m plane and a conductor with total length of 0.16 m. The source voltage is 1 Volt; termination is a 50 Ohm resistance. The magnetic field is measured in a height of 0.01 m over ground at 256 near-field points. The source parameters are computed with amplitude-data only. Simulated Annealing is used as optimization method for model parameter calculation.

In Table 2 near-field calculations from different models are shown and compared with a full field simulation (MoM) of the planar structure at a frequency of 100 MHz. Figure 7 shows the magnetic field at point $p_1 = (3,3,3)^T[m]$ in comparison with the full field simulation in a frequency range of 1 MHz to 500 MHz.

As presented, the field approximation shows better accuracy if the distribution of sources is matched with the current distribution depending on the conductor path. The result plots show also a smaller deviation if the phase correlation between sources is integrated in the model computation.

5 Conclusions

In this paper methods for source identification of radiating transmission lines are presented. The focus is on cables bundles and conducting traces on planar structures. Model accuracy is can be increased by considering correlation between the neighboring approximating sources.

The approaches were tested by simulations and first measurements and were compared with numerical full field simulation data.

References

CISPR 25 Ed.3: Vehicles, boats and internal combustion engines – Radio disturbance characteristics – Limits and methods of measurement for the protection of on-board receivers, 2008.

Rinas, D., Niedzwiedz, S., and Frei, S.: Far Field Estimations and Simulation Model Creation from Cable Bundle Scans, EMC Europe, Wroclaw, 203–208, 2010.

Balanis, C. A.: Antenna Theory Analysis & Design, Wiley, 1996.

Vives-Gilabert, Y.: Modélisation des emissions rayonées de composants électroniques, Université de Rouen, 2007.

Baudry, D., Kadi, M., Riah, Z., Arcambal, C., Vives-Gilabert, Y., Louis, A., and Mazari, B.: Plane wave spectrum theory applied to near-field measurements for electromagnetic compatibility investigations", IET Science, Measurement and Technology, 15 June 2008.

Isernia, T., Giovanni Leone, G. and Pierri, R.: Radiation Pattern Evaluation from Near-Field Intensities on Planes, IEEE Transaction on Antennas and Propagation, 44(5), 701–710, 1996.

Tong, X., Thomas, D. W. P., Nothofer, A., Sewell, P., and Christopoulos, C.: A Genetic Algorithm Based Method for Modeling Equivalent Emission Sources of Printed Circuits from Near-Field Measurements, APEMC Beijing, 2010.

Isernia, T., Leone, G., and Pierri, R.: Radiation pattern evaluation from near-field intensities on planes, IEEE Trans. Antennas Propogat., 44, 701–710, 1996.

Pierri, R., D'Elia, G., and Soldovieri, F.: A two probes scanning phaseless near-field far-field transformation technique, IEEE Trans. Antennas Propogat., 47, 792–802, 1999.

EMCoS Consulting and Software, www.emcos.com, January 2011.

6

An uncooled VGA-IRFPA with novel readout architecture

D. Würfel, M. Ruß, R. Lerch, D. Weiler, P. Yang, and H. Vogt

Fraunhofer Institute for Microelectronic Circuits and Systems, Duisburg, Germany

Abstract. An uncooled VGA Infrared Focal Plane Array (IRFPA) based on microbolometers with a pixel pitch of $25\,\mu$m for thermal imaging applications is presented. The IRFPA has a 16-bit digital video data output at a frame rate of 30 Hz. Thousands of Analog to Digital Converters (ADCs) are located under the microbolometer array. One ADC consists of a Sigma-Delta-Modulator (SDM) of 2nd order and a decimation filter. It is multiplexed for a certain amount of microbolometers arranged in a so called "cluster". In the 1st stage of the SDM the microbolometer current is integrated time-continuously. The feedback is applied using a switchable current source. First measurements of Noise Equivalent Temperature Difference (NETD) as a key parameter for IRFPAs will be presented.

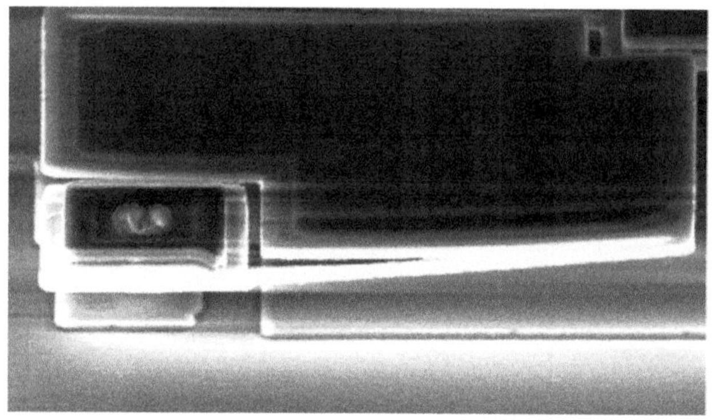

Fig. 1. SEM-photo of a microbolometer constructed at Fraunhofer IMS.

1 Introduction

Low-cost thermal imaging is dominated by uncooled infrared focal plane arrays (IRFPAs) using microbolometers based on vanadium oxide (Blackwell et al., 2008) or amorphous silicon (Durand et al., 2009) as the sensing material. According to Plank's law, each body with a temperature above absolute zero point emits electro-magnetic radiation. The wavelength and intensity depends on the temperature of the body. For example, a body at a temperature of 300 K has its maximum emittance at a wavelength of approx. $10\,\mu$m. FIR imager uses the IR-radiation of a warm body in the wavelength range between $8...14\,\mu$m. The FIR-sensitive element is the microbolometer which changes its resistance by absorbing the radiation of a warm body. Uncooled IRFPAs operate at ambient temperature and benefit from the abdication of a Stirling cooler in terms of low-cost, low power dissipation, low weight, and reduced volume. Typical applications for IRFPAs are thermal imaging and pedestrian detection for automotive driving assistance systems, fire fighting, biological imagery, or military applications like target recognition.

Correspondence to: D. Würfel
(daniel.wuerfel@ims.fraunhofer.de)

2 The microbolometer

The microbolometer is fabricated by post-processing on CMOS wafers in Fraunhofer IMS Microsystem Lab. The realization of a microbolometer is shown in Fig. 1. A micromachined membrane consisting of amorphous silicon is suspended by two via stacks of metal from the CMOS substrate. The membrane forms with two other layers a good interferometric structure for radiation absorption. On top of the membrane an antireflection layer with a sheet resistance of 377 Ω/sq is deposited. The bottom structure consists of a nearly perfect reflecting metal layer (Ruß et al., 2007). To increase the thermal resistance the membrane is fixed by two small legs. The distance between membrane and reflection metal reaches an optimum for one quarter of the radiation

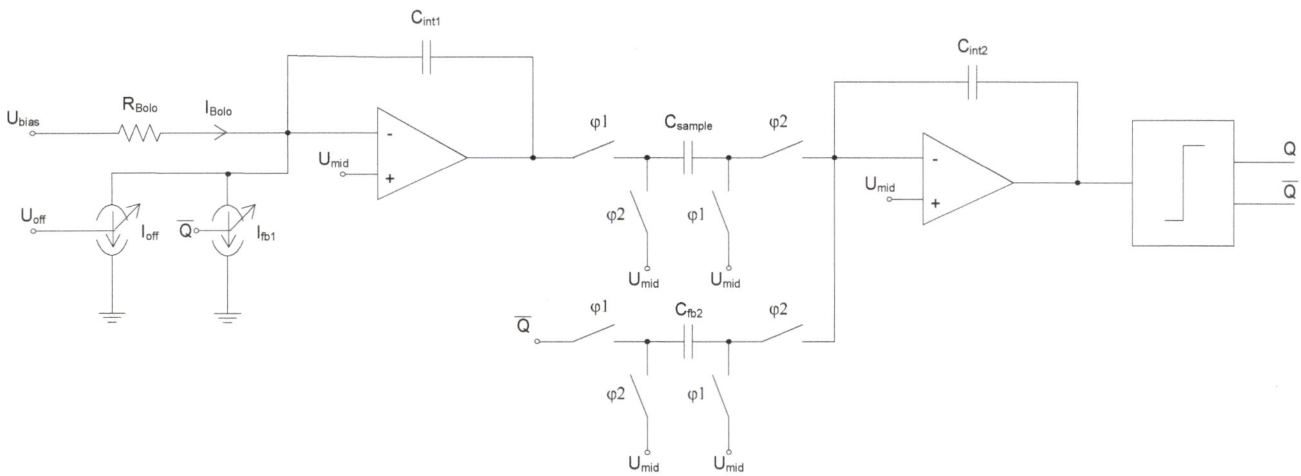

Fig. 2. Schematic of used 2nd Order SDM.

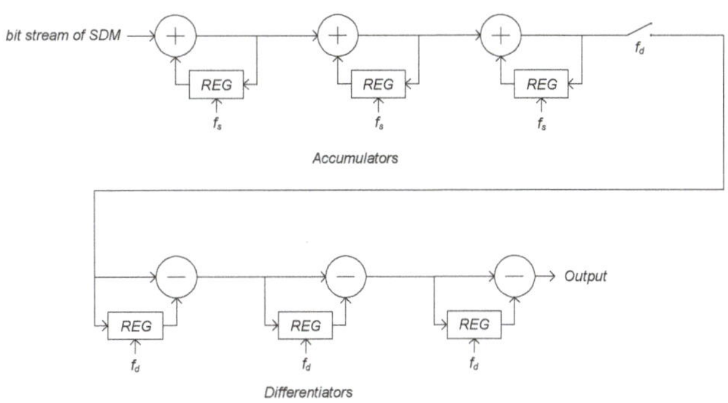

Fig. 3. Model of a 3rd Order SINC-Filter.

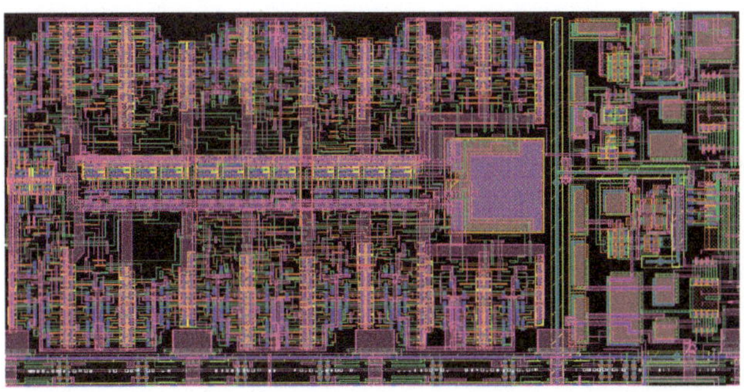

Fig. 4. Layout of a cluster (some metal layers are omitted for clarity).

wavelength. To reduce thermal losses by gas conduction a vacuum package is required. The microbolometer converts the infrared radiation into heat energy and this induces a temperature rise resulting in a change of the electrical resistance. The microbolometers are arranged in an array with a pixel pitch of $25\,\mu$m.

3 The novel readout architecture

The novel readout architecture depicted in Fig. 2 bases on a 2nd order SDM (Weiler et al., 2010). A SDM has the advantage of reaching high resolutions very easily. The price is sampling speed, because a SDM uses so called "oversampling". Quantisation noise is formed and moved out of the base band to higher frequencies. The output signal of a SDM is a digital bit stream which is fed into a decimation filter. This filter is a low-pass and suppresses the quantisation noise. The sampling speed is reduced while increasing the length of the output word. The SDM together with its decimation filter forms the ADC.

The for readout selected microbolometer is integrated directly into the SDM by closing the corresponding select-switch. Reference U_{bias} is a constant voltage. The microbolometer current consisting of an offset part and a signal part is integrated time-continuously. The feedback is applied by a switchable current source. A controllable offset current source is also included to compensate the offset part of the microbolometer current. The 2nd integrated stage works time-discretely. Because of the multiplexing of the readout circuit for 32 microbolometers a decimation filter with an finite settling time is necessary. So a 3rd order SINC-Filter was chosen for this function. Figure 3 shows a corresponding scheme based on (Norsworthy, 1997). It consists of 3 cascaded accumulators which are running at the sampling frequency f_s. Then there is a undersampling stage with frequency f_d. It follow 3 differentiators which are running at frequency f_d. The amount of 32 microbolometers which are read out by one ADC (consisting of the SDM and the decimation filter) is the result of the minimal area where one ADC can be placed. The ADC is arranged in a cluster consisting of 4 times 8 microbolometer areas. A layout of a cluster is depicted in Fig. 4. Three quarters of the area to the left are occupied by the digital SINC-Filter. The SDM is placed in

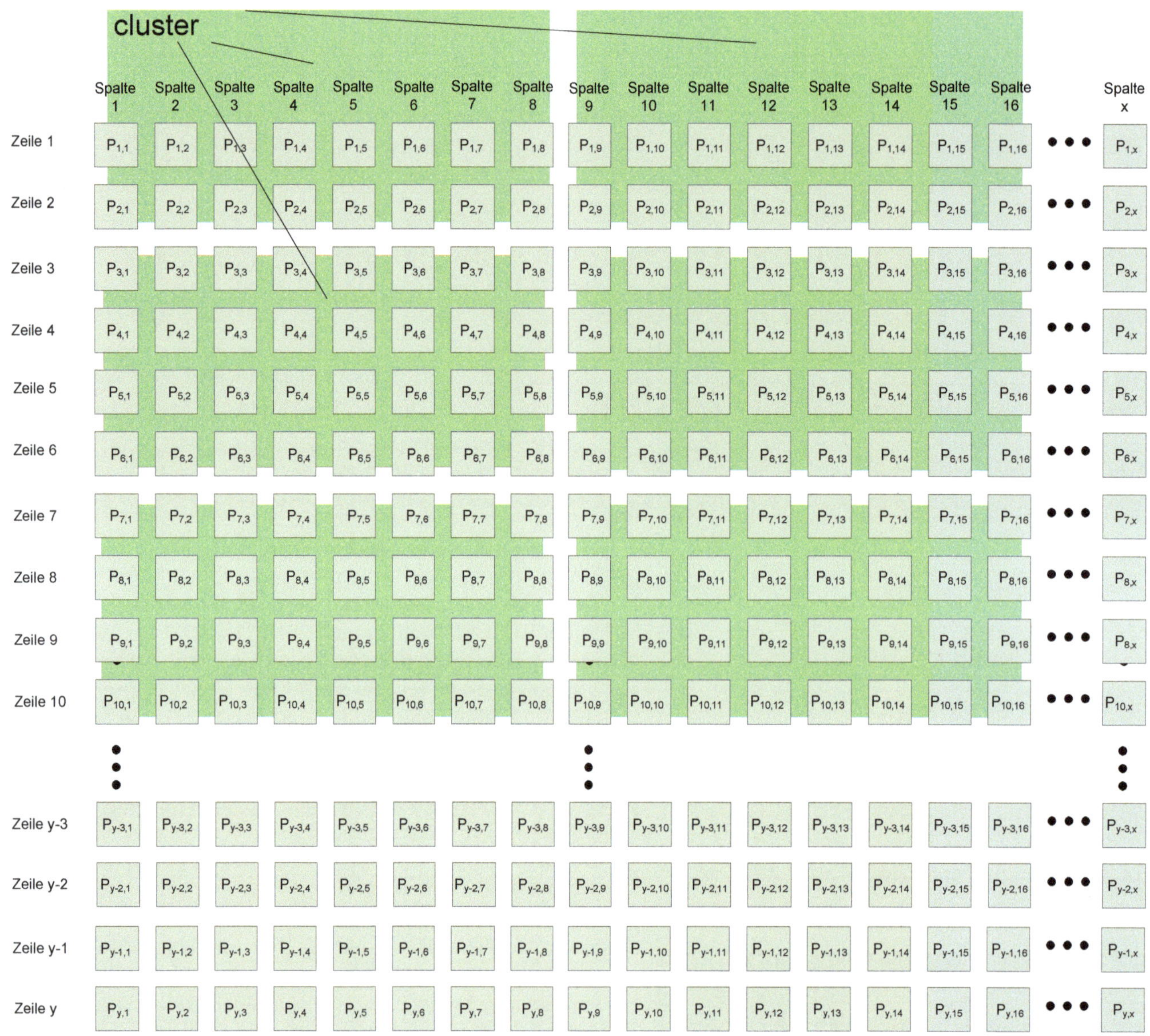

Fig. 5. Distribution of the clusters beneath the microbolometer array.

the right quarter. Beneath the whole microbolometer array approximately 10 000 ADCs are positioned. In Fig 5 in a cut-out the positions of some clusters beneath the array are shown. The digital output from the ADCs is multiplexed to a 16 digital video output interface. The frame rate is 30 Hz. The IRFPA has been fabricated in a 0.35 μm CMOS technology and bonded on a PCB-board (Fig. 6).

4 First measurement results

For first measurements the chip with microbolometers of the 1st generation was placed in a vacuum chamber with an FIR transparent window. A measured NETD histogram is shown in Fig. 7. The mean value of the NETD is approximately 150 mK. The value is higher than the design value but it has to be considered that it is the 1st generation of microbolometers and the microbolometer resistance failed the goal resistance extensively. But for all the VGA-IRFPA works promisingly. A picture of a hand has been taken (Fig. 8). It has to be remarked that beside an offset correction no further correction or filter is applied to the picture.

5 Conclusions

A VGA-IRFPA with a novel readout architecture has been presented. It is based on massively parallel use of

Fig. 6. Photo of the VGA-IRFPA bonded on a PCB-board.

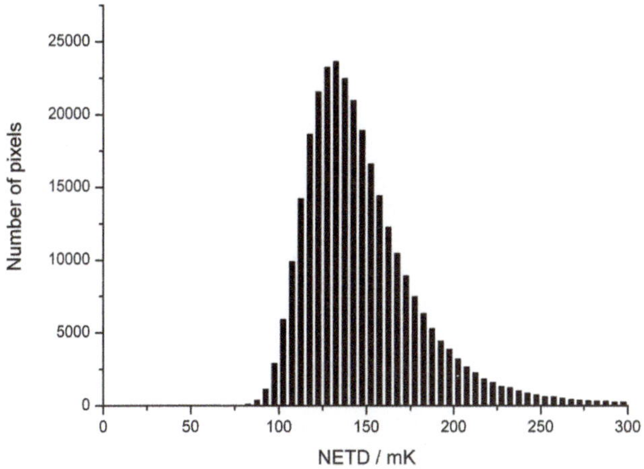

Fig. 7. NETD histogram.

SDM-readout followed by SINC-filters. The pixel-pitch of the microbolometers is 25 μm. First NETD measurement results are shown. In near future better results with the 2nd generation of microbolometers are expected.

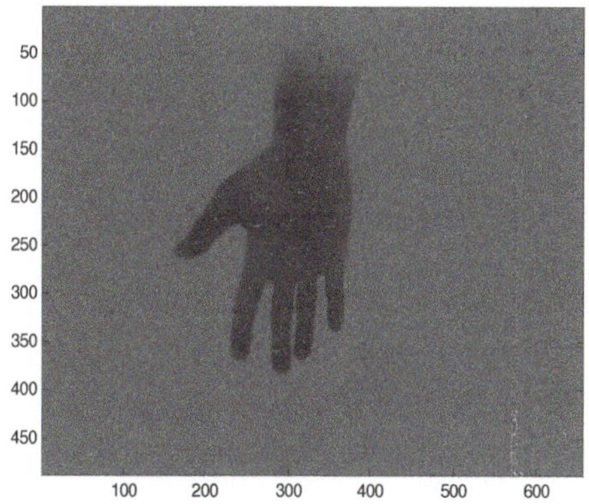

Fig. 8. Picture of a hand taken with the presented VGA-IRFPA.

Acknowledgements. The presented work is part of project "FIRKAM" funded by the German "Bundesministerium fur Bildung und Forschung BMBF". The authors would like to thank the program manager "VDI Technologiezentrum" and the FIRKAM project partners for their helpful information and discussions. We highly appreciate the support of the engineers and technicians of Fraunhofer IMS during design, fabrication, and characterization of the IRFPA.

References

Blackwell, R., Lacroix, D., Bach, T., Ishii, J., Hyland, S., Geneczko, J., Chan, S., Sujlana, B., and Joswick, M.: Uncooled VO_X thermal imaging systems at BAE Systems", Proc. SPIE, 6940, 694021, 1–8, 2008.

Durand, A., Minassian, C., Tissot, J. L., Vilain, M., Robert, P., Touvignon, A., Chiappa, J. M., and Pistre, C.: Uncooled amorphous silicon TEC-less 1/4 VGA IRFPA with 25 μm pixel-pitch for high volume applications, Proc. SPIE, 7298, 72982B, 1–7, 2009.

Russ, M., Bauer, J., and Vogt, H.: The geometric design of microbolometer elements for uncooled focal plane arrays, Proc. SPIE Conference Infrared Technology and Applications XXXIII, 6542, 1–10, 2007.

Weiler, D., Russ, M., Würfel, D., Lerch, R., Yang, P., Bauer, J., and Vogt, H.: A digital 25 μm pixel-pitch uncooled amorphous silicon TEC-less VGA-IRFPA with massive parallel Sigma-Delta-ADC readout, Proc. of SPIE, 7660, 76600S-1-7, 2010.

Norsworthy, S. R., Schreier, R., and Temes, G. C.: Delta-Sigma Data Converters, IEEE Press, 0-7803-1045-4, 1997.

Optimization of array geometry for direction-of-arrival estimation using a priori information

O. Lange[1] **and B. Yang**[2]

[1]Robert Bosch GmbH, Robert-Bosch-Strasse 200, 31139 Hildesheim, Germany
[2]Chair of System Theory and Signal Processing, Universität Stuttgart, Pfaffenwaldring 47, 70569 Stuttgart, Germany

Abstract. This paper focuses on the estimation of the direction-of-arrival (DOA) of signals impinging on a sensor array. A novel method of array geometry optimization is presented that improves the DOA estimation performance compared to the standard uniform linear array (ULA) with half wavelength element spacing. Typically, array optimization only affects the beam pattern of a specific steering direction. In this work, the proposed objective function incorporates, on the one hand, a priori knowledge about the signal's DOA in terms of a probability density function. By this means, the array can be adjusted to external conditions. On the other hand, a modified beam pattern expression that is valid for all possible signal directions is taken into account. By controlling the side lobe level and the beam width of this new function, DOA ambiguities, which lead to large DOA estimation errors, can be avoided. In addition, the DOA fine error variance is minimized. Using a globally convergent evolution strategy, the geometry optimization provides array geometries that significantly outperform the standard ULA with respect to DOA estimation performance. To show the quality of the algorithm, four optimum geometries are presented. Their DOA mean squared error is evaluated using the well known deterministic Maximum Likelihood estimator and compared to the standard ULA and theoretical lower bounds.

1 Introduction

This paper deals with the problem of estimating the direction-of-arrival (DOA) of signals impinging on an array of spatially distributed sensors. There exists a vast amount of DOA estimators, whose performance and accuracy compared to theoretical lower bounds like the Cramér-Rao lower bound have been intensively investigated. A good introduction to array signal processing and DOA estimation can be found in (Krim and Viberg, 1996) and the references therein. In this work, we focus on the deterministic Maximum Likelihood (DML) DOA estimator, which is asymptotically consistent and statistically efficient under certain regularity conditions (Stoica and Nehorai, 1989). Nevertheless, any other estimator that works with an arbitrary array geometry can be applied to the results of this work, as well.

Given a fixed number of sensor elements, the question of an optimum sensor placement with respect to DOA estimation performance naturally arises. In this context, it is intuitively clear that no "globally optimal" geometry for DOA estimation exists. Instead, an array geometry is always optimal with respect to certain presuppositions like the number of signals (Gershman and Böhme, 1997), the statistical properties of the DOA or the probability of large DOA estimation errors, etc.

Although there exist many different approaches to achieving an optimum geometry, basically all of them can be attributed directly or indirectly to the array beam pattern in terms of the side lobe level (SLL) and the shape of the main lobe. Because shape is a quantity that is difficult to characterize mathematically, the half power beam width (HPBW) is often used as a feature instead. It corresponds to the accuracy of the DOA estimates, i.e. the narrower the main beam, the lower the variance of the DOA estimates. Thus, the main beam width is directly related to the Cramér-Rao lower bound, which represents the minimum variance of an unbiased estimator (van Trees, 2002). Furthermore, a narrow beam width increases the possibility of angular signal separability in the multiple signal case.

It is well known in DOA estimation, that the mean squared error (MSE) departs from the Cramér-Rao lower bound when the signal-to-noise (SNR) ratio (or the sample size) falls below a specific limit. Beginning at this threshold, DOA estimates are dominated by large estimation errors, outliers

Correspondence to: O. Lange
(oliver.lange@de.bosch.com)

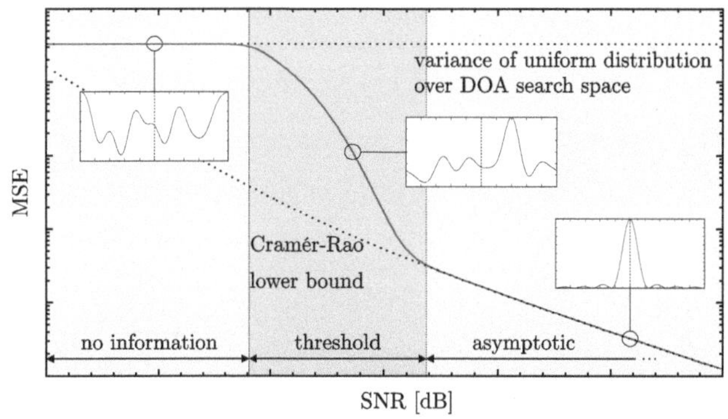

Fig. 1. Typical MSE curve of DOA estimation with three different regions of operation. In addition, likelihood functions are plotted corresponding to the asymptotic, the threshold and the no information region, respectively.

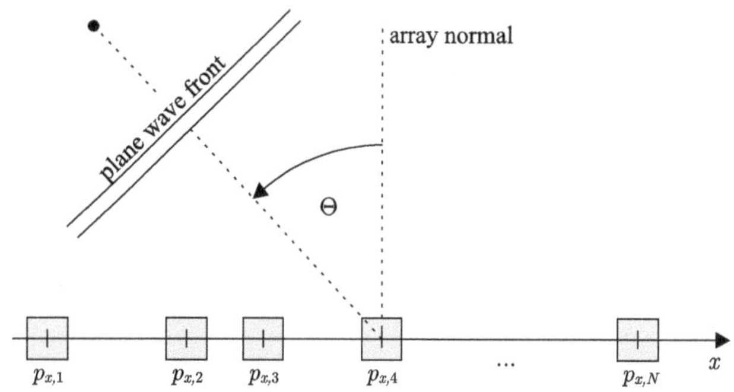

Fig. 2. Array geometry with N elements on the x-axis with nonuniform inter element spacings and one far field source at azimuth angle Θ.

whose probability of occurrence depends on the SLL of the beam pattern (Athley, 2005).

Moreover, in the multiple signal case, a low SLL prevents high SNR signals from overlapping low SNR signals.

By further reducing the SNR (or the sample size), the MSE goes to saturation, which can be approximated by the variance of a uniform distribution over the DOA search space (Richmond, 2006). Figure 1 quantitatively illustrates this typical DOA MSE curve and presents three regions of operation:

1. The *asymptotic region* is influenced by the shape of the main lobe, i.e. the HPBW; it can be approximated by the Cramër-Rao lower bound or the equivalent asymptotic variance of the estimator, respectively.

2. The *threshold region* is dominated by outliers, whose probability of occurrence is proportional to the SLL.

3. In the *no information region*, the SNR (or the sample size) is very low which leads to DOA estimates that are uniformly distributed over the search space.

All attempts of array geometry optimization that aim at DOA performance improvement have to take this MSE characteristic, which is based on a trade-off between SLL and beam width of the beam pattern, into account. In doing so, it should also be noted, that the beam pattern has to be unambiguous for all steering directions, where targets are to be expected. Ambiguities and high side lobes in this region should be avoided, as they lead to outliers in DOA estimation, even in high SNR regimes. As the usual beam pattern only represents the array response to a unit wave from a certain direction, we introduce a modified beam pattern (MBP), using a simple parameter transformation. By this means, the MBP allows for control of the HPBW and the SLL simultaneously for all valid steering directions.

To take "real world" external conditions into account, we further incorporate a probability density function (PDF), defined for all valid steering directions. It is affected, for instance, by the angular a priori distribution of the target's DOA, the element factor of the sensors, the transmit beam pattern and the effect of blinds or radomes. This a priori PDF is applied as an angular dependent weighting factor to the MBP. By this means, we expand an approach by Oktel and Moses (2005), how previous knowledge about the DOA and hardware characteristics can easily be transferred into the array design process.

Using a globally convergent optimization routine, the HPBW and the SLL of the weighted MBP can now be controlled resulting in array geometries that outperform the standard uniform linear array (ULA) with half wavelength element spacing in terms of DOA estimation accuracy.

This paper is organized as follows. The data model and the MBP are introduced in Sect. 2. We also derive the weighted MBP, which additionally uses a priori knowledge in terms of a beta distributed PDF. Furthermore, we introduce the deterministic Maximum Likelihood DOA estimator and show, how prior knowledge can improve DOA estimation. In Sect. 3, the objective function for array geometry optimization is defined. Some examples for optimum array geometries are presented and their DOA estimation performance is compared to the standard ULA using Monte Carlo simulations and the Cramér-Rao lower bound.

2 Data model and modified beam pattern

We consider a planar array geometry with N identical, omnidirectional sensors placed on a straight line, e.g. the x-axis (see Fig. 2). Q far-field sources that are orientated coplanar to the array are impinging on the array from azimuth directions $\boldsymbol{\Theta}_0 = [\Theta_{0,1}, \ldots, \Theta_{0,Q}]$. The narrow band assumption is

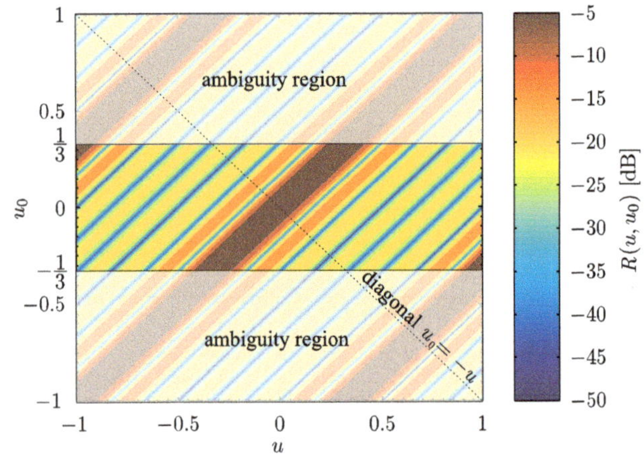

Fig. 3. ULA beam pattern $R(u, u_0)$ for varying target directions $u_0 = -1, \ldots, 1$. For $|u_0| > \frac{1}{3}$, DOA estimates are ambiguous due to grating lobes in the beam pattern.

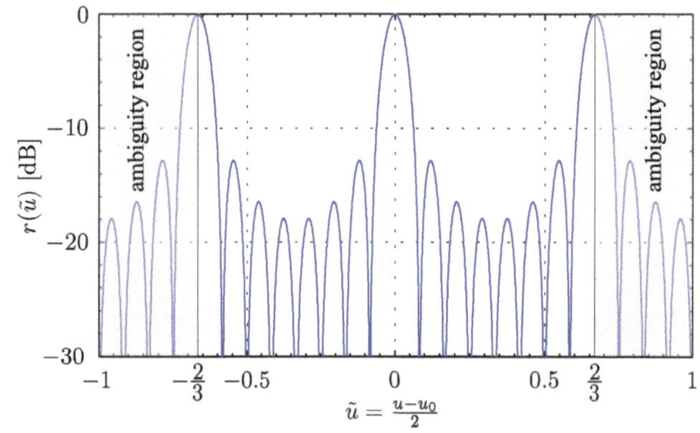

Fig. 4. Modified beam pattern (MBP) for an ULA with inter element spacing of $\frac{3}{4}\lambda$.

assumed to hold so that the complex baseband array output vector can be modeled by (van Trees, 2002)

$$\boldsymbol{x}(k) = \mathbf{A}(\boldsymbol{u}_0)\boldsymbol{s}(k) + \boldsymbol{n}(k), \qquad k = 1, \ldots, K \tag{1}$$

where $\mathbf{A}(\boldsymbol{u}_0) = [\boldsymbol{a}(u_{0,1}), \ldots, \boldsymbol{a}(u_{0,Q})]$ is the $(N \times Q)$ array steering matrix with $u_{0,q} = \sin\Theta_{0,q}$, $q = 1, \ldots, Q$. The complex baseband source signals are denoted by $\boldsymbol{s}(k) = [s_1(k), \ldots, s_Q(k)]^T$ where $(\cdot)^T$ means transpose, $\boldsymbol{n}(k)$ is the additive noise vector based on a Gaussian random process with zero mean and variance σ_n^2. The number of snapshots is denoted by K. The SNR of the q-th signal at each of the sensors is defined as

$$\mathrm{SNR}_q = \frac{\frac{1}{K}\sum_{k=1}^{K}|s_q(k)|^2}{\sigma_n^2}, \qquad q = 1, \ldots, Q. \tag{2}$$

The steering matrix $\mathbf{A}(\boldsymbol{u}_0)$ consists of Q steering vectors $\boldsymbol{a}(u_0)$ whose n-th element is defined by

$$a_n(u_0) = \mathrm{e}^{\mathrm{j}\frac{2\pi}{\lambda}p_{x,n}u_0}, \qquad n = 1, \ldots, N. \tag{3}$$

Here, the element positions are denoted by $p_{x,n}$ and λ is the wavelength.

2.1 Derivation of the modified beam pattern

The normalized array beam pattern $R(u, u_0)$ is defined as the squared magnitude of the response to a unit wave ($s(k) = 1\ \forall k$) from direction u_0 in the case of no noise, i.e.

$$\begin{aligned} R(u, u_0) &= \left|\tfrac{1}{N}\boldsymbol{a}^H(u)\boldsymbol{a}(u_0)\right|^2 \\ &= \left|\tfrac{1}{N}\textstyle\sum_{n=1}^{N}\mathrm{e}^{-\mathrm{j}\frac{2\pi}{\lambda}p_{x,n}(u-u_0)}\right|^2, \end{aligned} \tag{4}$$

where the superscript $(\cdot)^H$ denotes conjugate transpose. Figure 3 exemplarily shows the beam patterns corresponding to an 8-element ULA with element spacing $\frac{3}{4}\lambda$. The steering direction is varied from $u_0 = -1, \ldots, 1$. Due to violation of the spatial sampling theorem (van Trees, 2002), the region for unambiguous DOA estimates is limited to $|u_0| < \frac{1}{3}$ as indicated in Fig. 3.

It can be clearly seen from Fig. 3, that knowledge of

$$r(\tilde{u}) = R(\tilde{u}, u_0 = -\tilde{u}) = \left|\frac{1}{N}\sum_{n=1}^{N}\mathrm{e}^{-\mathrm{j}\frac{2\pi}{\lambda}p_{x,n}2\tilde{u}}\right|^2 \tag{5}$$

with $-1 \le \tilde{u} \le 1$, which represents the diagonal $u_0 = -u$ in Fig. 3, allows for perfect reconstruction of all beam patterns[1]

$$R(u, u_0) = r(\tilde{u})|_{\tilde{u} = \frac{u-u_0}{2}}. \tag{6}$$

This means, that $R(u, u_0)$ can be reduced to $r(\tilde{u})$ without any loss of information. Note that (Eq. 6) is only valid for omnidirectional sensor elements. Non uniform element characteristics are taken into account in the next subsection where prior knowledge is considered.

The function $r(\tilde{u})$ from (Eq. 5) is denoted as the modified beam pattern (MBP), because it can be obtained from the beam pattern by a simple rearrangement of the arguments u and u_0. Figure 4 plots the MBP for the same parameters as in Fig. 3. Note that the region of unambiguous DOA estimates can be identified as $-\frac{2}{3} < \tilde{u} = \frac{u-u_0}{2} < \frac{2}{3}$ which leads to the prior result $|u_0| < \frac{1}{3}$ for $-1 \le u \le 1$.

2.2 Definition of the a priori PDF

In many array signal processing applications, restrictive assumptions can be made concerning the statistical properties of the target's DOA. Hence, the probability of appearance or detection of a target at a certain direction u_0 is not uniform for all possible target directions $-1 \le u \le 1$. Often, a target is

[1] To distinguish between $R(u, u_0)$ and $r(\tilde{u})$, we use $\tilde{u}$ instead of u for notational convenience.

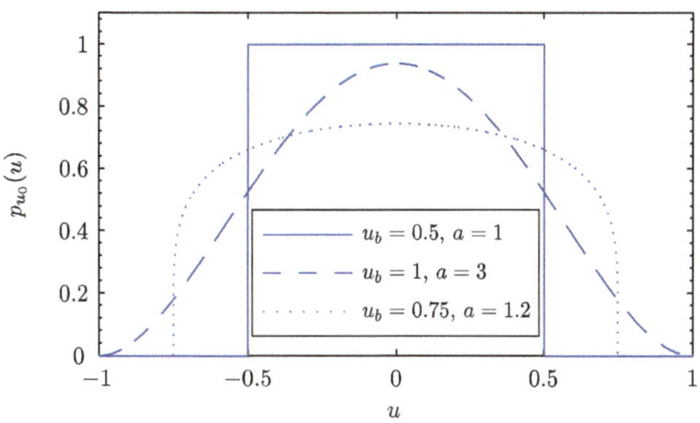

Fig. 5. Different beta probability densities, transformed to u space, from (Eq. 8) with $b = a$ for several values of u_b and a.

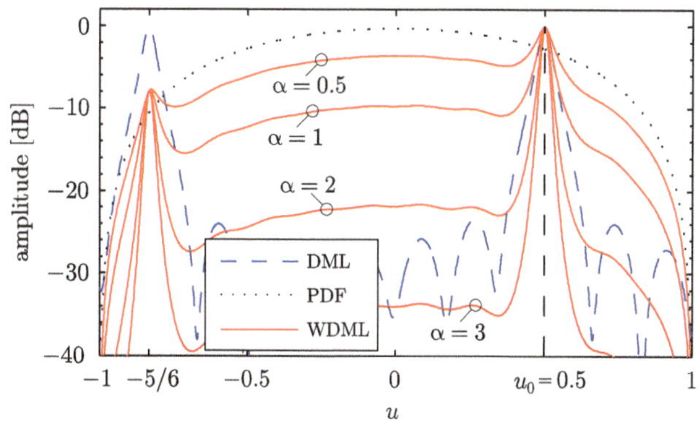

Fig. 6. Criterion functions for deterministic Maximum Likelihood (DML) and weighted deterministic Maximum Likelihood (WDML) DOA estimation with the normalized beta distribution from (Eq. 8) ($u_b = 1, a = 3$), which represents a priori knowledge about the target's DOA.

more likely to appear near the array boresight than laterally. Or there might be a certain region, e.g. in surveillance applications, where the possibility of occurrence of a target is near zero. In addition, incoming signals from certain DOA's can be damped by the use of hardware constraints like blinds or radomes or by the influence of the array element factor. Furthermore, there are applications like automotive radar, where targets are actively illuminated by a transmit antenna. Neglecting the influence of multi path scattering, the DOA angular distribution directly corresponds to the beam shape of the transmit antenna.

Thus, the mentioned characteristics directly or indirectly affect the statistical properties of the target's DOA. To account for the mentioned effects, we introduce the the PDF $p_{u_0}(u)$ which represents the angular a priori information. In this work, we focus (without loss of generality) on the beta distribution

$$p(v) = \begin{cases} \frac{1}{B(a,b)} v^{a-1}(1-v)^{b-1} & 0 \le v \le 1 \\ 0 & \text{elsewise,} \end{cases} \tag{7}$$

where $B(a,b) = \int_0^1 v^{a-1}(1-v)^{b-1} dv$ is the beta function. The parameters a and b are real positive constants. A transformation of the random variable $v \in [0,1]$ to u space with $u \in [-u_b, u_b], 0 < u_b \le 1$, results in

$$p_{u_0}(u) = \begin{cases} \frac{1}{2u_b B(a,a)} \left[\frac{u_b^2 - u^2}{4u_b^2} \right]^{a-1} & -u_b \le u \le u_b \\ 0 & \text{elsewise,} \end{cases} \tag{8}$$

where $a = b$ is chosen to assure a symmetric distribution with mean $\mu_{u_0} = 0$ and variance $\sigma_{u_0}^2 = \frac{1}{4(2a+1)}$. Figure 5 plots the PDF from (Eq. 8) for several values of u_b and a. For $a = 1$, the PDF is uniform, i.e., there is no prior information about the DOA u_0, except its restriction to the interval $[-u_b, u_b]$. As a increases, the PDF becomes narrower.

To be consistent with the standard parameter estimation theory, we still consider the target's DOA u_0 as a deterministic parameter. Namely, we regard it as a realization of a stochastic process with the PDF from (Eq. 8). This is an important assumption if theoretical bounds like the Cramér-Rao lower bound on DOA estimation are considered.

2.3 Advantage of prior knowledge for DOA estimation: weighted deterministic Maximum Likelihood estimator

The derivation of the deterministic Maximum Likelihood (DML) DOA estimator can be found, for instance, in (van Trees, 2002). In this work, we focus on the single signal case, i.e. the estimation of u_0. Hence, the DML estimate of u_0 is given by

$$\hat{u}_{0,\text{DML}} = \arg\min_u \{L(u)\} = \arg\max_u \left\{ L^{-1}(u) \right\} \tag{9}$$

with the likelihood function

$$L(u) = \text{tr}\left[\mathbf{P}_\mathbf{A}^\perp \hat{\mathbf{R}}_x \right], \quad u \in [-1, 1] \tag{10}$$

where $\text{tr}[\cdot]$ is the trace operator and $\mathbf{P}_\mathbf{A}^\perp$ is the projection matrix $\mathbf{P}_\mathbf{A}^\perp = \mathbf{I} - \mathbf{A}(u)\left[\mathbf{A}^H(u)\mathbf{A}(u)\right]^{-1}\mathbf{A}^H(u)$ onto the orthogonal complement of the column space of the steering matrix $\mathbf{A}(u)$. Furthermore,

$$\hat{\mathbf{R}}_x = \frac{1}{N} \sum_{k=1}^{K} \boldsymbol{x}(k)\boldsymbol{x}^H(k) \tag{11}$$

is the estimated spatial correlation matrix.

Any prior knowledge about the target's DOA statistical properties can improve the DML DOA estimation. By simply weighting the likelihood function from (Eq. 10) with the a priori PDF from (Eq. 8), we obtain the weighted deterministic Maximum Likelihood (WDML) estimate

$$\hat{u}_{0,\text{WDML}} = \arg\max_u \left\{ L^{-\alpha}(u) p_{u_0}(u) \right\}. \tag{12}$$

Here, the real valued parameter $\alpha \geq 0$ is used to control the ratio of the prior and the data: the larger the value of α, the lower the influence of the prior on the estimate. Note that (Eq. 12) is the Maximum a posteriori estimator, if α represents the amount of collected data, i.e. the product of snapshots and channels $\alpha = KN$ (van Trees and Bell, 2007).

We will demonstrate this advantageous behavior with a quantitative example (see Fig. 6 for a graphical interpretation). Again, we refer to the 8-element ULA with element spacing $\frac{3}{4}\lambda$ and assume one impinging source from $u_0 = \frac{1}{2}$. Due to the ambiguity at $u = -\frac{5}{6}$, the DML estimator (see dashed curve in Fig. 6) will probably provide a large estimation error. If, in addition, we assume, that the target's DOA follows a beta distribution as in (Eq. 8) with $a = 3$, the WDML estimator (see solid curves in Fig. 6) reduces the effect of the ambiguity and provides the correct DOA estimate $\hat{u}_{0,\mathrm{WDML}} = \frac{1}{2}$.

3 Optimization of the array geometry

In array geometry optimization, the objective function in general exhibits a multi modal character, i.e., local optimization routines might get stuck in local minima. Therefore, the optimization of the array geometry in this work uses the evolution strategy. A good survey of global optimization can be found in (Weise, 2007). Evolution strategy belongs to the family of evolutionary algorithms, which are based on biology-inspired methods like mutation, crossover, selection and survival of the fittest. In contrast to bit-encoded genetic algorithms, evolution strategies explore a real valued parameter space.

The aim of this array geometry optimization is to identify a better array geometry concerning DOA estimation for one target, underlying a given a priori DOA PDF, compared to the standard ULA with half wavelength element spacing.

3.1 Definition of the objective function

In the array geometry optimization task, an objective function $f(\boldsymbol{p}_x): \mathbb{R}^N \to \mathbb{R}$ has to be identified. For this purpose, we refer to the MBP (Eq. 5) and the target's DOA PDF (Eq. 8). We denote their product

$$g(\boldsymbol{p}_x, \tilde{u}) = r(\boldsymbol{p}_x, \tilde{u})\, p_{u_0}(\tilde{u}) \tag{13}$$

as weighted modified beam pattern (WMBP). To reduce the fine error variance, i.e. the Cramér-Rao lower bound, the main beam width has to be minimized. Therefore, the associated objective function $f(\boldsymbol{p}_x)$ calculates the HPBW of the WMBP from (Eq. 13):

$$f(\boldsymbol{p}_x) = \mathrm{HPBW}\left[g(\boldsymbol{p}_x, \tilde{u})\right], \tag{14}$$

which is evaluated numerically. To avoid an increased possibility of gross errors, care is taken to keep the SLL of the

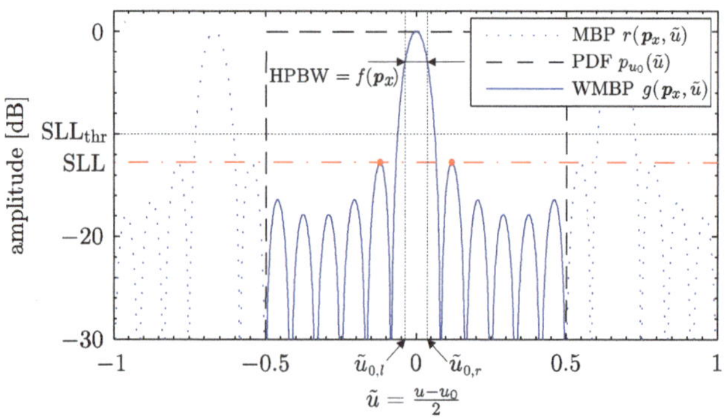

Fig. 7. Modified beam pattern (MBP) $r(\boldsymbol{p}_x, \tilde{u})$, beta PDF $p_{u_0}(\tilde{u})$ ($u_b = 0.5$, $a = 1$), weighted modified beam pattern (WMBP) $g(\boldsymbol{p}_x, \tilde{u})$ with marked half-power beam width, which is used as objective function $f(\boldsymbol{p}_x)$ for array geometry optimization.

WMBP below a threshold $\mathrm{SLL}_{\mathrm{thr}}$. Thus, the array geometry optimization results in the constrained minimization problem

$$\min_{\boldsymbol{p}_x} f(\boldsymbol{p}_x), \qquad \text{s.t.} \qquad \mathrm{SLL}\left[g(\boldsymbol{p}_x, \tilde{u})\right] \leq \mathrm{SLL}_{\mathrm{thr}}. \tag{15}$$

Figure 7 exemplarily illustrates the calculation of the objective function for the already mentioned 8-element ULA with sensor spacing $\frac{3}{4}\lambda$. At first, the product of the MBP and the PDF provides the WMBP, following (Eq. 13). Here, we use the beta PDF from (Eq. 8) with $u_b = 0.5$ and $a = 1$. Note that the SLL of the WMBP ($\mathrm{SLL} = -12.8$dB) is below the threshold $\mathrm{SLL}_{\mathrm{thr}} = -10$dB. Therefore, the HPBW of the WMBP is calculated using equation (Eq. 14) with $f(\boldsymbol{p}_x) = \tilde{u}_{0,r} - \tilde{u}_{0,l}$.

3.2 Optimization results

Three exemplary results of the geometry optimization will now be presented in order to illustrate the accuracy of the proposed objective function in conjunction with the evolution strategy. The DOA estimation performance for one signal in terms of the mean squared error (MSE) is evaluated for each generated geometry and compared to the standard ULA. The target's DOA u_0 is chosen to underlie the beta PDF, whose parameters are specified in the respective example. For DOA estimation, we use the WDML estimator from (Eq. 12) with $\alpha = 5$, i.e., a high influence of the prior compared to the Maximum a posteriori estimator is considered. Each of the MSE curves is based on $6 \cdot 10^4$ Monte Carlo runs per SNR simulation point. The number of snapshots is $K = 10$ and we consider array geometries with $N = 8$ sensor elements. The respective optimized array geometries, including the standard ULA, are shown in Fig. 8.

In addition, the MSE performance plots also include the asymptotic variance

$$\sigma^2_{\mathrm{DML}} = \frac{1}{2\pi^2 NU} \frac{1}{\mathrm{SNR}} \left(1 + \frac{1}{N} \frac{1}{\mathrm{SNR}}\right) \tag{16}$$

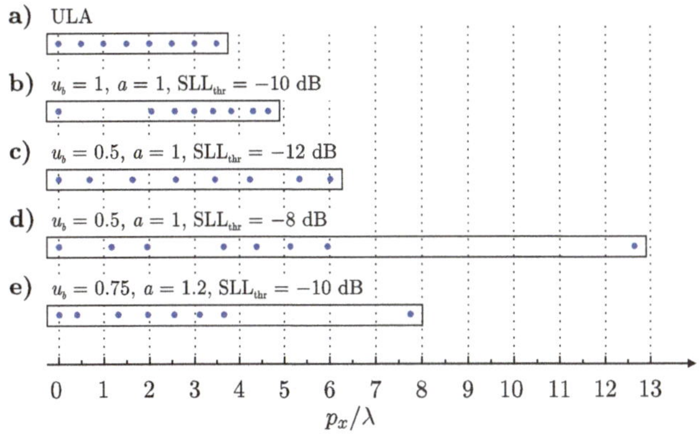

Fig. 8. ULA **(a)** and optimized array geometries with respect to **(b)** example 1, **(c)** and **(d)** example 2 and **(e)** example 3.

of the DML estimation error with the signal-to-noise ratio defined in (Eq. 2) and the variance of the element positions

$$U = \frac{1}{N}\sum_{n=1}^{N}\left(p_{x,n} - \frac{1}{N}\sum_{m=1}^{N} p_{x,m}\right)^2 \tag{17}$$

(Athley, 2005) and (van Trees, 2002). Note that in the single signal case, (Eq. 16) is identical to the Cramér-Rao lower bound for a stochastic signal $s(k)$ (Stoica and Nehorai, 1990).

3.2.1 Example 1

We choose the beta PDF with $u_b = 1$ and $a = 1$, i.e., the DOA estimation should be unambiguous for the half space $-1 \leq u \leq 1$. In addition, the SLL threshold is set to $SLL_{thr} = -10$dB. The optimized geometry is shown in Fig. 8b). Figure 9 plots the WMBP of the standard ULA and the optimized array. It can be seen that the ULA provides (near) ambiguous DOA estimates for $|\tilde{u}| > 0.9$. The optimized array, however, is unambiguous for the complete half space and exhibits a reduced HPBW at the expense of a slightly increased SLL. The associated DOA MSE curve is presented in the bottom of Fig. 9. It can be clearly seen, that in the case of $u_b = 1$, i.e. a uniform distributed DOA $u_0 \in [-1, 1]$, the optimized array significantly outperforms the ULA due to the effect of outliers. If the ULA's ambiguous regions are not taken into account, i.e. $u_b = 0.9$, its asymptotic MSE improves, but still does not reach the optimized array due to the marginal difference of the beam widths. Furthermore, Fig. 9 also shows, that the WDML estimator asymptotically approaches the asymptotic variance of the DML estimator, as it has been defined in (Eq. 16).

3.2.2 Example 2

In this example, we assume that the signal's DOA is limited to $-0.5 \leq u_0 \leq 0.5$. Therefore, we choose the beta PDF with

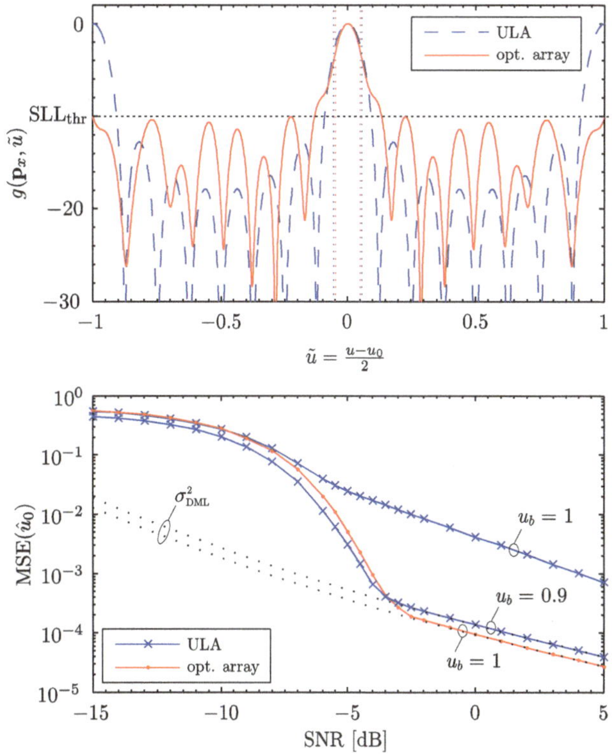

Fig. 9. Top: WMBP for example 1 for the standard ULA and the optimized array. Bottom: DOA MSE performance of the WDML estimator for example 1 for a single signal. The DOA u_0 follows the PDF from (Eq. 8) with $u_b = 1$ and $u_b = 0.9$, respectively, and $a = 1$.

$u_b = 0.5$ and $a = 1$ (see Fig. 5). To avoid ambiguities inside this region, the SLL is bounded above to $SLL_{thr} = -12$dB and $SLL_{thr} = -8$dB, respectively. The optimized geometries are shown in Fig. 8c) and d). Figure 10 plots the WMBP's for the ULA and the two optimized arrays. Due to the reduced DOA region, ambiguities outside of $-0.5 \leq u_0 \leq 0.5$ can now be accepted, as they have no impact on DOA estimation. This allows for an increased array aperture which leads to a smaller HPBW. The MSE curves in Fig. 10 show an improvement in terms of the SNR in the asymptotic region of 5dB and 10dB, respectively, compared to the ULA.

3.2.3 Example 3

Here, we assume a non-flat PDF with $u_b = 0.75$ and $a = 1.2$, which has already been plotted in Fig. 5. Thus, in contrast to the other examples, the distribution of the DOA $u_0 \in [-u_b, u_b]$ is no longer uniform. Again, the optimized array (see Fig. 8e) achieves a smaller HPBW at the expense of an increased SLL (see Fig. 11). Thus, the asymptotic MSE of the optimized array is reduced by 5.5dB.

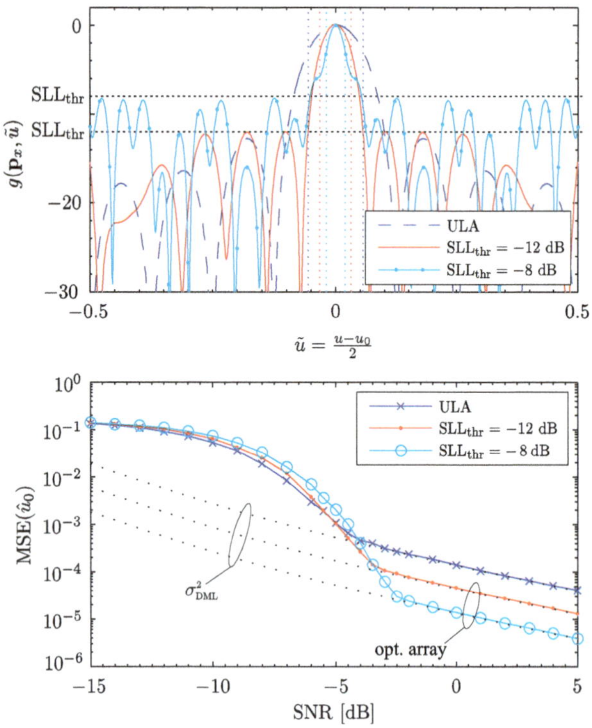

Fig. 10. Top: WMBP for example 2 for the standard ULA and the optimized arrays. Bottom: DOA MSE performance of the WDML estimator for example 2 for a single signal. The DOA u_0 follows the PDF from (Eq. 8) with $u_b = 0.5$ and $a = 1$.

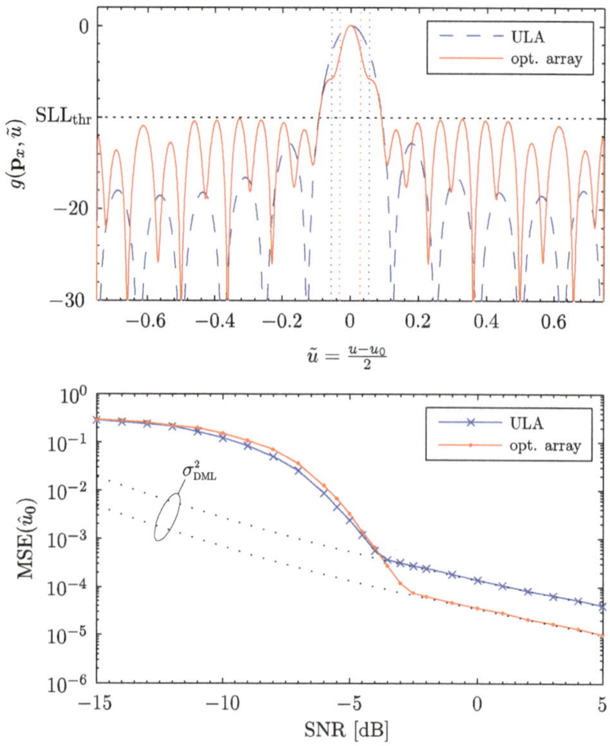

Fig. 11. Top: WMBP's for example 3 for the standard ULA and the optimized array. Bottom: DOA MSE performance of the WDML estimator for example 3 for a single signal. The DOA u_0 follows the PDF from (Eq. 8) with $u_b = 0.75$ and $a = 1.2$.

4 Conclusion

In this paper, a novel method of array geometry optimization with the aim of DOA estimation improvement has been presented. Based on a modification of the standard beam pattern and an angular a priori density function, an objective function has been derived. By controlling the beam width and the SLL of the modified beam pattern, the MSE curve of DOA estimation is affected, while ambiguities are avoided. By the use of the a priori PDF, we have demonstrated, how effects from the "real world" like the DOA's statistical properties, antenna characteristics like the element factor and the transmit beam pattern or hardware constraints like radomes or blinds can easily be taken into account in the optimization process. Using an evolution strategy, we have presented four optimum array geometries and evaluated their DOA estimation performance. Compared to the standard ULA, it has been shown, that the MSE can be significantly improved.

References

Athley, F.: Threshold region performance of Maximum Likelihood direction of arrival estimators, Signal Processing, IEEE Transactions on, 53, 1359–1373, doi:10.1109/TSP.2005.843717, 2005.

Gershman, A. B. and Böhme, J. F.: A note on most favorable array geometries for DOA estimation and array interpolation, Signal Processing Letters, IEEE, 4, 232–235, doi:10.1109/97.611287, 1997.

Krim, H. and Viberg, M.: Two decades of array signal processing research: The parametric approach, IEEE Signal Processing Magazine, 13, 67–94, 1996.

Oktel, U. and Moses, R. L.: A Bayesian approach to array geometry design, Signal Processing, IEEE Transactions on, 53, 1919–1923, doi:10.1109/TSP.2005.845487, 2005.

Richmond, C. D.: Mean-squared error and threshold SNR prediction of Maximum-Likelihood signal parameter estimation with estimated colored noise covariances, Information Theory, IEEE Transactions on, 52, 2146–2164, doi:10.1109/TIT.2006.872975, 2006.

Stoica, P. and Nehorai, A.: MUSIC, Maximum Likelihood, and Cramér-Rao bound, Acoustics, Speech and Signal Processing, IEEE Transactions on, 37, 720–741, doi:10.1109/29.17564, 1989.

Stoica, P. and Nehorai, A.: Mode, Maximum Likelihood and Cramér-Rao bound: conditional and unconditional results,

Acoustics, Speech, and Signal Processing, 1990. ICASSP-90., 1990 International Conference on, 5, 2715–2718, doi:10.1109/ICASSP.1990.116186, 1990.

van Trees, H. L.: Detection, Estimation, and Modulation Theory, Part IV, Optimum Array Processing, John Wiley & Sons, 2002.

van Trees, H. L. and Bell, K. L.: Bayesian Bounds for Parameter Estimation and Nonlinear Filtering/Tracking, Wiley-IEEE Press, 2007.

Weise, T.: Global Optimization Algorithms - Theory and Application, Weise, T., 2007-05-01 edn., http://www.it-weise.de/projects/book.pdf, 2007.

8

MIMO-OFDM equaliser for spatial multiplexing transmission modes

P. Beinschob and U. Zölzer

Department of Signal Processing and Communications, Helmut-Schmidt-Universität/University of the Federal Armed Forces Hamburg, Holstenhofweg 85, 22043 Hamburg, Germany

Abstract. In search for faster and more reliable communication, multiple-input multiple-output (MIMO) in conjuction with Orthogonal Frequency Division Multiplexing (OFDM) are subject of extensive research. In spatial multiplexing transmission an instantaneous rise of data rates governed by the number of transmit antennas can be realised. The system performance depends highly on signal-to-interference-plus-noise ratios (SINR) at the receiver. The receiver's equaliser is supposed to maximize the SINR by mitigating the spatial interference and thus separating the transmitted signals. For this problem several solutions exist such as linear and nonlinear, per subcarrier or OFDM symbol-based. An overview of common algorithms is given and complexity is discussed. Bit error rate (BER) performance evaluations are presented. Another aspect is the impact of the equalisation strategy on the performance of bit-interleaved soft information-based channel coding schemes. As a representative, LDPC codes are chosen. Simulation results show a significant BER performance loss for symbol decision-based equalisers compared to the uncoded performance. To overcome this problem a modification of the Maximum Likelihood algorithm is proposed which yields good performance for low SNR applications.

1 Introduction

In upcoming MIMO OFDM systems Spatial Multiplexing is an convenient way to linearly increase bandwidth efficiency with the number of transmit and receive antennas primary at the expense of higher SNR demands. Detection algorithms tend to vary in complexity, the lower the SNR requirements the higher the complexity.

Mobile scenarios are a particular challenge. This is partly due to outdated channel state information (CSI) as well as the fact that many receivers neglect the time-variant behaviour of the channel during a frame and even OFDM symbol. The multiple, independent OFDM send signals are superposed in the wireless channel, modelled by a channel matrix and additive, white gaussian noise (AWGN). The signals are convoluted by a time-varying channel matrix that results in a nonlinear system model. Thus, orthogonality of the OFDM subcarriers is disturbed. Simulation results based on 3GPP MIMO Spatial Channel Model show the detection performance over SNR for a range of velocities.

The rest of the paper is organized as follows. The underlying system model in time and frequency-domain, as well as a vectorised representation are presented in Sect. 2, followed by a brief review of per layer MIMO detection algorithms in Sect. 3. Joint detectors are discussed in Sect. 4. Furthermore the proposed time-domain interference mitigation is presented in Sect. 5. The paper is concluded by illustrating simulation results in Sect. 6 and a conclusion.

2 System model and structure

The vector of received values $\mathbf{r}$ at the time sample m of a MIMO system is the superposition of $L \cdot n_T$ previously sent samples and the current n_T samples, where $L+1$ is the length of the sampled channel impulse response and n_T is the number of transmit antennas. It is given by

$$\mathbf{r}[m] = \sum_{l=0}^{L} \mathbf{h}[l,m] \cdot \mathbf{s}[m-l] + \tilde{\mathbf{w}}[m], \quad (1)$$

where $\mathbf{s}[m]$ denotes the current vector of symbols of each of the transmit antenna, $\mathbf{w}$ is an identically, independently dis-

Correspondence to: P. Beinschob
(patric.beinschob@hsu-hamburg.de)

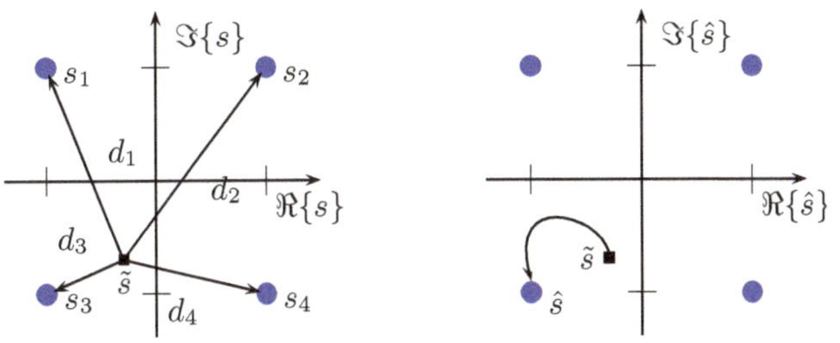

Fig. 1. Per layer maximum likelihood symbol decision and quantization.

tributed (iid) additive white Gaussian noise term and $\mathbf{h}[l,m]$ is the MIMO channel matrix in delay and time domain, indexed with l respectively m. The past sent samples are denoted by $\mathbf{s}[m-l]$, for $l \neq 0, l \leq L$. The data symbols of the K subcarriers are modulated by an inverse Fast Fourier Transform (IFFT). In simulations every value corresponding to a transmit antenna of the resulting vectors is transmitted using the formula above. The data symbols itself are drawn from an M-order QAM modulation alphabet $\mathcal{M}$. The mapping, denoted by $\mathcal{M}\{\cdot\}$, modulates $\kappa = \log_2 M$ bits to a QAM symbol. This is done consecutively for all n_T send streams/layer. The QAM constellations are considered power-normalized to simplify notation.

In frequency domain the system model in Eq. (1) can be described as

$$\mathbf{r}[n,k] = \mathbf{H}[n,k] \cdot \mathbf{s}[n,k] + \mathbf{w}[n,k], \tag{2}$$

where n denotes the time index of an OFDM symbol and k its subcarrier index. The vectors $\mathbf{r}[n,k]$ and $\mathbf{w}[n,k]$ are of dimension $n_R \times 1$, $\mathbf{s}[n,k]$ of $n_T \times 1$ and the matrix $\mathbf{H}[n,k]$ of $n_R \times n_T$, at which n_R is the number of transmit antennas. In simulations the MIMO channel coefficients $H_{r,t}[n,k]$, $r = 1,\ldots,n_R$, $t = 1,\ldots,n_T$ are modeled using the 3GPP spatial model which was developed to evaluate receiver algorithms in MIMO scenarios (3rd Generation Partnership Project 3GPP-TR25.996, 2008). The superposed received signals are transferred back into the frequency domain with the help of a FFT, resulting in the vectors $\mathbf{r}[n,k]$ of Eq. (2). A perfectly known channel matrix $\mathbf{H}$ is assumed for all detectors, as well as perfect synchronization and total avoidance of block interference, i.e. the cyclic prefix is longer than the maximum delay path. The MIMO detector's performance is evaluated in terms of soft information in form of channel log-likelihood ratios (LLRs), $L_C(x)$ and in direct form as hard decided bits.

2.1 Vectorised system model

The vectorised system description of Eq. (1) – for a block symbol based system where block interference is avoided, e.g. OFDM with cyclic prefix removed – can be denotes as

$$\mathbf{r}^{(\mathbf{v})} = \mathbf{h}^{(\mathbf{v})} \cdot \mathbf{s}^{(\mathbf{v})} + \mathbf{w}^{(\mathbf{v})}, \tag{3}$$

where the dimensions are $\mathbf{r}^{(\mathbf{v})}, \mathbf{w}^{(\mathbf{v})} \in \mathbb{C}^{K \cdot n_R}, \mathbf{s}^{(\mathbf{v})} \in \mathbb{C}^{K \cdot n_T}$ and $\mathbf{h}^{(\mathbf{v})} \in \mathbb{C}^{K \cdot n_R \times K \cdot n_T}$, (Stamoulis et al., 2002). The channel matrix structure is as follows:

$$\mathbf{h}^{(\mathbf{v})} = \begin{pmatrix} \mathbf{h}[0,0] & \mathbf{0} & & \cdots & \mathbf{0} \\ \mathbf{h}[1,0] & \mathbf{h}[0,1] & \mathbf{0} & \cdots & \\ \vdots & \vdots & \mathbf{h}[0,m] & \cdots & \\ \mathbf{h}[l,0] & \mathbf{h}[l-1,1] & \ddots & \ddots & \\ \vdots & \vdots & & \ddots & \\ \mathbf{h}[L+1,0] & \mathbf{h}[L,1] & & \ddots & \vdots \\ \mathbf{0} & \mathbf{h}[L+1,1] & \cdots & \ddots & \\ \vdots & \vdots & & \ddots & \mathbf{0} \\ \mathbf{0} & \mathbf{0} & \cdots & \cdots & \mathbf{h}[0,K] \end{pmatrix}, \tag{4}$$

where

$$\mathbf{h}[l,m] = \begin{pmatrix} h_{1,1}[l,m] & \ldots & h_{1,n_T}[l,m] \\ \vdots & \ddots & \vdots \\ h_{n_R,1}[l,m] & \ldots & h_{n_R,n_T}[l,m] \end{pmatrix}, \\ 0 \leq l \leq L, \quad 0 \leq m \leq K. \tag{5}$$

The vectorised formulation of the send vector is as follows

$$\mathbf{s}^{(\mathbf{v})} = \left(s_1[1],\ldots,s_{n_T}[1],\ldots,s_1[m],\ldots,s_{n_T}[K],\right)^T. \tag{6}$$

Analogue for the receive vector

$$\mathbf{r}^{(\mathbf{v})} = \left(r_1[1],\ldots,r_{n_R}[1],\ldots,r_1[m],\ldots,r_{n_R}[K]\right)^T. \tag{7}$$

3 Per layer detection

The idea of MIMO per layer detection is to map the problem back into the send space and decide upon the most likely symbol per send layer or send antenna stream $t = 1,\ldots,n_T$ – which is identical in spatial multiplexing – as depicted in Fig. 1. This is reduction in complexity but leads to suboptimal results compared to joint layer detection.

3.1 Linear receivers

Zero-Forcing is well known method, an optimal linear solution neglecting noise is given by

$$\tilde{\mathbf{s}} = (\mathbf{H}^H\mathbf{H})^{-1}\mathbf{H}^H\mathbf{r}. \tag{8}$$

High amplification of noise for low singular values leads to bad performance. The complexity for a block symbol is $\mathcal{O}(Kn_Rn_T^2)$. If noise is considered, the so-called linear MMSE solution is obtained,

$$\tilde{\mathbf{s}} = (\mathbf{H}^H\mathbf{H} + \sigma_w^2\mathbf{I}_{n_R})^{-1}\mathbf{H}^H\mathbf{r}. \tag{9}$$

It is limiting amplification, thus power, with a complexity of $\mathcal{O}(Kn_Rn_T^2)$.

3.2 Successive interference cancelation

Traditional Zero-Forcing approach is hard-decision only with a complexity of $\mathcal{O}(Kn_R^2 n_T^2)$ (Wolniansky et al., 1998). The following improved algorithms are also considered here: the VBLAST/MMSE variant, which utilize Eq. (9) for cancelation and a maximum a-posteriori order criterion is developed by Yapici (2005), called VBLAST/ZF/MAP.

4 Joint layer detection

4.1 Maximum likelihood symbol vector detection

Based upon **r** and channel matrix knowledge **H** optimal detection in a Maximum Likelihood (ML) sense is determining the most probable send symbol vector **s**, minimizing the distance:

$$\operatorname{argmin}_{\hat{\mathbf{s}}} \|\mathbf{r} - \mathbf{H} \cdot \hat{\mathbf{s}}\|^2 \quad \forall \hat{\mathbf{s}} \in \mathcal{S}, \tag{10}$$

where $\mathcal{S} = \mathcal{M}^{n_T}$ is the set of all possible send vectors. It contains M^{n_T} vectors. Testing all combinations and taking the most probable, yields optimal results at exponential complexity $O(KM^{n_T})$. A well investigated approach to reduce the complexity by maintaining Maximum-Likelihood performance is the so called Sphere Decoding (Hochwald and ten Brink, 2003).

4.2 Soft maximum likelihood detector

Realizing Eq. (10) straight forward, a list of all M^{n_T} distances is created:

$$d_i^2 = \|\mathbf{r} - \mathbf{H} \cdot \hat{\mathbf{s}}_i\|^2. \tag{11}$$

Instead of considering only $\hat{\mathbf{s}}_j$ with $d_j = \min_i d_i$ values, the neighborhood of the minimum is incorporated in the detection. Basic idea is to consider hypothetic vectors in an expectation value within a radius around **r** assuming Gaussian noise:

$$\tilde{\mathbf{s}} = \sum_{j \in S_\sigma} p_j \mathbf{s}_j, \; S_\sigma = \{i \in \{1, \ldots, M^{n_T}\} | d_i < c \cdot \sigma_w\},$$

$$p_j = \exp(-d_j^2/\sigma_w^2) / \sum_{S_\sigma} \exp(-d_j^2/\sigma_w^2)$$

The factor c of the noise deviation σ_w can be varied, it is set to $c := 6$ in the following. The complexity is the same as in the ML case.

5 Time domain interference mitigation

For time domain equalisation (TEQ), the vectorised system description Eq. (3) is considered. The Discrete Fourier Transform (DFT) matrix for the vectorised MIMO system can be written as $\mathbf{W} = \tilde{\mathbf{W}} \otimes \mathbf{I}_{n_T}$ with $\tilde{\mathbf{W}}_{k,m} = \frac{1}{\sqrt{K}} e^{-j(2\pi/K)km}$. The operator $\otimes$ denotes the Kronecker product. Following the MMSE criterion, an optimal interference mitigating filter matrix Ψ maximizing SINR (Stamoulis et al., 2002), is calculated by

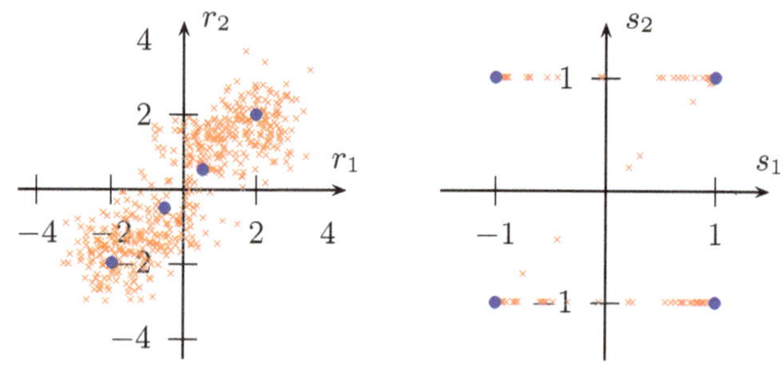

Fig. 2. Receive and send vector space for a BPSK 2×2 system with proposed SoftML equaliser.

$$\mathbf{R_{hh}} = \sigma_N \mathbf{I}_{K \cdot n_R} + \mathbf{h}^{(\mathbf{v})} \cdot (\mathbf{h}^{(\mathbf{v})})^H \tag{12}$$

$$\mathbf{h}_m = \mathbf{h}^{(\mathbf{v})} \cdot \mathbf{W}^H \mathbf{I}_{K \cdot n_T} \tag{13}$$

$$\Rightarrow \Phi = \mathbf{R}_{\mathbf{hh}}^{-1} \mathbf{h}_m, \tag{14}$$

with $m = 1, \ldots, K$. The solution is efficiently calculated by Cholesky decomposition of the covariance matrix $\mathbf{R_{hh}}$. Adding the identity matrix is for numerical stabilisation only (Haykin, 2001). Finally, the filter matrix is given by

$$\Psi = \mathbf{W}^H \Phi^H. \tag{15}$$

Applying the filter to the received signals the estimated n_T send signals for one OFDM symbol in time domain are calculated by $\tilde{\mathbf{s}}^{(\mathbf{v})} = \Psi \cdot \mathbf{r}^{(\mathbf{v})}$. The complexity is $O(K^3 \cdot n_R \cdot n_T^2)$. It is obviously higher than in the frequency domain case because of the K^3 dependence. However, for large systems approximations can reduce the complexity as shown by Hueske and Götze (2009).

6 Simulation results

First, results are presented for the frequency domain equaliser algorithms. A 4×4 system was considered with Rayleigh block fading. Uniform distributed bits were LDPC coded with an irregular LDPC code with design code rate of $1/2$, interleaved and 4-QAM modulated. LDPC codes are chosen because of high codeword distance and parallelisable decoder structure (Richardson et al., 2001). BER versus SNR for the uncoded system are depicted in Fig. 3, for the decoded case see Fig. 4. For the decoding the equalised symbols were LLR demapped per layer, which reduces complexity with respect to the joint detection. In case of uncoded transmission, there are approximate four orders of magnitude between the optimal detection algorithms based on maximum likelihood and the linear approaches (ZF, MMSE) at 22 dB. ZF and the MMSE approach show different behaviours for the coded

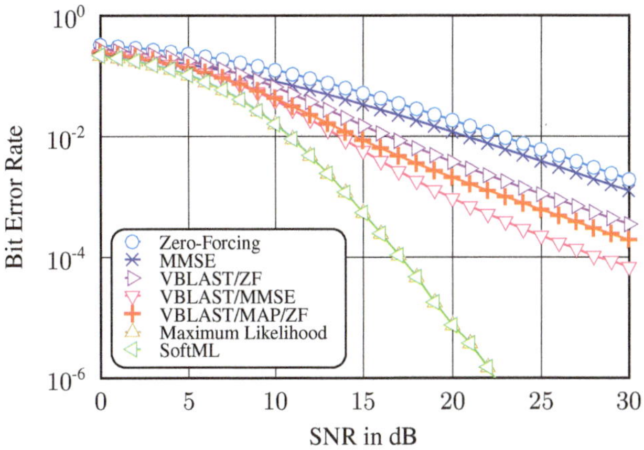

Fig. 3. BER vs. SNR for uncoded modulation over a Rayleigh block fading 4×4 MIMO channel and a selection of frequency domain equaliser.

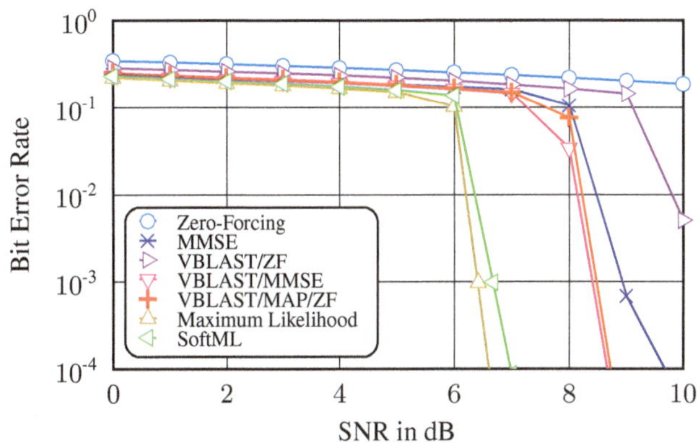

Fig. 4. BER vs. SNR for LDPC-coded bit-interleaved modulation over a Rayleigh block fading MIMO channel and the frequency domain equaliser from Fig. 3.

case. Comparison of VBLAST/ZF and VBLAST/MMSE demonstrates that neglecting noise leads to a significant performance loss. In the MMSE approach through the LLRs the uncertainty of an equalised symbol is carried to the decoder and can be reflected there. The simulations for comparison of FEQ resp. TEQ were performed on a 4×4 MIMO-OFDM system with $K = 128$ subcarriers and a cyclic prefix length $L = 6$. For a realistic channel the 3GPP Spatial Channel Model was used. The results are shown in Fig. 5. Note that for longer OFDM symbols e.g. $K' = 1024$ the normalised Doppler frequency increases with a factor 8, so results for the velocity region of 160m/s would translate into the region around 20 m/s. It can be seen that for high velocities of the mobile station TEQ shows significant gains in terms of symbol error rate (SER) which increases with growing SNR.

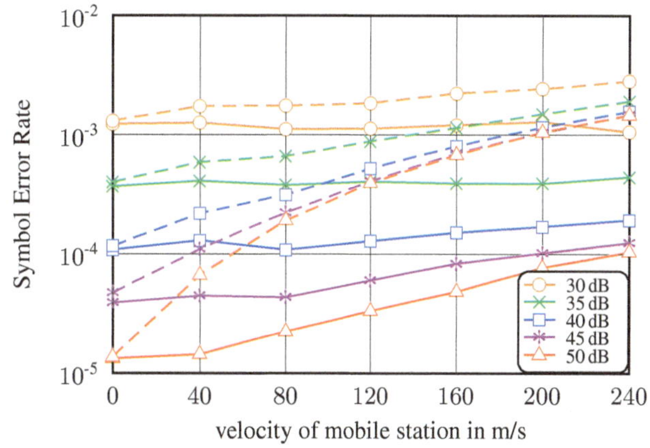

Fig. 5. Comparison of Time and Frequency Domain Equalisation performance TEQ (solid lines) resp. FEQ (dashed lines) for increasing Velocities and in high SNR regions, FEQ-method Zero-Forcing.

7 Conclusions

In this paper we have compared sub-optimal frequency domain equaliser with respect to coded performace in per layer LLR calculation and presented results for time domain equalisation mitigating intercarrier interference and spatial interference as well. We compared those results with the analogue equaliser in the frequency domain. For highly mobile terminals significant gains in terms of BER can be realized by employing time domain processing. When considering complexity, frequency domain equalisation is in favor, of course. To use those gains, future applications might switch to time domain equalisation at a certain velocity, depending on overall link quality and quality of service constrains.

References

3rd Generation Partnership Project (3GPP): Spatial channel model for Multiple Input Multiple Output (MIMO) simulations, Tech. Spec. Group Radio Access Network, Tech. Report 25.996, Rel. 8.0.0, 2008.

Haykin, S.: Adaptive Filter Theory, Prentice Hall, 4. edn., 2001.

Hochwald, B. and ten Brink, S.: Achieving near-capacity on a multiple-antenna channel, IEEE Trans. Commun., 51, 389–399, doi:10.1109/TCOMM.2003.809789, 2003.

Hueske, K. and Götze, J.: Ov-OFDM: A reduced PAPR and cyclic prefix free multicarrier transmission system, in: ISWCS 2009, 206–210, doi:10.1109/ISWCS.2009.5285357, 2009.

Richardson, T. J., Shokrollahi, M. A., and Urbanke, R. L.: Design of capacity-approaching irregular low-density parity-check codes, IEEE Trans. Inf. Theory, 47, 619–637, doi:http://dx.doi.org/10.1109/18.910578, 2001.

Stamoulis, A., Diggavi, S., and Al-Dhahir, N.: Intercarrier interference in MIMO OFDM, IEEE Trans. Signal Process., 50, 2451–2464, doi:10.1109/TSP.2002.803347, 2002.

Wolniansky, P. W., Foschini, G. J., Golden, G. D., and Valenzuela, R. A.: V-BLAST: An Architecture for Realizing Very High Data Rates Over the Rich-Scattering Wireless Channel, Proc. ISSSE, 295–300, 1998.

Yapici, Y.: V-Blast/MAP: A New Symbol Detection Algorithm for MIMO Channels, Master Thesis, Depart. of Electrical and Electronics Eng., Bilkent University, 2005.

9

Smoothing techniques for decision-directed MIMO OFDM channel estimation

P. Beinschob and U. Zölzer

Department of Signal Processing and Communications, Helmut-Schmidt-Universität/University of the Federal Armed Forces, Hamburg, Holstenhofweg 85, 22043 Hamburg, Germany

Abstract. With the purpose of supplying the demand of faster and more reliable communication, multiple-input multiple-output (MIMO) systems in conjunction with Orthogonal Frequency Division Multiplexing (OFDM) are subject of extensive research. Successful Decoding requires an accurate channel estimate at the receiver, which is gained either by evaluation of reference symbols which requires designated resources in the transmit signal or decision-directed approaches. The latter offers a convenient way to maximize bandwidth efficiency, but it suffers from error propagation due to the dependency between the decoding of the current data symbol and the calculation of the next channel estimate. In our contribution we consider linear smoothing techniques to mitigate error propagation by the introduction of backward dependencies in the decision-based channel estimation. Designed as a post-processing step, frame repeat requests can be lowered by applying this technique if the data is insensitive to latency. The problem of high memory requirements of FIR smoothing in the context of MIMO-OFDM is addressed with an recursive approach that acquires minimal resources with virtual no performance loss. Channel estimate normalized mean square error and bit error rate (BER) performance evaluations are presented. For reference, a median filtering technique is presented that operates on the MIMO time-frequency grids of channel coefficients to reduce the peak-like outliers produced by wrong decisions due to unsuccessful decoding. Performance in terms of Bit Error Rate is compared to the proposed smoothing techniques.

Correspondence to: P. Beinschob
(patric.beinschob@hsu-hamburg.de)

1 Introduction

MIMO OFDM systems are strong candidates for upcoming mobile networks. In Spatial Multiplexing transmission modes the channel capacity increases linearly with the number of transmit and receive antennas primary at the expense of higher SNR demands and algorithmic complexity in the receiver and detection algorithms (Foschini and Gans, 1998).

However, there can be observed large gaps between analytical derived capacity statements and realized data rates with proposed receiver designs. This is particularly true for mobile scenarios. Channel estimate errors tend to decrease the achievable rates because of the limited performance of the detection algorithms operating with channel estimates (Dall'Anese et al., 2009).

To acquire channel estimates with low errors a large number of reference symbols or pilots would be necessary. Exclusive bandwidth has to be dedicated to reference symbols and is not available for data transmission, therefore the effective data rate is reduced by the amount of pilots which is not desirable. In mobile receivers the channel estimate is quickly outdated depending on the relative radial velocity to the sender (Marzetta and Hochwald, 1999). After a time duration, commonly known as the coherence time T_C, the channel state must be assumed to be completely changed.

Cellular mobile networks in urban environment additionally experience multipath signal propagation often referred to street canyon scenarios and alike. So, Doppler shifts due to mobility vary on different propagation paths. Fading due to shadowing might occur suddenly and the channel estimation has to keep track.

The mentioned aspects and problems of mobile multi-antenna broadband communication and the consequences for the channel estimation is addressed in this paper. To mitigate influences we propose a post-processing – smoothing – algorithm to refine the channel estimates acquired by a decision-directed channel estimation (DDCE) and tracking algorithm.

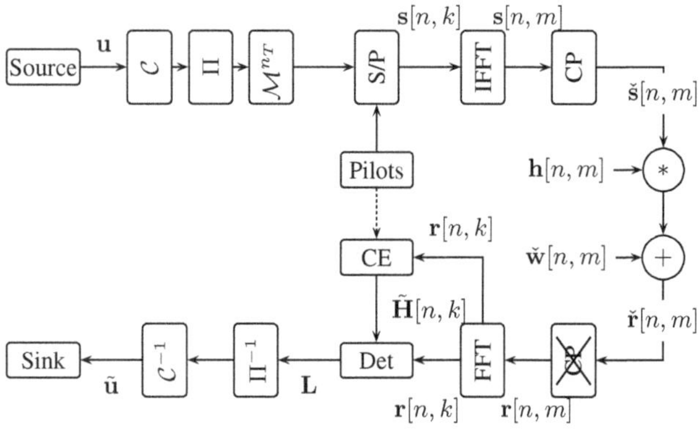

Fig. 1. MIMO-OFDM system model.

The DDCE based on Least Squares works well with a small number of pilot symbols in quasi-static scenarios.

The paper is organized as follows. The underlying system model in time and frequency-domain is presented in Sect. 2, followed by a brief review of the DDCE algorithm in Sect. 3. An approach to introduce linear smoothing techniques in the context of MIMO-OFDM channel estimation is taken in Sect. 4.1 and a memory-reduced version shown in Sect. 4.2. An alternative, non-linear filtering method is discussed in Sect. 4.3. Illustrating simulation results are given in Sect. 5 and finally a conclusion is drawn.

2 System model and structure

The vector of received values $\boldsymbol{r}$ at the time sample m of a MIMO system is the superposition of $L \cdot n_{\mathrm{T}}$ previously sent samples and the current n_{T} samples, where $L+1$ is the length of the sampled channel impulse response and n_{T} is the number of transmit antennas. It is given by

$$\boldsymbol{r}[m] = \sum_{l=0}^{L} \mathbf{h}[l,m] \cdot \boldsymbol{s}[m-l] + \tilde{\boldsymbol{w}}[m], \qquad (1)$$

where $\boldsymbol{s}[m]$ denotes the current vector of symbols of each transmit antenna, $\boldsymbol{w}$ is an identically, independently distributed (iid) additive white Gaussian noise term and $\mathbf{h}[l,m]$ is the MIMO channel matrix in delay and time domain, indexed with l respectively m. The past sent samples are denoted by $\boldsymbol{s}[m-l]$, for $l \neq 0, l \leq L$. The data symbols of the K subcarriers are modulated by an inverse Fast Fourier Transform (IFFT). In simulations every value corresponding to a transmit antenna of the resulting vectors is transmitted using the formula above. The data symbols itself are drawn from an M-order QAM modulation alphabet $\mathbb{S}$. The mapping, denoted by $\mathcal{M}\{\cdot\}$, modulates $\kappa = \log_2 M$ bits to a QAM symbol. This is done consecutively for all n_{T} send streams/layers. The QAM constellations are considered power-normalized to simplify notation.

In frequency domain the system model in Eq. (1) can be described as

$$\boldsymbol{r}[n,k] = \mathbf{H}[n,k] \cdot \boldsymbol{s}[n,k] + \boldsymbol{w}[n,k], \qquad (2)$$

where n denotes the time index of an OFDM symbol and k its subcarrier index. The vectors $\boldsymbol{r}[n,k]$ and $\boldsymbol{w}[n,k]$ are of dimension $n_{\mathrm{R}} \times 1$, $\boldsymbol{s}[n,k]$ of $n_{\mathrm{T}} \times 1$ and the matrix $\mathbf{H}[n,k]$ of $n_{\mathrm{R}} \times n_{\mathrm{T}}$, at which n_{R} is the number of receive antennas. In simulations the time domain MIMO channel coefficients $h_{r,t}[l,m]$, $r = 1,\ldots,n_{\mathrm{R}}$, $t = 1,\ldots,n_{\mathrm{T}}$ are modeled using the 3GPP spatial model which was developed to evaluate receiver algorithms in MIMO scenarios (3rd Generation Partnership Project , 3GPP). The superposed received signals are transferred back into the frequency domain with the help of a FFT, resulting in the vectors $\boldsymbol{r}[n,k]$ of Eq. (2). Perfect synchronization is assumed and total avoidance of block interference, i.e. the cyclic prefix is longer than the maximum delay path. The system's performance is evaluated in terms of bit error rates determined from hard decided channel decoder output. The channel decoder operates on soft information in form of channel log-likelihood ratios (LLRs), $\mathbf{L}$. A block diagram depicting the system model is given in Fig. 1.

3 Decision-directed Recursive Least Squares (RLS) channel estimation

MIMO-RLS algorithm estimates auto- and cross-correlation matrices $\boldsymbol{\Phi}$ resp. $\boldsymbol{\theta}$, time adaptive with forgetting factor ξ as described by Kay (1993). The channel estimation is done on each subcarrier k individually.

$$\boldsymbol{\Phi}[n,k] = \xi \cdot \boldsymbol{\Phi}[n-1,k] + \boldsymbol{s}[n,k] \cdot \boldsymbol{s}^{\mathsf{H}}[n,k] \qquad (3)$$

$$\boldsymbol{\theta}[n,k] = \xi \cdot \boldsymbol{\theta}[n-1,k] + \boldsymbol{s}[n,k] \cdot \boldsymbol{r}^{\mathsf{H}}[n,k] \qquad (4)$$

An estimate of the channel matrix is obtained as follows:

$$\tilde{\mathbf{H}}[n,k] = \left(\boldsymbol{\Phi}^{-1}[n,k] \cdot \boldsymbol{\theta}[n,k]\right)^{\mathsf{H}}. \qquad (5)$$

From the MIMO bit-wise log-likelihood ratio detection output, that is soft information of channel output $\mathbf{L}$, reconstructed send vectors are available. After the pilot sequence of length N_P in the preamble of the frame is processed, the reconstructed send vectors

$$\tilde{s}[n] = \mathcal{M}^{n_{\mathrm{T}}}\{\mathrm{sgn}\{\mathbf{L}\}\}, \quad \forall n > N_{\mathrm{P}}. \qquad (6)$$

are used to further refine the estimate of the auto- and cross-correlation matrices thus tracking the time-variable channel, as indicated by the switch in Fig. 2 (Akhtman and Hanzo, 2007). The coherence time of the channel determines the usability of the collected samples. Old samples describing an obsolete channel state should be omitted. This is accomplished by the introduction of the forgetting factor ξ. For the sake of notational simplicity the symbol $\boldsymbol{s}$ is used for pilot vectors as well as the reconstructed send vectors omitting the tilde.

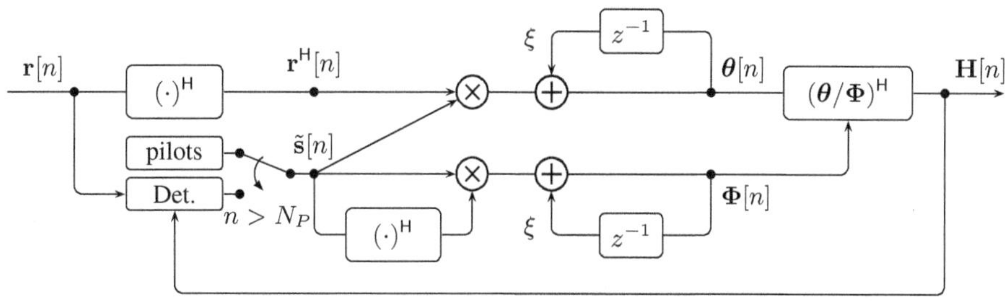

Fig. 2. Signal flow diagram for decision directed RLS channel estimation and tracking algorithm.

4 Smoothing techniques

4.1 RLS un-rolling

Post-processing after the RLS requires access to the spatial correlation samples for each OFDM symbol and subcarrier

$$\check{\boldsymbol{\Phi}}[n,k] = \boldsymbol{s}[n,k]\cdot\boldsymbol{s}^{\mathsf{H}}[n,k], \tag{7}$$

$$\check{\boldsymbol{\theta}}[n,k] = \boldsymbol{s}[n,k]\cdot\boldsymbol{r}^{\mathsf{H}}[n,k]. \tag{8}$$

Considering a $n_{\mathrm{T}} \times n_{\mathrm{R}}$ MIMO OFDM system with N_{S} OFDM symbols per frame and K subcarriers, the dimension of the arrays is $n_{\mathrm{T}} \times n_{\mathrm{R}} \times K \times N_{\mathrm{S}}$.

The weighted expectation value in Eqs. (3) and (4) is then

$$\boldsymbol{\Phi}[n,k] = \xi\cdot\boldsymbol{\Phi}[n-1,k] + \check{\boldsymbol{\Phi}}[n,k], \tag{9}$$

$$\boldsymbol{\theta}[n,k] = \xi\cdot\boldsymbol{\theta}[n-1,k] + \check{\boldsymbol{\theta}}[n,k]. \tag{10}$$

It can be re-formulated as a matrix vector-multiplication. The weighting factors are exponentially decreasing with increasing sample index n therefore they constitute a lower triangular matrix

$$\Xi = \begin{pmatrix} 1 & 0 & \cdots & & \\ \xi & 1 & 0 & \cdots & \\ \xi^2 & \xi & 1 & 0 & \cdots \\ \vdots & \vdots & \vdots & \ddots & 0 \\ \xi^{N_S-1} & \xi^{N_S-2} & \xi^{N_S-3} & \cdots & 1 \end{pmatrix}. \tag{11}$$

The weighting along the time dimension with the forgetting factor matrix can be applied independently on the other dimensions

$$\check{\boldsymbol{\Phi}}'_{t,r,k} = \Xi\cdot\check{\boldsymbol{\Phi}}_{t,r,k} \quad \forall t,r,k \tag{12}$$

and analogue

$$\check{\boldsymbol{\theta}}'_{t,r,k} = \Xi\cdot\check{\boldsymbol{\theta}}_{r,r,k}. \tag{13}$$

With matrices (12) and (13) the ordinary channel estimate can be calculated as well

$$\tilde{\mathbf{H}}[n,k] = \left(\check{\boldsymbol{\Phi}}'^{-1}[n,k]\cdot\check{\boldsymbol{\theta}}'[n,k]\right)^{\mathsf{H}}. \tag{14}$$

The smoothing algorithm can be derived by minimizing the modified cost function

$$\check{J}[n,k] = \sum_{\tilde{n}=1}^{N_S} \xi^{|N_{\mathrm{S}}-\tilde{n}|}\cdot\mathbf{e}^{\mathsf{H}}[\tilde{n},n,k]\cdot\mathbf{e}[\tilde{n},n,k] \tag{15}$$

with

$$\mathbf{e}[\tilde{n},n,k] = \boldsymbol{r}[\tilde{n},k] - \tilde{\mathbf{H}}[n,k]\cdot\boldsymbol{s}[\tilde{n},k]. \tag{16}$$

As a consequence it follows a transition for the weighting matrix to be fully occupied

$$\Xi' = \begin{pmatrix} 1 & \xi & \xi^2 & \xi^3 & \cdots & \xi^{N_{\mathrm{S}}-1} \\ \xi & 1 & \xi & \xi^2 & \cdots & \xi^{N_{\mathrm{S}}-2} \\ \xi^2 & \xi & 1 & \xi & \cdots & \xi^{N_{\mathrm{S}}-3} \\ \xi^3 & \xi^2 & \xi & 1 & \cdots & \xi^{N_{\mathrm{S}}-4} \\ \vdots & \vdots & \vdots & \vdots & \ddots & \vdots \\ \xi^{N_{\mathrm{S}}-1} & \xi^{N_{\mathrm{S}}-2} & \xi^{N_{\mathrm{S}}-3} & \xi^{N_{\mathrm{S}}-4} & \cdots & 1 \end{pmatrix}. \tag{17}$$

Smoothing weighting reflecting mobility induced channel time variance and effectively doubles the number of samples available for estimating the auto- and crosscorrelation matrices thus lowering the estimation error. Obviously, in this formulation arbitrary window functions can be applied as well instead of exponentially weighting factors in Ξ'. So smoothing weighting is applied by evaluating

$$\check{\boldsymbol{\Phi}}''_{t,r,k} = \Xi'\cdot\check{\boldsymbol{\Phi}}_{t,r,k}, \tag{18}$$

$$\check{\boldsymbol{\theta}}''_{t,r,k} = \Xi'\cdot\check{\boldsymbol{\theta}}_{t,r,k}, \tag{19}$$

The smoothed channel estimate for the unrolling (U) is yielded by

$$\tilde{\mathbf{H}}^{\mathrm{U}}_{S}[n,k] = \left(\check{\boldsymbol{\Phi}}''^{-1}[n,k]\cdot\check{\boldsymbol{\theta}}''[n,k]\right)^{\mathsf{H}}. \tag{20}$$

4.2 Recursive smoothing

The unrolling approach consumes a huge amount of memory impeding practical receiver implementations. Therefore a reduction of memory requirements is desirable. It is possible to reformulate the algorithm again into a recursive form (Kashima et al., 2006). Defining

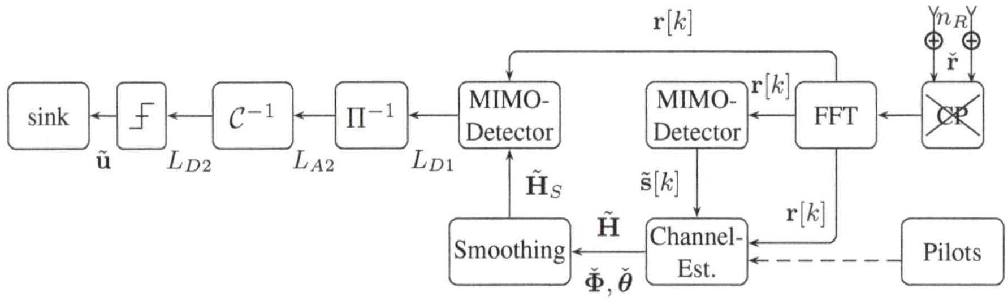

Fig. 3. Receiver structure with post processing of channel estimation with discussed smoothing techniques.

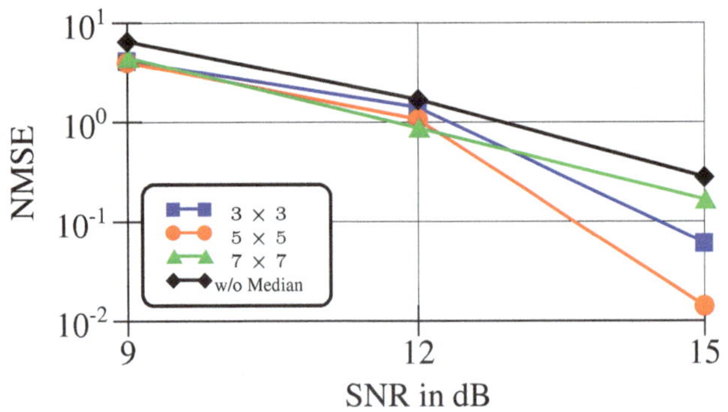

Fig. 4. NMSE comparison of different kernel sizes for Median Filtering in a time-variable channel with simulated mobile terminal velocity of $10\,\mathrm{m\,s^{-1}}$.

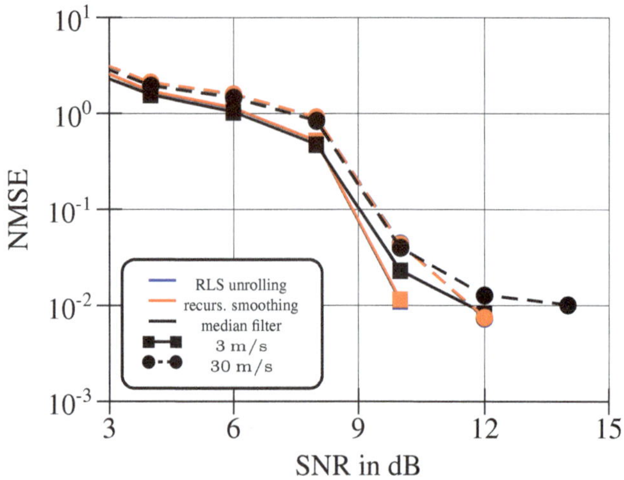

Fig. 5. NMSE comparison of different kernel sizes for Median Filtering in a time-variable channel with simulated mobile terminal velocity of $10\,\mathrm{m\,s^{-1}}$.

$$\mathbf{\Phi}_{\mathrm{S}}[n,k] = \sum_{\tilde{n}=1}^{N_{\mathrm{S}}} \xi^{|N_{\mathrm{S}}-\tilde{n}|} \cdot s[\tilde{n},k] \cdot s^{\mathrm{H}}[\tilde{n},k], \qquad (21)$$

$$\boldsymbol{\theta}_{\mathrm{S}}[n,k] = \sum_{\tilde{n}=1}^{N_{\mathrm{S}}} \xi^{|N_{\mathrm{S}}-\tilde{n}|} \cdot s[\tilde{n},k] \cdot r^{\mathrm{H}}[\tilde{n},k], \qquad (22)$$

it is possible to calculate the smoothed inverse autocorrelation matrix $\mathbf{P}_{\mathrm{S}} = \mathbf{\Phi}_{\mathrm{S}}^{-1}$ recursive using the already estimated inverse autocorrelation matrix $\mathbf{\Phi}$,

$$\mathbf{P}_{\mathrm{S}}[n,k] = \mathbf{P}[n,k] + \xi^2 \cdot \left[\mathbf{P}_{\mathrm{S}}[n+1,k] - \xi^{-1} \cdot \mathbf{P}[n,k]\right]. \qquad (23)$$

So a smoothed channel estimate can also be calculated using the previously estimated channel estimate $\tilde{\mathbf{H}}$ by evaluating

$$\tilde{\mathbf{H}}_{\mathrm{S}}^{\mathrm{R}}[n,k] = \tilde{\mathbf{H}}[n,k] + \xi \cdot [\tilde{\mathbf{H}}_{\mathrm{S}}[n+1,k] - \tilde{\mathbf{H}}[n,k]], \qquad (24)$$

where $n = N_{\mathrm{S}}, \ldots, 1$ and it is initialized by setting $\tilde{\mathbf{H}}_{\mathrm{S}}[N_{\mathrm{S}},k] = \tilde{\mathbf{H}}[N_{\mathrm{S}},k]$.

4.3 Time-frequency median filtering

Decision-directed channel estimation is prone to error propagation and wrong decisions may lead to impulse-like outliers in the channel estimate. As median filtering is especially suited for impulse noise it is investigated in this context as an alternative to the above discussed approaches to smoothing.

This non-linear filtering is applied on the $n_{\mathrm{T}} \cdot n_{\mathrm{R}}$ time-frequency grids resp. time-variant channel transfer functions. An important parameter is the filter kernel size. Often used sizes in image processing are 3×3, 5×5 or 7×7 samples. The optimal kernel size depends on the coherence time and coherence bandwidth respectively subcarriers.

5 Simulation results

The comparative simulations were conducted using a 4×4 MIMO-OFDM system configuration with $K = 128$ subcarriers and a cyclic prefix length $L = 6$ and two velocity setups ($3\,\mathrm{m\,s^{-1}}$ and $30\,\mathrm{m\,s^{-1}}$). For a realistic channel the 3GPP Spatial Channel Model was used with time-variant impulse responses. Uniformly distributed bits were coded using an irregular LDPC code with design code rate of $1/2$, interleaved and 4-QAM modulated. LDPC codes are chosen because of high codeword distance and parallelisable decoder structure (Richardson et al., 2001). They are also employed in the IEEE 802.11n Standard for MIMO systems.

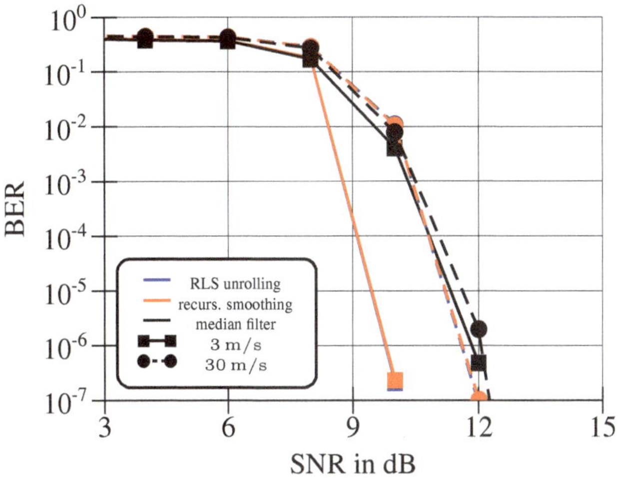

Fig. 6. Comparison of Bit Error Rates for discussed smoothing techniques.

For different median kernel sizes simulations had been conducted and the results in terms of NMSE are depicted in Fig. 4. A kernel size of 5×5 resulted in lowest average channel estimation errors which was then used in the overall comparative simulations.

Results in terms of channel estimation error measured in normalised mean squared error (NMSE) are presented in Fig. 5. A turbo-like cliff region (ten Brink, 2000) around 9 dB was observed. For SNR below this threshold the detector could not reproduce the send vectors reliable therefore error propagation in the decision-directed channel estimation algorithm started rendering complete frames useless and finally resulting in high channel estimation errors. Above the identified threshold the NMSE dramatically decreased. In this region performance differences for Median filtering are visible, whereas for the unrolling and the recursive Smoothing there were no significant differences in performance observable – the curves are almost identical, slightly better for the unrolling approach.

The Bit Error Rate (BER) versus SNR are depicted in Fig. 6. The results reflect the same tendency as the NMSE results did. In comparison to the linear smoothing techniques the performance of the Median filter is worse.

Reliable transmission is possible in the slowly time-variant case above 10 dB SNR employing one of the linear smoothing methods. It needs 12 dB or more with the Median filtering method. However, for the higher velocity the difference in minimum SNR is insignificant due to the high impact of intercarrier interference.

6 Conclusions

In this paper we have discussed smoothing techniques in the context of MIMO OFDM channel estimation. The output of a bandwidth-efficient decision-directed channel estimation and tracking algorithm is post-processed to improve the estimate's accuracy in mobile scenarios by effectively doubling the number of samples available for correlation matrix estimation. In a first approach, the modification of the RLS algorithm leads to a smoothing with high memory requirements. A memory efficient recursive smoothing algorithm has been presented having virtually no loss in NMSE or BER performance compared to the first approach in the conducted simulations with two velocity scenarios. As an alternative, median filtering is compared to the above mentioned. While it leads to comparable results for the high velocity case, a gap for the lower one reveals the disadvantage of the static kernel size in this method.

However, in future work these results has to be verified with measurement setups. Nevertheless the simulations indicate a potential to post-processing algorithms if the transmitted data is not sensitive to delays caused by the increased processing time and latency.

References

3rd Generation Partnership Project (3GPP): Spatial channel model for Multiple Input Multiple Output (MIMO) simulations, Tech. Spec. Group Radio Access Network, Tech. Report 25.996, Rel. 8.0.0, 2008.

Akhtman, J. and Hanzo, L.: Advanced Channel Estimation for MIMO-OFDM in Realistic Channel Conditions, IEEE International Conference on Communications, ICC '07, pp. 2528–2533, doi:10.1109/ICC.2007.418, 2007.

Dall'Anese, E., Assalini, A., and Pupolin, S.: On the Effect of Imperfect Channel Estimation upon the Capacity of Correlated MIMO Fading Channels, pp. 1–5, doi:10.1109/VETECS.2009.5073728, 2009.

Foschini, G. J. and Gans, M. J.: On Limits of Wireless Communications in a Fading Environment when Using Multiple Antennas, Wireless Personal Communications: An International Journal, 6, 311–335, 1998.

Kashima, T., Fukawa, K., and Suzuki, H.: Recursive Least Squares Channel Estimator with Smoothing and Removing for Iterative-MAP Receiver of MIMO-OFDM Mobile Communications, 7, 3052–3057, doi:10.1109/ICC.2006.255273, 2006.

Kay, S. M.: Estimation Theory, Fundamentals of statistical signal processing, vol. 1, Prentice Hall PTR, Upper Saddle River, NJ, 1993.

Marzetta, T. and Hochwald, B.: Capacity of a mobile multiple-antenna communication link in Rayleigh flat fading, IEEE Transactions on Information Theory, 45, 139–157, doi:10.1109/18.746779, 1999.

Richardson, T. J., Shokrollahi, M. A., and Urbanke, R. L.: Design of capacity-approaching irregular low-density parity-check codes, IEEE Trans. Inf. Theory, 47, 619–637, doi:http://dx.doi.org/10.1109/18.910578, 2001.

ten Brink, S.: Designing Iterative Decoding Schemes with the Extrinsic Information Transfer Chart, AEU Int. J. Electron. Commun., pp. 389–398, 2000.

10

Modeling automotive FlexRay transceivers for signal integrity and EMC simulations

H. Günther, U. Hilger, and S. Frei

Technische Universität Dortmund, Dortmund, Germany

Abstract. Automotive bus systems like FlexRay or CAN transmit safety critical data. To ensure correct functionality under all circumstances, extensive investigations about signal integrity and EMC have to be performed. To be able to use simulation in this validation process, suitable models for the components of the bus systems have to be developed. This paper shows how a combined transceiver model for signal integrity and EMC investigations can be created. The model shows good results in comparison to measurement data.

1 Introduction

Automotive bus systems are used to connect control units, intelligent sensors and actors in vehicles. From economical point of view extended cable networks are desirable. Systems are often operated close to their specification limits. To provide safety critical applications under these circumstances or electromagnetic disturbances, the functionality of a bus system must be ensured with sophisticated methods.

An approach to detect problematic behaviour and its causes is the computational investigation of the transmitted signal quality (signal integrity) of the physical layer of the bus system. Models for any component of a bus system have to be developed. Accuracy of models and methods has to be qualified by measurements with realistic systems.

To ensure a high immunity against electromagnetic disturbances, the transceivers should be qualified with special RF immunity tests. EMC simulations may help to investigate possibly critical cases. Therefore accurate EMC behavioural models are needed (Hilger et al., 2010).

This paper shows the capabilities of bus system signal integrity and EMC simulation and focuses on modelling of the transceiver device. SI transceiver models mainly deal with output characteristics while EMC models reflect the input behaviour of a device. However, intersections between those two fields exist and can be used to create a transceiver model to reflect both behaviours. As result the combined model is able to reflect the signal integrity behaviour of a real transceiver device in a wide range of input voltage and under electromagnetic disturbances and considers operation thresholds which limit normal operation ranges.

Correspondence to: H. Günther
(harald.guenther@tu-dortmund.de)

2 Simulation models for SI and EMC

2.1 Signal integrity simulation model of FlexRay transceiver

A FlexRay transceiver consists of general functional parts, e.g. for sending, receiving, and decoding data. Figure 1 shows a general structure of interconnection between those parts, which is the basis for the transceiver signal integrity simulation model (Günther et al., 2010).

For signal integrity investigations the most important part of the bus transceiver model is the output driver. The model is based on a physical approach which is shown schematically in Fig. 2. The three possible bus levels belonging to the internal transceiver states *idle*, *active low* and *active high* are realized by interconnection of two MOSFET transistors. They work as controlled switches based on the MOSFET equations and connect the bus output pin with ground or supply voltage, depending on the state of the transceiver. The adaption of the switching behavior and of the bus levels is done with a differentiable control function and the series connection of two diodes. When idle, both transistors are in high impedance mode and the transceiver is ready for signal reception. An additional ohmic voltage divider is used to model the correct impedance behavior in this case and provide the bias voltage. To ensure correct behavior in different load conditions and to model the slope correctly, capacitive effects of the transceiver device are modelled using output capacitors.

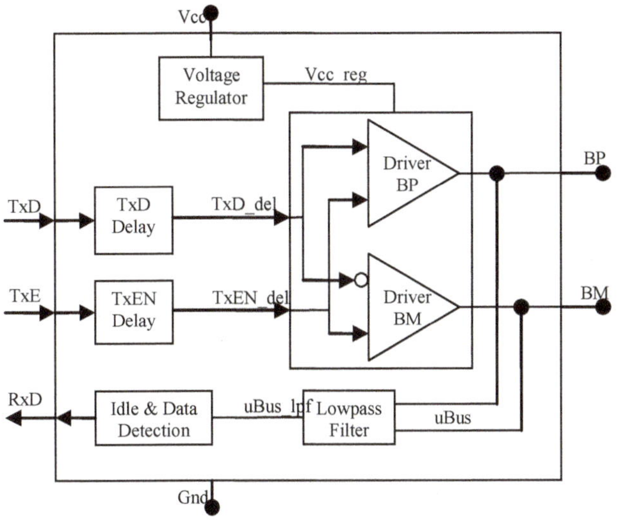

Fig. 1. Internal structure of output driver of FlexRay transceiver model.

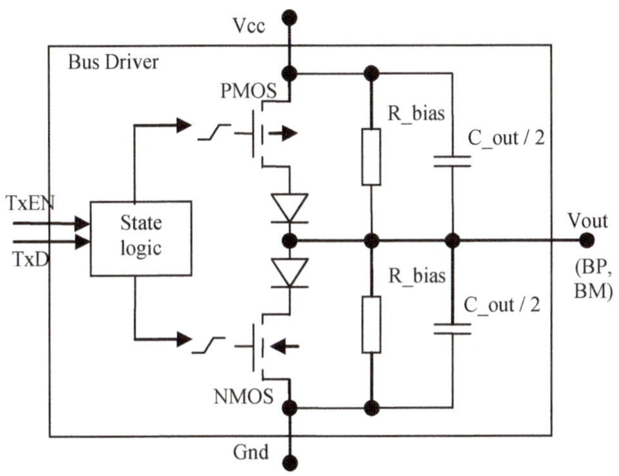

Fig. 2. Schematic structure of output driver of FlexRay transceiver model.

With several parameters like switching voltages of the MOSFETs and diodes and parameters of the control function it is possible to match details of signal integrity behavior of the model with the behavior of the real transceiver device. Datasheet information and measurement data is used for this.

An advantage of this behavioral modeling approach compared to behavioral modeling approaches simply using measured characteristic curves (IBIS Open Forum, 2009) is the possibility to integrate for example temperature and supply voltage dependencies of the transceiver device in the model quite easily. By using basic physical equations the influences of several quantities can be integrated directly.

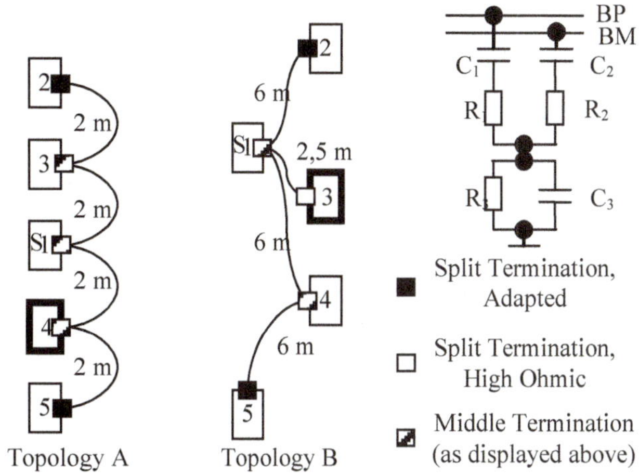

Fig. 3. Structure of FlexRay topologies.

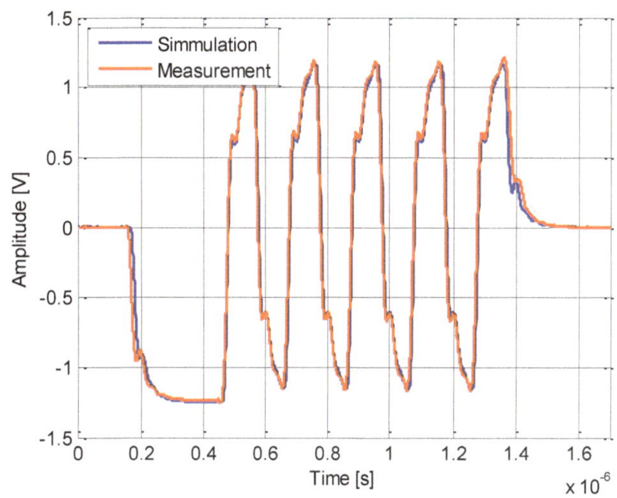

Fig. 4. Simulation and measurement results for topology A.

2.2 Behavior of SI model in typical FlexRay networks

The models for the bus system component were verified with measurement results for each component individually. The signal integrity behavior was validated at several load conditions. Good agreement between simulation and measurement data could be achieved.

Below the simulation results of two FlexRay bus system topologies shown in Fig. 3 are presented as example and compared with measurement data. It is shown that with the help of simulation based investigation conclusions about real world behavior of bus systems can be drawn.

Topology A is a chain connection of 5 bus nodes connected with bus lines with a length of 2 m each. Node 1 is sending. Figure 4 shows the comparison between simulation and measurement data at node 4. The agreement between model behavior and measurement values is very good.

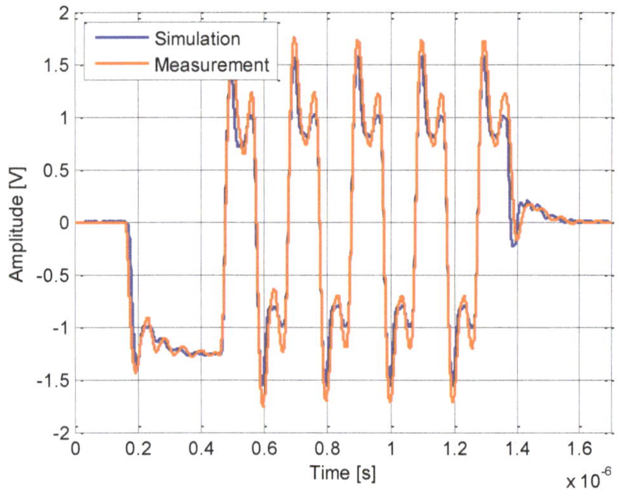

Fig. 5. Simulation and measurement results for topology B.

Topology B also includes 5 bus nodes connected with a network containing different line lengths. Node 1 is sending data onto the bus. Figure 5 shows the signal traces of simulation and measurement at node 3. Again very good agreement between simulation and measurement data can be seen.

2.3 EMC simulation model of FlexRay transceivers

Because of the nonlinear behavior of the transceivers for EMC modeling, not only the determination of the stationary impedance is important, but also the dynamic behavior in case of bus communication failure (Hilger and Frei, 2008a,b). In other investigations the input impedance of IC's were mostly determined by the reflection measurements of the input pins and potential ESD protection elements were implemented as simple diode models (Boyer et al., 2007). The nonlinear effects near or over critical values are not considered in these examinations. In Hilger and Frei (2008b) investigations on modeling of CAN and FlexRay transceivers were made and very simple equivalent impedance models for EMC simulations were presented. However the behavior in case of transient disturbances or the partial destruction of bits by amplitude modulated signals in combination with ESD protection circuits and the transceiver internal signal processing is not considered. The receiver unit of a differential bus transceiver can be modeled as ideal high impedance comparator with low pass filter. High frequencies or fast transient pulses may not affect the correct detection of the signal and pulse width. In Table 1 the equivalent RC-circuits of investigated bus transceivers are shown.

Figure 6 shows a simple behavioral model of the bus receiver with equivalent impedance bus load, comparator and low pass filter.

The comparator detects and amplifies the differential bus voltage. The connected low pass will filter all high frequency

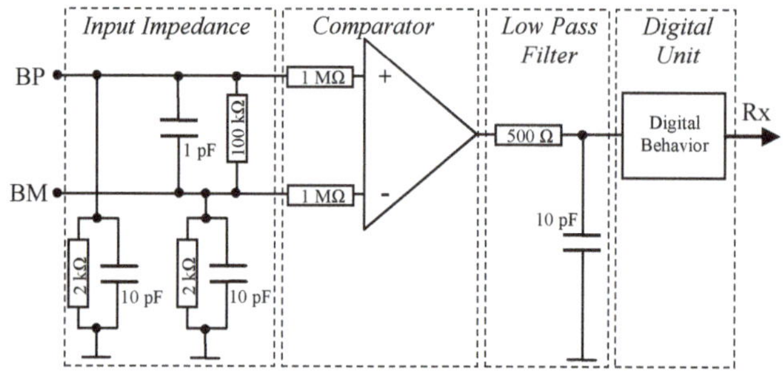

Fig. 6. Behavioral model of FlexRay receiver input.

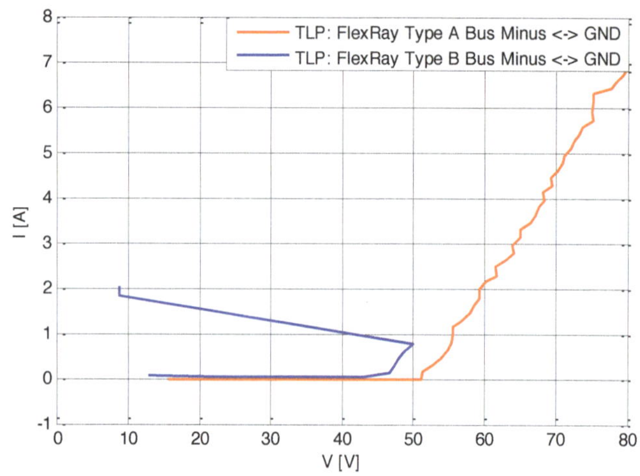

Fig. 7. TLP measurement, V/I characteristic of tested FlexRay transceivers.

parts of the signal and the digital unit is modeled as ideal A/D-D/A converter.

In VHDL-AMS the transceiver behavior can be modeled as combination of discrete elements and variable input impedances (V/I- and f/Z-tables) which can be found by measuring the DC V/I characteristic and the impedance in failure point in frequency domain.

Figure 7 shows sample V/I-measurement results done with HPPI TLP (HPPI Transmission Line Pulser, 2010) of two typical FlexRay transceivers. At levels over 45–50 V internal ESD circuits may switch and the current increases critically. Below 45 V both transceivers have high impedances.

In Fig. 8 the frequency dependent critical failure voltages of the FlexRay transceivers are presented. Because of constructive variations and internal signal processing the transceivers have different immunity levels.

The voltage change between 90 and 100 MHz are caused by resonances in the measurement setup.

Table 1. Equivalent impedances of investigated bus drivers.

Transceiver	CAN Type A	FlexRay Type A	FlexRay Type B
Equivalent Impedance	12 pF || 450 Ω	10 pF || 2 kΩ	42 pF || 20 kΩ

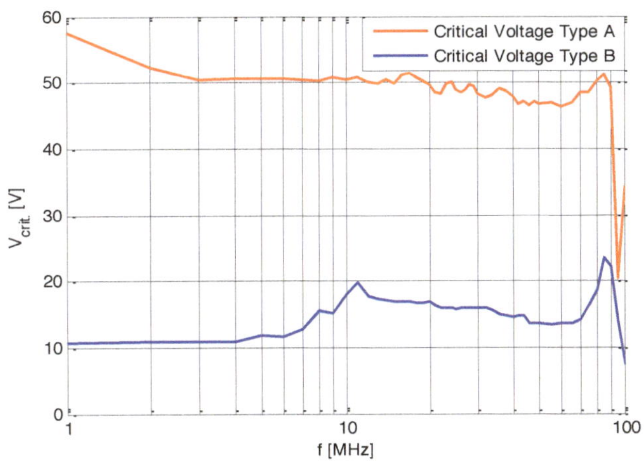

Fig. 8. Critical failure voltage of tested FlexRay transceivers.

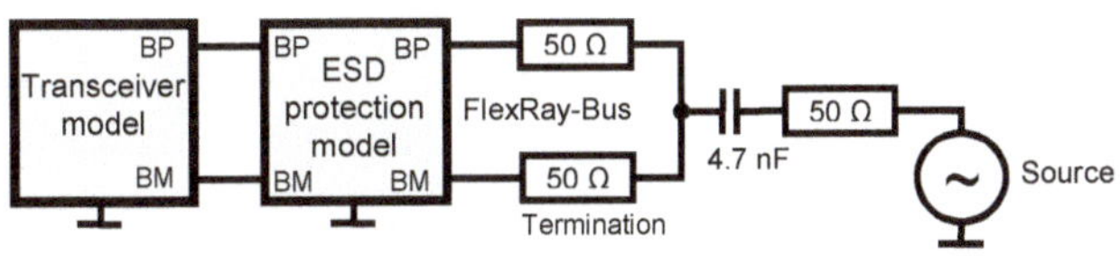

Fig. 9. Test setup incoupling with DPI method.

2.4 Investigations of FlexRay transceivers in combination with ESD protection circuits

To perform EMC simulations in combination with ESD protection circuits for signal integrity investigations the transceiver models can be modularly extended with VHDL-AMS models of protection elements.

ESD protection elements are intended for the protection against rarely occurring pulses. A permanent operation with CW-coupling above the breakdown voltage is not possible. The thermal power, generated by a coupled CW-current, can not be bypassed and will leads to destruction. The immunity simulations with sinusoidal disturbances were exemplarily done with a TVS diode array from Protek type GBLCS05C. The TVS array has a breakdown voltage of 6 V@1 mA and a parasitic impedance of 1.4 pF || 100 kΩ.

According to the DPI method (IEC 62132-4:2006, 2006) the noise voltage will be coupled to both bus lines through the termination resistors and a capacitor. In Fig. 9 the simulation and measurement setup is shown.

A comparison between simulation and measurement is given in Fig. 10. The voltages are limited by the protection element. At disturbance amplitude of about 8 V, the transceiver can not detect the differential voltage anymore. The disturbed differential bus signal with the resulting bit errors can be seen below. The correlation between simulation and measurement results is very good.

To verify the low pass filter characteristics and the digital behavioral model in combination with ESD protection circuits measurements at high and low frequencies were done.

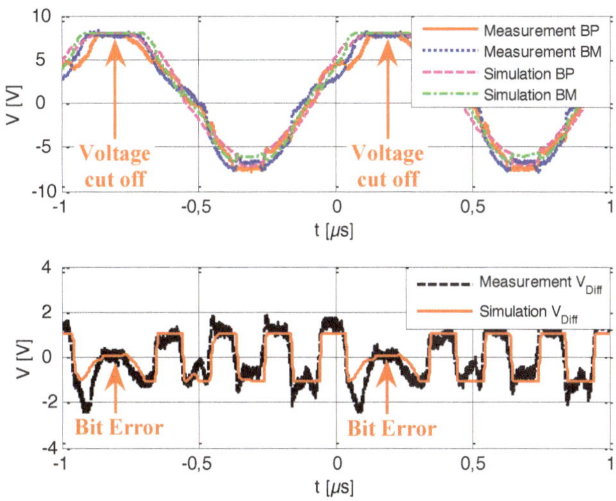

Fig. 10. Simulated and measured differential bus voltage, interference frequency 1 MHz with FlexRay Type A and Protek TVS Array.

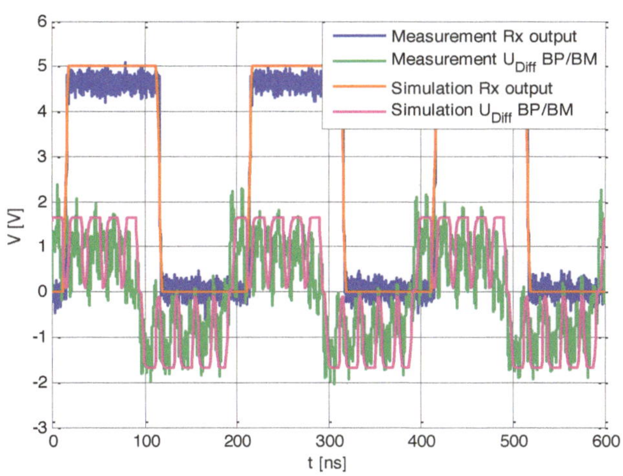

Fig. 11. Comparison between simulated and measured signal of a FlexRay transceiver.

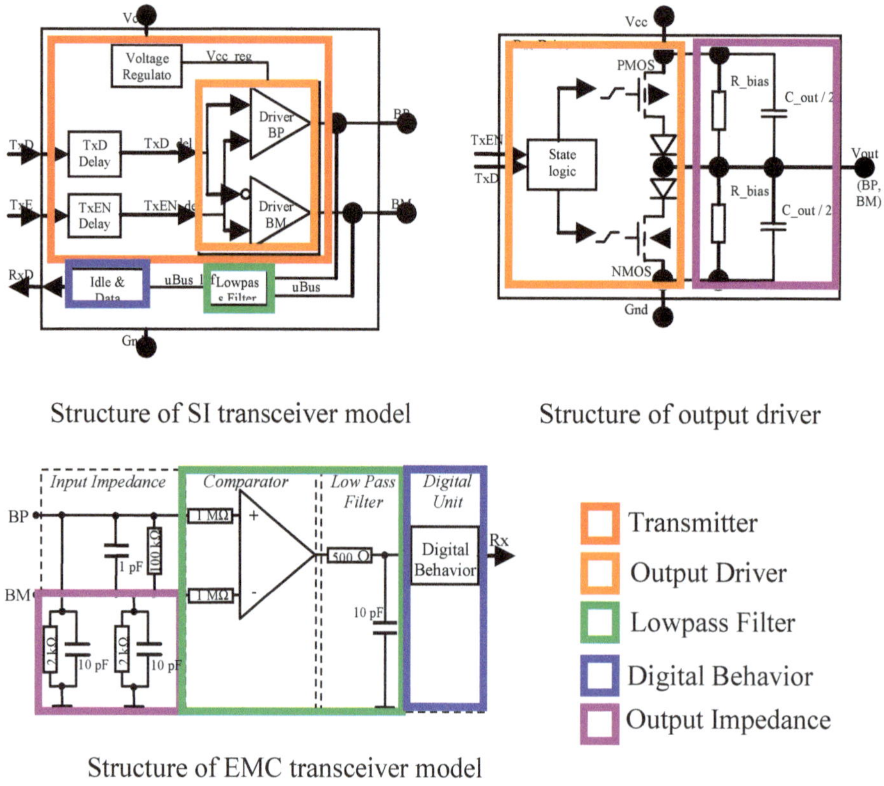

Fig. 12. Common parts of SI and EMC transceiver model.

In Fig. 11 a comparison between the simulation with the behavioral model and a measured signal of the FlexRay transceiver can be seen. The transceiver is combined with the Protek TVS array. Interference frequency is 50 MHz and the injected amplitude is 20 V. The correlation of presented curves is good.

3 A combined FlexRay transceiver model for SI and EMC simulations

The presented SI model of the FlexRay transceiver mainly focuses on the output behavior of the bus driver, while the presented EMC model reflects the input behavior of the circuit. Integration of the two models to a combined SI and EMC transceiver simulation model is possible. Figure 12 gives an overview of the common parts of the two models.

The transmitting part with the output drivers is present only in the SI model, thus it will be used for the combined model. Lowpass filter, digital behavior, and output impedances are present in both SI and EMC models, but show big similarities in general parts. The EMC model contains additional information about differential impedances and operation limits.

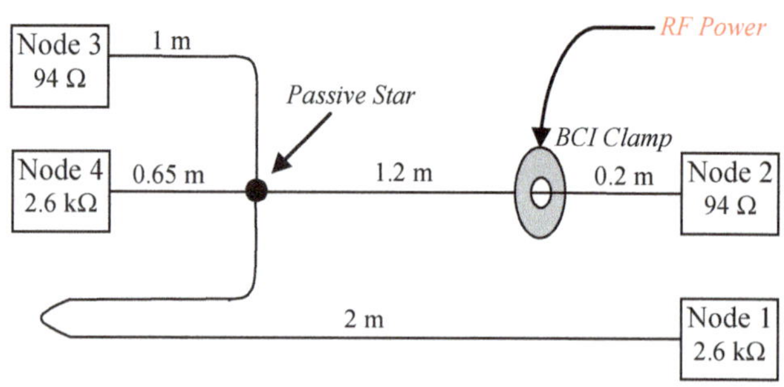

Fig. 13. BCI test setup with passive star and four nodes.

4 Simulation of BCI test with four node FlexRay network

With a VHDL-AMS model of a BCI clamp from Hilger et al. (2010) it is possible to make virtual CAN or FlexRay network tests. In combination with a cable model (Zhang et al., 2008) and the transceiver models an overall simulation model can be created. To investigate an extended network 4 FlexRay nodes are interconnected with cable models of twisted pair cables. In Fig. 13 the BCI test setup is shown. According to FlexRay Protocol Specification (2005) all nodes are split terminated. Nodes 1 and 4 with two 1.3 kΩ resistors, nodes 2 and 3 with two 47 Ω resistors and each

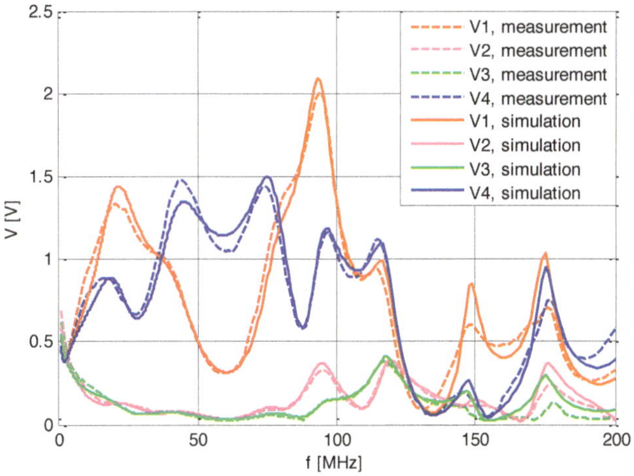

Fig. 14. Comparison between simulated and measured amplitudes at the four FlexRay nodes.

with 4.7 nF to GND. Common mode chokes or other protection elements are not used in this setup. The BCI clamp is positioned on the cable of node 2 in a distance of 0.2 m.

Figure 14 shows a comparison between the simulated and measured voltages at different FlexRay nodes. The measurements were done in the frequency domain with a network analyzer and in the time domain with a signal generator and oscilloscope. Here higher input power and active voltage probes were used. The BCI clamp was supplied with constant power of 10 dBm over all frequencies. The parasitic effects of connectors, the PCB and discrete elements, like resistors and capacitors, were considered in the model. The model shows very good correlation up to 200 MHz between simulated and measured results.

5 Conclusion

In order to investigate possible disturbances in bus system operation with simulation, behavioral models of bus transceivers for SI and EMC investigations were developed. SI model focus mainly on the output behavior of the transceivers, while EMC models reflect the input behavior. A detailed comparison of two developed models for these applications shows, that integration into a combined transceiver model for SI and EMC simulations is possible. Transmitter parts are taken from the SI model, since the EMC model gives detailed information about output impedances and operation limits. Digital receiving parts are present in both models. The combined model shows good matching with measurement results in an application example with disturbance signal injection through a BCI clamp.

References

Boyer, A., Bendhia, S., and Sicard, E.: Modeling of a Direct Power Injection Agression on a 16 Bit Microcontroler Input Buffer, EMC Compo 2007, 6th International Workshop on Electromagnetic Compatibility of Integrated Circuits, 2007.

FlexRay Protocol Specification: V2.1 Rev. A, http://www.flexray.com, 2005.

Günther, H., Frei, S., and Wenzel, T.: Simulation Methods for Signal Integrity of Automotive Bus Systems, APEMC, 2010.

Hilger, U. and Frei, S.: Modellierung von LIN-Transceivern für EMV-Simulationen im Kraftfahrzeug, EMV 2008, VDE Verlag, 2008a.

Hilger, U. and Frei, S.: Störfestigkeit von Bustransceivern gegen sinusförmiger Stögrößen in Verbindung mit nichtlinearen ESD-Schutzelementen, EMV 2008, VDE Verlag, 2008b.

Hilger, U., Miropolsky, S., and Frei, S.: Modeling of Automotive Bus Transceivers and ESD-Protection Circuits for Immunity Simulations of Extended Networks, EMC Europe, Wroclaw, 2010.

HPPI Transmission Line Pulser: Type TLP3010, http://www.hppi.de, 2010.

IEC 62132-4:2006: Integrated circuits – Measurement of electromagnetic immunity 150 kHz to 1 GHz – Part 4: Direct RF power injection method, 2006.

IBIS Open Forum: I/O Buffer Information Specification, Online, available: http://www.eigroup.org/ibis, 2009.

Zhang, H., Siebert, K., Frei, S., Wenzel, T., and Mickisch, W.: Multiconductor Transmission Line Modeling with VHDL-AMS for EMC Applications, IEEE-EMC 2008 Symposium, Detroit, 2008.

CMOS low noise amplifiers for 1.575 GHz GPS applications

M. Isikhan and A. Richter

IMMS gGmbH, Erfurt, Germany

Abstract. This paper presents Low Noise Amplifier (LNA) versions designed for 1.575 GHz L1 Band Global Positioning System (GPS) applications. A 0.35 μm standard CMOS process is used for implementation of these design versions. Different versions are designed to compare the results, analyze some effects and optimize some critical performance criteria. On-chip inductors with different quality factors and a slight topology change are utilized to achieve this variety. It is proven through both on-wafer and on-PCB measurements that the LNA versions operate at a supply voltage range varying from 2.1 V to 3.6 V drawing a current of 10 mA and achieve a gain of 13 dB to 17 dB with a Noise Figure (NF) of 1.5 dB. Input referred 1 dB compression point (ICP) is measured as −5.5 dBm and −10 dBm for different versions.

1 Introduction

The demand on current GPS applications forces the design of high performance, low cost L1 frequency band (1.575 GHz center frequency) receivers. Galileo/GPS combined system is under development and this will allow the realization of much higher precision applications (Alvarado et al., 2007). On the other hand, the Assisted GPS (A-GPS) concept necessitates the integration of GPS receiver into the mobile phone requiring improvement in receiver sensitivity, linearity and power consumption (Bokinge et al., 2006; van Diggelen, 2002). The demand is more intensive on highly integrated, compact solutions. Therefore, the GPS receiver building blocks should be convenient for SoC or possibly SiP implementation with a small amount of off-chip circuit components as possible. The target is to design an LNA with an optimum layout to be integrated in a receiver chip. The input and output of LNA is optimized for 50 Ω terminations making it suitable for integration into heterodyne receiver structures with external Image Reject Filter. A block diagram of a possible application is depicted in Fig. 1.

Correspondence to: M. Isikhan
(murat.isikhan@imms.de)

LNA is a critical building block on the received signal path. Important parameters of the performance of a receiver such as Noise Figure (NF), input impedance matching, reverse isolation, stability and linearity performance are mainly determined by the LNA (Razavi, 1998). The circuit topology and component selections to improve LNA performance in one criterion do not usually work in favor of all other performance criteria. Thus, a compromise is to be done during the decision for circuit topology and components. One of the performance criteria may take the priority depending on the application needs. Different versions are designed for this purpose through utilizing different Q on-chip inductors and a slight topology change. A highly integrated solution implemented in a standard CMOS technology with lower mask costs is a critical design target.

Different versions of LNA are designed and fabricated with a 0.35 μm standard CMOS technology and measurements are performed on-wafer and on-PCB. The design procedure and experimental results with comparisons are presented in this paper.

2 LNA circuit design

As mentioned in the introduction part, the NF of LNA has the most critical effect on the overall NF of the system. In order to have an overall NF of less than 2 dB for high precision GPS applications, the LNA NF should be kept well below 1.5 dB. 1.2 dB NF for the impedance-matched LNA in package could relax the noise requirements of cascaded components. A summary of the requirements are given in Table 1.

Different topologies are considered to achieve the performance targets. The LNA should provide sufficient gain at a bandwidth of 20.46 MHz for GPS L1 band applications. An amplifier topology with an inductive-capacitive load can achieve this requirement. Furthermore such a load selection keeps the NF at a low level. An ideal LNA should provide isolation of the signals at the input and output very well. The parasitic signals at the output of the LNA should not couple to the LNA input, possibly resulting in parasitic signal

Table 1. Target requirements.

Supply voltage	2.1 V ... 3.6 V
Supply current	10 mA
Center frequency	1.575 GHz
Input/output impedance	50 Ω
S11, S22	<−10 dB
S21	>15 dB
S12	<−25 dB
NF	<1.5 dB
ICP	−10 dBm

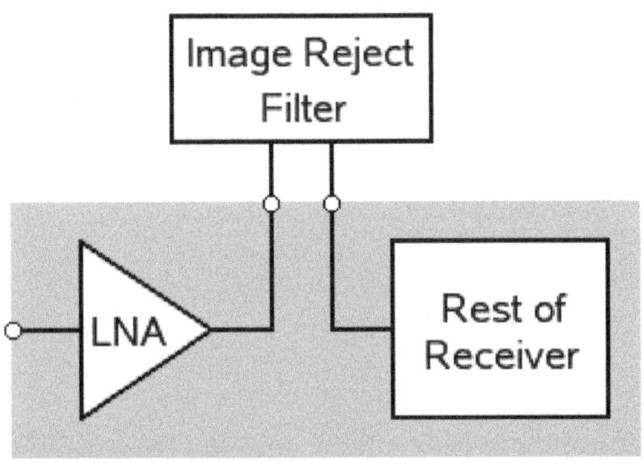

Fig. 1. Integration of LNA in a GPS receiver.

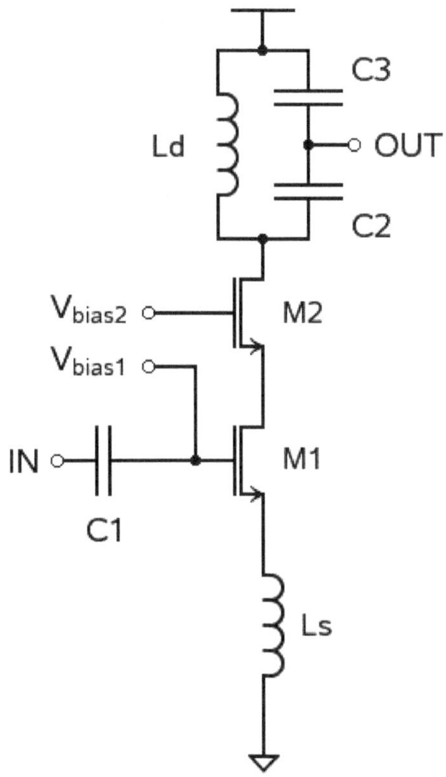

Fig. 2. Simplified core circuit of LNA.

transmissions from the antenna. This performance specification can be optimized through minimizing S12 parameter of the LNA. This attenuation in the reverse signal path, in other words reverse isolation is very critical for the overall receiver performance as well. In addition to this fact, a low S12 value provides the stable operation of an LNA. As well as providing reverse isolation and stability which are very critical, good isolation of input and output stage of the LNA brings the advantage of treating input and output impedance matching separately, performing some changes at the input stage of LNA for better impedance matching does not disturb the impedance seen at the output drastically. Using a cascode amplifier topology provides adequate reverse isolation and stability. The inductive degeneration topology provides the amplifier to achieve real input impedance. This is critical for input impedance matching with 50 Ω. Regarding these considerations, an amplifier topology with cascode n-MOS transistors, inductive capacitive load and a source degeneration inductor is selected for core circuit of LNA. This circuit is depicted in Fig. 2.

The inductive source degeneration proves good results when input impedance matching is considered. It provides creating a resistive part in addition to the reactive part seen at the gate of the input n-MOS transistor. This fact can be mathematically expressed with the following Eq. (1) where C_{gs} is the gate-source capacitance of input transistor M1 and g_m is its transconductance.

$$Zin = jwL_s + \frac{1}{jwC_{gs}} + \frac{g_m L_s}{C_{gs}} \tag{1}$$

It is a hard task to fix the input impedance exactly to 50 Ω using only a degeneration inductor. Trying to increase the transconductance through increasing the width of the transistor does not help to improve the resistive part of input impedance. In addition, the possibility of increasing transconductance through increasing the current is also limited since there is a maximum limit for the current consumption. Increasing the value of L_s helps to increase the resistive part of impedance, however, when small signal analysis is considered, it is obvious that increasing the source degeneration inductance results in lower gain. Due to technology and design limitations, the real part is lower than required 50 Ω. There is also a capacitive part of input impedance. To balance this and achieve a good input matching, an inductor should be placed on the signal path prior to the gate of input transistor. The Noise Figure considerations indicate that an inductor on the signal path should have a high quality factor in order to contribute less to the NF. On-chip inductors do not have a high enough quality factor for this application in the circuit with the given technology. Due to these facts, the inductor used for matching (L_{match}) is implemented as a high Q off-chip component. The bond wire inductance, effect of package and PCB parasitic circuit elements should

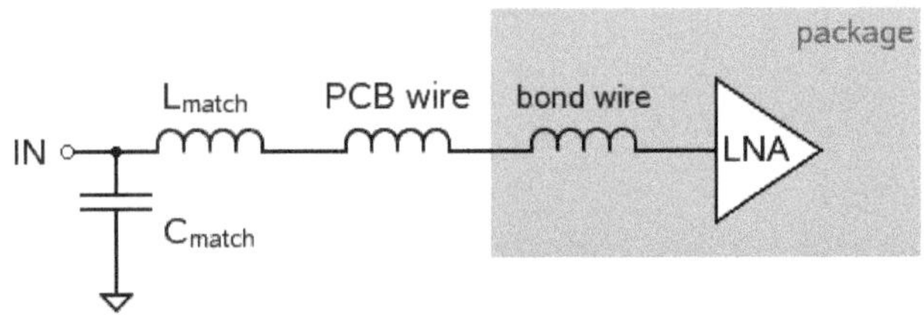

Fig. 3. Input impedance matching and parasitics of package and PCB.

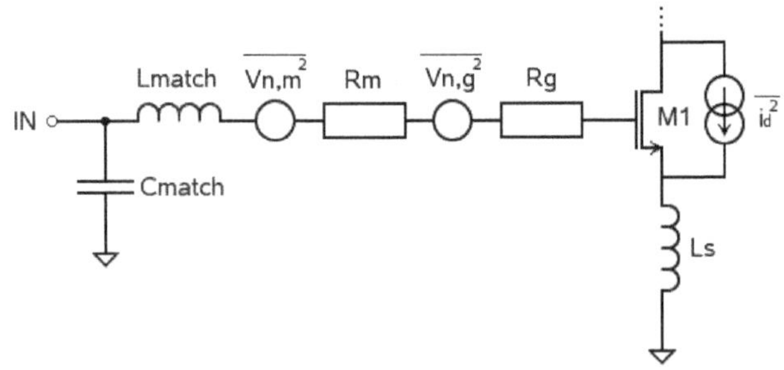

Fig. 4. Noise equivalent circuit of LNA.

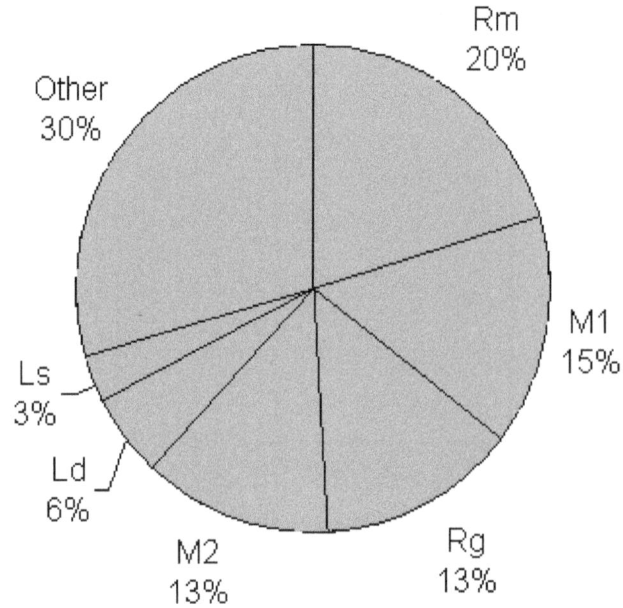

Fig. 5. Contribution of noise sources in LNA circuit.

be modeled accurate and taken into account for a good input impedance matching at the desired frequency range. To compensate the inductive impedance and to fix the resistive part exactly to 50 Ω, a matching capacitor C_{match} is also utilized. The circuit showing the off-chip components placed on the input signal path and the parasitic inductive effects of package and PCB are depicted in Fig. 3.

The main noise sources of a LNA with such inductive degeneration topology are the intrinsic thermal noise of input transistor M1, the gate resistance associated with it (R_g) and the resistance dependent to the Q of off-chip inductance for matching and the bond wire and conductor on PCB (R_m). The noise equivalent circuit for the LNA at the input transistor side is shown in Fig. 4.

As long as there is a good input matching, the LNA can be treated as a voltage amplifier for noise calculations and its noise factor F can be taken into consideration (Shaeffer and Lee, 1997). The noise considerations given in Shaeffer and Lee (1997) state the following Eq. (2) for the noise factor F;

$$F = 1 + \frac{R_m}{R_s} + \frac{R_g}{R_s} + \gamma \cdot R_s \cdot g_{d0} \cdot \left(\frac{w_0}{w_T}\right)^2 \tag{2}$$

This assumption is valid as long as the resonance of amplifier with the input circuit is achieved and w_0 is the resonant frequency. The resonant frequency should be fixed at the desired center frequency, 1.575 GHz. The n-MOS transition frequency is w_T which is equal to g_m/C_{gs}. R_s is the source resistance which is 50 Ω, R_m is the resistance due to off-chip inductor, bond wire and wiring on application board. R_g is the resistance associated with the gate of input n-MOS transistor. γ is a bias dependent factor and $g_{d0}$is zero bias drain conductance (Bokinge et al., 2006; Shaeffer and Lee, 1997).

Optimum values for zero bias drain conductance, gate resistance and the resistance of matching inductor should be achieved for a better noise performance. Gate width of n-MOS transistor and the inductance L_{match} are decided through a set of simulations to optimize noise parameters. The simulation results for the sources of noise in the LNA circuit are shown in the diagram above. The contribution of input transistor thermal noise and noise due to resistances on signal path can be compared.

The noise model for on-chip components and the estimated series resistance for off-chip discrete matching inductor are used for this simulation. It can be concluded that using an on-chip inductor instead would have contributed much more on the overall NF.

The load inductor L_d and the source degeneration inductor L_S are implemented as on-chip inductors to have a more compact solution. For such an implementation, the inductor modeling should be completed before the necessary simulations. The inductor values and their quality factors should be known and the parasitic extractions should be performed. The load inductor resonates with two capacitances and maximizes the gain at a resonant frequency. The output impedance seen at this frequency is ideally only resistive and 50 Ω for a good matching at the output. The capacitors, C_2 and C_3 have much higher quality factor compared to the inductor. Therefore, the overall Q of the load is mainly determined by the Q of the inductor. Utilization of a higher Q inductor will generate a sharper resonance characteristic and as a result a sharper gain (S21) characteristics at the frequency range of interest with a high maximum value of S21.

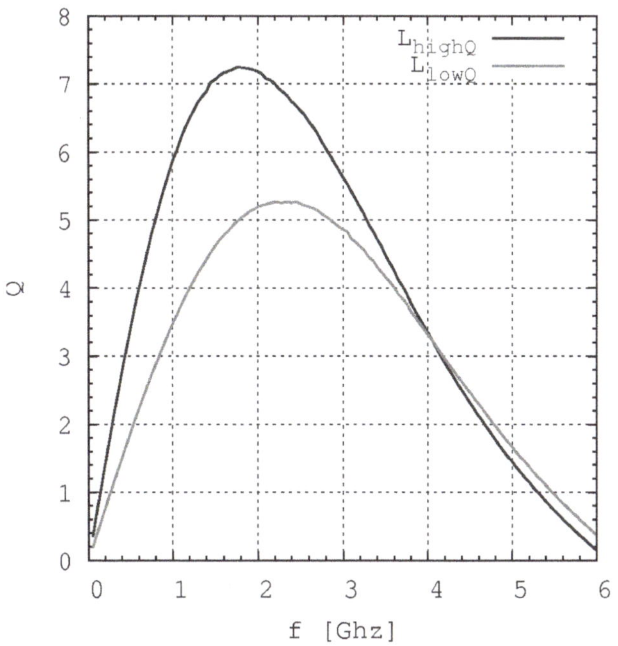

Fig. 6. Quality factors of 7 nH inductors L_d.

On the other hand an inductor with a low Q at the operation bandwidth will cause a smoother S21 characteristic, providing a higher available bandwidth but with a decreased value of maximum S21. The process variations should be taken into account during the simulations for deciding component values. A gain characteristic with a high maximum value and a bandwidth fitting the requirement can be fixed at the center frequency of 1.575 GHz. However, the process variations may alter the component values and shift the resonant frequency. The absence of a perfect package modeling may cause a more serious problem. Designing a higher bandwidth LNA with a decreased maximum gain provides more flexibility against these problems. It is decided to investigate the optimum Q of the available on-chip inductors to have an optimized S21 response. Two different 7 nH inductors with different quality factors are used for this purpose. Two different versions of LNA with the same topology are designed. The capacitor values to resonate with the inductor are also slightly different for the two versions as well to achieve resonance at the desired frequency and to obtain good output matching (S22 characteristic). The Q of two 7 nH inductors used for this design can be viewed in Fig. 6.

A third version of LNA for this investigation is also designed. The high Q inductor is used as load inductor. The source degeneration inductor L_s is not implemented on-chip. The gain can be increased by more than 3 dB by eliminating the source degeneration inductor. For this purpose, the source of input transistor is directly connected to the ground of the chip. The bond wire inductance acts as a degeneration inductor in this condition. A perfect modeling of such inductance realized by the bond wire effects in the package is difficult. However, it provides the presence of resistive part of input impedance. Different matching inductor and capacitor should be used on the application board compared to the first two versions. It is difficult to predetermine these values through simulations unless there is a very realistic model for package effects. This configuration can be utilized to maximize the gain despite having a complicated input matching and possibly stability problems. Small signal analysis on the circuit indicates that removing L_s results in a worse reverse isolation and hence stability problems.

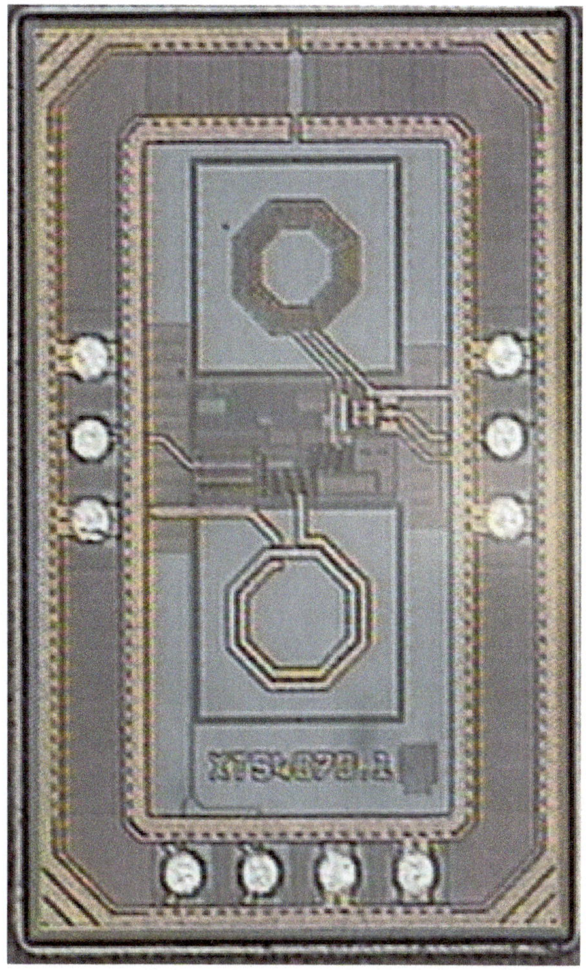

Fig. 7. IC photo of one LNA version.

3 Results

The LNA is encapsulated in a SOIC16 Ceramic Package. Measurements are performed on-wafer and on application board. Measurements are performed for the three design versions with different load and degeneration inductor implementations. The chip microphotograph of the design version

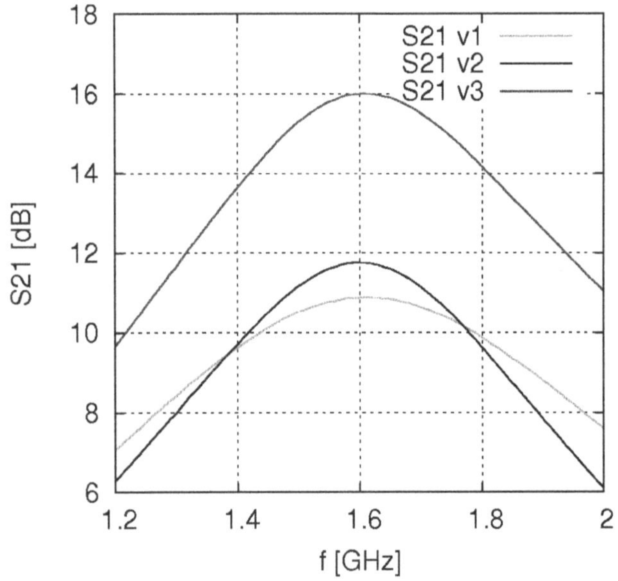

Fig. 8. S21 comparison of three design versions.

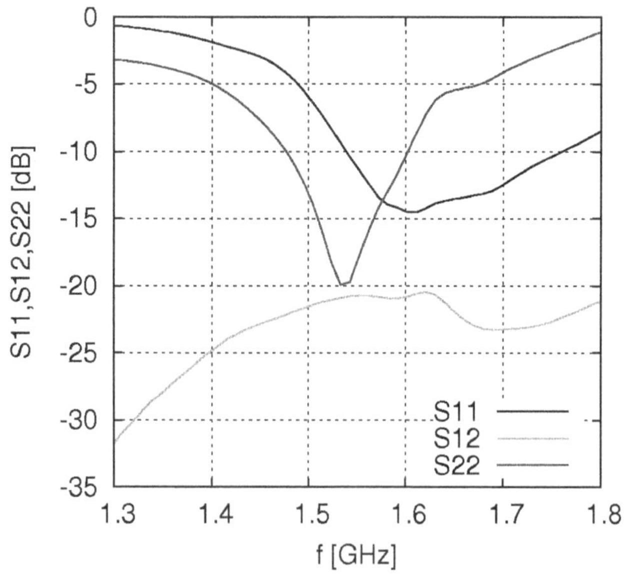

Fig. 10. S11, S12 and S22 of packaged LNA version 3.

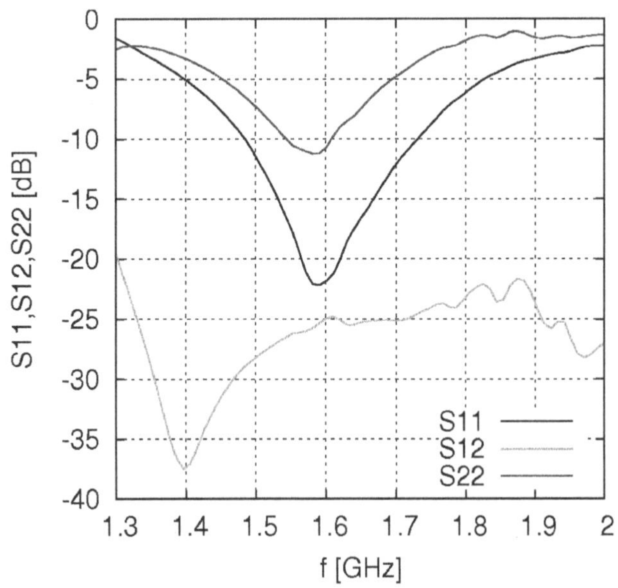

Fig. 9. S11, S12 and S22 of packaged LNA version 2.

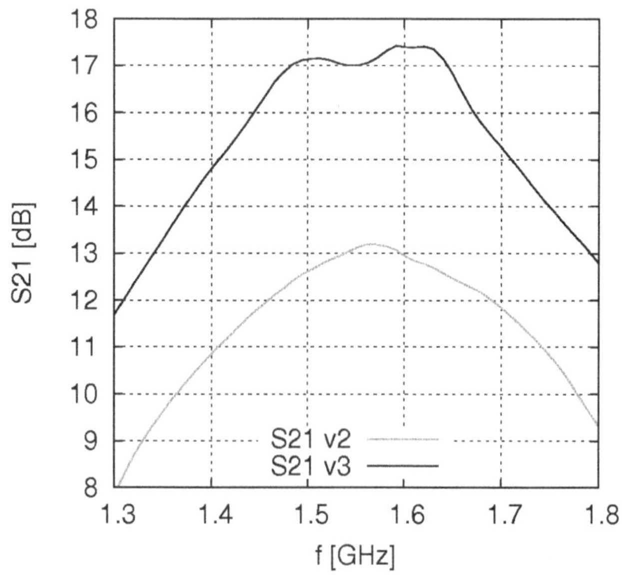

Fig. 11. S21 Comparison of packaged LNA versions 2 and 3.

with high Q on-chip inductor and on-chip degeneration inductor is shown in Fig. 7.

The set of investigations are performed on three different LNA versions with different load inductors and a slight topology difference. The comparison of S21 curves are shown in Fig. 8.

The figure shows the comparison of on-wafer measurements on the three versions. Version 2 with a high Q load inductor has a sharper response compared to first version as expected. In addition, the third version which includes no on-chip degeneration inductor has an upwards shifted response as expected. The output matching and noise figure measurements on-wafer fits to the simulated results well.

The S-parameter and NF measurements on the application PCB for version 2 and version 3 are shown in Figs. 9, 10, 11 and 12.

A good impedance matching at the input and output is achieved for both versions with S11 and S22 less than -10 dB at the frequency range of interest. The values of off-chip components are optimized to have a good input matching and noise figure. Although the on-wafer measurements fit the simulated results well, the measurement on the application board does not fit to the simulation results due to

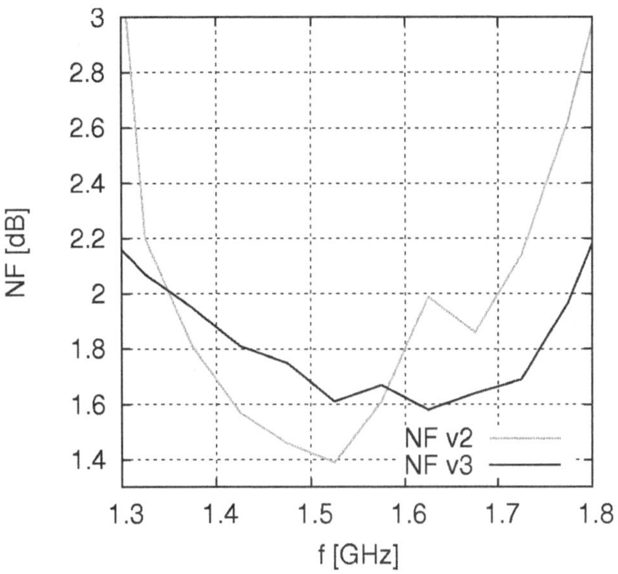

Fig. 12. NF Comparison of packaged LNA versions 2 and 3.

incomplete modeling of package and PCB parasitics. A reverse isolation (S12) better than 25 dB and 20 dB and S21 of 13.5 dB to 17 dB at the frequency range of interest is achieved. The NF is higher than expected 1.3 dB from the simulations. However the later narrowband noise measurements prove better results with a NF of 1.5 dB for both versions. Input referred 1 dB Compression Point (ICP) is measured as −5.5 dBm whereas the input referred third Intercept Point (IIP3) is measured as 3.3 dBm for version 2. ICP and IIP3 for version 3 are measured −10 dBm and 1 dBm, respectively. These results indicate a very good linearity performance of the device. The measurements are performed for supply voltage range varying from 2.1 V to 3.6 V and only negligible changes are observed. When stability issues are considered, it can be concluded that a stable operation for version 1 and version 2 is achieved. On the other hand version 3 has some stability problems at a frequency range of 200 MHz close to the frequency band of interest. This is again due to the incomplete modeling of parasitic effects.

A better performance for this application can be provided with more accurate modeling of package and PCB. The LNA IC can be integrated into a SiP environment. The matching elements can be implemented as components on a ceramic substrate and can be packaged together with LNA IC. A good performance can be achieved with an accurate 3-D substrate and package modeling for such application. Utilizing bond wire inductance as a circuit component in the design is a very strong motivation for these investigations since it obligates a prefect 3-D modeling for more realistic simulations.

4 Conclusion

Three 1.575 GHz low noise amplifier versions implemented with a 0.35 μm CMOS technology were presented. The experimental results indicate 17 dB gain and 1.5 dB noise figure. A high linearity with an IIP3 of 1 dBm is achieved. Further investigations to integrate this IC to a SiP environment are in progress. The performance can be improved with a SiP solution.

Acknowledgements. The presented work was partly funded by Thuringia TMWTA under No. B 509-03006. We also thank Björn Bieske and Uwe Baumann for their support with measurements.

References

Alvarado, U., Rodríguez, N., Mendizábal, J., Berenguer, R., and Bistué, G.: A Dual-Gain ESD-protected LNA with Integrated Antenna Sensor for a Combined Galileo and GPS front-end, Silicon Monolithic Integrated Circuits in RF Systems Topical Meeting on 10–12 January, pp. 99–102, 2007.

Bokinge, B., Einerman, W., Emericks, A., Grewing, C., Pettersson, O., Theil, D., and van Waasen, S.: A High Performance CMOS LNA for System-on-Chip GPS, Proceedings of Asia-Pacific Microwave Conference 2006.

van Diggelen, F.: Indoor GPS Theory and Implementation, IEEE Position Location and Navigation Symposium, Digest of Technical Papers, pp. 240, 2002.

Razavi, B.: RF Microelectronics, Prentice Hall, ISBN 0-13-887571-5, pp. 166–170, 1998.

Shaeffer, D. K. and Lee, T. H.: A 1.5-V, 1.5-GHz CMOS Low Noise Amplifier, IEEE Journal of Solid State Circuits, 32(5), 745–759, May 1997.

12

Pushing energy savings in adiabatic logic by carbon-nanotube field effect transistors

P. Teichmann[1,*], C. Friederich[1], and D. Schmitt-Landsiedel[1]

[1]Institute for Technical Electronics, Technische Universität München, Arcisstr. 21, 80333 München, Germany
[*]now at: Infineon Technologies AG, 81726 München, Germany

Abstract. For the first time carbon nanotube (CNT) transistor based adiabatic logic (AL) was analyzed in this work and compared to CNT based static CMOS (CCNT). Static CCNT inverters are used as a reference and compared to inverters in the AL families Efficient Charge Recovery Logic (ECRL) and Positive Feedback Adiabatic Logic (PFAL) in terms of energy dissipation. Energy savings by adiabatic logic in dependence of operating frequency, supply voltage and number of nanotubes per transistor are reviewed. It is shown that CNT based AL circuits provide high energy saving factors even for high frequencies compared to CNT based static CMOS circuits.

1 Motivation

Aggressive scaling of bulk MOSFETs has led to problems occurring due to increased short-channel-effects (SCE), leakage currents, and also due to fabrication limitations. One potential future replacement for bulk MOSFETs is the so-called Carbon Nanotube (CNT) FET. Made out of a wrapped around single-atom thick carbon cylinder, they promise to have excellent electrical as well as mechanical behavior. CNTs can be metallic or semiconductors (small-gap or moderate-gap), depending on the way they are rolled and the circumference of the tube (Hamada et al., 1992). So both, active devices and connections can be made out of CNTs. Single Wall Carbon Nanotubes (SWCNT) have a high mobility in the order of $100\,000\,\mathrm{cm}^{-2}\,\mathrm{Vs}^{-1}$ and a conductivity of up to $400\,000\,\mathrm{S\,cm}^{-1}$ (Hecht, 2007), due to the 1-D transport and thus reduced phase space for scattering leading to ballistic transport (Javey and Dai, 2006). Though these facts are very convincing, a lot of barriers have to be overcome before MOSFETs made out of SWCNTs can be produced on a large scale. One limitation is the Schottky Barrier (SB) that is formed by any metal-semiconductor connection due to the different work functions (Appenzeller et al., 2002). This limits the on-state current and thus the performance of the device. Additionally this leads to an ambipolar characteristic (Martel et al., 2001), the on-current is increased, when $|V_{\mathrm{GS}}|$ is increased. In digital circuits, this leads to problems in transistor stacks, where negative gate-to-source voltages appear and thus turn on a device that has a "0" input (Raychowdhury and Roy, 2007). One way to overcome this limitation is by connecting the nanotubes via Palladium, which is a metal with a high work-function (Javey et al., 2003). Thus the barrier can be greatly reduced or eliminated. Chirality cannot be controlled so far, nanotubes are to date a mix of 1/3 metallic and 2/3 semiconducting tubes (Hecht, 2007). The controlled assembly of nanotubes is another barrier that has to be overcome in order to allow for the integration of systems consisting of CNT devices with traditional metal interconnects, or CNT only. Derycke et al. (2001) propose a methodology to selectively produce p-type and n-type CNTs for inter-nanotube inverters. One of the tubes is covered with Polymethylmethacrylat (PMMA) and after thermal annealing and following exposure to oxygen, one of the tubes is n-type and the other is a p-type tube.

In contrast to dynamic losses in static CMOS

$$E_{\mathrm{CMOS}} = \frac{1}{2} \cdot C \cdot V_{\mathrm{DD}}^2 \tag{1}$$

the losses in adiabatic logic (AL) are described by

$$E_{\mathrm{AL}} = 8 \cdot \frac{RC}{T} \cdot C \cdot V_{\mathrm{DD}}^2 \tag{2}$$

The losses in AL do not only depend on the capacitance C and the voltage supply V_{DD} but also on the path resistance R and the period T of the operating frequency.

Due to their superior electrical characteristic, the high conductivity resulting from a (near-)ballistic transport, carbon nanotubes are interesting candidates for future large scale integration circuits; their capabilities allow for lowest consumption especially in Adiabatic Logic. In the following part of the work the principle of chirality and its meaning

Correspondence to:
D. Schmitt-Landsiedel
(dsl@tum.de)

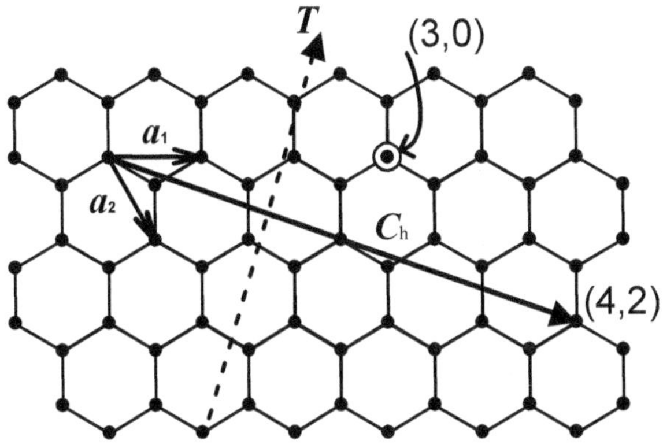

Fig. 1. The chiral vector $\boldsymbol{C}_\mathrm{h}$ is defined via unit vectors $\boldsymbol{a}_1$ and $\boldsymbol{a}_2$. It defines the electrical properties. As examples, the metallic (3,0) and a semiconducting (4,2) orientation is presented. Perpendicular to the chiral vector the translation vector $\boldsymbol{T}$ is found, which is the axis of the obtained nanotube.

for the properties of CNTs will briefly be presented, and later on, simulations are carried out with the Stanford CNT Hspice simulation model (Deng et al., 2008; Deng and Wong, 2007a,b) to see how Adiabatic Logic performs with those future devices.

2 The chirality of a CNT and the CNTFET

A CNT can be either metallic or semiconducting. The chiral vector $\boldsymbol{C}_\mathrm{h} = n_1\boldsymbol{a}_1 + n_2\boldsymbol{a}_2$ determines whether a metallic or semiconducting nanotube is formed. A graphical representation of the chirality vector can be observed in Fig. 1. It connects the origin with the point on the grid that is identical to the origin after wrapping up the carbon sheet. With the chiral number (n_1, n_2) the circumference of the nanotube can be determined as $|\boldsymbol{C}_\mathrm{h}| = a\sqrt{n_1^2 + n_2^2 + n_1 n_2}$, where $a = 2.49$ Å is the bond distance of two carbon atoms. Electrical characteristics are also determined via the chiral number. A nanotube with (n_1, n_2) is a metal, if $n_1 - n_2 = 3l$ $(l = 1, 2, 3...)$ and a semiconductor if $n_1 - n_2 \neq 3l$ (Dresselhaus et al., 2001). The bandgap energy E_gap is also dependent on the circumference of the tube. Additionally, tiny-gap and large-gap semiconducting CNT exists, also deterministic according to the chiral number (n_1, n_2) (Hamada et al., 1992). Two constellations are pictured in Fig. 1, the chiral number $(3,0)$ represents a metallic nanotube, whereas the $(4,2)$ constellation is a semiconducting nanotube. For tiny-gap and large-gap semiconducting CNTs the bandgap energy is proportional to $\frac{1}{R^2}$ and $\frac{1}{R}$, respectively (Dresselhaus et al., 2001), where R is the radius of the tube.

Figure 2 shows a schematic cross-section through a CNTFET device. The source and drain regions are connected by a semiconducting CNT. As in the bulk MOSFET device, the control gate of the CNTFET is separated from the semiconductor with an insulating layer.

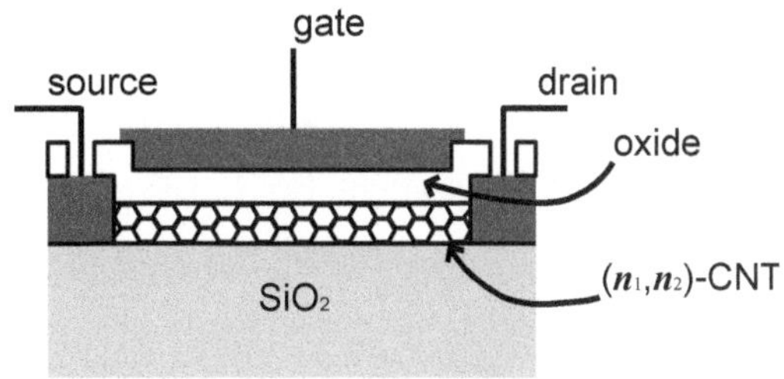

Fig. 2. Schematic cross-section of a Carbon Nanotube Field Effect Transistor (CNTFET) proposed by Marulanda (2008). A gate is separated via an insulating layer (oxide) from the CNT. Through the CNT a connecting channel between source and drain can be established.

Table 1. Parameter list for the applied Stanford CNT simulation model.

Parameter	Value	Explanation
V_DD	1.2 V	supply voltage
(n_1, n_2)	(19,0)	chirality of nanotube(s)
tubes	3	# of parallel tubes within a device
pitch	20 nm	distance between adjacent CNTs
L_ch	32 nm	physical channel length
$L_\mathrm{SS}, L_\mathrm{DD}$	32 nm	doped source/drain extension
t_ox	4 nm	top gate oxide thickness
k_ox	16	dielectric constant of oxide

3 Simulation results

Simulations are carried out with the HSpice CNT compact model provided by Deng and Wong (2007a,b) and Deng et al. (2008). The model includes non-idealities like near-ballistic transport, scattering, effects of the source/drain extension, and charge screening between tubes, when more than one tube is used for a CNTFET (Deng and Wong, 2007a). By inclusion of a transcapacitance network, good predictions of the transient behavior are expected (Deng and Wong, 2007a).

Based on the Stanford CNT model, static CMOS like (static CCNT) inverters as well as adiabatic inverters are compared with respect to the energy consumed per performed calculation. The standard parameter sets summarized in Table 1 are used if not stated otherwise. Three tubes in parallel are used in the CNTFET devices, each with a moderate-gap semiconductor tube of chiral number $(19,0)$. Pitches between the adjacent CNTs within one device are 20 nm. A channel of 32 nm physical length is separated via a 4 nm oxide. A chain of five inverters is used to determine the energy consumption and therewith the resulting energy saving factor ESF$=\frac{E_\mathrm{CMOS}}{E_\mathrm{AL}}$. The static CCNT structure is used as

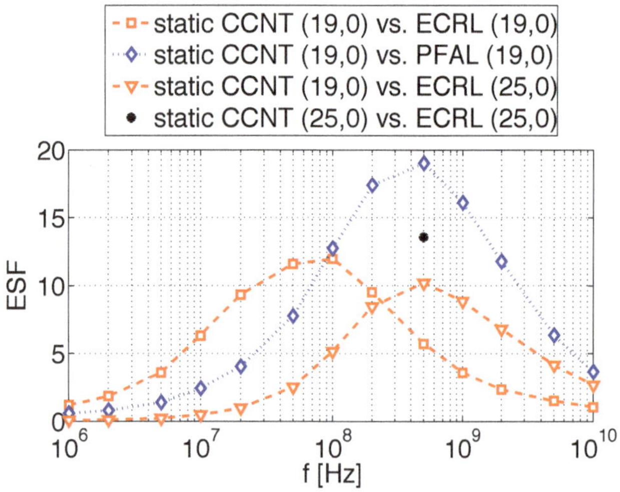

Fig. 3. The ESF gained by applying CNTFETs for ECRL and PFAL.

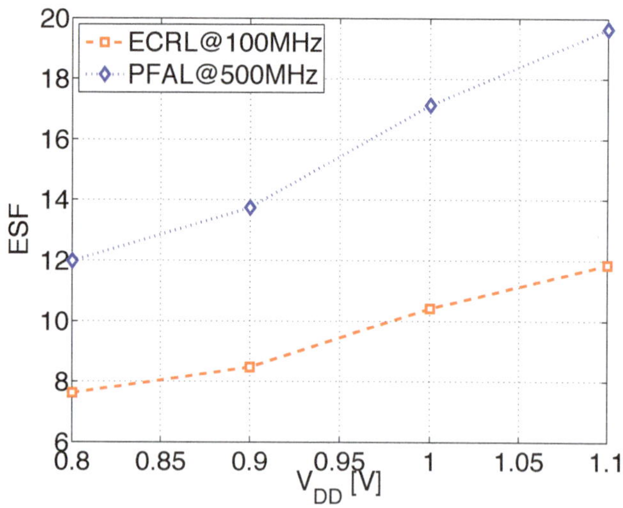

Fig. 4. Development of the ESF of an inverter circuit under the impact of voltage scaling.

a reference and compared to inverters in the AL families Efficient Charge Recovery Logic (ECRL, Moon and Jeong, 1996) and Positive Feedback Adiabatic Logic (PFAL, Blotti et al. 2000).

ESF versus the frequency can be observed in Fig. 3. A maximum ESF of 12 is found for ECRL for 100 MHz and 19 for PFAL at 500 MHz. A remarkable result is, that for a wide frequency range from 70 MHz to 2.4 GHz the ESF is above 10 for PFAL. If the chiral vector is changed, the bandgap of the semiconductor is influenced. A reduced bandgap energy leads to a reduced threshold-voltage V_{th}, as the barrier height from source/drain to the channel is reduced. The on-current is increased, but off-currents are also increased. Due to the reduced V_{th} leakage currents are increased, but the adiabatic losses are reduced, as the gate-overdrive $(V_{DD} - V_{th})$ is increased. ECRL is simulated for (25,0) and compared to static CCNT (19,0). For the frequency with maximum ESF also the static CCNT with (25,0) is simulated and compared to ECRL (25,0). For lower frequencies, increased leakage currents lead to an intersection of leakage losses and adiabatic losses for the (25,0) ECRL configuration, i.e. leakage dominates the losses. That leads to decreased ESF when compared to the case of (19,0) chirality. Reduced adiabatic losses lead to a shift of the optimum frequency f_{opt} (frequency with maximum ESF) to 500 MHz. At this frequency the ESF for (25,0) is higher compared to its (19,0) counterpart. Thus f_{opt} can be adopted by changing the chirality of the nanotube. At 500 MHz the static CCNT inverter was simulated with (25,0). An increase in the ESF is observed, that can be explained by increased leakage currents in the (25,0) static CCNT. For static CCNT circuits, performance is improved but dynamic losses are not directly impacted by the chirality.

In summary, the following values are derived from Fig. 3: f_{opt} is improved from 100 MHz to 500 MHz and the maximum ESF is decreased from 12 to 10 in case of replacing (19,0)-tubes in ECRL with (25,0)-tubes, and it is increased from 12 to 13.6 in case that (25,0)-tubes are used both in ECRL and static CCNT.

PFAL for the (19,0)-setup has already a high ESF with a maximum value of 19 at 500 MHz. Increasing the chirality for PFAL to (25,0) did not show any benefit, it rather worsens the energy behavior of PFAL. Thus it is not presented in the figures.

All previous investigations are carried out with a supply voltage of $V_{DD} = 1.2$ V. Supply voltage scaling allows to reduce the energy dissipation in static CCNT and AL. Simulations are carried out with reduced supply voltage. The according ESFs are presented in Fig. 4. It is reduced by 34% and 39% when going from 1.1 V to 0.8 V for ECRL and PFAL, respectively. But it has to be mentioned, that the performance for static CCNT is also decreased. As cascaded gates are used, the overall performance loss will decrease the maximum operating frequency or the effort for pipelining is increased. The first fact will determine a lower limit for voltage scaling (where safety margins for variations have to be regarded also), whereas the second fact leads to a trading of speed versus energy and area.

A high fan-out of a gate leads to the necessity of scaling the devices of the driving gate. Comparable to the width in planar CMOS technology, the number of tubes is a measure in CNTFETs for their driving capability. Equipping a gate with a higher number of tubes in the devices will impact the energy dissipation. Figure 5 shows the simulation results of ESF, when different numbers of tubes are used within the inverter chain. All gates in the chain are sized, and both devices, NCNT and PCNT. Again ECRL and static CCNT

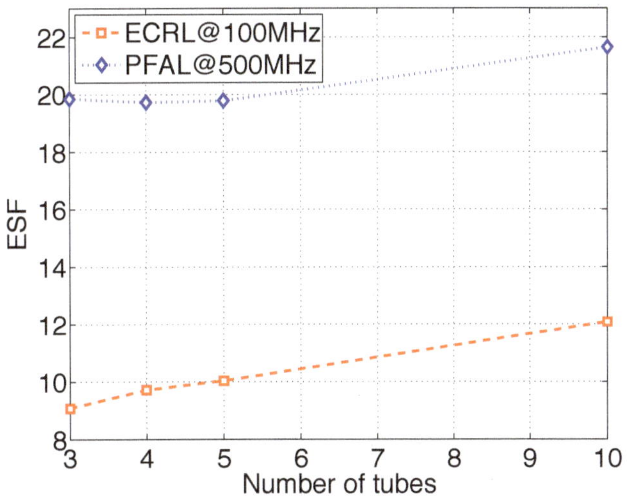

Fig. 5. ESF with respect to the number of parallel tubes used in a CNTFET device.

are simulated with a frequency of 100 MHz, and PFAL at 500 MHz, all with a supply voltage $V_{DD} = 1.2\,V$. The increased number of tubes obviously impacts the ESF. While the ESF for PFAL stays constant up to five tubes in parallel, and is increased from 20 to 22 when ten tubes are used, in ECRL it is continuously increased from 9 for three tubes to 12 in case of ten tubes. This is explained by the increased capacitance value and the reduced on-resistance due to parallel connection of more tubes. In static CMOS, the increased capacitance will increase the energy dissipation according to $E_{CMOS} \propto C$. In AL, also the reduced on-resistance impacts the energy dissipation with $E_{AL} \propto RC^2$. If the quadratic increase of the capacitance value is more than compensated by the reduced on-resistance, the energy in AL rises less than in static CCNT.

4 Summary and conclusions

For the first time carbon nanotube based adiabatic logic was analyzed in this work and compared to CNT based static CMOS. Simulation results show maximum energy saving factors of 12 and 19 for ECRL and PFAL, respectively. A remarkable result is, that for a wide frequency range from 70 MHz to 2.4 GHz the ESF is above 10 for PFAL. Changing the chirality to (25,0) will impact results for ECRL: ESF factors for frequencies greater than 100 MHz are increased with respect to the (19,0) results. The optimum frequency is increased from 100 to 500 MHz. The maximum ESF is slightly decreased. If carbon nanotubes become successors to planar CMOS transistors they offer a tremendous energy saving impact and improved performance especially in Adiabatic Logic. Due to their superior carrier transport they offer a small on-resistance and thus are exceptionally well applicable in Adiabatic Logic, where energy per operation depends not only on capacitance, as in static CMOS, but also on on-resistance.

Acknowledgements. This study was supported by DFG Grant SCHM 1478/6-1.

References

Appenzeller, J., Knoch, J., Martel, R., Derycke, V., Wind, S., and Avouris, P.: Carbon nanotube electronics, Nanotechnology, IEEE Transactions on, 1, 184–189, 2002.

Blotti, A., Di Pascoli, S., and Saletti, R.: Simple model for positive-feedback adiabatic logic power consumption estimation, Electronics Letters, 36, 116–118, 2000.

Deng, J. and Wong, H.-S.: A Compact SPICE Model for Carbon-Nanotube Field-Effect Transistors Including Nonidealities and Its Applications Part II: Full Device Model and Circuit Performance Benchmarking, Electron Devices, IEEE Transactions on, 54, 3195–3205, 2007a.

Deng, J. and Wong, H.-S.: A Compact SPICE Model for Carbon-Nanotube Field-Effect Transistors Including Nonidealities and Its Applications Part I: Model of the Intrinsic Channel Region, Electron Devices, IEEE Transactions on, 54, 3186–3194, 2007b.

Deng, J., Lin, A., Wan, G. C., and Wong, H.-S. P.: Carbon nanotube transistor compact model for circuit design and performance optimization, J. Emerg. Technol. Comput. Syst., 4, 1–20, 2008.

Derycke, V., Martel, R., Appenzeller, J., and Avouris, P.: Carbon Nanotube Inter- and Intramolecular Logic Gates, Nano Letters, 1, 453–456, 2001.

Dresselhaus, M. S., Dresselhaus, G., and Avouris, P. (Eds.): Carbon Nanotubes: Synthesis, Structure, Properties, and Applications, Springer Berlin/Heidelberg, 2001.

Hamada, N., Sawada, S., and Oshiyama, A.: New one-dimensional conductors: Graphitic microtubules, Phys. Rev. Lett., 68, 1579–1581, 1992.

Hecht, D.: Properties and Applications of Carbon Nanotube Films: A Revolutionary Material for Transparent and Flexible Electronics, Ph.D. thesis, University of California Los Angeles, 2007.

Javey, A. and Dai, H.: Carbon nanotube electronics, in: Proceedings of the 19th International Conference on VLSI Design, pp. 453–458, 2006.

Javey, A., Guo, J., Wang, Q., Lundstrom, M., and Dai, H.: Ballistic carbon nanotube field-effect transistors, Nature, 424, 654–657, 2003.

Martel, R., Derycke, V., Lavoie, C., Appenzeller, J., Chan, K. K., Tersoff, J., and Avouris, P.: Ambipolar Electrical Transport in Semiconducting Single-Wall Carbon Nanotubes, Phys. Rev. Lett., 87, 256 805, 2001.

Marulanda, J. M.: Current Transport Modeling of Carbon Nanotubes: Concepts, Analysis, and Design, Ph.D. thesis, Louisiana State University, 2008.

Moon, Y. and Jeong, D.-K.: An efficient charge recovery logic circuit, Solid-State Circuits, IEEE Journal of, 31, 514–522, 1996.

Raychowdhury, A. and Roy, K.: Carbon Nanotube Electronics: Design of High-Performance and Low-Power Digital Circuits, Circuits and Systems I: Regular Papers, IEEE Transactions on, 54, 2391–2401, 2007.

13

Localization of passive UHF RFID Labels with Kalman Filter

T. Nick and J. Götze

Information Processing Lab, TU Dortmund University, Otto-Hahn-Str. 4, 44227 Dortmund, Germany

Correspondence to: T. Nick (theresa.nick@tu-dortmund.de)

Abstract. Localization via Radio Frequency Identification (RFID) is frequently used in different applications nowadays. It has the advantage that next to its ostensible purpose of identifying objects via their unique IDs it can simultaneously be used for the localization of these objects. In this work it is shown how Received Signal Strength Indicator (RSSI) measurements at different antennae of a passive UHF RFID label can be combined for localization. The localization is only done based on the RSSI measurements and a Kalman Filter (KF). Because of non-linearities in the measurement function it is necessary to incorporate an Extended Kalman Filter (EKF) or an Unscented Kalman Filter (UKF) where simulations have shown that the UKF performs better than the EKF. Additionally to the selection of the filter there are different possibilities to increase the localization accuracy of the UKF: The advantages of using Reference Tags (RT) or more than one tag per trolley (relative positioning) in combination with an Unscented Kalman Filter are discussed and simulations results show that the localization error can be decreased significantly via these methods. Another possibility to increase the localization accuracy and in addition to achieve a more realistic simulation is the consideration of the angle between reader antenna and tag. Simulation results with the incorporation of different numbers of fixed antennae lead to the conclusion that this is a useful surplus in the localization.

1 Introduction

Radio Frequency Identification (RFID) Finkenzeller (2008) has become very popular in the last decades. Especially in the field of logistics many different use cases can be found for the identification as well as tracking of objects and goods Raza et al. (1999). A surplus of RFID is the possibility to gain knowledge about the object's identity as well as its position.

In this paper different approaches for the localization of a passive UHF RFID label (LT) with different types of Kalman Filters are shown. Next to the Extended Kalman Filter (EKF) and the Unscented Kalman Filter (UKF) additional information like relative positioning and the usage of Reference Tags (RTs) is used to improve the localization result.

A use case for the such a localization via RFID is the localization of trolleys coming inside and moving outside of a mail distribution center of Deutsche Post AG. With the help of the localization process a database can always track which trolley has been where at which point of time. This allows for missing trolleys to be found easier because possible actual locations are constrained. The problem of the localization of the trolleys though is that there is not only one trolley coming into the mail distribution center or leaving it, but that there are in most cases also trolleys parked on the side of the gates whose tags send their ID back. Those trolleys should not be detected as coming in or leaving because they are stationary. Figure 1 shows such a scenario where a trolley is just about ready to be moved out of the mail distribution center through the left gate and passing through the RFID localization area while many other trolleys are standing on both sides of that gate. In the front part of the picture it can be seen that there is also a trolley which is passing by the RFID area inside the mail distribution center. It is important to be able to distinguish between these trolleys and the ones coming in or going out because in the database the status of the trolleys remaining inside the mail distribution center is not changed. Through the usage of this RFID-based localization the number of missing trolleys should be reduced and therefore the necessity to replace them. Because there are about 80 mail distribution centers nationwide with about 30 gates each the hardware should be as inexpensive as possible.

Compared to existing localization techniques for RFID labels like "LANDMARC" Ni et al. (2004) or its expansion to work with passive tags and in 3-D Khan and Antiwal (2009) this paper's approach is aimed to achieve higher localization

Fig. 1. Test environment for RFID based trolley tracking.

accuracy. This aim should be reached with off-the-shelf hardware components for labels, reader and antennae. That is why a Kalman Filter is chosen which has proven to be well working for such an application Lee et al. (2010). The parameter which is available from almost all RFID readers is the Received Signal Strength Indicator (RSSI). But previous research has shown that RSSI measurements are noisy due to multiple effects of the wireless transmission (like multi-path propagation caused by reflections) Parameswaran et al. (2009); Elnahrawy et al. (2004). Therefore the combination of RSSI measurements with a Kalman Filter and additional information might lead to a more accurate localization result.

The paper is organized as follows: In Sec. 2 the mathematical basics about the Extended and Unscented Kalman Filter are explained. In the next section the approaches of differential and relative positioning are shown. Another option is to incorporate the angle between reader antenna and tag into the localization process which is described in Sect. 4. After that the Matlab simulation is explained and its results are given in Sect. 5. The paper closes by giving a conclusion and an outlook on future work in Sect. 6.

2 Extended and unscented Kalman Filter

In general a Kalman Filter uses a prediction and an update step to eliminate errors. Because of its structure only the result of the last step and a new measurement are needed to determine a function that describes how the old result is connected to the new one and how the measurement is connected with the prediction result. For a more detailed explanation of Kalman Filters see Thrun et al. (2006).

2.1 Extended Kalman Filter

The Extended Kalman Filter uses partial derivatives to be able to cope with non-linearities in the process or measurement function. Therefore it is necessary to calculate the Jacobian Matrices. The input into the filter is a Gaussian position estimate of the previous time step $t-1$ with mean $\boldsymbol{\mu}_{t-1}$ and its covariance $\mathbf{P}_{t-1}$ as well as the control $\mathbf{u}_t$ and the measurement $\mathbf{z}_t$ for updating the position.

In a first step a prediction of the new mean $\bar{\boldsymbol{\mu}}_t$ is given by

$$\bar{\mu}_t = g(\mathbf{u}_t, \mu_{t-1}), \tag{1}$$

where $g()$ represents the relation between the control of the present time step and the mean of the previous time step.

Next is the calculation of a prediction of the new covariance $\bar{\mathbf{P}}_t$

$$\bar{\mathbf{P}}_t = \mathbf{G}_t \mathbf{P}_{t-1} \mathbf{G}_t^T + \mathbf{R} \tag{2}$$

where $\mathbf{G}_t$ is the Jacobian matrix of the control function and $\mathbf{R}$ is the covariance matrix of the process noise.

With the help of this new covariance the Kalman Gain $\mathbf{K}_t$ can be calculated as

$$\mathbf{K}_t = \bar{\mathbf{P}}_t \mathbf{H}_t^T (\mathbf{H}_t \bar{\mathbf{P}}_t \mathbf{H}_t^T + \mathbf{Q})^{-1} \tag{3}$$

with $\mathbf{H}_t$ being the Jacobian matrix of the measurement function and $\mathbf{Q}$ being the covariance matrix of the measurement noise.

Now the new values for the mean μ_t and covariance $\mathbf{P}_t$ are computed as

$$\mu_t = \bar{\mu}_t + \mathbf{K}_t(\mathbf{z}_t - h(\bar{\mu}_t)) \tag{4}$$

and

$$\mathbf{P}_t = (\mathbf{I} - \mathbf{K}_t \mathbf{H}_t)\bar{\mathbf{P}}_t, \tag{5}$$

where $h()$ is the measurement prediction function. $\mathbf{x}_t$ and $\mathbf{P}_t$ are the outputs of the Extended Kalman Filter.

2.2 Unscented Kalman Filter

The Unscented Kalman Filter Wan and Merwe (2000) can also handle non-linearities in the filtering process. This is done trough stochastic linearization. So-called *Sigma Points* are used which are located at the mean and symmetrically along the axis of the covariance with two points for each dimension n leading to an overall number of $2n+1$ Sigma Points.

The input into the Unscented Kalman Filter is as well as for the Extended Kalman Filter a Gaussian position estimate of the previous time step $t-1$ with mean μ_{t-1} and covariance $\mathbf{P}_{t-1}$. Its first step is to calculate the Sigma Points of the previous step as

$$\chi_{t-1} = (\mu_{t-1} \quad \mu_{t-1} + \gamma\sqrt{\mathbf{P}_{t-1}} \quad \mu_{t-1} - \gamma\sqrt{\mathbf{P}_{t-1}}), \tag{6}$$

where $\gamma = \sqrt{n+\delta}$ and $\delta = \alpha^2(n+\kappa) - n$ with α and κ as scaling parameters and n as the dimension of the state space.

Next these points are propagated through the control function g

$$\bar{\chi}_t^* = g(\mathbf{u}_t, \chi_{t-1}) \tag{7}$$

and a predicted mean $\bar{\mu}_t$ and covariance $\bar{\mathbf{P}}_t$ are calculated as

$$\bar{\mu}_t = \sum_{i=0}^{2n} w_m^{[i]} \bar{\chi}_t^{*[i]} \tag{8}$$

and

$$\bar{\mathbf{P}}_t = \sum_{i=0}^{2n} w_c^{[i]} (\bar{\chi}_t^{*[i]} - \bar{\mu}_t)(\bar{\chi}_t^{*[i]} - \bar{\mu}_t)^T + \mathbf{R}, \tag{9}$$

where w_m and w_c are the weights according to

$$w_m^{[0]} = \frac{\delta}{n+\delta}$$

$$w_c^{[0]} = \frac{\delta}{n+\delta} + (1 - \alpha^2 + \beta) \text{ and}$$

$$w_m^{[i]} = w_c^{[i]} = \frac{1}{2(n+\delta)} \quad \text{for } i = 1, \ldots, 2n$$

with β as a scaling parameter and $\mathbf{R}$ as the covariance matrix of the process noise.

With the help of $\bar{\mu}_t$ and $\bar{\mathbf{P}}_t$ new Sigma Points $\bar{\chi}_t$ are calculated

$$\bar{\chi}_t = (\bar{\mu}_t \quad \bar{\mu}_t + \gamma\sqrt{\bar{\mathbf{P}}_t} \quad \bar{\mu}_t - \gamma\sqrt{\bar{\mathbf{P}}_t}), \tag{10}$$

which now capture the uncertainty after the prediction step.

For each of these Sigma Points a predicted observation point is computed through the measurement function h

$$\bar{\mathbf{Z}}_t = h(\bar{\chi}_t). \tag{11}$$

Now $\bar{\mathbf{Z}}_t$ is used to calculate the predicted observation $\hat{\mathbf{z}}_t$ and its uncertainty $\mathbf{S}_t$ according to

$$\hat{\mathbf{z}}_t = \sum_{i=0}^{2n} w_m^{[i]} \bar{\mathbf{Z}}_t^{[i]} \tag{12}$$

and

$$\mathbf{S}_t = \sum_{i=0}^{2n} w_c^{[i]} (\bar{\mathbf{Z}}_t^{[i]} - \hat{\mathbf{z}}_t)(\bar{\mathbf{Z}}_t^{[i]} - \hat{\mathbf{z}}_t)^T + \mathbf{Q}, \tag{13}$$

where $\mathbf{Q}$ is the covariance of the additive measurement noise.

For the calculation of the Kalman Gain $\mathbf{K}_t$ the cross-covariance $\bar{\mathbf{P}}_t^{x,z}$ between state and observation is needed

$$\bar{\mathbf{P}}_t^{x,z} = \sum_{i=0}^{2n} w_c^{[i]} (\bar{\chi}_t^{[i]} - \bar{\mu}_t)(\bar{\mathbf{Z}}_t^{[i]} - \hat{\mathbf{z}}_t)^T, \tag{14}$$

which leads to

$$\mathbf{K}_t = \bar{\mathbf{P}}_t^{x,y} \mathbf{S}_t^{-1}. \tag{15}$$

Now the outputs μ_t and $\mathbf{P}_t$ of the Unscented Kalman Filter can be computed with the help of the measurement $\mathbf{z}_t$ as

$$\mu_t = \bar{\mu}_t + \mathbf{K}_t(\mathbf{z}_t - \hat{\mathbf{z}}_t) \tag{16}$$

and

$$\mathbf{P}_t = \bar{\mathbf{P}}_t - \mathbf{K}_t \mathbf{S}_t \mathbf{K}_t^T. \tag{17}$$

3 Differential and relative positioning

In Nick et al., 2011a,b approaches are shown how to localize a passive UHF RFID label with an Unscented Kalman Filter and the help of relative positioning or reference tags, respectively. The basic principles of both methods are given in the following.

3.1 Relative positioning

For the relative positioning the object is not only equipped with one ($N = 1$) UHF RFID tag, but with $N > 1$ tags that have a fixed relative position to each other. In most cases it is sufficient to equip the object with $N = 2$ tags. This hardly adds cost to the localization method and achieves the highest increase in localization accuracy due to the relative position information between the two labels.

When needed in the measurement function h (see Eq. 11) of the Unscented Kalman Filter the RSSI value is transformed into a distance. With the help of these distances d_i from each reader antenna to each tag the x-, y- and z-coordinates of the RFID tags can be computed. For the distances d_i holds:

$$d_i = \sqrt{(x_i - x)^2 + (y_i - y)^2 + (z_i - z)^2}, \tag{18}$$

where x_i, y_i and z_i are the coordinates of the reader antenna i ($i = 1, \ldots,$ M, M = number of antennae) and x, y and z are the coordinates of the label that has to be located.

When the object that should get localized is equipped with two RFID labels additional information next to the RSSI values can be used for the localization. The two tags always have the same relative position to each other (the same constant distance between each other in all three coordinates). As a second information you know that these two tags need to be moving with the same velocity. These data can be given to the measurement function h of the Kalman Filter and the measurement vector $\mathbf{z}_t$ has to be build accordingly to these parameters as well

$$\mathbf{z}_t = \begin{bmatrix} d_{i1} \\ d_{i2} \\ dist_x \\ dist_y \\ dist_z \\ 0 \\ 0 \\ 0 \end{bmatrix}. \tag{19}$$

The first $2 \cdot M$ components d_{i1} and d_{i2} of the vector are the distances calculated from the RSSI measurements. The next three values are the known fixed distances of the two tags and the last three values are the differences in velocities in all three directions which should be zero because the tags are fixed to the same object.

The additional information about the constant distances and equal velocitites transform the UKF into a Constrained Unscented Kalman Filter (CUKF) Simon (2010). The type of CUKF used here is called "Perfect Measurement" because it augments the measurement vector by components which are known.

3.2 Differential Positioning with Reference Tags (RTs)

By placing Reference Tags (RTs) with known location in the area in which another tag shall be located more information is available for the localization process. The RTs answer to the reader queries just like every other tag. That way more RSSI values are collected. This information can be used for localizing the tagged object.

RSSI values of the RTs and the k nearest neighbor (kNN) method Duda et al. (2000) are used to achieve a better localization result. For the unknown tag (LT) all RSSI measurements t_i to the M antennae are written in vector $\mathbf{t}$

$$\mathbf{t} = [t_1 \quad t_2 \ldots t_M]. \tag{20}$$

The RSSI values from the N RTs to all M antennae are inserted into matrix $\mathbf{\Theta}$

$$\mathbf{\Theta} = \begin{bmatrix} \Theta_{11} & \Theta_{12} & \ldots & \Theta_{1M} \\ \vdots & & & \vdots \\ \Theta_{N1} & \Theta_{N2} & \ldots & \Theta_{NM} \end{bmatrix}. \tag{21}$$

Based on these RSSI values the Euclidean distance between each RT and the LT can be calculated as

$$e_j = \sqrt{\sum_{i=1}^{M} (\Theta_{ji} - t_i)^2} \tag{22}$$

where $j \in (1, N)$ and e_j is the distance between the jth RT and the LT. For all N RTs this leads to a vector $\mathbf{e}$ which contains the distances between the LT and all RTs in descending order.

Based on the measure of the Euclidean distance the k nearest neighbors are chosen. The value of k can be set to best fit to the problem that is solved with the algorithm in different application scenarios. In a next step the weights $w_i (i = 1, \ldots, k)$ of these k nearest neighbors are calculated according to

$$w_i = \frac{1/e_i^2}{\sum_{i=1}^{k} (1/e_i^2)}. \tag{23}$$

This form of computing the weights implies that the nearer a RT is to the LT the higher its weight is.

If the kNN algorithm would be used for the localization of the LT its position in x-, y- and z-coordinates could now be calculated like

$$(x, y, z) = \sum_{i=1}^{k} w_i (x_{RT_i}, y_{RT_i}, z_{RT_i}) \tag{24}$$

like it is done in Ni et al. (2004).

However here the kNN method is used as an additional information for the Unscented Kalman Filter. Therefore the kNN result will be handled differently here.

Usually the measurement function h contains the distance information $d_i (i = 1, \ldots, M)$ from the LT to all M antennae (from Eq. 18) and the matching measurement vector $\mathbf{z}_t$ contains the distances based on the measured RSSI values for the LT. Now with additional information through the RTs h also contains k-times the actual position estimates in x-, y- and z-coordinate before incorporating the measurement.

Those position estimates can be said to be equal to the positions of the RTs plus additive noise which is comprised in the covariance matrix of the measurement noise $\mathbf{Q}$. The inverse of the weights calculated in Eq. (23) are used here. This form of choosing the noise for the lower part of the matrix leads to a higher noise when the weight is smaller which is due to the fact that the RT is further away from the LT.

To fit to the measurement function h the measurement vector $\mathbf{z}_t$ consists of two parts as well. The first part of the vector comprises the distance measurements $d_i (i \in (1, M))$ based on the RSSI values from the LT to all M antennae. The rest of the vector contains the x-, y- and z-coordinates of those RTs which are the k nearest to the LT.

4 Angle-dependent RSSI measurements

As described in Nick et al. (2011c) the angle between reader antenna and tag has an influence on the measured RSSI value. Therefore it is useful to take this angle into consideration when localizing an object which is equipped with an RFID label.

Based on the RSSI measurement in the directivity of the antenna a distance D between antenna and tag can be calculated. With this distance D, the position of the antenna O and the directivity a circle can be set up as can be seen in Fig. 2. On every point of this circle with radius $r = D/2$ the same RSSI value is measured. Based on the radius r_i of every of those circles for each antenna the localization can be done via the UKF.

The description via circles is only possible in this case because it reflects the antenna's properties. If an antenna with a differently formed beam is used it is necessary to employ another geometric form. For example it might be suitable for some antennae to use an ellipse instead of the circle to describe the points where the same RSSI value is measured.

The measurement function h of the Unscented Kalman Filter only contains the distance measurements in this case. But

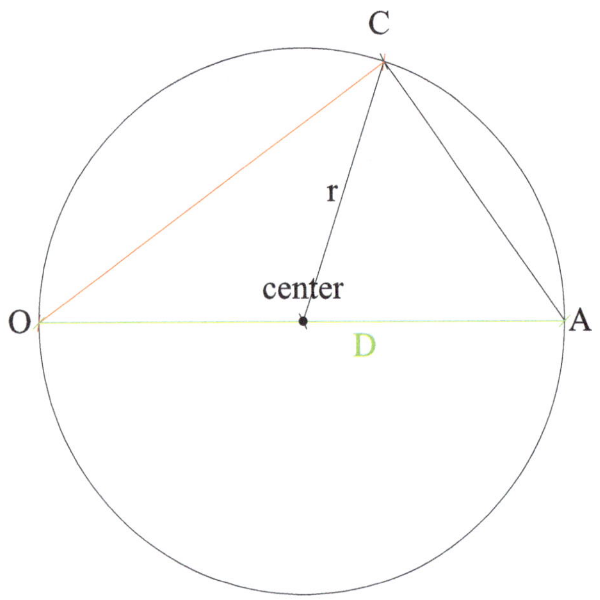

Fig. 2. Method with circle.

Table 1. Results of 3-D localization for EKF and UKF.

# of Ant	Error EKF	Error UKF
2	86.18 cm	69.06 cm
3	66.75 cm	49.30 cm
4	43.45 cm	27.18 cm

Table 2. Results of 3-D localization of UKF_RP and UKF_RT.

# of Ant	Error UKF_RP	Error UKF_RT
2	65.84 cm	68.41 cm (3)
3	39.81 cm	45.90 cm (2)
4	25.60 cm	26.82 cm (2)

compared to the relative or differential methods, in this case the distance is not calculated between reader antenna and tag. For the angle-dependent approach the distance between the center of the circle and the tag is calculated based on the computed radii r_i from all $i = 1, \dots, M$ antennae used.

5 Simulations

The simulation tool of the 3D localization of passive UHF RFID labels is written in Matlab. The considered space is 5m×5m×5m in which the antennae (and the RTs) are placed and the LT is moving For the movement of the tag some way points are given between which other points are interpolated based on the time interval dt and the velocity v. At every way point the RSSI values to all used antennae are calculated. For simulating more realistic scenarios noise is added to these computed RSSI values to take into account disturbances in real world scenarios. These calculated noisy RSSI values are then used as measurement inputs for the Kalman Filter.

The result of the simulation shows the localization error of the Kalman Filter over the desired path. It is the Root Mean Square Error (RMSE) over the estimates of the filter used for the localization. The first 10 results are left out since the filter needs an adjustment period which will take about this number of position estimates (depends on the chosen initial values).

For all methods described in the previous sections the simulation of the localization is carried out with a different number of antennae. The results of the simulations can be seen in Table 1 for the Extended and the Unscented Kalman Filter. Table 2 shows the results for the relative positioning (UKF_RP) as well as the differential positioning (UKF_RT). The number in brackets behind the localization results of the UKF_RT denotes the number of reference tags used for the kNN positioning. The reference tags are placed with a distance of $1m$ from each other in all three dimensions with a total of 36 RTs placed in the considered localization area.

It becomes obvious that the Extended Kalman Filter is outperformed by the UKF for all different number of antennae used. When incorporating additional information into the localization process the UKF_RP achieves better results than the UKF_RT, but both approaches are able to improve the localization accuracy compared to the UKF. Figure 3 shows the results of the different methods for an exemplary path. The colored markers denote the positions of the reader antennae of which three are used for this example.

These simulations have not taken the angle between reader antenna and tag into account so far. When considering the angle for the localization with an Unscented Kalman Filter (UKF_wa) the results as shown in Table 3 are achieved.

The results of the UKF_wa have almost the same localization error compared to the UKF_RT which performed best so far. This leads to the conclusion that the incorporation of the angle is important for a good localization result and it makes the simulation more realistic as well because it reflects the antenna's properties. An important parameter to achieve these good localization results with the UKF_wa is the directivity of the antenna. That means it is of importance to choose wisely how each of the antennae used is aligned according to the path on which the tagged object is moving. Slight changes in the alignment may result in huge variations of the localization error.

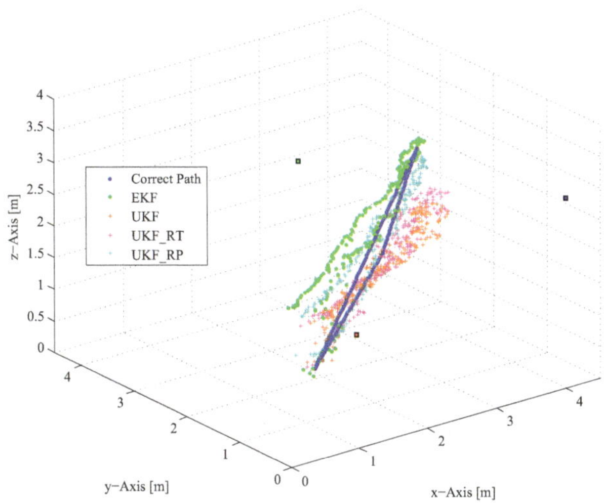

Fig. 3. Localization results of different Kalman Filter with 3 antennae.

Table 3. Results of 3-D localization of UKF_wa.

# of Ant	Error UKF_wa
2	65.96 cm
3	38.79 cm
4	26.61 cm

6 Conclusion and future work

In this paper it is shown through simulations that it is possible to localize a passive UHF RFID label solely with RSSI measurements and additional information combined in a Kalman Filter. The localization results of the Unscented Kalman Filter are better than these of the Extended Kalman Filter. The accuracy of the localization result can further be improved by the usage of reference tags or an additional tag on the object. When incorporating the angles between reader antennae and the tag into the localization process the simulation becomes more realistic and an accurate localization is possible as well.

For the future these simulation results will be verified through real world measurements. The results gained from these measurements might differ from the simulation results due to multipath propagations in the localization environment. Therefore it can be useful to incorporate a camera into the localization process which is independent of these effects. To be able to view the results of the RFID as well as the camera-assisted localization easily it might be useful to implement a demonstration tool to show the different localization results.

Acknowledgements. The reported R + D work was carried out in the frame of the BMBF-Project smaRTI (Smart ReUsable Transport Items); the smaRTI project is carried out in the frame of the Efficiency Cluster Logistic Ruhr (part of the Leading-Edge Cluster – High-Tech Strategy for Germany). This particular research was supported by the BMBF (Bundesministerium fuer Bildung und Forschung) of Federal Republic of Germany under grant 01IC10L10H (Sichere, flexible Detektion und Lokalisierung von UHF RFID-Labeln in rauer Umgebung). The responsibility for this publication is held by the authors only. In particular we have to thank our industrial partners Infineon Technologies AG and Deutsche Post AG supporting us by the required information regarding the RFID test environment and tag data.

INVENTED IN GERMANY

SPONSORED BY THE
Federal Ministry of Education and Research

References

Duda, R., Hart, P., and Stork, D.: Pattern Classification, Wiley-Interscience, 174–191, 2000.

Elnahrawy, E., Li, X., and Martin, R.: The Limits of Localization Using Signal Strength: A Comparative Study, in: Proceedings of IEEE Sensor and AdHoc Communications and Networks (SECON), 2004.

Finkenzeller, K.: RFID Handbuch, Hanser Fachbuchverlag, 5th edn., 2008.

Khan, M. and Antiwal, V.: Location Estimation Technique using Extended 3-D LANDMARC Algorithm for Passive RFID Tag, in: Proceedings of 2009 IEEE International Advance Computing Converence (IACC 2009), 2009.

Lee, K.-C., Oka, A., Pollakis, E., and Lampe, L.: A Comparison between Unscented Kalman Filtering and Particle Filtering for RSSI-Based Tracking, in: Proceedings of WPNC, 2010.

Ni, L., Liu, Y., Lau, Y., and Patil, A.: LANDMARC: Indoor Location Sensing Using Active RFID, Wireless Networks, 10, 701–710, 2004.

Nick, T., Goetze, J., John, W., and Stoenner, G.: Localization of Passive UHF RFID Labels Using An Unscented Kalman Filter with Relative Position Information, in: Proceedings of European Workshop on Smart Objects: Systems, Technologies and Applications (RFID-SysTech 2011), 2011a.

Nick, T., Goetze, J., John, W., and Stoenner, G.: Localization of UHF RFID Labels with Reference Tags and Unscented Kalman Filter, in: Proceedings of IEEE International Conference on RFID-Technologies and Applications (RFID-TA 2011), 2011b.

Nick, T., Khan, M. Z. I., Goetze, J., and John, W.: Localization of Passive UHF RFID Label with Unscented Kalman Filter Based on Angle-Dependent RSSI Measurements, in: Proceedings of IEEE 5th International Conference on Signal Processing and Communication Systems (IEEE ICSPCS), 2011c.

Parameswaran, A., Husain, M., and Upadhyaya, S.: Is RSSI a Reliable Parameter in Sensor Localization Algorithms – An Experimental Study, in: Proceedings of Field Failure Data Analysis Workshop (F2DA'09), 2009.

Raza, N., Bradshaw, V., and Hague, M.: Applications of RFID Technology, in: Proceedings of IEE Colloquium on RFID Technology, 1999.

Simon, D.: Kalman filering with state constraints: a survey of linear and nonlinear algorithms, IET Control Theory and Applications, 4, no. 8, 1303–1318, 2010.

Thrun, S., Burgard, W., and Fox, D.: Probabilistic Robotics, The MIT Press, 39–71 2006.

Wan, E. A. and Merwe, R. V. D.: The Unscented Kalman filter for nonlinear estimation, in: Proceedings of IEEE 2000 Adaptive Systems for Signal Processing, Communicationa and Control Symposium, 2000.

Iterative least squares method for global positioning system

Y. He[1] **and A. Bilgic**[2]

[1]Institute for Integrated Systems, Ruhr University Bochum, 44780 Bochum, Germany
[2]KROHNE Messtechnik GmbH, Ludwig-Krohne-Str. 5, 47058 Duisburg, Germany

Abstract. The efficient implementation of positioning algorithms is investigated for Global Positioning System (GPS). In order to do the positioning, the pseudoranges between the receiver and the satellites are required. The most commonly used algorithm for position computation from pseudoranges is non-linear Least Squares (LS) method. Linearization is done to convert the non-linear system of equations into an iterative procedure, which requires the solution of a linear system of equations in each iteration, i.e. linear LS method is applied iteratively. CORDIC-based approximate rotations are used while computing the QR decomposition for solving the LS problem in each iteration. By choosing accuracy of the approximation, e.g. with a chosen number of optimal CORDIC angles per rotation, the LS computation can be simplified. The accuracy of the positioning results is compared for various numbers of required iterations and various approximation accuracies using real GPS data. The results show that very coarse approximations are sufficient for reasonable positioning accuracy. Therefore, the presented method reduces the computational complexity significantly and is highly suited for hardware implementation.

1 Introduction

Location Based Services (LBS) (Monsmondo et al., 2006; Perusco, 2002; He and Bilgic, 2009; Schiller and Voisard, 2004) are wireless "mobile content" services which are used to provide location-specific information to mobile users moving from location to location. They utilize the ability to make use of the geographical position of the mobile device. Currently, GPS technique (Schreiner, 2007; Djuknic and Richton, 2002), network positioning methods (Drane et al., 1998) as well as other positioning methods (He et al., 2008; Chan and Ho, 1994) are commonly used positioning methods for location estimation for LBS.

Correspondence to: Y. He
(yuheng.he@ruhr-uni-bochum.de)

The result of the free availability of satellite positioning parameters has led to wide adoption of the GPS systems. Building GPS devices in commercially available cell phones has been achieved by mobile device providers, such that the number of cell phones equipped with GPS functionality has rapidly grown in the last few years.

In order to do the positioning, an initial set of pseudoranges between the receiver and the satellites is needed. Non-linear LS is the most common method to determine the receiver's position from the pseudoranges. Usually, linearization is done to convert the non-linear problem into an iterative algorithm, which requires the solution of an overdetermined system of linear equations in each iteration step *itr*, i.e. linear LS method is applied in each iteration step *itr*. For solving the linear LS problems in each iteration step an iterative version of the QR decomposition (QRD) (Götze, 1994) is applied in this paper. Instead of annihilating the lower diagonal elements during the QRD, CORDIC-based approximate rotations are used. By choosing the accuracy of the approximation, e.g. by choosing *itg* optimal CORDIC angles per rotation, the LS computation can be simplified. However, we only obtain an approximate solution to the LS problem, whose accuracy depends on *itg*. The accuracy of the positioning results of GPS method is compared for varying numbers of iterations *itr* of the positioning algorithms and varying numbers of iterations *itg* of the iterative QRD using real GPS data. The results show that very coarse approximations (small *itg*) are sufficient for obtaining a reasonable position estimate. Therefore the presented methods reduce the computational complexity and the required power consumption significantly.

GPS positioning is introduced in Sect. 2 resulting in an algorithm, which requires the solution of LS problems in each iteration *itr* ($itr = 1, 2, \cdots, itr_{\max}$). For solving these LS problems an iteration version of the QRD is presented in Sect. 3 using CORDIC-based approximate rotations, where *itg* denotes the approximation accuracy ($itg = 1, 2, \cdots, itg_{\max}$). The trade-off between *itr* (iteration of the positioning method) and *itg* (iteration of the iterative QRD) is investigated in Sect. 4, where experimental results are given

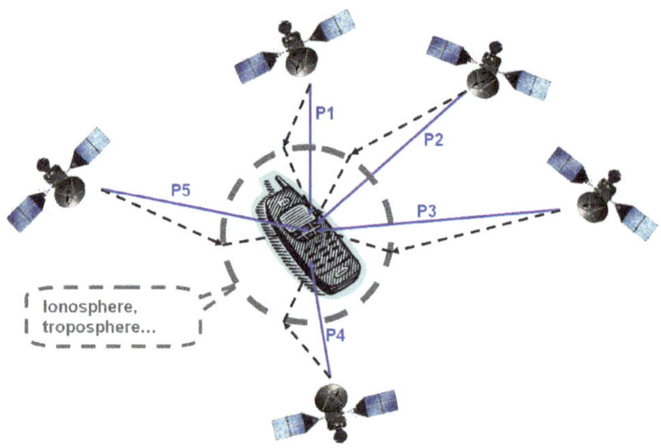

Fig. 1. Pseudoranges: the distance from satellites to GPS receiver.

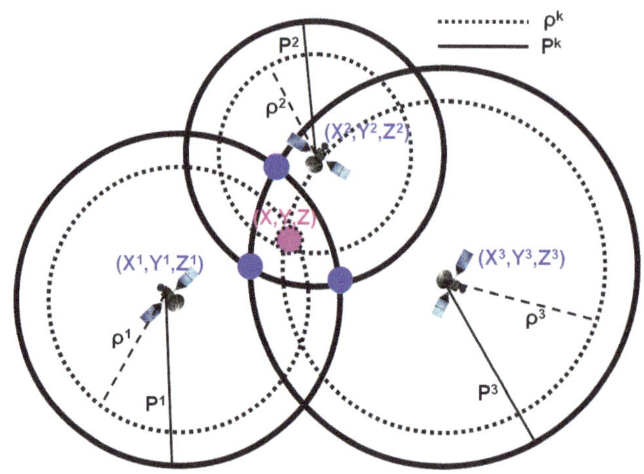

Fig. 2. Observed pseudoranges P^k and geometrical pseudoranges ρ^k.

using real GPS data. The paper finishes with a conclusion and an outlook to the future work in Sect. 5.

2 GPS positioning

The whole GPS positioning procedure includes three tasks: acquisition, tracking and positioning (Borre et al., 2007). The acquisition tries to find satellites and to get their positions. It gives rough estimates of signal parameters. Tracking keeps track of these parameters as the signal properties change over time. After tracking, the navigation data can be extracted and pseudoranges (measured distance from satellites to GPS receiver) can be computed. The final task of the receiver is to compute the user position.

The GPS satellites' arrangement ensures that every point on our planet is in contact with at least six satellites at all times. Each satellite k continuously broadcasts a digital radio signal that includes its position $(X^k,\ Y^k,\ Z^k)$ and its time t^k. On board atomic clocks ensure an accurate time to a billionth of a second. The radio signal of satellite spreads with $c = 3 \times 10^8$ in universe, the velocity of light in vacuum. GPS receivers measure the time delay τ^k of the signal from each satellite k to the receiver, so $\tau^k = t - t^k$, where t is time of receiver. The measurement of t in the receiver is not very accurate (as compared to the satellite time t^k). Furthermore the speed of radio signal from the satellites cuts down because of ionosphere and troposphere. Signal propagation duration from satellites to the receiver is longer than expected. Therefore the measured distance from the satellite to the receiver P^k, measured by $P^k = \tau^k \cdot c$, is a rough distance estimate called "Pseudorange" (see Fig. 1). The receiver simultaneously collects these measurements from at least four satellites and processes them to solve for position and time measurement error.

2.1 Observation equation

The most commonly used algorithm for position computation from pseudoranges is based on the LS method.

This method is used to find the receiver position from pseudoranges to four or more satellites.

The basic observation equation for the pseudorange P^k is

$$P^k = \rho^k + c(dt - dt^k) + T^k + \ell^k + e^k. \tag{1}$$

ρ^k is the geometrical range between satellite k and receiver, which can be computed as:

$$\rho^k = \sqrt{(X^k - X)^2 + (Y^k - Y)^2 + (Z^k - Z)^2} \tag{2}$$

where $(X,\ Y,\ Z)$ is the position of receiver (see Fig. 2). dt denotes the receiver clock offset and dt^k is the satellite clock offset. From the ephemerids, which also include information on the satellite clock offset dt^k, the position of the satellite $(X^k,\ Y^k,\ Z^k)$ can be computed. T^k is the tropospheric error and ℓ^k is the ionospheric error. These two errors are computed from a priori models, whose coefficients are part of the broadcast ephemerids. e^k is the observation error of the pseudorange. Therefore, Eq. (1) contains four unknowns X, Y, Z and dt. The error terms are minimized by using the LS method.

Equation (2) is nonlinear with respect to the receiver position $(X,\ Y,\ Z)$, so the equation is linearized before using the LS method. The nonlinear term in Eq. (2):

$$f(X,\ Y,\ Z) = \sqrt{(X^k - X)^2 + (Y^k - Y)^2 + (Z^k - Z)^2} \tag{3}$$

is linearized. Starting from an initial position for the receiver $(X_1,\ Y_1,\ Z_1)$, the position estimate is improved iteratively. The center of the Earth (0, 0, 0) can be chosen as the initial point, if no a-priori-information (as e.g. from AGPS) is available. Let *itr* be the number of the iterations

($itr = 1, 2, \cdots, itr_{\max}$). The increments $\Delta X_{itr}, \Delta Y_{itr}, \Delta Z_{itr}$ update the receiver coordinates as follows:

$$\begin{aligned} X_{itr+1} &= X_{itr} + \Delta X_{itr}, \\ Y_{itr+1} &= Y_{itr} + \Delta Y_{itr}, \\ Z_{itr+1} &= Z_{itr} + \Delta Z_{itr}. \end{aligned} \tag{4}$$

The Taylor expansion of $f(X_{itr} + \Delta X_{itr},\ Y_{itr} + \Delta Y_{itr},\ Z_{itr} + \Delta Z_{itr})$ is

$$\begin{aligned} f(X_{itr+1}, Y_{itr+1}, Z_{itr+1}) &= f(X_{itr}, Y_{itr}, Z_{itr}) \\ &+ \frac{\partial f(X_{itr}, Y_{itr}, Z_{itr})}{\partial X_{itr}} \Delta X_{itr} \\ &+ \frac{\partial f(X_{itr}, Y_{itr}, Z_{itr})}{\partial Y_{itr}} \Delta Y_{itr} \\ &+ \frac{\partial f(X_{itr}, Y_{itr}, Z_{itr})}{\partial Z_{itr}} \Delta Z_{itr} \end{aligned} \tag{5}$$

Equation (5) includes only first-order terms, and the partial derivatives are

$$\frac{\partial f(X_{itr}, Y_{itr}, Z_{itr})}{\partial X_{itr}} = -\frac{X^k - X_{itr}}{\rho^k_{itr}}$$

$$\frac{\partial f(X_{itr}, Y_{itr}, Z_{itr})}{\partial Y_{itr}} = -\frac{Y^k - Y_{itr}}{\rho^k_{itr}}$$

$$\frac{\partial f(X_{itr}, Y_{itr}, Z_{itr})}{\partial Z_{itr}} = -\frac{Z^k - Z_{itr}}{\rho^k_{itr}}$$

Let $\rho^k_{itr} = \sqrt{(X^k - X_{itr})^2 + (Y^k - Y_{itr})^2 + (Z^k - Z_{itr})^2}$ be the range computed from the satellite position $(X^k,\ Y^k,\ Z^k)$ to the approximate receiver position $(X_{itr},\ Y_{itr},\ Z_{itr})$, so the first-order linearized observation equation becomes

$$\begin{aligned} P^k_{itr} &= \rho^k_{itr} - \frac{X^k - X_{itr}}{\rho^k_{itr}} \Delta X_{itr} - \frac{Y^k - Y_{itr}}{\rho^k_{itr}} \Delta Y_{itr} \\ &- \frac{Z^k - Z_{itr}}{\rho^k_{itr}} \Delta Z_{itr} + c(dt_{itr} - dt^k) + T^k_{itr} + \ell^k_{itr} + e^k_{itr}. \end{aligned} \tag{6}$$

where dt_{itr} is the estimated clock error at the receiver.

2.2 Applying least-squares method

Rearranging Eq. (6), it can be rewritten as

$$\begin{aligned} &\begin{bmatrix} -\frac{X^k - X_{itr}}{\rho^k_{itr}} & -\frac{Y^k - Y_{itr}}{\rho^k_{itr}} & -\frac{Z^k - Z_{itr}}{\rho^k_{itr}} & +1 \end{bmatrix} \begin{bmatrix} \Delta X_{itr} \\ \Delta Y_{itr} \\ \Delta Z_{itr} \\ cdt_i \end{bmatrix} \\ = &\quad P^k_{itr} - \rho^k_{itr} + cdt^k - T^k_{itr} - \ell^k_{itr} - e^k_{itr}. \end{aligned} \tag{7}$$

A unique solution can not be found from a single equation. Therefore $m \geq 4$ satellites are required to form a system of linear equations (usually $m \geq 6$ satellites are available). Let $b^k_{itr} = P^k_{itr} - \rho^k_{itr} + cdt^k - T^k_{itr} - \ell^k_{itr} - e^k_{itr}$ and $\mathbf{b}_{itr} = [b^1_{itr},\ b^2_{itr}, \cdots,\ b^m_{itr}]^T$. Then we obtain the LS problem:

$$\min_{\mathbf{x}_{itr}} ||\mathbf{A}_{itr}\mathbf{x}_{itr} - \mathbf{b}_{itr}||_2, \tag{8}$$

where

$$\mathbf{A}_{itr} = \begin{bmatrix} -\frac{X^1 - X_{itr}}{\rho^1_{itr}} & -\frac{Y^1 - Y_{itr}}{\rho^1_{itr}} & -\frac{Z^1 - Z_{itr}}{\rho^1_{itr}} & 1 \\ -\frac{X^2 - X_{itr}}{\rho^2_{itr}} & -\frac{Y^2 - Y_{itr}}{\rho^2_{itr}} & -\frac{Z^2 - Z_{itr}}{\rho^2_{itr}} & 1 \\ -\frac{X^3 - X_{itr}}{\rho^3_{itr}} & -\frac{Y^3 - Y_{itr}}{\rho^3_{itr}} & -\frac{Z^3 - Z_{itr}}{\rho^3_{itr}} & 1 \\ \vdots & \vdots & \vdots & \vdots \\ -\frac{X^m - X_{itr}}{\rho^m_{itr}} & -\frac{Y^m - Y_{itr}}{\rho^m_{itr}} & -\frac{Z^m - Z_{itr}}{\rho^m_{itr}} & 1 \end{bmatrix}$$

and $\mathbf{x}_{itr} = [\Delta X_{itr}\ \ \Delta Y_{itr}\ \ \Delta Z_{itr}\ \ cdt_{itr}]^T$. The solution $\Delta X_{itr},\ \Delta Y_{itr},\ \Delta Z_{itr}$ has to be added to the approximate receiver position to get the next approximate position as in Eq. (4). These iterations continue until the solution $\Delta X_{itr},\ \Delta Y_{itr},\ \Delta Z_{itr}$ is at meter level.

For solving Eq. (8) we need to find $\hat{\mathbf{x}}_{itr}$ which minimizes the length of the error vector $\hat{\mathbf{e}}_{itr} = \mathbf{b}_{itr} - \mathbf{A}_{itr}\hat{\mathbf{x}}_{itr}$ with $||\mathbf{e}_{itr}||^2 = (\mathbf{b}_{itr} - \mathbf{A}_{itr}\mathbf{x}_{itr})^T(\mathbf{b}_{itr} - \mathbf{A}_{itr}\mathbf{x}_{itr})$ the sum of squares of the m separate errors. Minimizing this quadratic gives the normal equations

$$\mathbf{A}^T_{itr}\mathbf{A}_{itr}\hat{\mathbf{x}}_{itr} = \mathbf{A}^T_{itr}\mathbf{b}_{itr} \Rightarrow \hat{\mathbf{x}}_{itr} = (\mathbf{A}^T_{itr}\mathbf{A}_{itr})^{-1}\mathbf{A}^T_{itr}\mathbf{b}_{itr} \tag{9}$$

and the error vector is

$$\hat{\mathbf{e}}_{itr} = \mathbf{b}_{itr} - \mathbf{A}_{itr}\hat{\mathbf{x}}_{itr}. \tag{10}$$

There are various algorithms for solving LS problems (Eq. 8) (Golub and Loan, 1996). In the subsequent section, we use the QRD by Givens rotations.

3 Iterative solution of LS

In each iteration of the positioning algorithm a LS problem must be computed. Since the pseudoranges are subject to measurement errors and the convergence (number of required iterations) of the algorithm depends on the accuracy of these LS solutions, it is worthwhile to investigate the use of an iterative LS solver and the trade-off between the number of iterations of the positioning method and the number of iterations of the iterative LS solver.

The iterative version of the QR decomposition (QRD) presented in Götze (1994) is used for the iterative solution of Eq. (8), since it is well suited for hardware implementation, suitable for the adaption to measurement errors (pseudoranges), and yields a solution vector which converges linearly to the exact solution. Here we will briefly review this iterative QRD.

The QR decomposition of a $m \times n$ matrix $\mathbf{A} = Q \cdot R$ can be computed by applying a sequence of Givens rotations $\mathbf{G}(\phi_{ij})$

Table 1. Mean value of position-accuracy results (in meter) by GPS with exact LS method and CORDIC-based approximate rotations.

itr	*itg*	GPS-itr	GPS-itg
	1		1282.8
	2		1256.6
2	3	1241.4	1246.1
	8		1241.4
	9		1241.4
	1		85.756
	2		47.744
3	3	38.484	39.020
	8		38.484
	9		38.484
	1		81.495
	2		35.452
4	3	32.472	32.985
	8		32.472
	9		32.472
	1		81.495
	2		35.452
5	3	32.472	32.985
	8		32.472
	9		32.472

($\phi_{ij} = \arctan(a_{ij}/a_{jj})$) to the matrix such that the matrix entries below the diagonal of $\mathbf{A}$ are annihilated, i.e. generate $a'_{ij} = 0$ for

$$(i,j) = \{(2,1)(3,1)\ldots(m,1)(3,2)\ldots(m,2) \ldots(n+1,n)\ldots(m,n)\}. \tag{11}$$

The resulting matrix is upper triangular and denoted by $\mathbf{R}$. The product of all required Givens rotations forms the orthogonal matrix $\mathbf{Q}^T = \prod_{j=1}^{n} \prod_{i=j+1}^{m} \mathbf{G}(\phi_{ij})$ such that

$$\min_{\mathbf{w}} \|\mathbf{Aw} - \mathbf{b}\|_2 = \min_{\mathbf{w}} \|\mathbf{Rw} - \mathbf{Q}^T\mathbf{b}\|_2 \tag{12}$$

and the solution $\mathbf{w}$ can be computed by back substitution.

One iteration of the iterative version of the QRD works exactly like the QRD but instead of using exact rotations $\mathbf{G}(\phi_{ij})$ that annihilate a_{ij} ($a'_{ij} = 0$) CORDIC–based approximate rotations are used resulting in $|a'_{ij}| = |d||a_{ij}|$ with $0 \leq |d| < 1$. The CORDIC–based approximate Givens rotations are computed by determining the shift value ℓ, $\ell \in \{0,1,2,\ldots,b\}$ (b wordlength) corresponding to the CORDIC angle $\phi_{ij}(\ell) = \arctan 2^{-\ell}$ which is closest to the exact rotation angle ϕ_{ij}. Instead of annihilating the matrix elements during the course of the QRD an approximate rotation will only reduce the matrix elements by the maximal factor possible with itg specific CORDIC angles $\phi_{ij}(\ell)$. This CORDIC–based approximate rotation is applied to the QRD for solving LS problems.

Since the matrix elements below the diagonal are no longer annihilated the QRD procedure using CORDIC–based approximate rotations must iteratively be applied until the matrix is ultimately upper triangular. One obtains *iterative* versions of the QRD distinguished by *itg*, which determines the accuracy of the approximation of the rotations.

Defining the lower diagonal quantity for iteration *itg*:

$$S^{(itg)} = \sum_{j=1}^{n} \sum_{i=j+1}^{m} \left(a_{ij}^{(itg)}\right)^2 \tag{13}$$

the above algorithm guarantees

$$\lim_{itg\to\infty} S^{(itg)} \to 0$$

which is equivalent to

$$\lim_{itg\to\infty} \mathbf{A}^{(itg)} \to \mathbf{R} \text{ and } \lim_{itg\to\infty} \mathbf{Q}^{T(itg)} \to \mathbf{Q}^T$$

When we apply this method to the LS problems, which must be solved for positioning in Eq. (8), very few steps, i.e. $itg \ll b$, of the CORDIC-based approximate rotation are sufficient to obtain similar results as using exact rotations. This yields a significant reduction in computational complexity by itg/b (we will show in the following section for real GPS data that even $itg = 1$ gives reasonable results). Furthermore, this method only requiring shift and add operations is very well suited for hardware implementation.

4 Experimental results

4.1 GPS raw data collection

For GPS positioning, a SiGe GN3S Sampler v2 is used to capture the raw GPS data, which are low level signal data (raw intermediate frequency samples) being delivered by the GPS satellite network and processed by the SiGe radio front end (GN3S, 2009). Each GPS satellite continuously broadcasts a navigation message at a rate of 50 bits per second.

The obtained raw GPS data at each measurement point includes information of $m = 7$ satellites with matched pseudoranges. 78 raw GPS data records are gained. Afterwards position calculation is done for GPS method for various numbers of required iterations *itr* (see Sect. 2) and various approximation accuracies *itg* (see Sect. 3).

4.2 Experimental results

The mean values of the positioning results for the 78 measurements are presented in Table 1. The calculated positions are compared with the exact positions (known coordinates at the measure points from land surveying office in Bochum Germany) and the accuracy of the positioning results of GPS method for varying *itr* and *itg* is compared. The number of required iterations *itr* is at least two. The second column *itg*

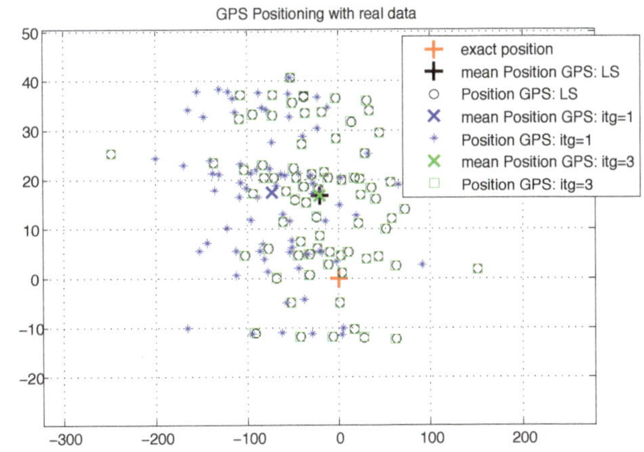

Fig. 3. Positioning results with GPS method.

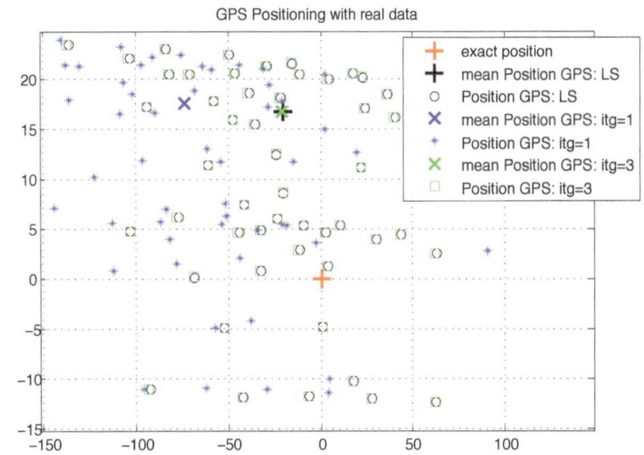

Fig. 4. Enlarged positioning results with GPS method.

in Table 1 is the approximation accuracy in QRD. Figures 3 and 4 show the position estimates of GPS positioning. In Table 1 the 3rd column GPS-itr is the accuracy of GPS positioning with exact LS method and the 4th column GPS-itg is the accuracy of GPS positioning using QRD with CORDIC-based approximate rotations for solving LS problems. The results show that with the increasing number of iterations in QRD, the accuracy of GPS-itg also increases. If $itg \geq 8$, the accuracy of the exact LS method is achieved. Especially, if itr is big enough, e.g. $itr \geq 3$, only one iteration in QRD $itg = 1$ is required to compute the position with reasonable positioning accuracy, which means very coarse approximations are sufficient.

The position estimates of 78 measurements are shown in Fig. 3. GPS position with exact LS (black circles), its mean value (black cross), GPS position using iterative QRD with $itg = 1$ (blue stars), its mean value (blue cross), GPS position using iterative QRD with $itg = 3$ (green squares), its mean value (green cross) and the exact position (red cross) are shown in the figure. All the position results are considered as accuracy of position, i.e. the position estimates subtracted by the exact position (so the exact position is set to (0, 0)).

Figure 4 is an enlarged part of Fig. 3. It is obvious to notice that GPS position using QRD with $itg = 3$ (the green squares) are more closer to GPS position with exact LS (black circles) than GPS position using QRD with $itg = 1$ (blue stars). If the number of iterations in QRD increases, i.e. itg increases, the position results will become more and more similar to the results of exact LS. The mean value of $itg = 3$ is almost the same as the mean value of exact LS (green cross is almost at the same position as black cross in Fig. 4).

5 Conclusion

An iterative LS approach and an iterative version of the QRD using CORDIC-based approximate rotation are applied to position computation. The accuracy of the positioning results of GPS method is compared for various numbers of required iterations itr and various approximation accuracies itg of CORDIC-based approximate rotations by using real GPS data. It is shown that a significant reduction concerning computational complexity and hardware requirements can be obtained. Furthermore, this method only requiring shift and add operations is very well suited for hardware implementation.

The presented method is very efficient for the implementation of the standard triangulation method based on non-linear LS. In future work we apply this idea to *recursive* computations of the position estimates using Orthogonal DGPS (Chang and Paige, 2003) and Kalman Filter based recursive GPS algorithms (Grewall et al., 2001). In the first case no iterative LS method is required for positioning which leads to further reduction of the computational effort and the power consumption. In the second case using the square root version of the Kalman filter (Merwe and Wan, 2001) QRD can be applied and the required approximation accuracy (number of itg) can be investigated to obtain desired position accuracy.

References

Borre, K., Akos, D., Bertelsen, N., Rinder, P., and Jensen, S.: A Software Defined GPS and Galileo Receiver, Birkhäuser, 2007.

Chan, Y. and Ho, K.: A Simple and Efficient Estimator for Hyperbolic Location, Signal Processing, IEEE Transactions, 42(8), 1905–1915, 1994.

Chang, X. and Paige, C.: An Orthogonal Transformation Algorithm for GPS Positioning, Society for Industrial and Applied Mathematics, 24(5), 1710–1732, 2003.

Djuknic, G. M. and Richton, R. E.: Geolocation and Assisted GPS, IEEE Computer Society, 34(2), 123–125, 2002.

Drane, C., Macnaughtan, M., and Scott, C.: Positioning GSM telephones, IEEE Communications Magazine,, 36(4), 46–54, 1998.

GN3S Sampler: http://www.sparkfun.com/commerce/product_info.php?products_id=8238, 2009.

Golub, G. and Loan, C.: Matrix Computations (Johns Hopkins Studies in Mathematical Sciences), The Johns Hopkins University Press, ISBN-13: 978-0801854149, 1996.

Götze, J.: Iterative Version of the QRD for Adaptive RLS Filtering, SPIE Conference on "Advanced Signal Processing: Algorithms, Architectures and Implementations", 438–450, 1994.

Grewal, M., Weill, L., and Andrews, A.: Global Positioning Systems, Inertial Navigation and Integration, Wiley & Sons, ISBN-13, 978-0471350323, 2001.

He, Y. and Bilgic, A.: Integrating Location Based Services in Vehicle-to-IMS, Fachgespraech der GI-Fachgruppe KuVS: Application and Services of Location Based Services, Bonn, September 2009.

He, Y., Hueske, K., Gotze, J., and Coersmeier, E.: Efficient Computation of Joint Direction-Of-Arrival and Frequency Estimation, Signal Processing and Information Technology, IEEE International Symposium, 144–149, 16–19 December 2008.

Merwe, R. and Wan, E.: The square-root unscented Kalman filter for state and parameter estimation, in: Proc. Int. Conf. Acous. Speech, Signal Process. (ICASSP), 6, 3461–3464, May 2001.

Monsmondo, M., Kapov, L., and Kovacic, M.: Bringing Location Based Services to IP Multimedia Subsystem, IEEE Melecon 2006, 16–19 May, 746–749, 2006.

Perusco, L. and Michael, K.: Control, Trust, Privacy, and Security: Evaluating Location-Based Services, IEEE Technology and society magazine, 26(1), 4–16, 2007.

Schiller, J. and Voisard, A.: Location-Bases Services, Morgan Kaufmann by Elservier, ISBN: 1-55860-929-6, 2004.

Schreiner, K.: Where We At? Mobile Phones Bring GPS to the Masses, IEEE Computer Graphics and Applications, 27(3), 6–11, 2007.

An efficient architecture for the integration of sensor and actuator networks into the future internet

J. Schneider, A. Klein, C. Mannweiler, and H. D. Schotten

University of Kaiserslautern, Institute for Wireless Communication and Navigation, Paul-Ehrlich-Straße – Building 11, 67663 Kaiserslautern, Germany

Abstract. In the future, sensors will enable a large variety of new services in different domains. Important application areas are service adaptations in fixed and mobile environments, ambient assisted living, home automation, traffic management, as well as management of smart grids. All these applications will share a common property, the usage of networked sensors and actuators. To ensure an efficient deployment of such sensor-actuator networks, concepts and frameworks for managing and distributing sensor data as well as for triggering actuators need to be developed. In this paper, we present an architecture for integrating sensors and actuators into the future Internet. In our concept, all sensors and actuators are connected via gateways to the Internet, that will be used as comprehensive transport medium. Additionally, an entity is needed for registering all sensors and actuators, and managing sensor data requests. We decided to use a hierarchical structure, comparable to the Domain Name Service. This approach realizes a cost-efficient architecture disposing of "plug and play" capabilities and accounting for privacy issues.

1 Introduction

Today, most sensors are restricted to a local platform and cannot be accessed from external systems or applications. This drawback causes high redundancy in terms of hardware and hence high capital expenditures. In this paper, we present an architecture for integrating different sensors and actuators into the Internet. Our concept enables efficient management and access to sensors, actuators, and additional meta information via open and standardized interfaces, and provides means for ID and access management, while using the Internet as comprehensive transport medium. Important requirements and objectives, that need to be addressed, are:

- Privacy and Security
- Manufacturer Independence
- Dynamic System Adaptations
- Scalability
- Plug and Play Capability
- Open and Standardized Interfaces
- Low implementation Costs

All these issues are related to the question: "How will such a concept be accepted by users?". In order to gain the users' confidence, the architecture must take care of all these aspects. The market acceptance is not only dependent on users' confidence, moreover the system needs to be composed of mass market products to achieve low-priced products and increase the system's attractiveness. Since our selected transport medium is the Internet, the users benefit from low installation costs and high reusability. Our proposed architecture is not limited to any special application. Moreover, we introduce an architecture which is able to deal with several application fields like home automation, ambient assisted living, home security, smart grid, etc. The paper is organized as follows: Sect. 2 briefly lists the state of the art. In Sect. 3, we present the system architecture and sensor integration. Finally, the paper concludes with a summary and an outlook on future work.

2 State of the art

Today, related to sensors and sensor networks a lot of research work has been done in the field of energy management (Wang, 2006; Emmert and Staehle, 2007; Wang and

Correspondence to: J. Schneider
(schneider@eit.uni-kl.de)

Yang, 2007). In particular, wireless sensor networks have become a popular target for several energy management concepts. Although, our proposed concept has the potential to reduce energy consumption in wireless sensor networks, the focus of our approach is on easily integrating different sensors and actuators in the present and future Internet structure. The project C-Cast (C-Cast, 2010) introduced a centralized context management architecture. This architecture consists of Context Providers (e.g. wireless sensor networks), Context Consumers (e.g. actuators for system adaptations) and a Context Broker. The Context Broker acts as a central agent. Therefore, all Context Providers have to send a registration to the Context Broker with their capabilities (available sensor values). However, this architecture does not account for privacy issues. Zuniga (Zuniga and Krishnamachari, 2003) proposed two different solutions to integrate sensors into the Internet. In the first approach, a direct connection is used, i.e. each sensor is directly connected to the Internet and can be reached via its IP address. In the second approach, an indirect connection is used; this means a gateway is responsible for providing an interface between the Internet and each sensor. Especially for home automation, a special EIB/KNX (KNX, 2010) bus is used to connect different actuators with each other. Communication is performed via a serial bus and each actuator is assigned a unique address. Main drawback is the enclosed system communication, since bus specification is restricted to KNX Association members and license fees are expensive. For managing sensor data, a distributed Context Service Architecture has been published in (Mannweiler, 2010) and describes the design as well as the implementation of a highly scalable solution for a context service registry in mobile environments. This overcomes the bottleneck of centralized context management systems. Additionally, this system has successfully been tested with five small sensor networks, each disposing of three to five individual sensors. Our designed architecture also aims at cooperating with such context service architectures and service delivery platforms.

3 System architecture and sensor integration

Sensor networks are an emerging research topic and have a wide range of potential applications. Hence, we present a comprehensive architecture to cover multiple application fields. Our introduced system is capable of supporting applications and services which use sensor data for controlling actuators and system adaptations. Services are able to access sensor information of different SCPs via open and standardized interfaces, that also enable to integrate new sensors, irrespective of manufacturer. Thus, the system is extensible in a "Plug and Play" manner. Figure 1 illustrates the generic system structure.

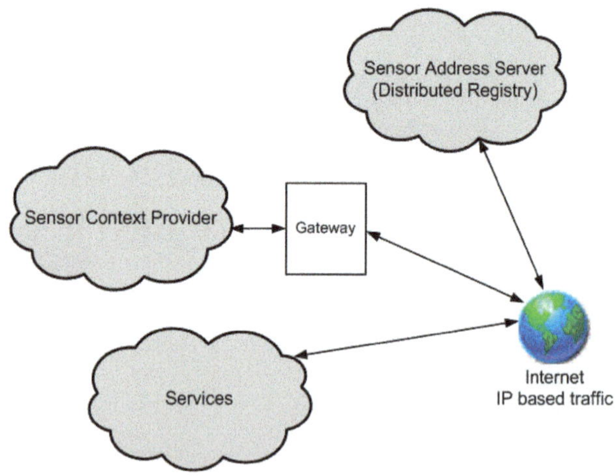

Fig. 1. Generic system structure.

Our system consists of the following functional components:

- Sensor Context Providers (SCP): These entities provide sensor context information which is gathered by sensors or sensor networks.
- Gateway: The gateway implements the interface between SCP(s) and the Internet.
- Sensor Address Server (SAS): All SCPs are registered at this entity and the SAS will choose an appropriate SCP on sensor data requests.
- Service: These entities are able to request and subscribe for sensor data. Logical functions are used for combining these values and exploiting them e.g. for system adaptations. All services need to be advertised to a node of the registry located in the Local Area Network or to the distributed registry of the Wide Area Network, respectively.
- Actuators (AC): These entities are used to control external devices like heaters, light, etc.

To achieve a high reusability of deployed hardware, each SCP and AC requires a network connection to a Local Area Network (LAN). Multimedia data, sensor context data as well as other Internet services can use the same infrastructure. Additionally, installation costs for large buildings will be reduced by using wireless LAN. For sensor data abstraction and exchange, our concept applies a three layer sensor context model as described in Table 1.

Raw sensor data (layer 0) is directly gathered by sensors and provided without any data processing. Layer 1 context data is derived from one or more layers, i.e. layer 0 and/or layer 1. The minimum requirement for a SCP is to provide sensor context data of layer 0. In the following subsection,

Table 1. Three layer context model.

Context layer	Description
0	Raw sensor context data without any processing
1	Abstracted context data with additional meta information
2	Service-level abstraction

more detailed information on SCPs will be depicted. In contrast, services (layer 2) use sensor context data from layer 0, 1 and 2 for their actual purpose.

3.1 Sensor Context Provider (SCP)

The Sensor Context Provider (SCP) can be seen as a sensor context source, since this entity is responsible for gathering and provisioning raw sensor data. A SCP is not restricted to deliver only values of one sensor, rather most of these entities will provide a set of sensor values and parameters. However, each SCP must be capable of supporting the respective bus or hardware interface of its attached sensors and sensor networks. For example, a sensor network can also be seen as a SCP and directly connected to a gateway. For provisioning of sensor data, we decided to use an indirect connection, i.e. a gateway is responsible for implementing an interface between the Internet and each SCP. This solution provides more flexibility, since sensor data processing is accomplished in the gateway.

3.2 Gateway

The gateway implements an open interface for providing raw or abstracted sensor data to external services and applications via the Internet. By means of a standardized XML-derivative, different access rights, and security concepts, sensor data information can be reliably exchanged, taking privacy issues into account. Each gateway must provide the following functionalities:

- Registration of all attached sensor nodes with their capabilities (available values, accuracy, sensor location, etc.)
- Detection of arrival and departure of sensors
- Announcement and request of sensor values to the Sensor Address Server via corresponding node
- Provision of sensor data (Context Layer 0)
- Optional: sensor data processing to derive higher level context data (Context Layer 1)

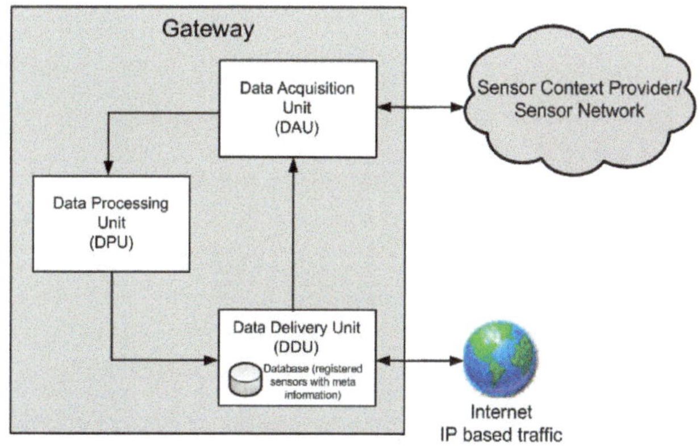

Fig. 2. Generic gateway model.

A generic gateway model is shown in 2.
The gateway can be split into three main functional blocks:

- Data Acquisition Unit (DAU): This unit implements the hardware interface between sensors and gateway, and is responsible for acquiring raw sensor data from SCP(s).
- Data Processing Unit (DPU): The complexity of this unit depends on the Context Layer. To provide layer 0 context, sensor data will only be formatted in an applicable XML structure and delivered. In contrast, if layer 1 context can be requested, sensor context data will be generated via combining and reasoning of current and previous sensor values. For this procedure, only sensor data from the respective sensor network or SCP can be used. Gateways cannot request sensor data from other gateways. Moreover, gateways, that provide layer 1 context, can also provide their raw sensor data, used for derivation of layer 1 context, as layer 0 context.
- Data Delivery Unit (DDU): The DDU module provides the interface between gateway and Internet, and is responsible for registration with SAS. It implements a "request/provide" as well as an asynchronous "publish/subscribe" mode. Both control and sensor data have to pass this module. Moreover, it disposes of a database with all currently connected sensors and their meta-information. Additionally, the DDU is responsible for mapping and transferring a sensor data request to the according SCP and managing all sensor data subscriptions, taking access rights and privacy issues into account.

3.3 Sensor Address Server (SAS)

An important entity is the so-called Sensor Address Server (SAS). It is responsible for administrating a list of registered gateways, and for coordinating and mediating sensor data requests. The SAS is implemented as a hierarchical, zone-oriented structure comparable to the Domain Name Service

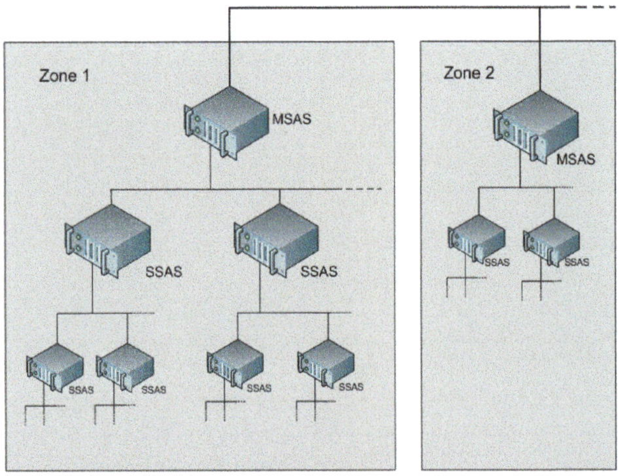

Fig. 3. Structure of Sensor Address Server (SAS).

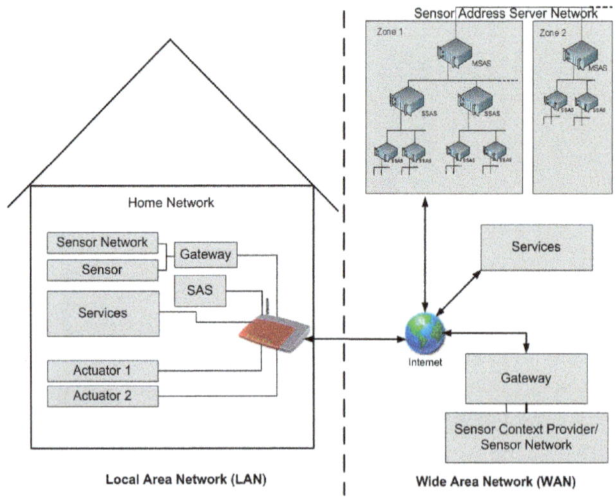

Fig. 4. Generic model of the system architecture.

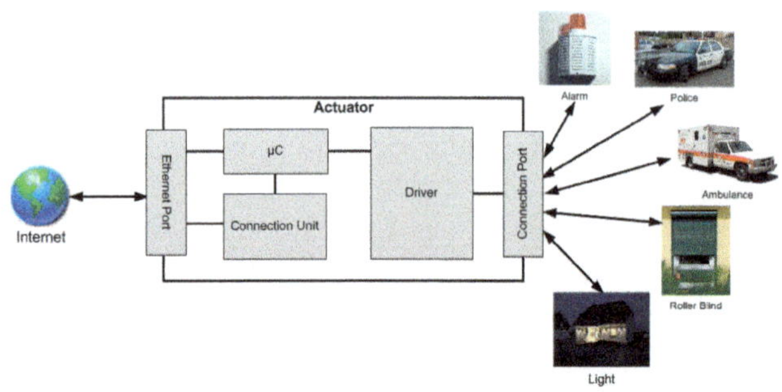

Fig. 5. Actuator block diagram.

(DNS) (Mockapetris, 1987). It groups sensors and SCPs depending on their location and allocates suitable sensor gateways to requests (e.g. by services). Figure 3 shows the system structure of the SAS. A more detailed description is given in (Schneider, 2010).

At system start-up, each component in the network allocates a co-located SAS. Therefore, each client or gateway sends a "getSAS" request to a known broadcast address. This address includes the current location of the client and a parameter "mobile", indicating whether the client is mobile or not. In case of a mobile client, an appropriate SAS is chosen with respect to the current position of the client. To ensure an up-to-date allocation of the SAS, each mobile client has to renew its announcement after a certain time interval. With the allocation of a SAS to a client, sensor data can be announced to as well as requested from the SAS. This includes sensor data and meta-information. Each announcement is valid for a certain time interval and has to be renewed. To ensure privacy issues, each SCP or gateway, respectively, can choose between two different privacy modes. In the "private-mode", advertised sensor information is restricted to the zone-internal SAS. Hence, sensor data can only be requested inside the local sub-network (LAN). If a gateway chooses the "public-mode", advertised sensor data is forwarded to an appropriate SAS in the Internet. Hence, sensor data and context, respectively, can be requested inside the sub-network as well as from the Internet (WAN). For managing the dynamic IP addresses of the clients, a mapper is used to assign the IP address to a fixed URL. Appropriate management services are e.g. DynDNS (dyndns, 2010), 2myDNS (2mydns, 2010), no-IP (no-ip, 2010). A generic model of the system architecture is shown in Fig. 4.

3.4 Actuators (AC)

Actuators (ACs) execute commands of smart control systems, like alarm systems, for security or ambient assisted living, home automation, health care, etc. Each AC establishes an interface for connecting systems without bus/network capabilities. Figure 5 shows a block diagram of an actuator including examples for connected entities.

Actuators can only execute commands on request, where two different classes of commands are possible:

- Action Command: Actuator triggers e.g. an alarm.
- Control Command: Actuator receives a control command from a service and replies with its current actuator status or an error message.

An action command triggers an event on the connection port of the actuator. Usually, these ports are connected to other entities like alarm generator, roller blinds, etc.

3.5 Services

Services are the controlling entities in our architecture. With the capability to request and subscribe for sensor data, as well as to control actuators, they act as middleware. Logical functions such as data combining and reasoning will be applied in these entities. As mentioned in Sect. 2, Mannweiler (Mannweiler, 2010) introduced a design for a distributed service registry. To extend this system to a distributed service architecture, we include a privacy concept. Therefore, an additional registry node is added to the Local Area Network (LAN), for registering all services inside the LAN. If a service is marked as public, its registration gets forwarded to the public distributed registry node, located either in the same SAS zone as the LAN's SAS or a different SAS zone. This concept allows the user to restrict services to the local subnetwork. If a user decides to make it public, the service registration is forwarded to the Wide Area Network's (WAN) SAS.

4 Conclusions

This paper presented the design of a highly scalable and extensible solution for the integration of sensors and actuators into the future Internet, that overcomes the weakness of locally restricted sensor platforms or centralized sensor and actuator management architectures by means of open and standardized interfaces, and Sensor Address Servers. Further, privacy modes allow users to decide if their sensor data and/or services are available only for private or public use. In combination with security concepts, the proposed system architecture ensures reliable exchange of sensor data and meta-information for many important application areas, such as service adaptations in fixed and mobile environments, ambient assisted living, home automation, traffic management, as well as management of smart grids. Additionally, the partitioning into SCPs, gateways, services, and actuators allows for eased integration of new and rather low-cost components, since complex data processing and reasoning algorithms are shifted to service components. Thus, many components of the architecture, e.g. actuators and SCPs, can be constituted by mass market products. Furthermore, the use of the Internet as comprehensive transport medium ensures low installation costs, too. Therefore, these features will increase the attractiveness of the presented approach and hence the market acceptance.

Future work will include three areas: First, implementation of a Sensor Address Server network and the introduced privacy concept. Second, combining of the SAS network and the Distributed Registry. Finally, large scale tests with a high number of sensors, actuators, and services with different privacy levels will be performed.

Acknowledgements. This work has been supported by the Federal Ministry of Education and Research of the Federal Republic of Germany (Foerderkennzeichen 01 BK 0808, GLab; german-lab, 2010).The authors alone are responsible for the content of the paper.

References

Wang, Q., Hempstead, M., and Yang, W.: A Realistic Power Consumption Model for Wireless Sensor Network Devices, SECON '06 3th Annual IEEE Communications Society Conference on Sensor, Mesh and Ad Hoc Communications and Networks, 2006.

Emmert, B. and Staehle, D.: Impact of Energy Models on Energy Efficient Sensor Network Routing, University of W—rzburg, Institute of Computer Science, Technical Report No. 415, April 2007.

Wang, Q. and Yang, W.: Energy Consumption Model for Power Management in Wireless Sensor Networks, SECON '07 4th Annual IEEE Communications Society Conference on Sensor, Mesh and Ad Hoc Communications and Networks, 2007.

C-Cast Context Casting, http://www.ict-ccast.eu/, last access: March 2010.

KNX Association, http://www.knx.org/, last access: Mai 2010.

Zúniga, M. and Krishnamachari, B.: Integrating Future Large-scale Wireless Sensor Networks with the Internet, USC Computer Science Technical Report CS 03-792, 2003.

Mannweiler, C., Amann, B., Schneider, J., Klein, A., and Schotten, H. D.: A Distributed Context Service Architecture for Heterogeneous Radio Environments, ITG-Fachbericht der 15, ITG Fachtagung Mobilkommunikation, Osnabrück, 2010.

Schneider, J., Mannweiler, C., Klein, A., and Schotten, H. D.: Einbindung von Sensoren und Sensornetzwerken in das Future Internet, ITG-Fachbericht der 15. ITG Fachtagung Mobilkommunikation, Osnabrück, 2010.

Mockapetris, P.: Domain Names – Concepts and Facilities, IETF, RFC 1034, http://tools.ietf.org/rfc/rfc1034.txt, November 1987.

Mockapetris, P.: Domain Names – Implementation and Specification, IETF, RFC 1035, http://tools.ietf.org/rfc/rfc1035.txt, November 1987.

http://www.dyndns.com, last access: January 2010.

http://www.2mydns.com, last access: January 2010.

http://www.no-ip.com, last access: January 2010.

http://www.german-lab.de, last access: January 2010.

Analysis and simulations of a frequency synthesizer with internal and external noise sources

G. S. Sangha and M. H. W. Hoffmann

Department of Microwave Techniques, University of Ulm, Germany

Abstract. In this paper, a novel technique to analyse the noise behaviour of a second-order-PLL based synthesizer is demonstrated. Its non-linear behaviour is analyzed with independent noise-sources introduced at different points of the circuit. Phase plane trajectories will be used to demonstrate that noise sources at different locations in the loop produce qualitatively different contributions to the output signal of the synthesizer.

1 Introduction

The PLL based synthesizer has become an important component for a wide range of communications systems. The increasing requirements concerning stability of synthesizers are requiring deeper insight into the non-linear effects of different noise sources in these systems. These affect the phase noise of the output signal. Previous papers have been published on different aspects of phase noise in synthesizer signals, (e.g. Kundert, 2001; Curtin and O'Brien, 1999; Heydari, 2004; Sancho and Suàrez, 2001; Lee and Hajimiri, 2000; Blanchard, 1976; Drucker, 2000), yet a systematic overview on the effects of different noise sources on the phase noise and its spectral forming by non-linearities is still missing. This paper attempts to contribute to close that gap.

2 Analysis methodology

A PLL based frequency synthesizer is used for the analysis. It is shown in Fig. 1 as a model in terms of phases. In order to take into consideration a more complete set of parameters influencing the noise behaviour of the synthesizer, also the fluctuations of the reference-signal amplitude R and local-oscillator amplitude n_4 are considered in the phase detector function $f(R, n_4, \varphi)$. With these terms, the PLL might be described by the following non-linear differential equation:

$$\begin{aligned}\tau_1\ddot{\varphi} &+ \left\{b + K_o\tau_2\frac{\partial f(R, n_4, \varphi)}{\partial\varphi}\right\}\dot{\varphi} + K_o f(R, n_4, \varphi)\\ &+ K_o\tau_2\frac{\partial f(R, n_4, \varphi)}{\partial R}\dot{R} + K_o\tau_2\frac{\partial f(R, n_4, \varphi)}{\partial n_4}\dot{n}_4\\ &= \tau_1\ddot{\theta}_i + b\dot{\theta}_i - b\omega_o \qquad (1)\end{aligned}$$

In this equation, φ is the phase difference between the reference-phase θ_i and the local-oscillator phase θ_o.

In this paper a non-standard noise modelling technique following Rice (1944,1945) will be used.

If a linear analysis is sufficient, the phase detector function might be linearised using a first order Taylor series approach. This leads to a linear differential equation, which is different from what is found in literature. This is due to the above mentioned introduction of new sources of noise, which are usually not taken into account. A closer analysis of the linearised behaviour has been presented by the authors in Sangha and Hoffmann (2005).

The noise sources considered in the following analysis are the following:

- The noise at input of the PLL including phase noise as well as amplitude fluctuations
- The amplitude variations of the local oscillator, since they influence the phase-detector output
- Noise source included in the phase-detector, and in the control filter.

2.1 Rice's noise models

Rice (1944,1945) has developed two different models to describe the measured noise mathematically. The first model is

Correspondence to: M. H. W. Hoffmann
(michael.hoffmann@ieee.org)

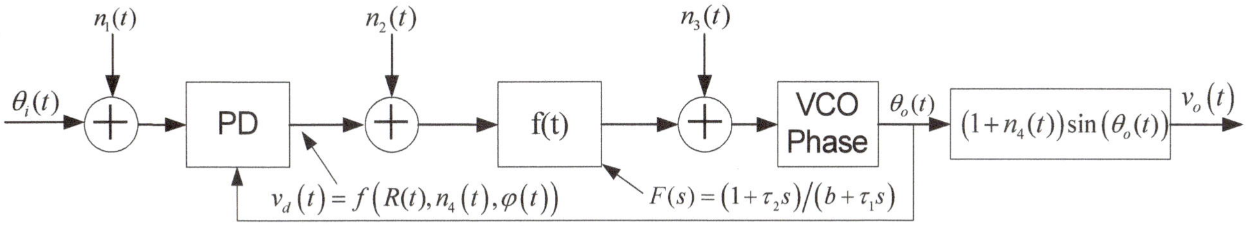

Fig. 1. Synthesizer model used for the analysis

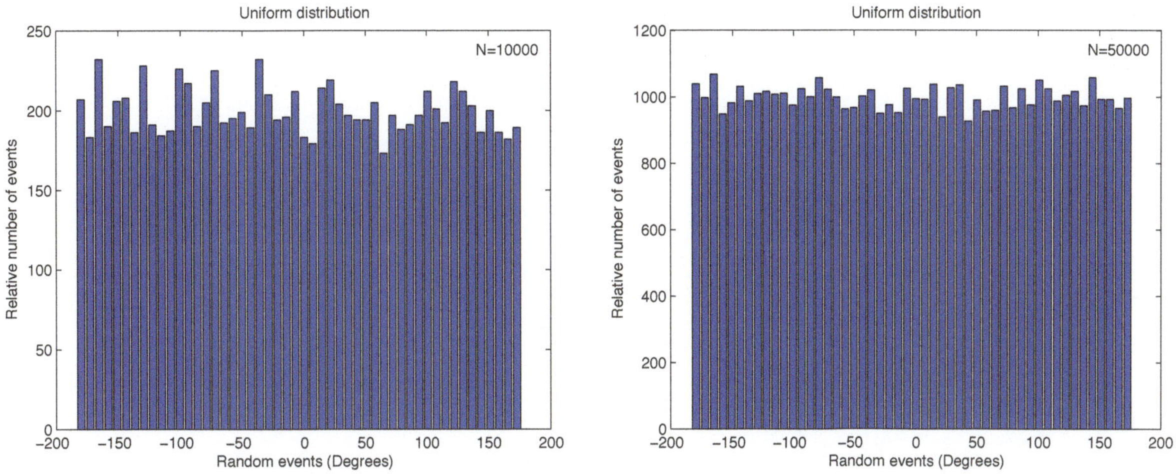

Fig. 2. Uniform distribution for different number of events.

well known and widely used:

$$u(t) = V_0/2 + \sum_{n=1}^{N} \{V_n \cos(n\Delta\omega t + \varphi_n)\}$$
$$= A_0/2 + \sum_{n=1}^{N} \{A_n \cos(n\Delta\omega t) + B_n \sin(n\Delta\omega t)\} \quad (2)$$

In this model both V_n and φ_n are pair-wise independent random variables with V_n being Rayleigh-distributed and φ_n being uniformly distributed in the interval $[0, 2\pi)$, (Rice, 1944, 1945). This is equivalent to saying that parameters A_n and B_n are normally distributed. Due to these distributions the mathematical analysis is rather complex.

The not so well know Rice's second model is the heart of the noise analysis that was presented in Sangha and Hoffmann (2005). The other way of approaching the measured noise can be given as:

$$u(t) = A_0/2 + \sum_{n=1}^{N} A_n \cos(n\Delta\omega t + \varphi_n) \quad (3)$$

Parameters A_n are deterministic. Parameters φ_n are random variables uniformly distributed in the interval $[0, 2\pi)$, (Rice, 1944, 1945). Due to the uniform distributions of φ_n's, this model is mathematically much simpler than the first model.

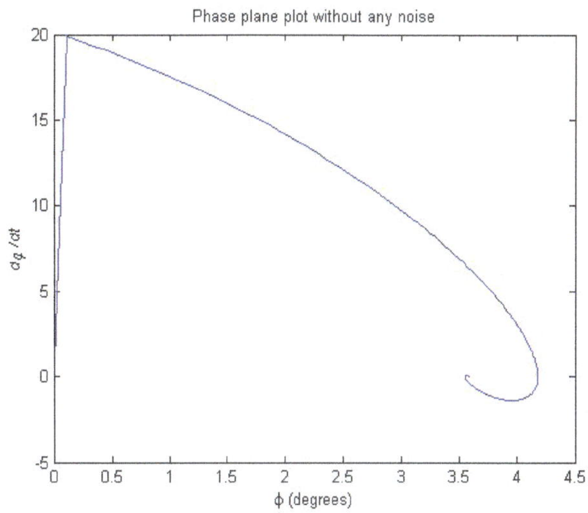

Fig. 3. Phase plane trajectory of the synthesizer without noise.

By virtue of the central limit theorem, it leads to the same results as the first model.

The accuracy of the model increases as the number $N$of sum terms is increased. Fig. 2 demonstrates this fact by comparing the uniform distribution obtained using the Rice's second model for a value of N equal to 10 000 and 50 000, re-

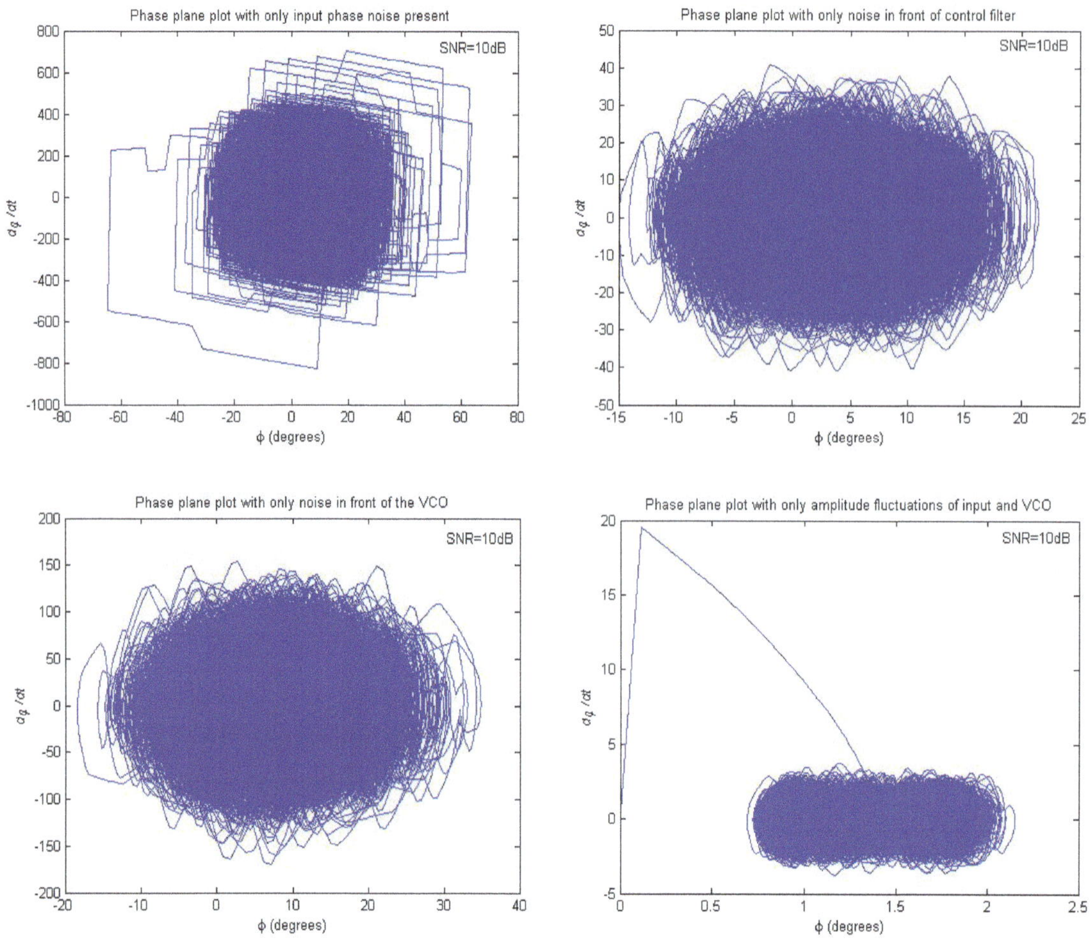

Fig. 4. Phase plane trajectories for a synthesizer with noise introduced at different points of the circuit.

spectively. The horizontal axis in this figure is divided into 50 bins in the interval (-180°, 180°).

Note that the Rice's second model can easily be used to generate differently coloured noises. E.g. all the A_n's would be equal and constant for white noise. In a similar manner, $1/f$ noise can be defined using the relation $A_n/\sqrt{n}$ for amplitude terms from Rice's model.

3 Dynamic behaviour of a synthesizer with noise

The behaviour of a PLL might not be described by a linear model when it is far from lock, for example during switching transients, and when responding to large signal inputs. Important parameters such as settling time, switching speed and capture range are often dominated by non-linear behaviour.

In Sangha and Hoffmann (2005) the behaviour of the loop with respect to noise was presented. Here we will concentrate on the nonlinear aspects of the PLL with noise. With the noise present at different points in the loop, the locking behaviour can be best demonstrated by observing the phase plane trajectories. Simulations are performed in SIMULINK using a nonlinear model (Fig. 1) of the PLL. It does not use any of the standard VCO or PLL models found in SIMULINK.

The input noise was designed to be white noise that is divided into two components affecting the amplitude and phase respectively. The amplitude fluctuations of VCO were designed in a way that comes close to reality. Noise in front of the CF and VCO was 1/f noise superimposed by white noise.

All the random noise sources in this simulation are designed for SNR of 10 dB. The angular frequency of the input signal is chosen to be $1.5 \cdot 10^5$ rad/sec and the damping factor ζ is chosen to be 0.7.

Figure 3 shows the settling behaviour of the synthesizer without noise, while Fig. 4 shows how the PLL settling behaviour is modified when a noise source at one of the points shown in Fig. 1 is introduced. Comparing these curves, it is concluded that the input phase noise is most critical for the settling behaviour of the synthesizer. Furthermore, Fig. 5 shows that with increasing noise level, there is a probability of cycle slipping. Fig. 4 also shows how the amplitude fluctuations of input and VCO signals affect the settling be-

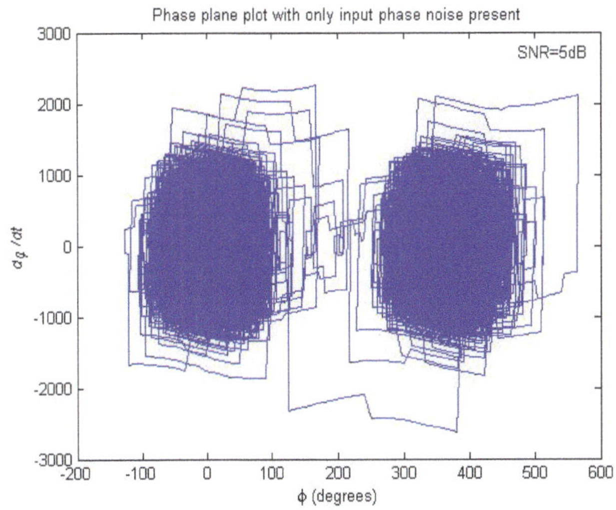

Fig. 5. Phase plane trajectory of the synthesizer with input phase noise having SNR = 5 dB.

haviour of the synthesizer. The amplitude fluctuations, which are usually ignored, are taken into account for completeness.

4 Conclusions

In this paper we have demonstrated a useful technique for non-linear noise analysis of a PLL based synthesizer taking into account a more complete combination of noise sources. We showed how the noise at different points in the loop can affect the settling behaviour of the synthesizer.

References

Blanchard, A.: Phase-Locked Loops Applications to Coherent Receiver Design, John Wiley and Sons, 1976.

Curtin, M. and O'Brien, P.: Phase-Locked Loops for High-Frequency Receivers and Transmitters-Part 2, Analog Devices, Analog Dialogue, 33-5, 1999.

Drucker, E.: Model PLL Dynamics and Phase-Noise Performance, Microwaves and RF Magazine, 73–82, 2000,. http://www.mwrf.com.

Heydari, P.: Analysis of the PLL Jitter Due to Power/Ground and Substrate Noise, IEEE Transactions on Circuits and Systems-I, 51, 2404–2416, 2004.

Kundert, K.: Modelling and Simulation of Jitter in PLL Frequency Synthesizers, Cadence Design Systems, Inc., 2001. www.cadence.com.

Lee, T. H. and Hajimiri, A.: Oscillator Phase Noise: A Tutorial, IEEE J. of Solid-state circuits, 35, 326–336, 2000.

Rice, S. O.: Mathematical Analysis of Random Noise, Bell Syst. Techn. J., 23, 282–332, 1944.

Rice, S. O.: Mathematical Analysis of Random Noise, Bell Syst. Techn. J., 24, 46–156, 1945.

Sancho, S. and Suàrez, A.: Nonlinear Dynamics of Microwave Synthesizers- Stability and Noise, IEEE Transactions on Microwave Theory and Techn., 49, 1792–1803, 2001.

Sangha, G. S.and Hoffmann, M. H. W.: Analysis and Design of a Frequency Synthesizer with Internal and External Noise Sources, in Proc. 18th Intern. Conf. on Noise and Fluctuations-ICNF2005, 521–524, 2005.

17

A VLSI design concept for parallel iterative algorithms

C. C. Sun and J. Götze

Dortmund University of Technology, Information Processing Lab, Otto-Hahn-Str. 4, 44227 Dortmund, Germany

Abstract. Modern VLSI manufacturing technology has kept shrinking down to the nanoscale level with a very fast trend. Integration with the advanced nano-technology now makes it possible to realize advanced parallel iterative algorithms directly which was almost impossible 10 years ago. In this paper, we want to discuss the influences of evolving VLSI technologies for iterative algorithms and present design strategies from an algorithmic and architectural point of view. Implementing an iterative algorithm on a multiprocessor array, there is a trade-off between the performance/complexity of processors and the load/throughput of interconnects. This is due to the behavior of iterative algorithms. For example, we could simplify the parallel implementation of the iterative algorithm (i.e., processor elements of the multiprocessor array) in any way as long as the convergence is guaranteed. However, the modification of the algorithm (processors) usually increases the number of required iterations which also means that the switch activity of interconnects is increasing. As an example we show that a 25×25 full Jacobi EVD array could be realized into one single FPGA device with the simplified μ-rotation CORDIC architecture.

1 Introduction

Modern VLSI manufacturing technology has kept shrinking down to Deep Sub-Micron (DSM) with a very fast trend and Moore's law is expected to hold for the next 10 years (Gelsinger, 2008). Now, since the DSM nano-technology allows the integration of an ever-increasing number of IP macro-cells on a single silicon die, parallel multiprocessor platforms have received great attention and have been realized into several state-of-the-art applications (e.g., Dual-Core CPU, MPSoC and Parallel Computing) (Vangal et al., 2007; Wolf, 2004; Vitullo et al., 2008).

10 years ago, for 0.35 μm technology, design engineers were focusing on reducing the area size. Later, when it came to 0.13 μm technology they paid huge efforts to improve the signal delay and reduce the power consumption. As the VLSI manufacturing technology keeps shrinking down into 65 nm, the design methodology for nano-circuits poses new challenges: area requirements of the wire interconnections are increasing explosively in relation to the area of processor elements, bus transmission bottleneck in the million transistors SoC designs, and leakage current is now dominating the power consumption (Sainarayanan et al., 2007; Stine et al., 2007).

These changes bring us to analyze the impacts on parallel iterative algorithms as VLSI technology keeps evolving. As long as the convergence properties of the iterative algorithms are guaranteed, it is possible to modify/simplify the architecture during the iteration steps and reduce the computational complexity significantly with regard to the implementation. However, this simplification will usually cause an increased number of iterations for convergence. Therefore, it actually becomes a trade-off problem between the performance/complexity of the hardware, the load/throughput of interconnects and the overall energy/power consumption due to the behavior of parallel iterative algorithms.

Computing the Eigenvalue Decomposition (EVD) with the parallel Jacobi method is used as an example since the convergence of this methodology is very robust to modification of the processor elements. Finally, a VLSI design concept for parallel iterative algorithms is presented which takes into account the influence of the modifications on area, timing delay and power consumption.

The paper is organized as follows: in Sect. 2 we will first describe the design concepts for parallel iterative algorithms. After that, we will further clarify the definition of the serial and parallel Jacobi method, respectively, in Sect. 3. Then, in Sect. 4 the design issues of the Jacobi EVD array and their suitability for different hardware implementations are discussed, which lead to the simplified μ-rotation CORDIC processor. Section 5 shows the experimental and syntheses results. Section 6 concludes this paper.

Correspondence to: C. C. Sun
(chichia.sun@tu-dortmund.de)

2 Design concept and implementation issues

A design concept for parallel iterative algorithms, is presented taking into consideration the influences of different VLSI technologies in terms of area, power and timing delay. Implementing an iterative algorithm on a multiprocessor array, there is a trade-off between the complexity of an iteration step (assuming that the convergence of the algorithm is retained) and the number of required iteration steps. For example, suppose we have a hardware platform, which requires an iteration step of the iterative algorithm to be executed K times in order to obtain the convergence. The iteration step is executed in parallel on the platform. If we simplify the processors in order to improve the logical utilization of the platform, the number of required iterations usually increase from K to $K{+}L$. It also means that the switch activity of interconnects between these processor elements is increasing due to the behavior of iterative algorithm. How to find a superior solution to balance the design criteria is the major issue of this paper, especially for low-power or limited-area devices.

In this paper, we selected the Jacobi EVD method as a typical iterative algorithm since the convergence of this methodology is very robust to modification of the processor elements (Brent and Luk, 1985; Gotze et al., 1993; Goetze and Hekstra, 1995; Klauke and Goetze, 2001). We have investigated the influences in DSM design with different sizes of multiprocessor arrays (i.e., 4×4, 16×16 and 25×25). After that, several modifications of the algorithm/processor were studied and their impacts on different FPGA devices were investigated (e.g., Xilinx Virtex series in 0.22 μm, 0.15 μm and 65 nm). According to these analyses, we present an efficient strategy to comply with the design criteria, especially in balancing the number of iterations and the computational complexity.

3 Eigenvalue decomposition

An Eigenvalue decomposition of a real symmertric $n{\times}n$ matrix **A** is obtained by factorizing **A** into three matrices $A{=}Q \wedge Q^T$, where Q is an orthogonal matrix ($QQ^T{=}I$) and $\wedge$ is a diagonal matrix which contains the eigenvalues of **A**.

3.1 Jacobi method

The cyclic-by-row Jacobi method computes the EVD of a $n{\times}n$ symmetric matrix iteratively by applying a sequence of orthonormal rotations to the left and the right of the matrix **A**, as shown in the following:

$$A_{k+1} = Q_k A_k Q_k^T, \quad \text{with} \quad k = 0, 1, 2, \ldots, \tag{1}$$

where Q_k is an orthonormal rotation by the angle θ in the (i, j) plane:

$$\begin{array}{rcl} & \begin{array}{cccccc} & & \text{col i} & & \text{col j} & \\ & & \downarrow & & \downarrow & \end{array} & \\ Q_k = & \left[\begin{array}{cccccc} 1 & 0 & & \cdots & & 0 \\ 0 & \ddots & & & & \\ & & \cos\theta_k & & \sin\theta_k & \\ \vdots & & & \ddots & & \vdots \\ & & -\sin\theta_k & & \cos\theta_k & \\ & & & & \ddots & 0 \\ 0 & & & \cdots & & 1 \end{array}\right] & \begin{array}{l} \\ \\ \leftarrow \text{row i} \\ \\ \leftarrow \text{row j} \\ \\ \\ \end{array} \end{array} \tag{2}$$

The order of sequential plan rotations $\{Q_k\}$ is called cyclic-by-row manner, if (i, j) is chosen as follows:

$$(i, j) = (1, 2)(1, 3) \ldots (1, n)(2, 3) \ldots (2, n) \ldots (n-1, n)\,. \tag{3}$$

The execution of all $N{=}n(n{-}1)/2$ index pairs (i, j) is called a sweep. After several sweeps are applied, the matrix **A** will converge into a diagonal matrix $\wedge$, which contains the eigenvalues:

$$\lim_{k\to\infty} A_k = diag[\lambda_1, \lambda_2, \ldots, \lambda_n] = \begin{bmatrix} \lambda_1 & 0 & \cdots & 0 \\ 0 & \lambda_2 & & \vdots \\ \vdots & & \ddots & 0 \\ 0 & \cdots & 0 & \lambda_n \end{bmatrix}. \tag{4}$$

In practice we can observe the Frobenius norm of the off-diagonal elements until it is close to zero or perform a predefined number of sweeps which depends on the size of matrix **A**.

We have to choose the rotation angle in order to annihilate the off-diagonal elements of Matrix **A** by solving a 2×2 symmetric EVD subproblem as shown in the following:

$$\begin{bmatrix} a'_{ii} & a'_{ij} \\ a'_{ji} & a'_{jj} \end{bmatrix} = \begin{bmatrix} \cos\theta & -\sin\theta \\ \sin\theta & \cos\theta \end{bmatrix} \begin{bmatrix} a_{ii} & a_{ij} \\ a_{ji} & a_{jj} \end{bmatrix} \begin{bmatrix} \cos\theta & -\sin\theta \\ \sin\theta & \cos\theta \end{bmatrix}^T. \tag{5}$$

We can solve the subproblem and cause the maximal reduction $\{a_{i,j}, a_{j,i}\}{=}0$ by applying an optimal angle of rotation θ_{opt}:

$$\theta_{\text{opt}} = \frac{1}{2}\arctan(\tau)\,, \tag{6}$$

where $\tau{=}\frac{2a_{ij}}{a_{jj}-a_{ii}}$, and the range of θ_{opt} is limited to $|\theta_{\text{opt}}| {\le} \frac{\pi}{4}$.

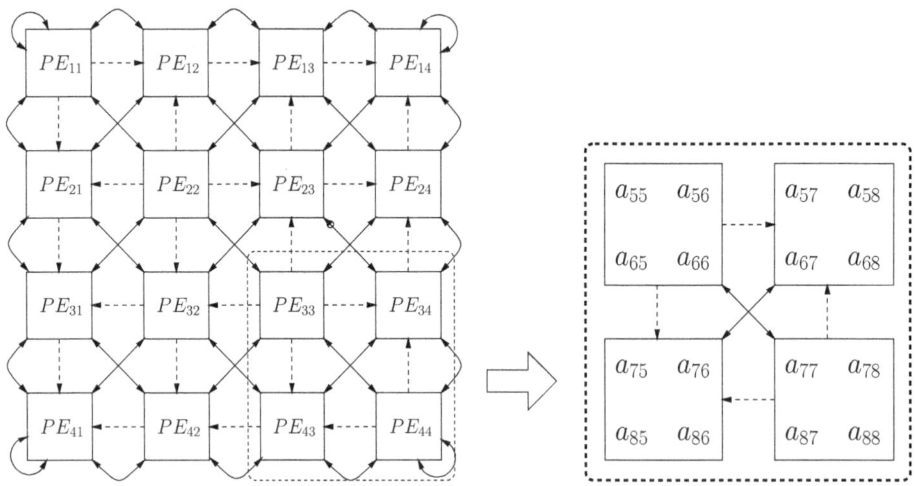

Fig. 1. A 4×4 EVD array, n=8.

3.2 Parallel Jacobi EVD array

The parallel array presented by Brent and Luk consists of $\frac{n}{2} \times \frac{n}{2}$ Processor Elements (PEs) and each PE contains a 2×2 sub-block of the matrix to be decomposed (Brent and Luk, 1985). Figure 1 shows a typical 4×4 EVD array with 16 PEs. This systolic Jacobi array can perform $\frac{n}{2}$ subproblem in parallel and each sweep requires $n-1$ steps. Initially each PE holds a 2×2 sub-matrix of A:

$$PE_{pq} = \begin{pmatrix} a_{2p-1,2q-1} & a_{2p-1,2q} \\ a_{2p,2q-1} & a_{2p,2q} \end{pmatrix}, \quad \text{where } p \text{ and } q = 1, 2, \cdots, \tfrac{n}{2}. \tag{7}$$

The optimal angel θ_{opt}, which is able to annihilate the off-diagonal elements ($a_{2p-1,2q}$ and $a_{2p,2q-1}$), is computed by diagonal PEs (i.e., PE_{11}, PE_{22}, PE_{33} and PE_{44}) using Eq. (6). After these rotation angles are computed, they will be sent to the off-diagonal PEs. This transmission is indicated by the dashed lines in Fig. 1. All PEs will perform a two-sided rotation with the corresponding row (θ_r) and column (θ_c) rotation angles.

$$PE'_{pq} = Q(\theta_r) \cdot PE_{pq} \cdot Q(\theta_c)^{\mathrm{T}}, \quad \text{where } Q(\theta) = \begin{bmatrix} \cos\theta & -\sin\theta \\ \sin\theta & \cos\theta \end{bmatrix}. \tag{8}$$

One sweep needs to perform $n-1$ parallel rotation steps. After these rotations are applied, the local matrices are interchanged between processors as indicated by the solid lines in Fig. 1 for execution of the next sweep. We can use the CORDIC processor to realize the BLV EVD array (Walther, 1971; Volder, 1959; Parhi and Nishitani, 1999). It should be noticed that since we selected the CORDIC processor to approximate the rotation, we can transmit the $\tan\theta_{\text{opt}}$ directly instead of the angles (see Sect. 4). In this way, we can improve efficiency of the communication bus and make this systolic array more suitable for VLSI implementation.

4 Architecture considerations

In this section we will show the reasons why it is necessary to simplify the CORDIC architecture and how to achieve this goal. As the evaluation of the VLSI technology keeps shrinking down to the nanoscale, it is possible to implement a full Jacobi EVD array into one single FPGA device (Ahmedsaid et al., 2003). However if we still use the original full CORDIC processor which is generated by the Xilinx IP-CORE library (www.xilinx.com), only moderate parallelism can be obtained due to the limited FPGA configuration resources. For example, we could only realize a 6×6 multicore array at most in the biggest Xilinx FPGA device as shown in Table 2. Therefore, we must simplify the CORDIC architecture in order to fit the design criteria.

At first we have slightly modified a simplified scaling free μ-rotation CORDIC which was presented in Goetze and Hekstra (1995) as shown in Fig. 2. It is able to perform the single inner iteration efficiently. This simplified PE has 2 adders, 2 shifters and 4 multiplexers, and it reduces the number of inner iterations from 16 or 32 times for a full CORDIC with word length 16 and 32 bits, respectively, to only one or 6 inner iterations with the CORDIC circular rotation mode. However, decreasing the inner iterations will cause an increased number of outer sweeps because of the imprecise inner iterations. Therefore, the simplified CORDIC architecture can reduce the size of area but requires more sweeps. On the other hand, the full CORDIC architecture needs fewer sweeps but requires more area.

Table 1 gives a set $\mathcal{A}$ approximated rotation angles for a simplified 32-bits scaling free μ-rotation CORDIC PE. For

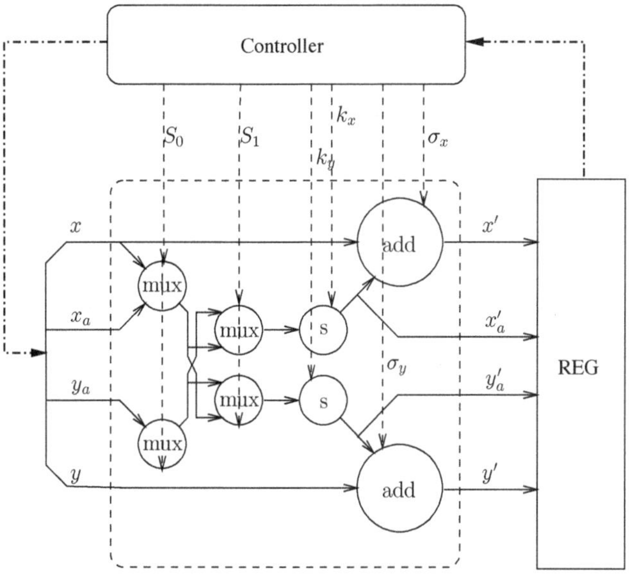

Fig. 2. The block diagram of a simplified CORDIC PE, including 2 adders, 2 shifters and 4 multiplexers.

a given accuracy n_m, this look-up table is constructed using the aforementioned four approximation methods in Goetze and Hekstra (1995). These orthonormal μ-rotations are chosen such that they satisfy a predefined accuracy condition in order to approximate the original rotation angles and are constructed by the cheapest possible method. It should be noticed that we have slightly modified the look-up table. First, since we only need the $\tan\theta$ for searching the optimal angle in Eq. (6), we can store $2\times\tan\theta$ instead of performing arctan operation to reverse the rotation angle in the look-up table. Second, we look into the critical path in Table 1. For angle index k=-1, it requires six cycles per iteration. In fact, the global clock in synchronous circuit is usually determined by the critical path, which also means that the maximum timing delay per iteration is 6 cycles. Therefore, in order to improve the computational balance, we repeat the inner iteration steps of the angles until they are close to the critical one. For example, when an optimal rotation angle index k=-8, it will repeat three times from the index -8 to the index -10. In this way, we can balance the overall computing overhead and improve the computational efficiency.

Figure 3 shows a block diagram of a 4×4 full Jacobi EVD array including one controller and 16 PEs. The shaded diagonal processors will first search the optimal rotation angle and then forward these angles to the off-diagonal PEs.

5 Experimental results

In this work, we have simulated four different cases of the cyclic–by–row parallel Jacobi EVD method in Matlab and on Xilinx FPGA respectively:

Table 1. The set $\mathcal{A}$ of μ-rotations for 32-bit accuracy, showing the method used, the tan θ angle and the cost of rotation and scaling.

angle index	method	angle	cost (shift-add operations)		cycle	repeat
k		$2\times\tan\theta_k$	rot.	scl.	count	
−1	IV	1.49070	4	8	6	1
−2	IV	0.54296	4	6	5	1
−3	IV	0.25501	4	6	5	1
−4	IV	0.12561	4	4	4	1
−5	III	6.25841×10^{-2}	6	0	3	2
−6	III	3.12606×10^{-2}	6	0	3	2
−7	III	1.56263×10^{-2}	6	0	3	2
−8	II	7.81266×10^{-3}	4	0	2	3
−9	II	3.90627×10^{-3}	4	0	2	3
−10	II	1.95313×10^{-3}	4	0	2	3
−11	II	9.76563×10^{-4}	4	0	2	3
−12	II	4.88281×10^{-4}	4	0	2	3
−13	II	2.44141×10^{-4}	4	0	2	3
−14	II	1.22070×10^{-4}	4	0	2	4
−15	II	6.10352×10^{-5}	4	0	2	5
−16	I	3.05176×10^{-5}	2	0	1	6
−17	I	1.52588×10^{-5}	2	0	1	6
−18	I	7.62939×10^{-6}	2	0	1	6
−19	I	3.81470×10^{-6}	2	0	1	6
−20	I	1.90735×10^{-6}	2	0	1	6
−21	I	9.53674×10^{-7}	2	0	1	6
−22	I	4.76837×10^{-7}	2	0	1	6
−23	I	2.38419×10^{-7}	2	0	1	6
−24	I	1.19209×10^{-7}	2	0	1	6
−25	I	5.96046×10^{-8}	2	0	1	6
−26	I	2.98023×10^{-8}	2	0	1	6
−27	I	1.49012×10^{-8}	2	0	1	6
−28	I	7.45058×10^{-9}	2	0	1	5
−29	I	3.72529×10^{-9}	2	0	1	4
−30	I	1.86265×10^{-9}	2	0	1	3
−31	I	9.31323×10^{-10}	2	0	1	2
−32	I	4.65661×10^{-10}	2	0	1	1

1. Full rotation CORDIC with 32 iteration steps.
2. Half rotation CORDIC with 16 iteration steps.
3. Simplified μ-rotation CORDIC with one single inner iteration step (μ-CORDIC).
4. Simplified μ-rotation CORDIC with 6 inner iteration steps (6-CORDIC).

5.1 Matlab simulation

At present we have tested with numerous random symmetric matrices **A** of size 4×4 to 50×50. Figure 4 shows the average number of Shift-Add operations needed to compute the eigenvalues for each size of EVD array. Apparently, both Full and Half CORDIC require much more effort than the simplified CORDIC. The 6-CORDIC requires a little more than the μ-CORDIC in average. On the other hand, we have

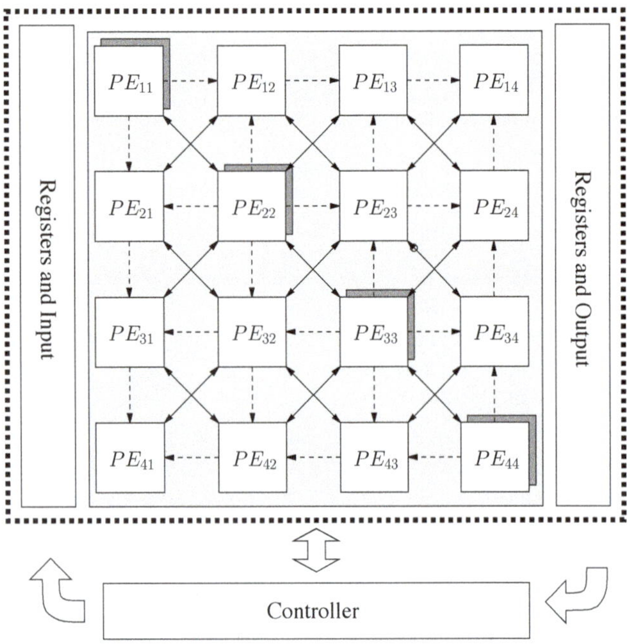

Fig. 3. The block diagram of a 4×4 Jacobi EVD array with 16 μ-rotation elements for FPGA implementation.

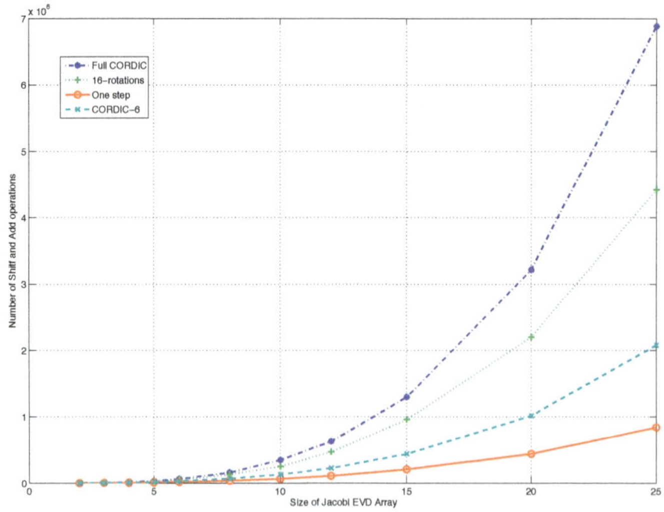

Fig. 4. Number of Shift-Add operations vs. Jacobi EVD array sizes for different CORDIC solutions.

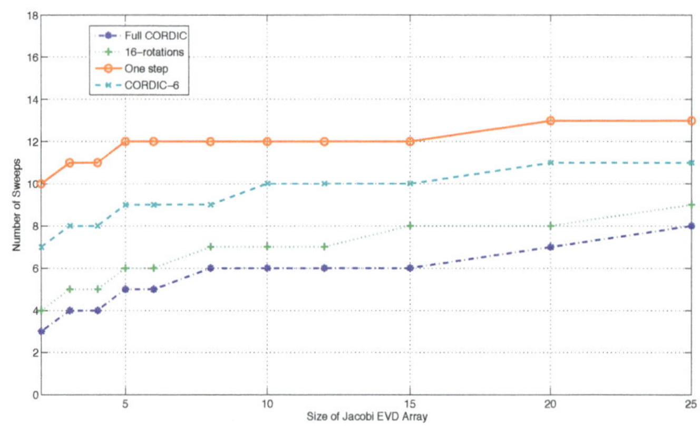

Fig. 5. The required number of sweeps vs. Jacobi EVD array sizes for different CORDIC solutions.

also simulated the number of the sweeps as shown in Fig. 5. Here, when the Jacobi EVD array's size is 20×20, the μ-CORDIC requires 13 sweeps which is almost twice than the Full CORDIC. Although the simplified μ-rotation CORDIC PE can improve the computational efficiency, it also increases the timing delay. The simplified 6-CORDIC not only requires less sweeps than the μ-CORDIC but also reduces the timing delay. Therefore, the simplified 6-CORDIC is actually a good compromise between the timing delay and the computational effort.

Consequently, from an algorithmic point of view, there is no doubt that we would rather realize the Jacobi method by utilizing the orthonormal simplified μ-rotation CORDIC method. However, when it comes to the VLSI circuit design (i.e., here we use VHDL for RTL design), things become totally different.

5.2 FPGA implementation

We have modeled a μ-rotation CORDIC PE in VHDL and compared with a full-pipeline CORDIC which is generated by the Xilinx Coregen automatically. Later, we synthesized these two CORDIC processors by Xilinx ISE into three different FPGA devices. It should be noticed that the word-length is 32 bits. Table 2 shows the syntheses results for Area, Timing Delay and the size of EVD array for each FPGA device (e.g., XCV1000-6FG680 0.22 μm, XC2V8000-5FF1517 0.15 μm and XC5VL330-2FF1760 65 nm). There are some important points that can be observed.

First of all, for the XCV1000-6FG680 0.22 μm, we are not able to implement a full EVD array directly because of the FPGA device could not provide enough configuration resources for implementation. Second, when the VLSI technology came to 0.15 μm, the FPGA device still can not provide enough hardware resources for regular CORDIC implementation. However, it is possible to implement a 14×14 EVD array with the presented μ-rotation CORDIC architecture. Although it needs more sweeps than the regular one, it enables significantly increased parallelism compared to the full CORDIC method. Finally, when the VLSI design keeps shrinking down into 65nm we are able to realize a 25×25 EVD array for solving the Eigenvalue problem of a 50×50 symmetric matrix **A** with the simplified μ-rotation CORDIC PE. Using the 6-CORDIC method allows three times the

Table 2. Area, Delay and the maximal size of EVD array of different Xilinx FPGA devices (i.e., XCV1000-6FG680, XC2V8000-5FF1517 and XC5VL330-2FF1760).

		XCV1000, 0.22 μm	XC2V8000, 0.15 μm	XC5VL330, 65 nm
6-CORDIC×3	Area	454/24.576 LUTs	464/93.184 LUTs	332/207.360 LUTs
	Delay	12.506 ns (79.9 MHz)	8.802 ns (113.6 MHz)	3.934 ns (254.2 MHz)
	EVD	7×7	14×14	25×25
	Matrix	14×14	28×28	50×50
Full CORDIC	Area	5.938/24.576 LUTs	5.938/93.184 LUTs	5.938/207.360 LUTs
	Delay	14.977 ns (66.8 MHz)	7.295 ns (137.1 MHz)	3.52 ns (284 MHz)
	EVD	2×2	4×4	6×6
	Matrix	4×4	8×8	12×12

matrix size of the full CORDIC. Therefore, utilizing the Full CORDIC would cause a partition problem and the processor array would require handling the partition sequentially. This requires an external memory and a more complicated control routine.

6 Conclusions

In this paper, we presented a design concept for parallel iterative algorithms when the VLSI design keeps evolving into nanoscale. For iterative algorithms we are able to simplify/modify the PEs as long as the convergence is guaranteed, such that the parallelism of the implementation can be increased. This is paid for by an increased number of iterations. Computing the EVD by the parallel Jacobi algorithm was used as an example. We have synthesized it into three different Xilinx FPGA devices. The experimental results show that we can realize a 25×25 full Jacobi EVD array into Xilinx XC5VL330 65nm FPGA device. In future work we will investigate the influences of the interconnects, i.e., with advancing VLSI technology the simplified PEs become smaller and smaller in comparison with the interconnection structure of the processor array. This fact requires that the varying importance of interconnects must be incorporated into the design concept.

References

Ahmedsaid, A., Amira, A., and Bouridane, A.: Improved SVD systolic array and implementation on FPGA, in: IEEE International Conference on Field-Programmable Technology (FPT), pp. 3–42, 2003.

Brent, R. P. and Luk, F. T.: The Solution of Singular-Value and Symmetric Eigenvalue Problems on Multiprocessor Arrays, SIAM Journal on Scientific and Statistical Computing, 6, 69–84, 1985.

Gelsinger, P.: Moore's Law: "We See No End in Sight,", Tech. rep., Intel Chief Technology Officer, http://websphere.sys-con.com/node/557154, 2008.

Goetze, J. and Hekstra, G.: An Algorithm and Architecture Based on Orthonormal Micro-Rotations for Computing the Symmetric EVD, INTEGRATION – The VLSI Journal, 20, 21–39, 1995.

Gotze, J., Paul, S., and Sauer, M.: An Efficient Jacobi-Like Algorithm for Parallel Eigenvalue Computation, IEEE Transactions on Computers, 42, 1058–1065, 1993.

Klauke, S. and Goetze, J.: Low Power Enhancements for Parallel Algorithms, in: IEEE International Symopsium on Circuits and Systems, 2001.

Parhi, K. K. and Nishitani, T.: Digial Signal Processing for Multimedia Systems, MARCEL DEKKER, New York, 1999.

Sainarayanan, K. S., Raghunandan, C., and Srinivas, M.: Delay and Power Minimization in VLSI Interconnects with Spatio-Temporal Bus-Encoding Scheme, in: IEEE Computer Society Annual Symposium on VLSI, pp. 401–408, 2007.

Stine, J. E., Castellanos, I., Wood, M., Henson, J., Love, F., Davis, W. R., Franzon, P. D., Bucher, M., and Basavarajaiah, S.: FreePDK: An Open-Source Variation-Aware Design Kit, in: IEEE International Conference on Microelectronic Systems Education, pp. 173–174, 2007.

Vangal, S., Howard, J., Ruhl, G., Dighe, S., Wilson, H., Tschanz, J., Finan, D., Iyer, P., Singh, A., Jacob, T., Jain, S., Venkataraman, S., Hoskote, Y., and Borkar, N.: An 80-Tile 1.28TFLOPS Network-on-Chip in 65 nm CMOS, Solid-State Circuits Conference, 2007. ISSCC 2007. Digest of Technical Papers. IEEE International, pp. 98–589, 2007.

Vitullo, F., L'Insalata, N. E., Petri, E., Saponara, S., Fanucci, L., Casula, M., Locatelli, R., and Coppola, M.: Low-Complexity Link Microarchitecture for Mesochronous Communication in Networks-on-Chip, IEEE Transactions on Computer, 57, 1196–1201, 2008.

Volder, J.: The CORDIC trigonometric computing technique, IRE Trans. Electron. Comput., EC-8, 330–334, 1959.

Walther, J.: A unified algorithm for elementary functions, in: Proc. Spring Joint Comput. Conf., vol. 38, pp. 379–385, 1971.

Wolf, W.: The future of multiprocessor systems-on-chips, in: Annual ACM IEEE Design Automation Conference, pp. 681–685, 2004.

18

Comparison of order reduction algorithms for application to electrical networks

Lj. Radić-Weissenfeld[1,2], **S. Ludwig**[1,2], **and W. Mathis**[1]

[1]Leibniz University Hannover, Institute of Electromagnetic Theory, Hannover, Germany
[2]Fraunhofer Research Institution for Electronic Nanosystems, Advanced System Engineering, Paderborn, Germany

Abstract. This paper addresses issues related to the minimization of the computational burden in terms of both memory and speed during the simulation of electrical models. In order to achieve a simple and computational fast model the order reduction of its reducible part is proposed. In this paper the overview of the order reduction algorithms and their application are discussed.

1 Introduction

Electrical models suggested for order reduction are electrical networks that represent the electromagnetic behaviour and performance of IC interface, electrical representation of signal integrity or power integrity. The electrical representations are huge electrical networks especially if they describe complex structures up to frequencies in the giga range. Therefore smaller electrical networks are desired in order to minimize computational costs in terms of both memory and speed during time and frequency simulation.

There are several ways to reach a smaller model. A smaller model can be created in the modelling process by considering restricted properties of a restricted interface (Kazemzadeh et al., 2008). If the model precisely describes properties of the whole IC interface for a high frequency range the smaller model can be obtained by utilizing order reduction algorithms (Antoulas and Sorensen, 2001; Ludwig et al., 2008; Ludwig and Mathis, 2009; Odabasioglu et al., 1998). Since created electrical models in the modeling phase are passive electrical networks, which usually consist of only time invariant elements, the order reduction algorithms based on the use of the Krylov subspace and Lyapunov balancing are suggested. These order reduction algorithms reduce the number of unknowns i.e. degrees of freedom as much as possible without sacrificing accuracy at the observation points.

In this paper order reduction algorithms based on the system projection on the Krylov subspace and Lyapunov balancing are compared. Moreover the prerequisite for combining these order reduction algorithms is discussed. The focus is particularly on the multipoint frequency-domain projection of the original model on the Krylov subspace, projection-based truncated balanced method and the combination of those.

Correspondence to:
Lj. Radić-Weissenfeld
(radic@tet.uni-hannover.de)

2 Description of electrical models

Electrical models are defined either as single-input single-output (SISO) systems or as multiple-input multiple-output (MIMO) systems. Systems are parts of a model connected to the input/output ports. Therefore inputs and outputs of a system are ports of an electrical model and a system is a reducible part of an electrical model as it is shown in Fig. 1.

The electrical model in Fig. 1 can be described by the following system equations

$$\mathbf{C}s\mathbf{X} = \mathbf{G}\mathbf{X} + \mathbf{B}\mathbf{U} \tag{1}$$

$$\mathbf{V} = \mathbf{L}\mathbf{X}, \tag{2}$$

where $\mathbf{U}, \mathbf{V} \in \mathbb{R}^p$ are input and output vectors, p is the number of ports, while $\mathbf{C}, \mathbf{G} \in \mathbb{R}^{N\times N}$, $\mathbf{B}, \mathbf{L}^T \in \mathbb{R}^{N\times p}$ are system matrices of the order N. The matrices and vectors in Eq. (2) result from the modified nodal analysis of the electrical network.

After the order reduction, order reduced system matrices $\mathbf{C}_{\text{red}}, \mathbf{G}_{\text{red}} \in \mathbb{R}^{q\times q}$, $\mathbf{B}_{\text{red}}, \mathbf{L}_{\text{red}}^T \in \mathbb{R}^{q\times p}$ are of the order $q \ll N$, while the order reduced model has the same number of inputs and outputs p. Now the order reduced system is described by following equations

$$\begin{aligned} \mathbf{C}_{\text{red}} s\mathbf{X}_{\text{red}} &= \mathbf{G}_{\text{red}}\mathbf{X}_{\text{red}} + \mathbf{B}_{\text{red}}\mathbf{U} \\ \widetilde{\mathbf{V}} &= \mathbf{L}_{\text{red}}\mathbf{X}_{\text{red}}, \end{aligned} \tag{3}$$

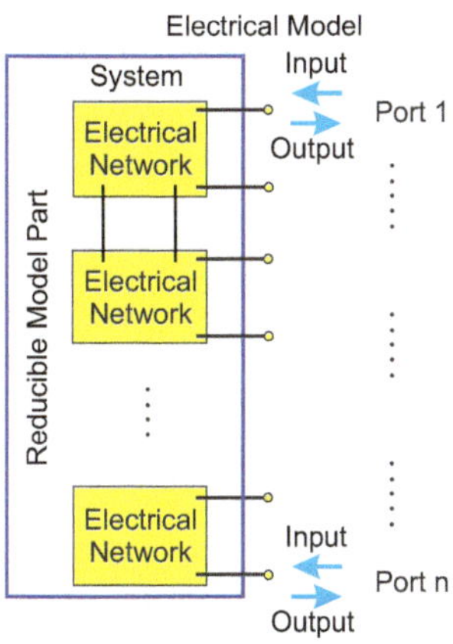

Fig. 1. A block diagram of an electrical model.

After the order reduction vector $\mathbf{V}$ is slightly changed and in Eq. (3) $\widetilde{\mathbf{V}}\approx\mathbf{V}$ holds. The change of the output vector emerges because the vector depends on the system. Therefore the change of the vector is a result of the order reduction of the system. Figure 2 illustrates the consequences of the order reduction of a linear system. The only difference between the original and reduced system is their description by system matrices, while input and output vectors are almost preserved.

The order reduced system can be synthesized as a reduced electrical model for time and frequency analyses and simulations with a circuit simulator (Ludwig et al., 2008; Yang et al., 2007).

3 Order reduction algorithms

The combination of order reduction algorithms, derived from a Krylov subspace projection and singular value based methods are successfully used for the order reduction of passive linear electrical networks (Antoulas and Sorensen, 2001). The Krylov subspace projection method is commonly used for models with a low number of input/output ports and a high order. The singular value based methods are successfully used for models of a modest order, i.e. a few hundred and a high number of ports.

3.1 Krylov-subspace projection

The basic principle of order reduction algorithms based on the moment matching is the projection of the system on the Krylov subspace (Odabasioglu et al., 1998; Freund, 2004). These order reduction algorithms are based on a mere projection utilizing the projection matrix that contains only the most dominant system part. Since only the most dominant system part is projected, the created system is small having the same dominant properties as the original. The Krylov

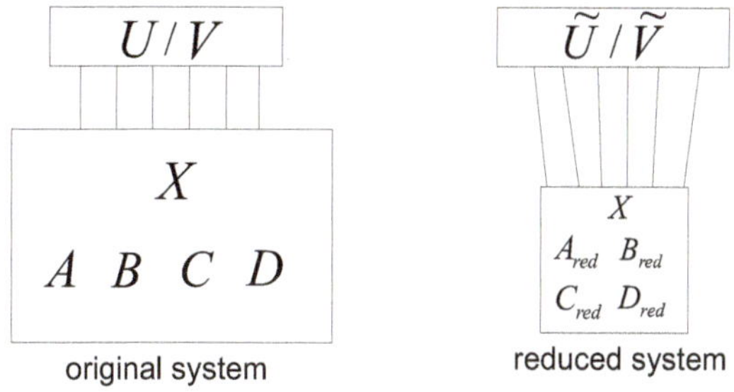

Fig. 2. A block diagram of an original and order reduced system.

subspace used in the order reduction is generated by matrices from Eq. (2)

$$\begin{aligned} Kr(\mathbf{R},\mathbf{E},q) \equiv colsp[&\mathbf{R},\mathbf{A}\mathbf{R},\mathbf{A}^2\mathbf{R},\ldots,\\ &\mathbf{A}^{k-1}\mathbf{R},\mathbf{A}^k\mathbf{r}_0,\mathbf{A}^k\mathbf{r}_1,\ldots,\mathbf{A}^k\mathbf{r}_l], \end{aligned} \tag{4}$$

where $k=\lfloor q/N \rfloor$ and $l=q-kN$ (Odabasioglu et al., 1998). q is the order to which the system of the order N is reduced. The order reduction is based on the moment matching, i.e. preserving the coefficients of the Taylor expansion of a transfer function

$$\mathbf{Y}(s) = \mathbf{M}_0 + \mathbf{M}_1 s + \mathbf{M}_2 s^2 + \cdots, \tag{5}$$

where $\mathbf{M}_k\in\mathbb{R}^N$ is a matched moment. The system matrices are reduced, while preserving the passivity of the reduced system (Odabasioglu et al., 1998). The similarity between original and order reduced system depends on the selected expansion points as well as on the dimension of the Krylov subspace. This order reduction does not enable neither an error prediction nor the preservation of the dominant controllability and observability system part.

3.2 Lyapunov balancing for order reduction

Another group of order reduction algorithms is built by algorithms based on the dominant controllability and observability system part. The controllability and observability of a system is defined by controllability $\mathbf{P}$ and observability $\mathbf{Q}$ Gramians as solutions of the Lyapunov equations

$$\begin{aligned} &\mathbf{GPC} + \mathbf{CPA}^T + \mathbf{BB}^T = \mathbf{0}\\ &\mathbf{CQG} + \mathbf{G}^T\mathbf{QC} + \mathbf{LL}^T = \mathbf{0}, \end{aligned} \tag{6}$$

where $\mathbf{G},\mathbf{B},\mathbf{L}\in\mathbb{R}^{N\times N}$ are state space matrices of an original system given in Eqs. (1) and (2).

The order reduction of system described by Eq. (6) can be done by truncation of the weak system part recognized in small singular values of the product of observability and controllability Gramians (Antoulas and Sorensen, 2001). These singular values can be used to determine the error of the order reduction as shown in (Obinata and Anderson, 2001; Kokotović et al., 1976) or to indicate the convergence of the absolute magnitude error as in Phillips and Silveira (2005).

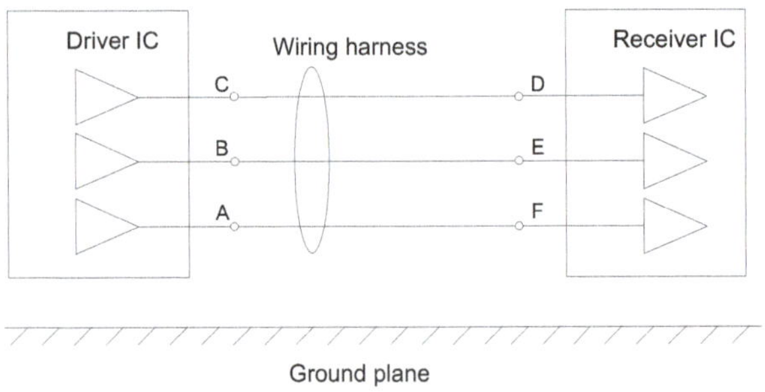

Fig. 3. Schematic view of a multiconductor transmission line model.

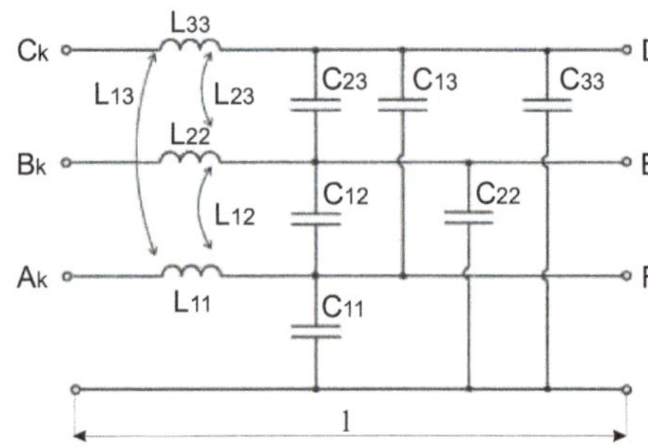

Fig. 4. Per-unit-length cable harness model (Musolino, 2007).

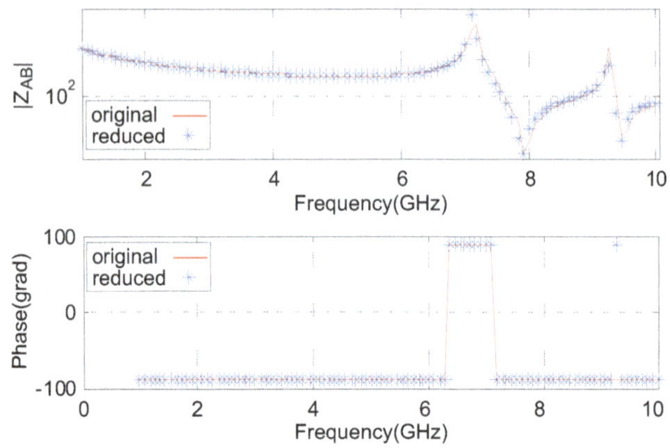

Fig. 5. Amplitude and phase of original and reduced impedance Z_{AB}, between ports A and B.

3.3 Combination of order reduction algorithms

For very high order systems, it is not easy to select the dominant system part in mean of its singular values because the Lyapunov equations are usually not solvable. Therefore it is suggested to use the projection on Krylov subspace as a first step of the order reduction and in a second step to reject the redundant system part using controllability and observability Gramians (Radić-Weissenfeld et al., 2008a,b). For some very high order systems almost the whole redundant system part is rejected in the first step already (Radić-Weissenfeld et al., 2008c). Indeed for such systems the projection on Krylov subspace immediately gives a small system that contains the dominant part only. This issue is discussed within the following example.

4 Example

As an example for the order reduction the electrical model of a cable harness is used. The model published in Musolino (2007) is developed for the electrical fast transient (EFT) immunity characterization of ICs. The whole model includes a driver and a receiver IC connected to each other by a cable harness and comprises an EFT source which induces a transient disturbance at IC terminals. In Fig. 3 the model is given as a schematic view of a multiconductor transmission line.

The cable harness contains three cables on the same radial distance to each other. The Fig. 4 shows the per-unit-length cable harness model. For 1000 segments from Fig. 4 the electrical model has 9000 elements and six ports. The size of matrices from Eq. (2) and the order of the system N is 6003.

The high order of the system matrices is reduced by the projection on the Krylov subspace. By using three expansion points and two-dimensional Krylov subspaces for six external ports the order reduced system has an order of 36 that is 0.5% of the original system order. The high order reduction is achieved since the number of external ports is much lower than the size of the matrices. The good matching of the properties of the original system in the whole frequency range of interest is shown in Figs. 5 and 6. The magnitude and phase of the original and order reduced transfer functions are very similar, since good expansion points are selected. Further order reduction based on Lyapunov equations (Eq. 6) is not suggested, because the whole redundant system part is rejected as well as a part of the system that can be defined as a dominant part. This is supported by results gained from the order reduction by using two expansion points and two-dimensional Krylov subspace. In this case the system of the order 24 does not approximate the original system at all, because the dominant system part is rejected. Indeed, Fig. 7 shows that beside the first 36 singular values the following singular values can be considered as related to the dominant system part. The difference between the magnitude of the 4th and 24th singular values is as big as the difference between the 24th and 43th singular values. Therefore the additional order reduction would reduce the system part of importance and using more expansion points would not drastically reduce the error of the order reduction.

While the original system has sparse matrices, the order reduced system is described by dense matrices. The density

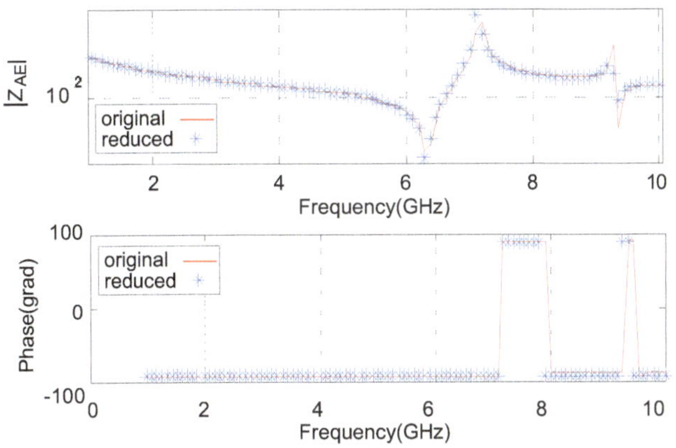

Fig. 6. Amplitude and phase of original and reduced impedance Z_{AE}, between ports A and E.

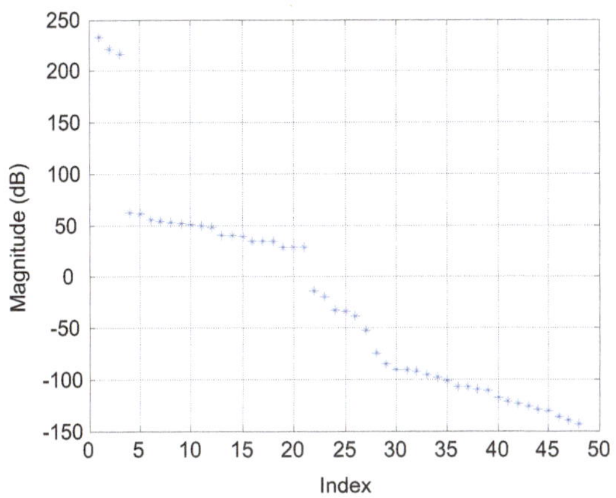

Fig. 7. Singular values of the system.

of the order reduced matrices influences the duration of the time simulation. This is due to the fact that the electrical circuit build from the dense matrices has the number of the nodes equal to the matrix size and it is almost fully connected network, if the synthesis method from Ludwig et al. (2008); Yang et al. (2007) is utilized.

5 Conclusions

In this paper an overview of order reduction techniques for electrical networks is given. The most successfully applied order reduction algorithm is a mere projection of a system on the Krylov subspace. Often this algorithm is used as a first step of the order reduction. Further order reduction can be done if the projection does not select a whole redundant system part. For the implementation of the second-step of the algorithm the Lyapunov balanced based algorithm is suggested, which selects the redundant system part.

Acknowledgements. Represented research and development work is carried out in the frame of the MEDEA+ project A701 PARACHUTE project (Parasitic Extraction and Optimization for Efficient Microelectronic System Design and Application). This particular research was supported by the BMBF (Bundesministerium fuer Bildung und Forschung) Federal Republic of Germany under 01M 3169 A, 01M 3169 D and 01M 3169 E. The responsibility for this publication is held by authors only.

References

Antoulas, A. C. and Sorensen, D. C.: Lyapunov, Lanczos, and Inertia, Linear Algebra and its Applications, 326, 137–150, 2001.

Freund, R. W.: SPRIM: structure-preserving reduced-order interconnect macromodeling, Proceedings of the IEEE/ACM International conference on Computer-aided design, pp. 80–87, 2004.

Kazemzadeh, R., Ludwig, S., Radić-Weissenfeld, Lj., and Mathis, W.: Efficient Modelling of IC Conducted Emission for Power Integrity Analysis, Adv. Radio Sci., in press, 2009.

Kokotović, P. V., O'Malley, R. E., and Sannuti, P.: Singular perturbation and order reduction in control theory – An overview, Automatica, 12, 123–132, 1976.

Ludwig, S., Radić-Weissenfeld, Lj., Mathis, W., and John, W.: Efficient Model Reduction of Passive Electrical Networks with a Large Number of Independent Sources, Proceedings of the IEEE International Symposium on Circuit and Systems – ISCAS, pp. 1280–1283, Seattle, USA, 2008.

Ludwig, S. and Mathis, W.: Transformation of Passive Electrical Networks with Distributed PWL-Sources using a Model Order Reduction, Adv. Radio Sci., in press, 2009.

Musolino, F.: EFT immunity characterization of ICs, Proceedings of the EMC Compo, pp. 163–167, 2007.

Obinata, G. and Anderson, B. D. O.: Model Reduction for Control System Design, Springer-Verlag, London, Great Britain, 2001.

Odabasioglu, A., Celik, M., and Pillegi, L.: PRIMA: Passive Reduced-Order. Interconnect Macromodeling Algorithm, IEEE Transactions on computer-aided design of integrated circuits and systems, 17(8), 645–654, 1998.

Phillips, J. R. and Silveira, M.: Poor Man's TBR: A Simple Model Reduction Scheme, IEEE Transactions on Computer-Aided Design of Integrated Circuits and Systems, 24(1), 43–55, 2005.

Radić-Weissenfeld, Lj., Ludwig, S., Mathis, W., and John, W.: Model order reduction of linear time invariant systems, Adv. Radio Sci., 6, 129–132, 2008a, http://www.adv-radio-sci.net/6/129/2008/.

Radić-Weissenfeld, Lj., Ludwig, S., Mathis, W., and John, W.: Two-step Order Reduction of IC Conducted Emission Models, Asia-Pacific EMC Symposium, Singapur, 2008b.

Radić-Weissenfeld, Lj., Ludwig, S., Mathis, W., and John, W.: Comparison of order reduction algorithms for IC emission models, EMC Symposium, Wroclaw, 2008c.

Yang, F., Zeng, X., Su, Y., and Zhou, D.: RLCSYN: RLC Equivalent Circuit Synthesis for Structure-Preserved Reduced-order Model of Interconnect, IEEE International Symposium on Circuit and Systems – ISCAS, New Orleans, USA, 2007.

Signal processing for plane wave actuators

T. Corbach, M. Holters, and U. Zölzer

Helmut-Schmidt-University, University of the Federal Armed Forces Hamburg, Germany

Abstract. Plane wave actuators without an enclosure per se have a forward and backward radiation. The backward radiation is unwanted in many applications when a single direction radiation is desired. To avoid the disadvantages of an enclosure a system is proposed, which provides a high suppression of the unwanted backward radiation using a pair of plane wave actuators. This is achieved by adapted input signal filters. The influences of the second plane wave actuator to the forward radiated signal are suppressed as well. Additionally, the system also provides for- and backward radiation of different signals with a high suppression of the radiation directions crosstalk. The required power for the signal suppression depends on the physical damping of the plane wave actuators and the space in between. The first realized prototype is designed for flat panel dipole loudspeakers to deal with the mentioned problems in the acoustic domain. The filter design and a calibration algorithm for any given pairs of dipole loudspeaker are explained. The good performance of the developed system is proven by measurement results with the prototype system.

1 Introduction

Directed radiation is wanted in many applications to lower the effect of interference or to concentrate the radiated intensity to the target area. Most point source actuators per se radiate omnidirectional, whereas actuators of nearly plane waves mainly radiate to the forward and backward. Approximating a plane wave actuator by a two-dimensional array of point sources achieves similar directivity including the backward radiation. The backward radiation of both, plane wave actuators and its point source approximation, is unwanted in many applications when a single direction radiation is desired. However, an additional enclosure effectively suppressing the backward radiation increases the size and hence lowers the good structural integration properties. The first realized prototype is designed for electrostatic flat panel dipole loudspeakers to deal with the mentioned problems in the acoustic domain. This paper presents alternative approaches to achieve a one sided radiation, describes the theory of the developed novel approach, demonstrates a practical realization and shows the results achieved with a DSP-based prototype.

Correspondence to: T. Corbach
(corbach@hsu-hh.de)

2 Alternative approaches

Different approaches to suppress acoustical or electromagnetic traveling waves are available. Regardless the physical domain one cancellation method suppressing backward travelling wave is to use a 180° phase delayed second wave. The destructive superposition leads to a cancellation.

To suppress the backward radiation of an electrostatic flat panel loudspeaker by active measures a second parallel similar panel is used. A 180° phase delay and an additional delay according to the panel distance cancels out one radiation direction (Olson, 1973). This approach only grants the suppression of the backward radiation but does not take the influences on the forward radiation into account. Due to the dipole character of both panels the forward radiation is inevitably disturbed by the signal of the second panel.

A further approach of suppressing the backward radiation is the use of passive measures like an enclosure. But a high degree of suppression with passive measures is achieved at the cost of the structural integration properties of the radiator. Furthermore, the enclosure itself causes reflections disturbing the forward radiation. It has to be noted that these reflections can be suppressed with the proposed approach as well.

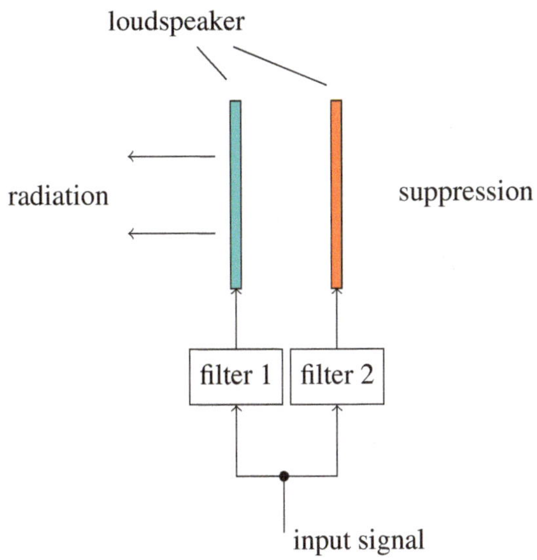

Fig. 1. Scheme of the proposed system with the desired radiation.

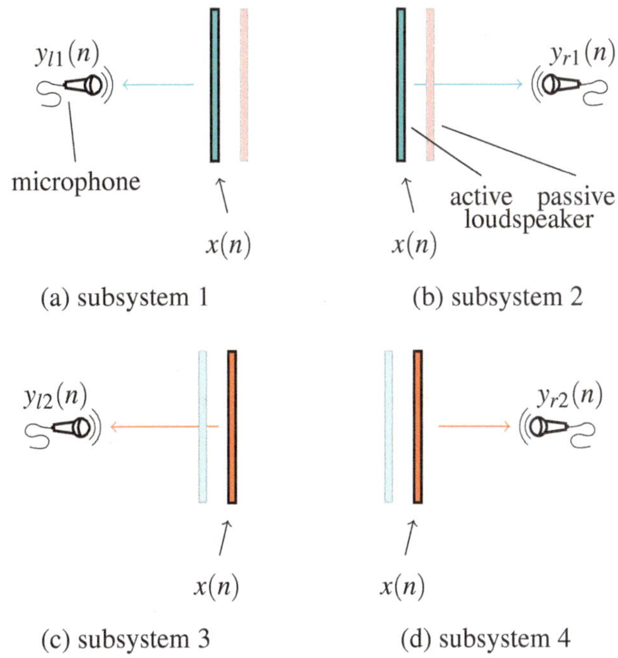

Fig. 2. Partition of the system into its subsystems **(a)** to **(d)** according the actuated panel and the observed side of the system.

3 Functionality

In this section the theory and functionality of the single-sided radiation is explained first, followed by a short explanation of the mentioned stereo expansion. Unlike the alternative approaches the developed system is able to treat forward and backward radiation as independent channels with a low crosstalk.

The proposed system is based on a parallel double electrostatic dipole loudspeaker with a digital signal processing of the input signal. The desired digital filters are gained by transfer functions of the system which have to be determined. These transfer functions represent the behavior of every single subsystem which the system can be divided into. Once the subsystems are described and determined it is possible to compute the desired filter transfer functions as well as the filter impulse responses in time domain (Corbach et al., 2009).

3.1 Subsystem transfer function

The system is divided into subsystems according to the active panel $s \in \{1,2\}$ controlled by the discrete input signal $x(n)$ and the occurring discrete output signal $y_{sk}(n)$ at a reference position at the observed radiation direction $k \in \{l,r\}$. Due to the assumption of plane waves one reference position on axis is sufficient. For 2 panels and 2 possible radiation directions of the whole system, there is a total of 4 subsystems as depicted in Fig. 2.

In time domain the occurring output signals of the subsystems are

$$y_{l1}(n) = h_{l1}(n) \cdot x(n) \,, \tag{1}$$

$$y_{r1}(n) = h_{r1}(n) \cdot x(n) \,, \tag{2}$$

$$y_{l2}(n) = h_{l2}(n) \cdot x(n) \text{ , and} \tag{3}$$

$$y_{r2}(n) = h_{r2}(n) \cdot x(n) \,. \tag{4}$$

For example the impulse response $h_{l1}(n)$ of subsystem 1 (Fig. 2a) describes all influences to the input signal $x(n)$ radiated by loudspeaker 1 and the output signal $y_{l1}(n)$ is observed on the left side of the double dipole. The impulse response $h_{r1}(n)$ (Fig. 2b) describes all influences to the input signal $x(n)$ radiated by the same loudspeaker and the output signal $y_{r1}(n)$ is observed on the opposite side of the double dipole. All reflections and transmission through the passive and active panels are included in the particular impulse response.

It has to be noted that this special case of a symmetric configuration of similar panels leads to equal impulse responses for the subsystems 1 and 4 (Fig. 2a and d), as well as for the subsystems 2 and 3 (Fig. 2b and c). To keep the filter computation general this simplification is not done here.

The subsystem transfer functions can be determined by common measuring techniques like swept sine or noise sequences. Here the logarithmic swept sine technique (Farina, 2007) is used. The desired subsystem impulse responses are defined by

$$h_{l1}(N-n) = y_{l1}(n)\cdot x_{\mathrm{swp}}^{-1}(n)\ , \tag{5}$$

$$h_{r1}(N-n) = y_{r1}(n)\cdot x_{\mathrm{swp}}^{-1}(n)\ , \tag{6}$$

$$h_{l2}(N-n) = y_{l2}(n)\cdot x_{\mathrm{swp}}^{-1}(n)\ , \text{and} \tag{7}$$

$$h_{r2}(N-n) = y_{r2}(n)\cdot x_{\mathrm{swp}}^{-1}(n)\ , \tag{8}$$

where $x_{\mathrm{swp}}(n)$ denotes the sweep test signal, $x_{\mathrm{swp}}^{-1}(n)$ denotes the inverse sweep and $y_{s,k}(n)$ the measured sequence at the observed side. The test signal $x_{\mathrm{swp}}(n)$ and its inverse $x_{\mathrm{swp}}^{-1}(n)$ are both of length N.

As the measurements take place at just one point, the functionality of the system is guaranteed in exactly this measuring point. The high directivity and the low distance between the panels enlarge the area of controlled radiation nearly to the full sphere.

3.2 Filter impulse response

To determine the desired filter impulse responses the whole system is described by the subsystems' impulse responses. This is achieved by expressing all appearing signal parts by the input signal, the desired filter impulse response, and the subsystems' impulse responses. The signal part on the left side, created by input signal $x_{\mathrm{in}}(n)$ convolved with the unknown filter impulse response $h_{f1}(n)$ and the impulse response h_{l1} of subsystem 1, can be denoted by

$$x_{l1}(n) = h_{l1}(n)\cdot \underbrace{h_{f1}(n)\cdot x_{\mathrm{in}}(n)}_{x_1(n)}\ . \tag{9}$$

Similarly, the other signal parts on both radiation sides of the double dipole loudspeaker can be denoted by

$$x_{r1}(n) = h_{r1}(n)\cdot \underbrace{h_{f1}(n)\cdot x_{\mathrm{in}}(n)}_{x_1(n)}\ , \tag{10}$$

$$x_{l2}(n) = h_{l2}(n)\cdot \underbrace{h_{f2}(n)\cdot x_{\mathrm{in}}(n)}_{x_2(n)}\ \text{and} \tag{11}$$

$$x_{r2}(n) = h_{r2}(n)\cdot \underbrace{h_{f2}(n)\cdot x_{\mathrm{in}}(n)}_{x_2(n)}\ . \tag{12}$$

In the further equations the substitutions

$$x_1(n) = h_{f1}(n)\cdot x_{\mathrm{in}}(n)\ \text{and} \tag{13}$$

$$x_2(n) = h_{f2}(n)\cdot x_{\mathrm{in}}(n) \tag{14}$$

are done.

Figure 3 shows the whole system with its signal parts $x_{l1}(n), x_{r1}(n), x_{l2}(n)$ and $x_{r2}(n)$ of the different subsystems. To realize a single-sided radiation by the filtered excitation, the signal parts have to result in the behavior of a single panel

$$x_{l1}(n)+x_{l2}(n) = h_{l1}(n)\cdot x_{\mathrm{in}}(n) \tag{15}$$

for the desired radiation side.

To achieve the suppression of the second, unwanted radiation direction the filtered excitation has to result in signal parts cancelling out each other. By equation this yields

$$x_{r1}(n)+x_{r2}(n) = 0\ . \tag{16}$$

Inserting Eqs. (9) and (11) into Eq. (15), and Eqs. (10) and (12) into Eq. (16) results in the required equations to describe the desired radiation characteristic by the subsystems and its signal parts:

$$h_{l1}(n)\cdot x_1(n)+h_{l2}(n)\cdot x_2(n) = h_{l1}(n)\cdot x_{\mathrm{in}}(n) \tag{17}$$

$$h_{r1}(n)\cdot x_1(n)+h_{r2}(n)\cdot x_2(n) = 0\ . \tag{18}$$

The final computation of the filter impulse responses $h_{f1}(n)$ and $h_{f2}(n)$ is done in the frequency domain. Therefore the time signals of Eqs. (17) and (18) are transformed into frequency domain by the Fourier transform with adequate time domain zero padding:

$$H_{l1}(e^{j\Omega})X_1(e^{j\Omega})+H_{l2}(e^{j\Omega})X_2(e^{j\Omega}) = H_{l1}(e^{j\Omega})X_{\mathrm{in}}(e^{j\Omega}) \tag{19}$$

$$H_{r1}(e^{j\Omega})X_1(e^{j\Omega})+H_{r2}(e^{j\Omega})X_2(e^{j\Omega}) = 0\ . \tag{20}$$

The back substitutions in frequency domain

$$X_1(e^{j\Omega}) = H_{f1}(e^{j\Omega})X_{\mathrm{in}}(e^{j\Omega})\ \text{and} \tag{21}$$

$$X_2(e^{j\Omega}) = H_{f2}(e^{j\Omega})X_{\mathrm{in}}(e^{j\Omega})\ , \tag{22}$$

as well as solving the Eqs. (19) and (20) for the filter transfer functions $H_{f1}(e^{j\Omega})$ and $H_{f2}(e^{j\Omega})$ gives

$$H_{f1}(e^{j\Omega}) = \frac{H_{l1}(e^{j\Omega})H_{r2}(e^{j\Omega})}{H_{l1}(e^{j\Omega})H_{r2}(e^{j\Omega}) - H_{r1}(e^{j\Omega})H_{l2}(e^{j\Omega})}\ \text{and} \tag{23}$$

$$H_{f2}(e^{j\Omega}) = \frac{-H_{r1}(e^{j\Omega})}{H_{r2}(e^{j\Omega})}H_{f1}(e^{j\Omega})\ . \tag{24}$$

Finally the inverse DFT is applied to the filter transfer functions to obtain the desired filter impulse responses $h_{f1}(n)$ and $h_{f2}(n)$:

$$H_{f1}(e^{j\Omega}) \bullet\!\!-\!\!\circ\, h_{f1}(n) \quad H_{f2}(e^{j\Omega}) \bullet\!\!-\!\!\circ\, h_{f2}(n)\ . \tag{25}$$

3.3 Signal computation

The filtered input signals $x_{f1}(n)$ and $x_{f2}(n)$ actuating the panels are computed by a linear convolution with the corresponding filter impulse response $h_{f1}(n)$ respectively $h_{f2}(n)$:

$$x_{f1}(n) = h_{f1}(n)\cdot x_{\mathrm{in}}(n) \tag{26}$$

$$x_{f2}(n) = h_{f2}(n)\cdot x_{\mathrm{in}}(n) \tag{27}$$

Depending on the given setup of loudspeakers and its distances, the filter transfer functions may show gains of several dB for certain frequencies. For input signals with amplitudes using the full quantization range, clipping will occur leading to worse performance and decreased audio quality. To avoid this, the filter transfer function has to be normalized to 0 dB or an equivalent damping factor should be introduced in the time domain.

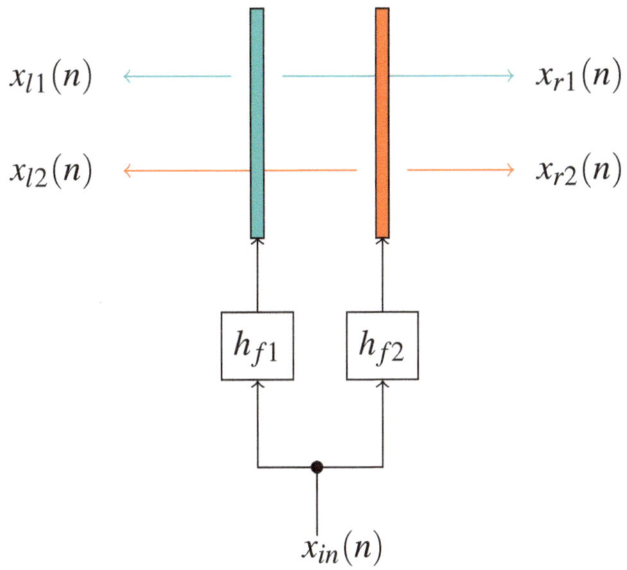

Fig. 3. Scheme of the entire system including all signal parts radiated by the single panels.

3.4 Synthetic filter impulse response

To clarify the systems functionality a filter computation with the use of synthetic subsystem impulse responses is done. Therefore, the plane wave sources are assumed to be ideal and equal at a sample-distance of d, and the wave transmission through the source is assumed to have a frequency independent attenuation of g.

Due to the symmetry this leads to equal impulse responses

$$h_d(n) = h_{l1}(n) = h_{r2}(n) = \delta(n - d_d) \tag{28}$$

for the subsystems 1 and 4, with d_d representing the samples needed from the source to the reference position. Since the subsystems 2 and 3 describe the transmission through a source to the reference point, a d_t-delayed and attenuated unit impulse yields their impulse response

$$h_t(n) = h_{l2}(n) = h_{r1}(n) = g \cdot \delta(n - d_t) \ . \tag{29}$$

With direct delay of $d_d = 5$, transmission delay of $d_t = 9$, and an attenuation of $g = 0.8$ the subsystems are described by the impulse responses $h_d(n)$ and $h_t(n)$ depicted in Figs. 4 and 5.

The Fourier transform yields the subsystems' transfer function $H_{l1}(e^{j\Omega})$, $H_{r2}(e^{j\Omega})$, $H_{l2}(e^{j\Omega})$, and $H_{r1}(e^{j\Omega})$. The desired filter transfer functions $H_{f1}(e^{j\Omega})$ and $H_{f2}(e^{j\Omega})$ are computed using Eqs. (23) and (24). The inverse Fourier transform then gives the desired filter impulse responses $h_{f1}(n)$ and $h_{f2}(n)$ depicted in Figs. 6 and 7. The sample-distance d between each impulse in the filter impulse response exactly amounts

$$d = d_t - d_d = 9 - 5 = 4 \ , \tag{30}$$

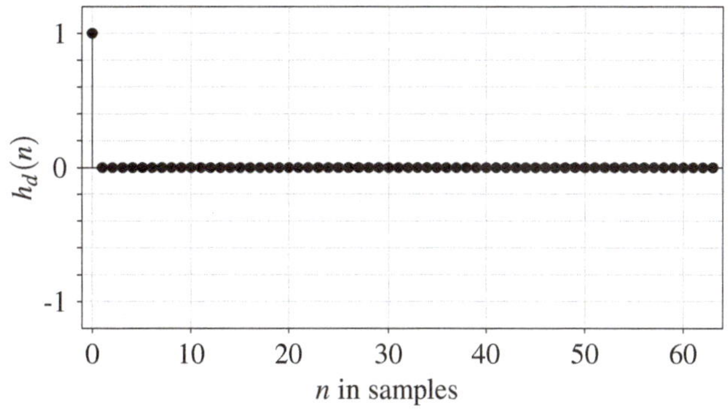

Fig. 4. Synthetic impulse response of the sub systems 1 and 4.

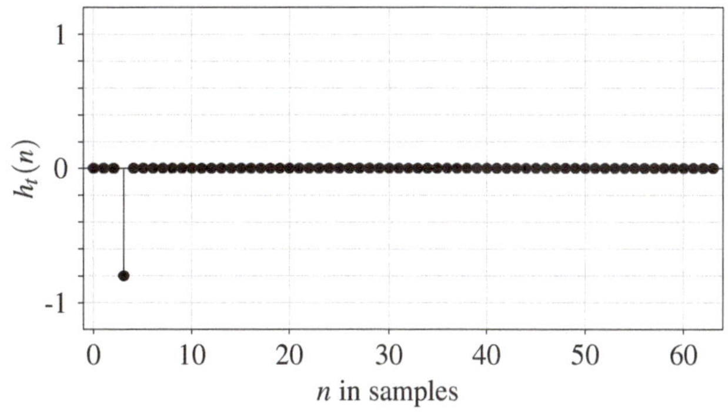

Fig. 5. Synthetic impulse response of the sub systems 2 and 3.

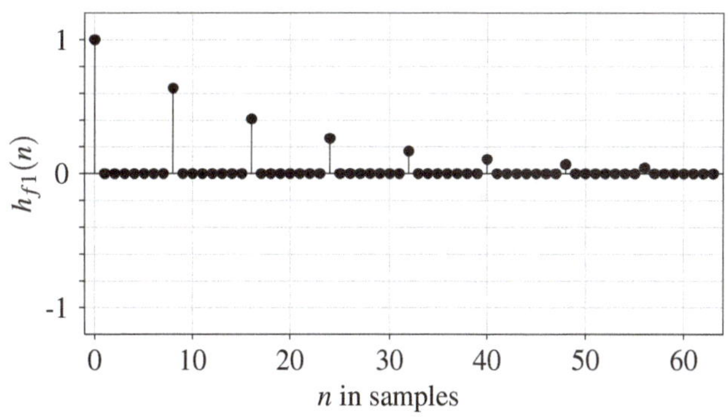

Fig. 6. Computed impulse response $h_{f1}(n)$.

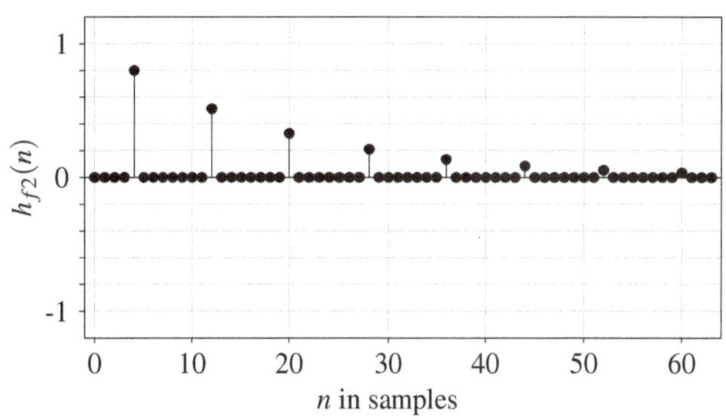

Fig. 7. Computed impulse response $h_{f2}(n)$.

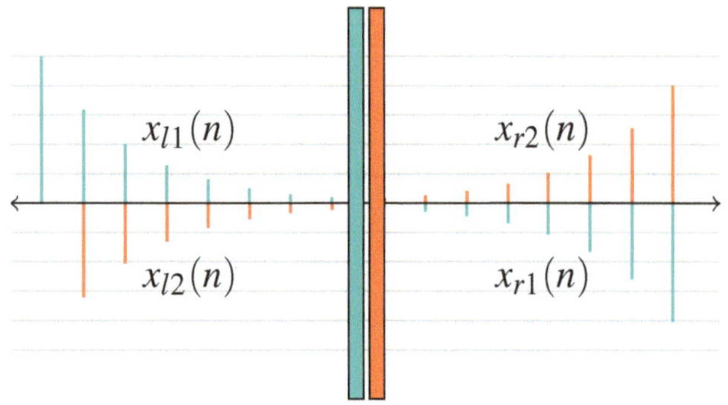

Fig. 8. Signal parts of each loudspeaker after about 70 samples with an unit impulse input signal $x_{\text{in}}(n) = \delta(n)$.

and each impulse decreases by a factor of g. This gains an equation for the filter impulse responses

$$h_d(n) = \sum_{k=0}^{N} g^{2k} \cdot \delta(n - d \cdot 2k) \tag{31}$$

$$h_t(n) = \sum_{k=0}^{N} g^{2k+1} \cdot \delta(n - d(2k+1)) \ . \tag{32}$$

For this example the impulse response is cut after 64 samples. A closer look at $h_{f1}(n)$ and $h_{f2}(n)$ explains the behavior of the system. Let us assume a single unit impulse as input signal $x_{\text{in}}(n) = \delta(n)$, which has to be radiated to the radiation direction. The first impulse $h_{f1}(0) = 1$ represents this desired radiation of the untouched signal to the radiation direction. The first impulse at the second radiator $h_{f2}(d) = g$ gains the needed 180° phase shifted, attenuated wave to suppress the unwanted backward radiation. Note, that there already is a 180° phase shift between forward and backward radiated signal of each source due to its dipole character. Since the second radiator itself has a backward radiation, too, a further, again attenuated impulse $h_{f1}(2d) = g^2$ is needed to keep the desired radiation uninfluenced. The attenuation leads to an exponential decay of needed signal “echoes” to achieve the single sided radiation.

Figure 8 shows the signal parts of the sources after about 70 time steps. This shows the direct dependency of the additionally needed power and the physical attenuation from the origin of the wave to origin of the 180° phase shifted wave.

3.5 Stereo radiation

Besides the single-sided radiation this approach allows the radiation of independent signals to each direction. The suppression of the crosstalk between the directions is equal to the suppression of backward radiation of the single-sided radiation. The stereo radiation is easily achieved by applying the filter unit $(h_{f1}(n), h_{f2}(n))$ to both input signals $x_{\text{in}1}(n)$ and $x_{\text{in}2}(n)$. The filter outputs are summed up by choosing different radiation directions for each filter unit as shown in Fig. 9. Now the sum of the output of filter $h_{f1}(n)$ driven by input signal $x_{\text{in}1}(n)$ and the output of filter $h_{f2}(n)$ driven by input signal $x_{\text{n}2}(n)$ has to actuate panel 1. Panel 2 has to be actuated by the sum of the output of filter $h_{f1}(n)$ driven by $x_{\text{in}2}(n)$ and output of filter $h_{f2}(n)$ driven by $x_{\text{in}1}(n)$.

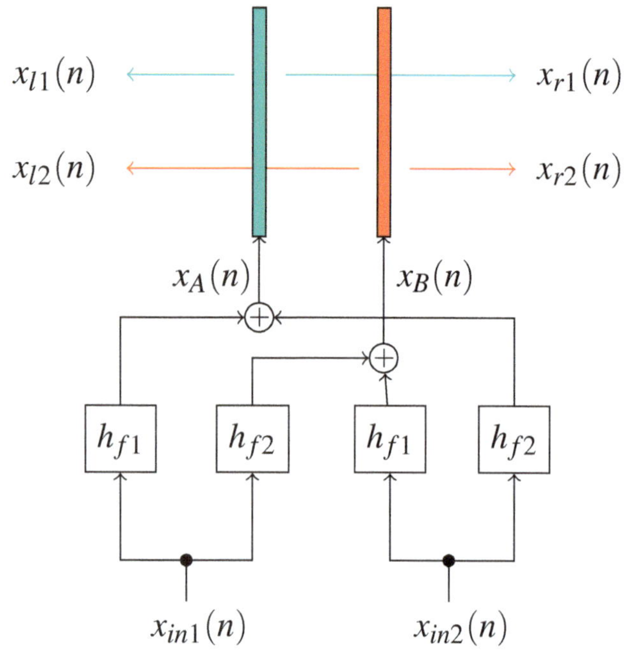

Fig. 9. Scheme of the system with dedicated signal processing for every independent input signal summed up to stereo radiation.

$$x_A(n) = h_{f1}(n) \cdot x_{\text{in}1}(n) + h_{f2}(n) \cdot x_{\text{in}2}(n) \tag{33}$$

$$x_B(n) = h_{f1}(n) \cdot x_{\text{in}2}(n) + h_{f2}(n) \cdot x_{\text{in}1}(n) \tag{34}$$

Certainly this causes higher restriction to the needed damping factor to avoid clipping (see Sect. 3.3).

4 Prototype system

The developed prototype is based on a Texas Instruments floating-point DSP built in the PADK evaluation board of Lyrtech with all needed periphery. The real-time filtering of any input signal for single-sided radiation and stereo-radiation, as well as a calibration algorithm to determine the filter impulse responses were implemented. The calibration algorithm makes use of the exponential sine sweep measurement technique. The used length of the filter impulse responses $h_{f1}(n)$ and $h_{f2}(n)$ amounts to 512 samples. The sampling frequency was set to 44.1 kHz. As loudspeakers the Panphonics S60 audio elements (Pan, 2009) with a size of $600 \times 600 \times 4$ ($w \times h \times d$ in mm) are used. The distance between the loudspeakers is set to 10 mm. The microphone-loudspeaker distance during the automatic calibration is 5 cm.

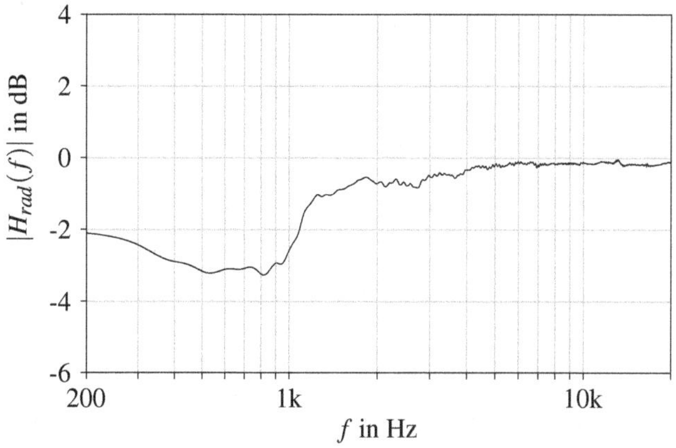

Fig. 10. Difference of the compared frequency responses at the radiation side between enabled and disabled signal processing.

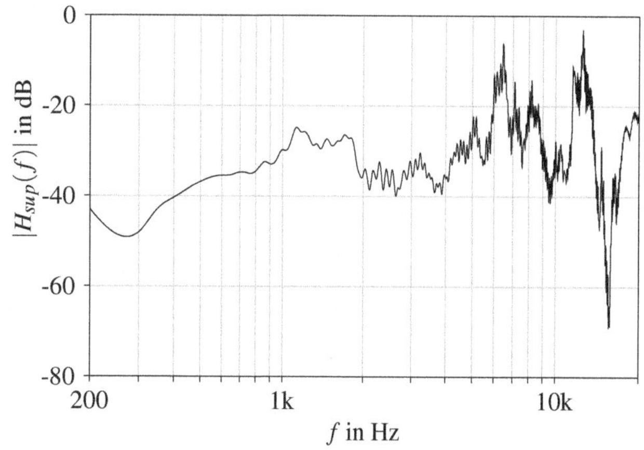

Fig. 11. Difference of the compared frequency responses at the suppression side between enabled and disabled signal processing.

5 Measurement results

The measurements are done on axis with microphones on both sides at the same distance of about 20 cm. Once again the exponential sine sweep method is applied to get the impulse responses and transfer functions of the entire system. The reflections of the walls at the measurement location are removed in the computed transfer functions by windowing its impulse responses.

The transfer functions of the entire system are measured twice for enabled and disabled signal processing. That means for the case of disabled signal processing the panel located to the radiation side is actuated, whereas the panel to the suppression side stays passive. The achieved differences between enabled and disabled signal processing at both sides $|H_{\text{rad}}(f)|$ and $|H_{\text{sup}}(f)|$ are depicted in Figs. 10 and 11. The behavior on the radiation side of the double dipole with activated signal processing was desired to have low differences to a single radiating dipole loudspeaker.

The measured result in Fig. 10 shows only minor differences between the enabled and disabled signal processing of less than 4 dB. For most audio signals like speech or music the auditory impression approves nearly no perceivable differences when activating the signal processing.

However, the second radiation direction was desired to be suppressed by the system to achieve a single-sided radiation by dipoles. Figure 11 shows the difference of the transfer function for disabled and enabled signal processing of the system. A high broadband active attenuation is observable. For increased frequencies (> 5 kHz) the attenuation values show higher fluctuations. Explanations for this behavior could be the used passive amplifiers providing the required bias voltage for the electrostatic panels. For higher frequencies the capacitive behavior of the panels causes a high current. The used combination of the passive adapter and a common hi-fi amplifier is not sufficient to provide accurate signals.

6 Discussion

A limitation of this approach is the additional power needed to suppress all occurring "echoes". To avoid quantization clipping the filters have to be normalized. As countermeasure to the digital normalization a higher amplification is needed to get a similar effective output compared to single source radiation. This additional power needed directly depends on the attenuation of radiating through an actuator and requires amplification headroom of the sources for low attenuation values. The decay of the filter impulse response, hence the required filter length, depends on the attenuation as well. The attenuation through a typical actuator is usually sufficient in the audio domain, leading to reasonable ampflications and filter lengths.

In principle the proposed approach of controlling the radiation of plane wave actuators with forward and backward radiation should work independently of the physical domain of the wave. The advantages of this approach compared to passive methods have to be investigated separately for each physical domain. However, it makes use of the cancellation by destructive superposition of waves, hence it is applicable wherever sources of travelling waves radiate into 2 opposite directions.

7 Conclusion

The results of the comparative measurements demonstrate the functionality of the developed approach of filter computation for the audio domain. With its practical realization a high broadband active attenuation of nearly 40 dB is achieved at the suppression side of the double dipole loudspeaker. The influences to the desired radiation side are very low and produce nearly no audible changes. Hence, an effective system was developed that achieves a high active attenuation of the unwanted radiation side without influencing the desired radiation direction. This approach of suppressing the second radiation direction of dipole loudspeakers is applicable wherever loudspeakers have to be integrated and no further passive efforts to suppress the backward radiation is wanted or possible.

According to this developed system further applications can be found beyond the single-sided radiation. For example the extension to a stereo radiation of different independent signals to both sides is already realized.

References

Panphonics Oy, Audio Element Manual v1.1, Panphonics, http://www.panphonics.com/pdf/AudioElementManualv1-1.pdf, 2009.

Corbach, T., Holters, M., and Zölzer, U.: Flat loudspeaker radiation characteristics controlling method, involves compensating sound signals that are emitted from loudspeaker by phase shifted compensation signal in moment of passage via another loudspeaker, German Patent – DE102007049407A1, 2009.

Farina, A.: Advancements in impulse response measurements by sine sweeps, in: 122nd AES Convention, Vienna, Austria, 2007.

Olson, H. F.: Gradient Loudspeakers, Journal of the AES, 21, 86–93, 1973.

Transformation of passive electrical networks with distributed PWL-sources using a Model Order Reduction

S. Ludwig[1,2] and W. Mathis[1]

[1]Leibniz University Hannover, Institute of Electromagnetic Theory, Hannover, Germany
[2]Fraunhofer Research Institution for Electronic Nano Systems, Advanced System Engineering, Paderborn, Germany

Abstract. In modeling of distributed systems with distributed sources large networks with RLC-elements and independent sources arise. This high complexity leads to a high effort in simulations. Therefore model reduction can be used to reduce these networks, preserving the behavior at the observed nodes in the networks. For the reduction of networks with a large number of independent sources only a weak reduction is enabled with standard model reduction techniques. In this paper an efficient reduction of networks with a large number of sources with piece-wise-linear waveforms is presented, using the decomposition of piece-wise-linear functions. With the proposed method a higher reduction of the network and/or a higher accuracy can be achieved with model reduction. The validity and efficiency of the proposed method is shown by reducing a RCI-Grid model.

1 Introduction

Since the early beginnings of circuit theory similarity and equivalence transformation of electrical networks are of strong interest. In the modeling of distributed systems, such as electromagnetic or thermal fields, with distributed sources very large networks are created. For example in electromagnetic fields in ICs the switching currents act like distributed sources or in thermal simulations of substrates the transistors act like distributed heat sources. For these very large networks a smaller network with approximated port behavior is required to allow for fast simulations. With the help of model order reduction and network synthesis it is possible to find such a reduced network (Antoulas, 2005; Ionutiu et al., 2007; Ludwig et al., 2008a; Radic et al., 2008). Replacing the original network with the reduced network enables investigations with a lower effort for simulations.

In our former work we have developed an efficient reduction method for networks with groups of distributed sources which have equal or proportional amplitudes (Ludwig et al., 2008a,b,c). In this paper we extend this approach to sources with piece-wise-linear (PWL) waveforms (PWL-sources), which do not necessarily have equal or proportional amplitudes.

In this work the design flow for the approximative transformation of large passive electrical networks, which especially contain a high number of PWL-sources, is presented in Sect. 2. Firstly, the network is transformed with an equivalence-transformation in Sect. 2.1 to reduce the number of independent sources and to enhance the reduction process. Secondly, the network is reduced with an approximative transformation using model order reduction algorithms in Sect. 2.2. The efficiency of the complete reduction process is shown by reducing an example network in Sect. 3 and the paper is concluded in Sect. 4.

2 Transformation of networks with distributed sources

For the transformation of electrical networks the invariant as well as the approximative properties of the network have to be defined. Invariant properties are for example stability and passivity. Properties that can be approximated are the behavior at the nodes for the connection with other networks and electrical values that should be observed or measured. Also the behavior at the nodes where independent sources are connected can be approximated. Other properties, such as the number of elements or nodes or the behavior at nodes which are not of interest can be disregarded.

Distributed systems with distributed sources are modeled with passive RLC-elements a high number of independent sources. In state-of-the-art model reduction these

Correspondence to: S. Ludwig
(ludwig@tet.uni-hannover.de)

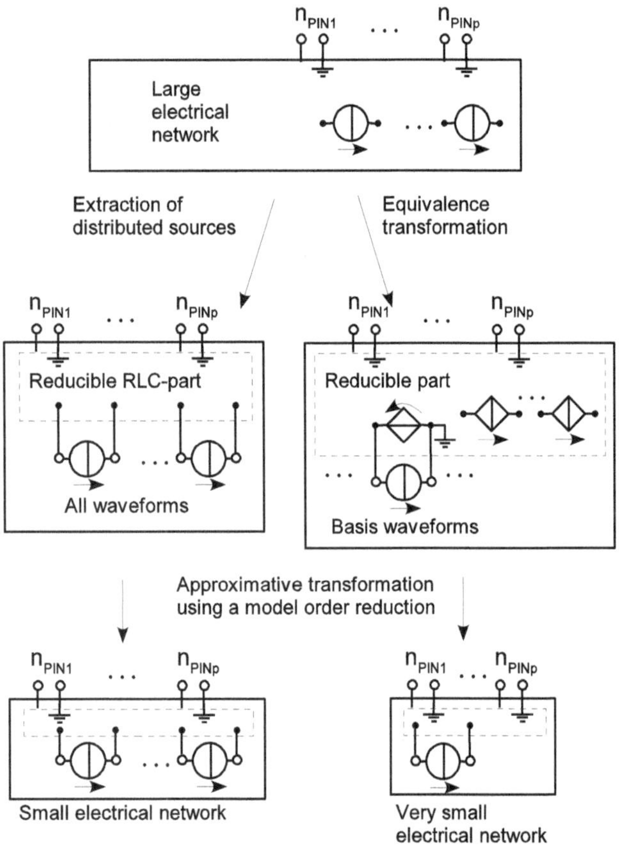

Fig. 1. Transformation flow for networks with distributed sources.

independent sources are extracted from the RLC-part of the network and connected through ports and only the RLC-part is reduced (left path in Fig. 1). The advantage of this extraction is that only passive RLC-elements are used in the reducible model and the passivity of the whole model can be preserved during a model order reduction (Odabasioglu et al., 1998). This leads to a passive reduced model and prevents unstable time-domain simulations. For networks with a large number of independent sources, this extraction produces a large number of ports. A large number of ports is a strong limitation for the model reduction (Silva et al., 2007b; Feldmann, 2004). In many distributed systems the waveforms of the sources are not completely independent of each other. For equal or proportional waveforms of the sources, which are widely used in modeling, we have developed a method to describe these sources efficiently for model reduction (Ludwig et al., 2008a,b,c). The idea is to reduce the number of independent sources in order to enable stronger reduction of the reducible part (dotted lines in Fig. 1) of the network. For these networks we propose an equivalence transformation (right path in Fig. 1) instead of the extraction of the sources. The basis-waveforms are given by one representative of each group and are realized with additional independent sources. The waveforms are copied into the network by replacing the distributed sources with controlled sources with a specific gain. As only the independent sources, representing a group of sources, have to be extracted and connected through ports, a network with a reduced number of ports is generated. The resulting network can be more efficiently reduced than the network with extracted sources (Ludwig et al., 2008a). The network generated in this way is generally not passive. The connection of non-passive systems can lead to unstable time-domain simulations and therefore a passive system is preferable. Controlled sources connected with the additional independent sources are added. These sources act like dissipative elements that dissipate exactly the amount of energy generated by the controlled sources in the network and vice a versa. Thereby the equivalent network is passive (Ludwig et al., 2008a,c) and passivity can be preserved in the reduced model with the description proposed in Ludwig et al. (2008c).

In models that are generated from measurements of distributed systems the sources in the model have PWL-waveforms, which are not necessarily equal or proportional. In the next section the described approach for an equivalence-transformation of the network is extended to networks with PWL-sources.

2.1 Equivalence transformation for the reduction of the number of PWL-sources

In this section we propose a method to describe PWL-sources in a network for an efficient reduction. We consider networks with sources, where the waveforms of the sources are described with continuous PWL-functions $f_q(t)$ on a closed interval $[0, T]$ for which holds

$$f_q(t) = m_i t + o_i \quad \text{for all} \quad a_{i-1} \leq t \leq a_i \tag{1}$$

for pairs of real numbers (m_i, o_i), i=1, ..., k with

$$0 < a_0 < a_1 < ... < a_k < T \tag{2}$$

according to the definition in Aliprantis et al. (2005).

Every PWL-function describing a waveform of a source in the network is decomposed into PWL-basis-functions. The union of all basis-functions gives the complete basis-functions space of the size b. The waveforms of all sources are now the linear combination of all b basis-functions multiplied with a corresponding weighting factor w_i.

$$f_q(t) = \sum_{i=1}^{b} w_{q,i} f_i(t). \tag{3}$$

The basis-functions can be found for example with the method proposed in Aliprantis et al. (2005).

After their decomposition the PWL-functions have to be realized in the network. In the network every basis-function is realized with an independent PWL-source. The distributed sources in the network are replaced with controlled sources, with the controlling values and gains corresponding to the linear combination of the basis-functions as in Eq. (3). With this method a network with equivalent behavior at the pins of the network is generated, hence this modification is an equivalence-transformation as shown in the right path in Fig. 1.

If the number of basis-functions is lower than the number of distributed PWL-sources in the original network, with this method a network with a lower number of independent PWL-sources can be generated. By extracting the lower number of independent PWL-sources a reducible network with a lower number of ports can be generated, which can be more efficiently reduced. Additionally, the reducible part of the network is still passive and reciprocal by adding appropriate controlled sources to the additional independent sources for the basis-functions, as described earlier in this section.

The difference compared to the former method for groups of waveforms with equal or proportional amplitude is that not for every group of waveforms an independent source is added but every basis-function is realized with an independent source. The distributed sources inside the network are again replaced by controlled sources, which are now not only controlled by the representative source for each group but by all additional independent sources.

2.2 Approximative transformation using a Model Order Reduction (MOR)

Model order reduction algorithms based on projection are commonly used for the reduction of large systems of electrical networks (Ionutiu et al., 2007). The advantage of the projection methods is the simple implementation and the low computational effort for the construction of the reduced model.

For MOR the electrical network of the model has to be described as a system of differential algebraic equations in the form of

$$(s\mathbf{C}+\mathbf{G})\mathbf{x} = \mathbf{B}\mathbf{u}, \qquad \mathbf{y} = \mathbf{L}^T\mathbf{x} \tag{4}$$

using modified nodal analysis, where the number of equations N is defined as the order of the system. p is the number of ports. The positive definite system matrices $\mathbf{C}, \mathbf{G} \in \mathbb{R}^{N\times N}$ contain the stamps for capacitors, inductors and resistors. The input and output vectors $\mathbf{u}, \mathbf{y}$ are related with the system vector $\mathbf{x} \in \mathbb{R}^N$ through ports with the help of the matrices $\mathbf{B}, \mathbf{L} \in \mathbb{R}^{N\times p}$, $\mathbf{B}=\mathbf{L}$. The matrix transfer function $\mathbf{H} \in \mathbb{C}^{p\times p}$ of the system is given by

$$\mathbf{H}(s) = \mathbf{L}^{\mathbf{T}}(s\mathbf{C}+\mathbf{G})^{-1}\mathbf{B}. \tag{5}$$

Circuit equations of typical networks have system matrices that are very sparse, even though the order N is quite large. For the reduction we use the approach proposed in Phillips and Silveira (2005). Firstly a number of real frequency points s_i is defined and for every frequency point

$$\mathbf{z}_k = (s_k\mathbf{C}+\mathbf{G})^{-1}\,\mathbf{B} \tag{6}$$

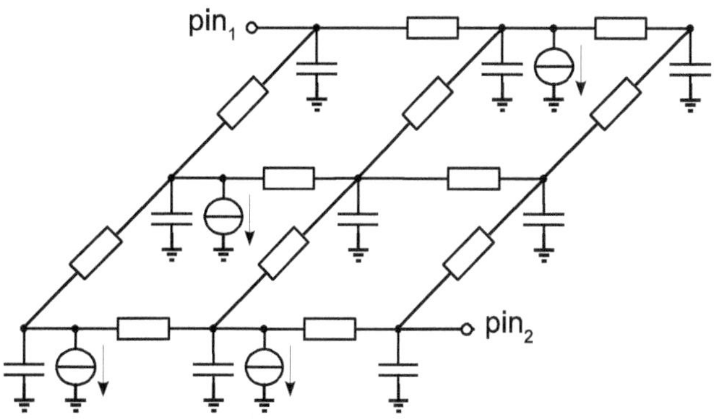

Fig. 2. Example of an 3×3 RCI-Grid model with 4 PWL-sources.

is calculated. By combining all $\mathbf{z}_i$ in $\mathbf{Z} = [\mathbf{z}_1, \mathbf{z}_2, ..., \mathbf{z}_i]$ and using a singular value decomposition

$$[\mathbf{U}, \mathbf{S}, \mathbf{V}] = SVD(\mathbf{Z}) \tag{7}$$

the projection matrix $\mathbf{T}$ is the orthogonal column span of $\mathbf{U}(:, 1:q)$ and the singular values in $\mathbf{S}$ can be used for choosing the reduced order and for error estimation (Phillips and Silveira, 2005), as only the dominant singular vectors corresponding to the q largest singular values are used for the projection matrix. With the projection matrix $\mathbf{T}$ the system vector is changed with $\widetilde{\mathbf{x}}=\mathbf{T}\mathbf{x}$. The reduced system matrices are calculated with

$$\widetilde{\mathbf{C}} = \mathbf{T}^T\mathbf{C}\mathbf{T} \qquad \widetilde{\mathbf{G}} = \mathbf{T}^T\mathbf{G}\mathbf{T}$$
$$\widetilde{\mathbf{B}} = \mathbf{T}^T\mathbf{B} \qquad \widetilde{\mathbf{L}} = \mathbf{T}^T\mathbf{L}. \tag{8}$$

If the projection matrix is real, this method preserves the passivity of suitably described systems (Odabasioglu et al., 1998). The order reduced transfer function

$$\widetilde{\mathbf{H}}(s) = \widetilde{\mathbf{L}}^{\mathbf{T}}(s\widetilde{\mathbf{C}}+\widetilde{\mathbf{G}})^{-1}\widetilde{\mathbf{B}} \tag{9}$$

with $\widetilde{\mathbf{H}} \in \mathbb{C}^{p\times p}$ is generated, with an order $n \ll N$ of the system matrices. The reduced system closely matches the input-output-behavior of the original model, $\widetilde{\mathbf{H}} \approx \mathbf{H}$ (Phillips and Silveira, 2005).

In the last step for the approximative transformation of the network, the reduced system is synthesized as an electrical network, for example with a method described in Palenius and Roos (2004); Yang et al. (2007); Ludwig et al. (2008a).

With this transformation a smaller network can be found. In the resulting network, properties like passivity are preserved while the behavior at the pins is approximately the same as in the original network.

3 Results for an example RCI-grid

As an example a network grid as shown in Fig. 2 is used. The example grid contains resistors, capacitors and independent current PWL-sources. The 4900 resistors of our example are connected as a two-dimensional grid with 50×50

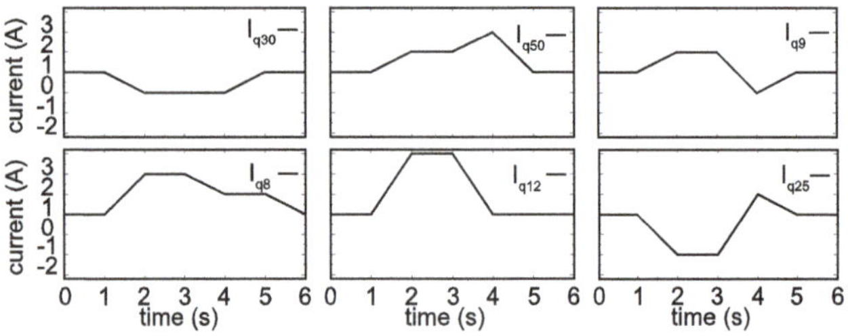

Fig. 3. Some waveforms of the current sources of the 50×50 RCI-Grid model.

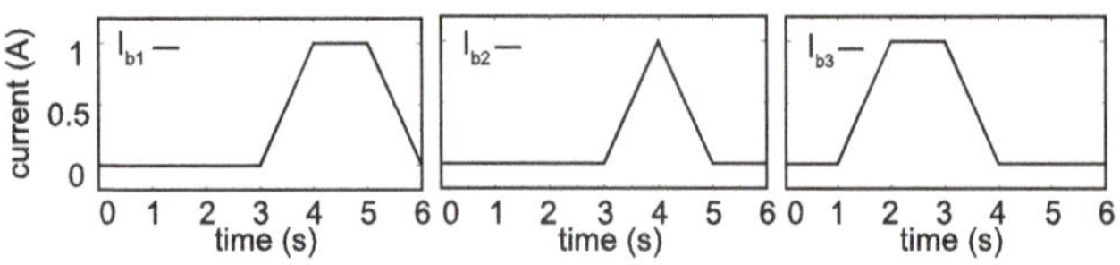

Fig. 4. Basis-waveforms for the 50×50 RCI-Grid model using decomposition of the original PWL-waveforms.

nodes and pins at two opposite corners of the network. Between every node in the 50×50 grid and the ground node a capacitor is connected, which amounts to 2500 capacitors. The values of the elements are randomly chosen in the ranges 1±0.5 ohm and 0.1±0.05 F for the resistors and capacitors, respectively. Between 55 arbitrary chosen nodes in the network and the ground node independent current sources with PWL-waveforms are connected. Some examples for the PWL-waveforms used in the example are shown in Fig. 3. All waveforms can be decomposed into the three basis-waveforms shown in Fig. 4.

With the standard method of extracting all independent sources the system has 57 ports. If the waveforms of the independent sources are decomposed into the three basis-waveforms and the equivalence-transformation in Sect. 2.1 is used, only three independent sources are necessary. This results in only five ports of the reducible part of the network. With this description the behavior at the two pins of the model is not changed. Notably the system of the equivalence-transformed network can be, already before the reduction with MOR, calculated almost four times faster due to the lower number of ports (Table 1). Both system descriptions are reduced with the PMTBR algorithm as described in Sect. 2.2. One hundred expansion points s_i are equally distributed in the frequency range range of interest $10^{-5} \leq f \leq 10^{5}$ and used for the generation of the projection matrix. In Fig. 5 the singular values **S** of the system with extracted sources and the equivalence-transformed system are shown. It can be seen that the singular values of the system after equivalence-transformation decay faster, which means that the reduced systems will be smaller and more accurate. For comparison both systems are first reduced to the same order of n=40. Secondly both systems are reduced to a maximal singular

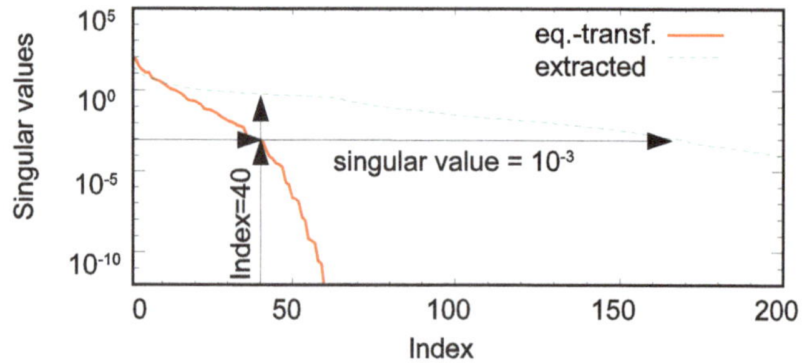

Fig. 5. Hankel singular values of the 50×50 RCI-grid.

Table 1. Comparison of reduction results of the system.

Original order N=2500	Ports p	Reduced order n (Reduction)	MATLAB Speed-up	Accuracy
Extracted sources	57	2500 –	1×	Exact at the pins
Extracted reduced with PMTBR	57	165 (93.4%)	9.4×	Very Good
Extracted reduced with PMTBR	57	40 (98.4%)	42×	Bad
Equivalence-transformed	5	2500 –	3.8×	Exact at the pins
Equivalent reduced with PMTBR	5	40 (98.4%)	52×	Very Good

value of 10^{-1}, resulting in an order of n=40 for the system of the equivalence-transformed network and n=165 for the system of the network with extracted sources. The possible MATLAB-speed-up of the reduction of both systems is shown in Table 1. The transfer functions of the reduced systems with the maximal singular value of 10^{-1} show a good agreement with the original system, as shown for the transfer function Z_{11} in Fig. 6. However, since the reduced system of the equivalence-transformed network is smaller, it can be faster simulated than the reduced system of the network with extracted sources (52× in comparison to 9.4×, Table 1). For the same order, the speed-up of the system with extracted sources is almost as high as for the equivalence-transformed system (42× and 52×, Table 1). Nevertheless, the reduced system of the equivalence-transformed network is more accurate (Figs. 6 and 7). With this example it is shown, that the proposed method of equivalence-transformation with the decomposition of the waveforms as an additional step before the reduction leads to smaller and therefore faster models than the standard method of extracting all PWL-sources. Also it was shown, that for the same size of the reduced system our equivalence-transformation before reduction leads to more accurate models.

For investigations with SPICE-like simulators the smallest reduced system that is close enough to the original, which is

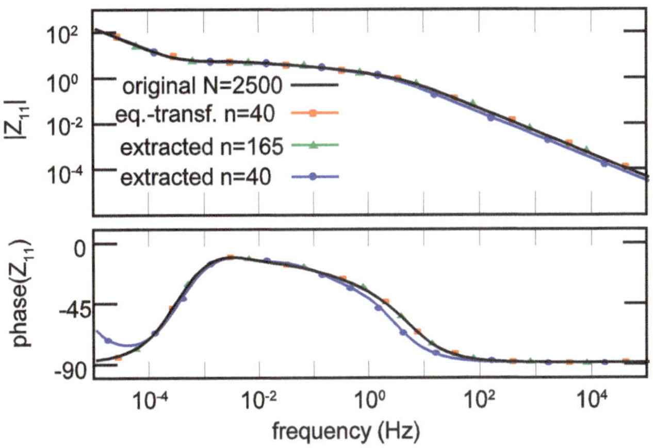

Fig. 6. Magnitude and phase of transfer function Z_{11} of the 50×50 RCI-grid (MATLAB).

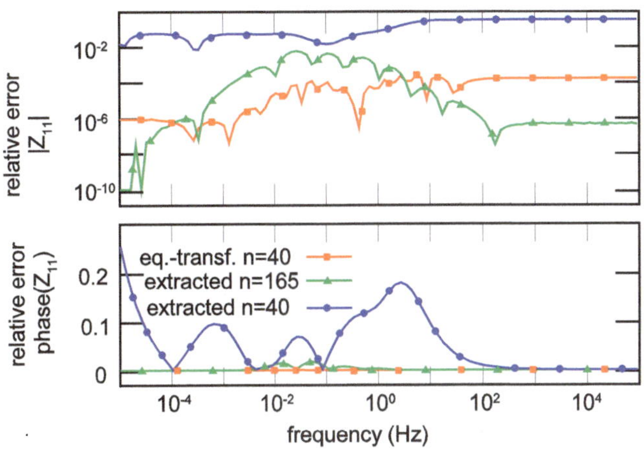

Fig. 7. Magnitude- and phase-error of transfer function Z_{11} 50x50 of the RCI-grid (MATLAB).

the equivalence-transformed network reduced with PMTBR (last line in Table 1) is synthesized with the method described in (Ludwig et al., 2008a). The resulting network is smaller (Table 2) than the original network and simulations are around ten times faster. As an example the good agreement of the network behavior at the pins can be seen for a simulation of the voltage fluctuation at pin1 in Fig. 8.

4 Conclusions

In this paper, an efficient reduction process of networks with a large number of PWL-sources is presented. The former developed method of an equivalence-transformation instead of extracting sources is extended to PWL-waveforms which are decomposed into basis-functions. The validity and efficiency of the proposed method is shown by reducing a RCI-Grid model. With the proposed reduction process a higher reduction and a higher accuracy can be achieved. This results in smaller and/or more accurate networks generated with the reduction process.

Table 2. Comparison of reduction results of the network.

	Original network	Reduced network
Network elements	7450	1685
Network nodes	2500	45
SPICE parsing time	18 s	0.9 s
SPICE simulation time	27 s	2.6 s
SPICE memory	16.7 mbytes	2.1 mbytes

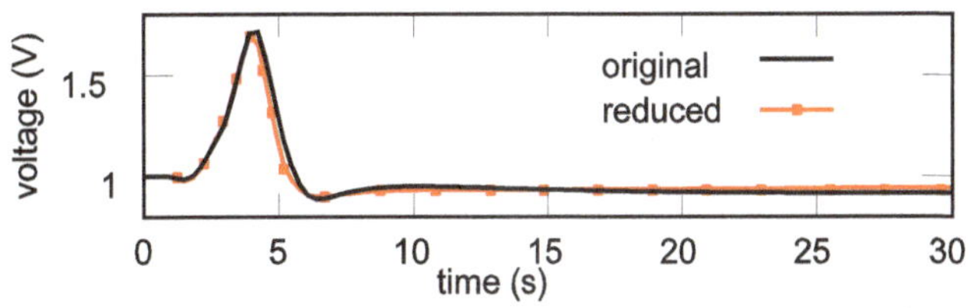

Fig. 8. Voltage drop at pin1 of the 50×50 RCI-grid (HSPICE).

Acknowledgements. S. Ludwig wants to acknowledge Lj. Radic-Weissenfeld and R. Kazemzadeh of the Institute of Electromagnetic Theory at the Leibniz University of Hannover for many useful discussions. Represented research and development work is carried out in the frame of the MEDEA+ project A701 PARACHUTE project This particular research was supported by the BMBF under 01M 3169 A, 01M 3169 D and 01M 3169 E. The responsibility for this publication is held by the authors only.

References

Aliprantis, C. D., Harris, D., and Tourkey, R.: Continous Piecewise Linear Functions, Macroeconomic Dynamics, 10(1), 77–99, 2005.

Antoulas, A. C.: Approximation of Large-Scale Dynamical Systems, Adv. Design Control, SIAM, 2005.

Feldmann, P.: Model order reduction techniques for linear systems with large number of terminals, DATE'2004 – Design, Automation and Test in Europe, Exhibition Conference, vol. 2, Paris, France, 2004.

Ionutiu, R., Lefteriu, S., and Antoulas, A. C.: Comparison of Model Reduction Methods with Applications to Circuit Simulation, Scientific Computing in Electrical Engineering SCEE 2006, Mathematics in Industry, Springer, vol. 11, 2007.

Ludwig, S., Radić-Weissenfeld, Lj., Mathis, W., and John, W.: Efficient Model Reduction of Passive Electrical Networks with a Large Number of Independent Sources, IEEE International Symposium on Circuits and Systems ISCAS, Seattle, USA, 2008a.

Ludwig, S., Radić-Weissenfeld, Lj., Mathis, W., and John, W.: Efficient passive network description of IC conducted emission models for model reduction, Adv. Radio Sci., 6, 133–137, 2008b, http://www.adv-radio-sci.net/6/133/2008/.

Ludwig, S., Radić-Weissenfeld, Lj., Mathis, W., and John, W.: Passive and Reciprocal Network Description of Independent

Sources for Efficient Model Reduction, International Conference on Signals and Electronic Systems – ICSES, Krakow, Poland, 2008c.

Odabasioglu, A., Celik, M., and Pileggi, L. T.: PRIMA: Passive Reduced- Order Interconnect Macromodeling Algorithm, IEEE Transactions on Computer-Aided Design of Integrated Circuits and Systems, 17(8), 645–654, 1998.

Palenius, T. and Roos, J.: Comparison of reduced-order interconnect macromodels for time-domain simulation, IEEE Transactions on Microwave Theory and Techniques, 52, 2240–2250, 2004.

Phillips, J. R. and Silveira, L. M.: Poor Mans TBR: A Simple Model Reduction Scheme, IEEE Transactions on Computer-Aided Design of Integrated Circuits and Systems, 24(1), 43–55, 2005.

Radić-Weissenfeld, Lj., Ludwig, S., Mathis, W., and John, W.: Model order reduction of linear time invariant systems, Adv. Radio Sci., 6, 129–132, 2008a, http://www.adv-radio-sci.net/6/129/2008/.

Silva, J. M. S., Villena, J. F., Flores, P., and Silveira, L. M.: Outstanding Issues in Model Order Reduction, Scientific Computing in Electrical Engineering, Springer, Berlin Heidelberg, Germany, 2007.

Yang, F., Zeng, X., Su, Y., and Zhou, D.: RLCSYN: RLC Equivalent Circuit Synthesis for Structure-Preserved Reduced-order Model of Interconnect, IEEE International Symposium on Circuits and Systems ISCAS, New Orleans, USA, 2007.

A MIMO FMCW radar approach to HFSWR

J. O. Hinz and U. Zölzer

Department of Signal Processing and Communications, Helmut-Schmidt-Universität/University of the Federal Armed Forces Hamburg, Holstenhofweg 85, 22043 Hamburg, Germany

Abstract. In this paper we propose one possible approach how to apply the concept of multiple-input multiple-output (MIMO) to monostatic Frequency Modulated Continuous Wave (FMCW) High-Frequency Surface Wave Radar (HFSWR) in a maritime environment. Common tasks for a HFSWR are sea-state monitoring and ship detection, where our focus is on ship detection.

A limiting factor in HFSWR is the available bandwidth, which is inversely proportional to the range resolution capability of the radar and typical below 100 kHz. The question is how to extend or combine a conventional single-input multiple-output (SIMO) FMCW phased-array type radar with stretch processing and the colocated MIMO concept to "reuse" the very limited HF radar band resources. Another important question to answer is how MIMO FMCW waveforms can be separated at the receiver.

1 Introduction

Typical HFSWR systems are used for sea-state monitoring or ship detection, operating in the HF radar frequency band of 3 to 30 MHz. These (quasi-)monostatic shore-based systems consist of a "floodlight" transmit antenna with a beamwidth of around 100° directed at the sea in conjunction with a receiver array of up to 16 elements (Gurgel et al., 1999). Depending on the operating frequency and assuming a receiver antenna element spacing of $\lambda/2$ the receive array can easily occupy several hundreds meters of coastline. Due to the highly conductive ocean sea surface in combination with a vertical polarization the waves propagate up to few hundreds of kilometres in a so-called ground-wave or surface-wave and experience less attenuation than conventional microwave radars. Due to the low frequency it even allows propagation along the curvature of the earth.

In case of ship detection the worst case is the detection of ships in sea clutter, otherwise detection capability is usually limited by the external noise level. Another important aspect of HFSWR is the high spectrum occupancy by other users, such as radio amateurs or radio stations. Thus it is difficult to obtain a radio frequency interference (RFI) free continuous bandwidth. This is the reason why some authors first monitor the frequency occupancy to select an RFI free or less occupied frequency band (Gurgel et al., 2007) to avoid RFI whenever possible. Other authors propose to improve target detection in sea clutter by using a dual frequency approach (Leong and Ponsford, 2004). Still both approaches will suffer from degraded performance in case the operating frequency does not match the receiver antenna element spacing.

Due to the high dynamic requirements in HFSWR most radar systems use a FMCW waveform in combination with stretch processing to limit the sampling requirements of the analog-to-digital converter (Richards, 2005). This is in contrast to most MIMO approaches where correlation processing is used. This makes it interesting to investigate how a potential MIMO FMCW system could look like and if it is possible to separate the different transmitted signals at the receiver. The rest of the paper will be structured as following. Section 2 is going to introduce the FMCW waveform and the signals at transmitter and receiver as well as stretch processing. After that Sect. 3 gives a deeper insight into HFSWR. Then the paper will proceed with MIMO radar in Sect. 4 and depict the combination of MIMO FMCW and HFSWR in Sect. 5. Finally a conclusion is drawn.

Correspondence to: J. O. Hinz
(jan.hinz@hsu-hamburg.de)

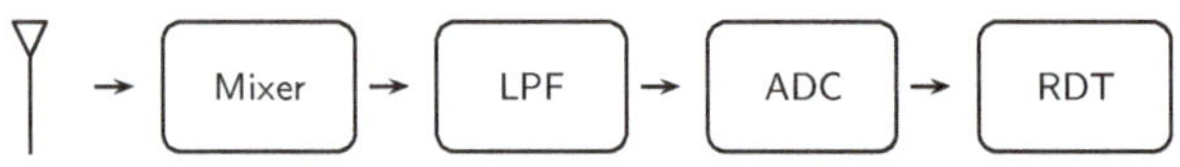

Fig. 1. FMCW receiver structure.

2 FMCW waveform and Stretch Processing

The FMCW waveform has three important advantages as compared to a pulse radar system, which are a constant power envelope, a high Doppler shift tolerance and the ability to use a simplified receiver processing called stretch processing. The main disadvantages are the signal leakage from the transmitter into the receiver and range-Doppler coupling. The following equation shows the mathematical description from (Meta et al., 2007) of a single chirp FMCW signal at the transmitter

$$s_{tx}(t) = \exp\left(j2\pi\left(f_s t + \frac{1}{2}\alpha t^2\right)\right), \tag{1}$$

in which t is the time-variable within the single chirp period, f_s denotes the start frequency of the chirp and α denotes the sweep-rate, which is defined as the ratio of the bandwidth of the chirp (B) divided by the chirp duration (T).

The received signal from a single target is an attenuated, delayed and Doppler-shifted version of the transmitted chirp. In the following only a single non-moving point target is assumed

$$s_{rx}(t) = A \cdot \exp\left(j2\pi\left(f_s(t-\tau) + \frac{1}{2}\alpha(t-\tau)^2\right)\right), \tag{2}$$

in which A is used to represent the attenuation factor and τ is the round-trip time for the signal to propagate from radar system to the target and back. A multiple target case can be seen as a superposition of the different echoes at the receiver.

A typical FMCW single channel receiver structure is illustrated in Fig. 1.

The illustrated mixer in Fig. 1 demodulates the received signal (attenuated, delayed and Doppler-shifted) with the transmitted signal, which is sometimes referred to as reference waveform. This results in a direct demodulation of the received signal into baseband. To eliminate the upper modulation products a low-pass filter is applied. This process is commonly known as stretch processing but sometimes also called deramp processing, deramp on receive, dechirp or one-pass processing (Richards, 2005). It is a technique to reduce the sampling-rate requirements of the following analogue-to-digital converter (ADC), followed by the range-Doppler transform (RDT). This reduction in sampling-rate is needed due to the high dynamic range. A graphical interpretation of stretch processing for one target can be found in Fig. 2, where the round-trip delay τ of the receive signal results in a difference frequency Δf, the so-called beat frequency. Stretch processing can be applied if the target round-trip delay τ is small as compared to the chirp duration T and if the beat signal frequency offset due to the delay is large as compared to the Doppler-shift due to the radial velocity of the targets.

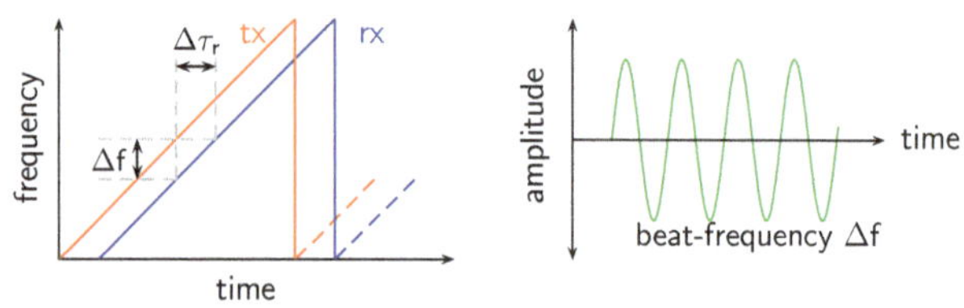

Fig. 2. Stretch processing and resulting beat signal.

In mathematical terms the beat signal of a static point target is expressed in the following equation

$$s_{if}(t) = A \cdot \exp\left(j2\pi\left(f_s\tau + \alpha t\tau - \frac{1}{2}\alpha\tau^2\right)\right). \tag{3}$$

The range resolution Δr and Doppler resolution Δf_D capabilities (McGregor et al., 1994) of the FMCW waveform are given by

$$\Delta r = \frac{c}{2B} \tag{4}$$

and

$$\Delta f_D = \frac{1}{NT}, \tag{5}$$

in which the parameter c and N are used to denote the speed of light and the number of chirps considered in one coherent integration time (CIT), respectively. Strongly related to the range resolution and the Doppler resolution is the range-Doppler transform (RDT). It is a 2-dimensional fast Fourier transform (FFT) used to produce the range-Doppler map (RD map). Sometimes the terms fast time and slow time are used to indicate that the first domain of the transform are the time samples within a single chirp (range transform), while the second domain is across the number of chirps (N) to form the Doppler transform. The purpose of the RDT can be seen as a set of two filter banks used to split the targets according to their range frequency and their Doppler frequency.

3 HFSWR

In addition to estimating the range and Doppler shift of a target one is usually also interested in determining the target's azimuth. Due to the limited number of antenna elements and the use of stretch processing in most HFSWRs this task can be performed by digital beamforming as in (Gurgel et al., 1999).

In Fig. 3 the block diagram of a typical SIMO FMCW HFSWR is illustrated. The abbreviation SP following each antenna element is used to indicate stretch processing, thus the combination of mixer and low-pass filtering as described

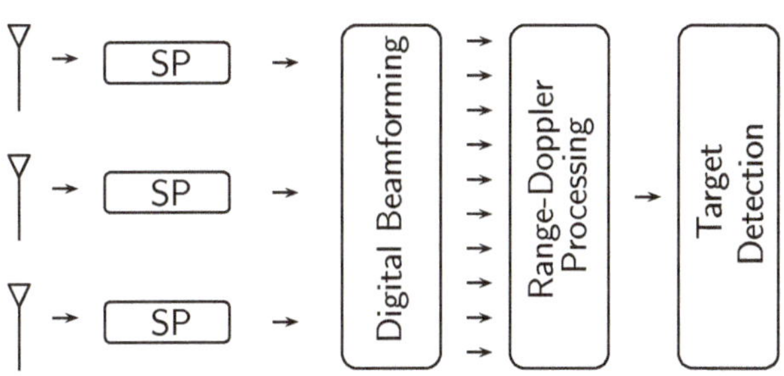

Fig. 3. SIMO FMCW HFSWR receiver structure.

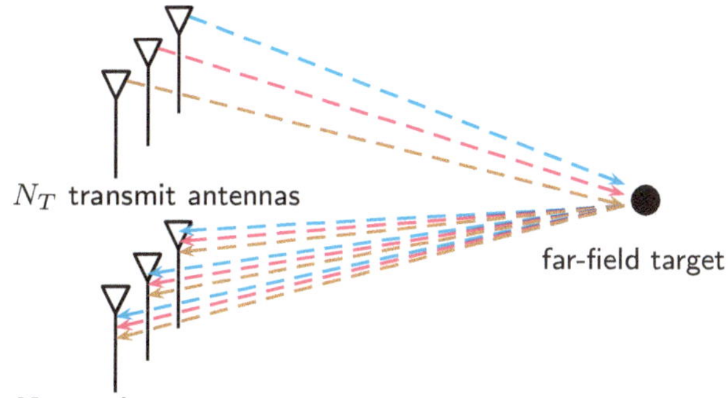

Fig. 4. Colocated MIMO Radar.

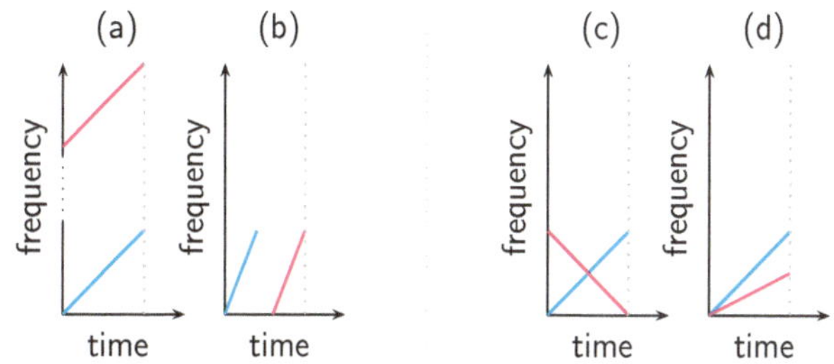

Fig. 5. MIMO FMCW – four possible signalling concepts.

in Sect. 2. The additional bandpass-filters, low-noise amplifiers and ADCs are omitted for simplicity. The low-pass filtered and digitalized baseband antenna elements signals are then fed into a digital beamformer, which in most cases is a Dolph-Chebychev beamformer used to produce a number of formed beams (M) in the angular domain of $-60°$ to $+60°$. For each of the beams a RDT is applied. The final part is the target detection (TD) in a 2- or 3-dimensional domain.

4 MIMO

The term MIMO has become a popular technique in communications and is used to increase the data rate or the reliability of a given link by the use of multiple antennas at the transmitter and the receiver side. Important here is the fact that in this approach all transmitters share the same frequency band, but transmit different signals. It is furthermore required that the receiver is able to separate the signals from each transmit antenna.

For the term MIMO radar one has to distinguish between colocated MIMO radar and statistical MIMO radar. What we consider here is the colocated MIMO radar case, which can be roughly characterized by the condition that the distance between the different transmitter/receiver elements is small as compared to the distance between the radar and the potential targets. The statistical MIMO radar assumes widely spaced transmitter and receiver elements to exploit spatial diversity. Still in both cases each transmitter element is assumed to transmit a different signal, so each receiver element receives a superposition of each transmitter element signal and each target as illustrated in Fig. 4 for the single target case in the far-field.

5 MIMO FMCW Radar Approach to HFSWR

The main question is how the concept of colocated MIMO radar can be combined with the linear FMCW waveform and receiver stretch processing in a typical HFSWR environment, possibly even added to an existing SIMO FMCW HFSWR. This implies the selection of appropriate FMCW waveforms to enable the receiver to separate them. In Fig. 5 four possible signalling concepts (a to d) for two transmit elements are shown.

The two first approaches (a and b) can be summarized as dual-frequency operation (Leong and Ponsford, 2004) and time-staggered operation (Frazer et al., 2007). The first approach enables the receiver to separate the different transmitter FMCW waveforms due to an offset in frequency or in time. The main disadvantage of the dual-frequency operation or even multiple-frequency operation is that with increasing frequency separation between the two transmitted waveform the receiver antenna-element spacing becomes suboptimal for receive beamforming and will either result in a small aperture or the introduction of grating lobes. In addition there are additional licenses required to operate at two or more frequency bands.

The time-staggered approach (b) is an option, where the receiver is able to distinguish the different transmitter elements by their time offset. Still to enable the receiver to distinguish the different waveforms the time offset between the elements has to be larger than the maximum round-trip time τ.

The third option (c) is to use two FMCW waveforms with opposite slope, occupying the same frequency band and having the same time duration. In approach (d) FMCW waveforms with the same start frequency and same duration but different bandwidth and thus different slopes are considered. The RF chain and the antenna element spacing are thus optimal for both last mentioned waveforms. Furthermore no additional licence is required. The main disadvantage of ap-

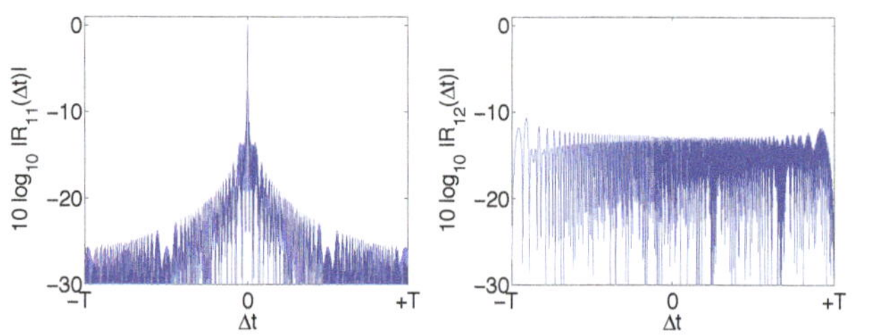

Fig. 6. Normalized auto- and crosscorrelation for opposite slope MIMO FMCW approach (c).

proach (d) is the worsening of the range resolution of the second waveform. Even if in Fig. 5 approach (d) only two different transmit waveforms are shown the approach can easily be extended to more than two waveforms. In the following we are going to investigate the possible application of the MIMO FMCW approach (c) and (d).

5.1 Correlation properties

In this section the autocorrelation and crosscorrelation properties of the waveforms are being investigated. The normalized auto- and crosscorrelation of the opposite slope approach (c) are presented in Fig. 6. Here R_{11} is used to denote the normalized autocorrelation of the first waveform while R_{12} indicates the normalized crosscorrelation of waveform one and waveform two. Due to symmetry same applies to the comparison of R_{22} and R_{21}, which is not shown here.

Good correlation properties can be characterized by a distinct peak in the autocorrelation function and a low interference level in the crosscorrelation function. In case a correlation-processor based range resolution is used the correlator output with the return of one or several targets will result in one or more distinct peaks.

The detection of these targets is thus limited by the level of noise and clutter as well as additional crosscorrelation residues from other waveforms. In the worst case the additional crosscorrelation will even result in additional peaks in correlation output, which could be misinterpreted as targets. In the presented example of Fig. 6 with two chirps of opposite slopes (c) the signals show good autocorrelation as well as good crosscorrelation properties. Thus the peak of the autocorrelation and average crosscorrelation level differ from each by a level of 14 dB, with no windowing applied. Approach (d) also offers good correlation properties, even if the maximum crosscorrelation level is higher and is not homogeneously distributed. This can be explained by the fact that the instantaneous frequencies over time vary linear in the same direction of higher frequencies. In general the correlation properties are a good indicator if a particular waveform set is suitable for a MIMO approach.

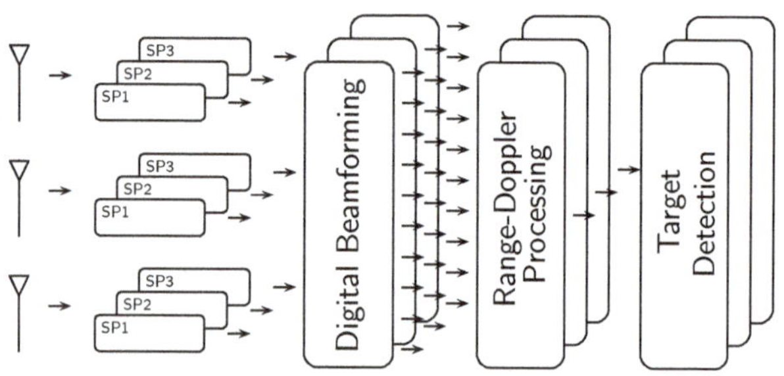

Fig. 7. HFSWR MIMO FMCW receiver structure.

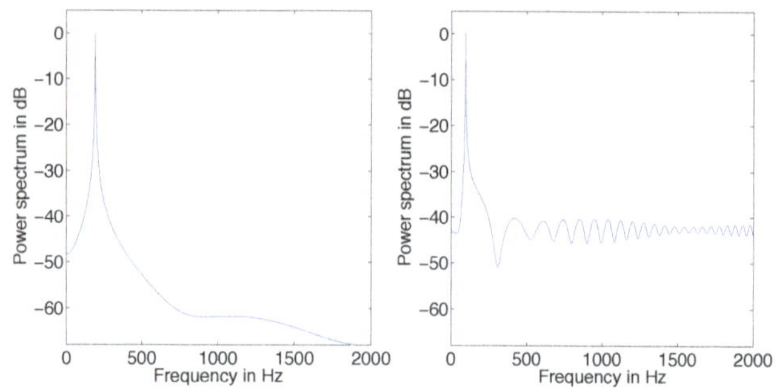

Fig. 8. Stretch processed FMCW for SIMO and MIMO.

5.2 Proposed receiver structure

The next step is to investigate if these good properties are remained for the case of stretch processed FMCW. To evaluate this a potential MIMO receiver structure is presented in Fig. 7, which can be seen as an extension of Fig. 3.

Here the abbreviations SP1 to SPN_T are used to denote stretch processing at each receiver element with the N_T respective reference transmit waveforms, where in the illustrated example the number N_T of transmit waveform is chosen to be 3.

Following the stretch processing are several beamformer, where each one operates across all antenna elements but only takes one particular stretch processed output. After the beamforming the data is range-Doppler transformed, followed by the target detection. Thus the number of receive chains scales with the number of transmit waveforms N_T. This receiver structure is highly motivated by the goal of a potential extension of an existing SIMO system. More advanced receiver concepts might be investigated in the future.

5.3 SIMO and MIMO stretch processing

To show the difference in stretch processing of SIMO and MIMO FMCW signals Fig. 8 is presented. It contains the received, demodulated, low-pass filtered and range transformed signal of a single target for the SIMO (only transmitter with 100 kHz bandwidth active) and the MIMO case (transmitters with 100 kHz and 50 kHz bandwidth active) of

MIMO FMCW approach (d). The formal description of the respective signals can be found as part of Sect. 2. The transmitted waveforms differ by their bandwidth but share the same start frequency f_s and the same time duration T of 0.5 s.

If one compares the two parts of Fig. 8 one can see that for the MIMO case the interference level of the additional waveform is increased to a level of −40 dB. This can be easily explained by the fact that the lower mixing products of two chirps with linear but different slope do not result in one constant frequency difference but in a distribution across the presented frequency range. Still it is clearly possible to determine the characteristic frequency peak of the beat signal of the given target. Same applies if one looks at the other waveform, where the different slope α results in a different beat frequency of the target. Still one has to keep in mind that the range resolution capability of the 50 kHz waveform is worsened as compared to the 100 kHz waveform. By applying range-Doppler transform and target detection independently for both waveforms the target's range and radial velocity can be estimated. Thus both detection results can be used to reassure the existence of a potential target.

6 Conclusions

It has been shown that a stretch processed MIMO FMCW approach in HFSWR is possible, in which the transmitter elements operate in the same or partially same frequency band simultaneously and thus the frequency band is reused. This is an important point in HFSWR as a "free" continuous frequency bandwidth is difficult to find.

The presented approach can be seen as an alternative to the dual frequency or the time staggered concepts. Approaches to avoid radio frequency interference and selection of the cleanest frequency can be incorporated into the concept, still limited by the static receiver interelement antenna distance. Additional simulations including attenuation factors and sea clutter and investigation about the allowed number of active transmitters should be performed.

The proposed receiver structure allows the modification of existing SIMO FMCW HFSWR systems and enables the reassurance of a potential target detection. Another modification might be to consider a distributed transmitter system such as a bistatic or multistatic system as the space occupied by an additional "floodlight" transmit antenna is not the limiting factor and is small as compared to the receiver array.

References

Frazer, G., Abramovich, Y., and Johnson, B.: Spatially Waveform Diverse Radar: Perspectives for High Frequency OTHR, in: IEEE Radar Conference 2007, 385–390, doi:10.1109/RADAR.2007.374247, 2007.

Gurgel, K. W., Essen, H. H., and Kingsley, S. P.: High-frequency radars: physical limitations and recent developments, Coastal Engineering, 37, 201–218, doi:10.1016/S0378-3839(99)00026-5, 1999.

Gurgel, K.-W., Barbin, Y., and Schlick, T.: Radio Frequency Interference Suppression Techniques in FMCW Modulated HF Radars, in: OCEANS 2007 – Europe, 1–4, doi:10.1109/OCEANSE.2007.4302289, 2007.

Leong, H. and Ponsford, A.: The Advantage of Dual-Frequency Operation in Ship Tracking by HF Surface Wave Radar, in: RADAR 2004 – Toulouse, France, 2004.

McGregor, J., Poulter, E., and Smith, M.: Switching system for single antenna operation of an S-band FMCW radar, IEE Proceedings - Radar, Sonar and Navigation, 141, 241–248, doi:10.1049/ip-rsn:19941231, 1994.

Meta, A., Hoogeboom, P., and Ligthart, L.: Signal Processing for FMCW SAR, IEEE Transactions on Geoscience and Remote Sensing, 45, 3519–3532, doi:10.1109/TGRS.2007.906140, 2007.

Richards, M. A.: Fundamentals of Radar Signal Processing, McGraw-Hill, 1. edn., 2005.

Fast beampattern evaluation by polynomial rooting

P. Häcker, S. Uhlich, and B. Yang

Chair of System Theory and Signal Processing, Universität Stuttgart, Pfaffenwaldring 47, 70550 Stuttgart, Germany

Abstract. Current automotive radar systems measure the distance, the relative velocity and the direction of objects in their environment. This information enables the car to support the driver.

The direction estimation capabilities of a sensor array depend on its beampattern. To find the array configuration leading to the best angle estimation by a global optimization algorithm, a huge amount of beampatterns have to be calculated to detect their maxima. In this paper, a novel algorithm is proposed to find all maxima of an array's beampattern fast and reliably, leading to accelerated array optimizations. The algorithm works for arrays having the sensors on a uniformly spaced grid. We use a general version of the gcd (greatest common divisor) function in order to write the problem as a polynomial. We differentiate and root the polynomial to get the extrema of the beampattern. In addition, we show a method to reduce the computational burden even more by decreasing the order of the polynomial.

1 Introduction

A common problem in signal estimation theory is the estimation of objects' direction (DOA, direction of arrival) using an array of sensors. In far range automotive radar systems, we are facing narrow-band and far field conditions. In addition, a Pulse-Doppler or frequency modulated continuous wave radar deals most of the time with the single object case, because all objects are first separated in range and velocity. Only objects with the same range and velocity have to be separated in DOA. We consider the case of azimuth estimation only since it is much more important than elevation in automotive applications. Together with the distance estimates, the azimuth angle determines the position of the relevant object uniquely relative to the vehicle. Using this information, the car can act in an intelligent way. Two examples of this behavior are ACC (Jurgen, 2006) (Adaptive Cruise Control) and LCA (Ruder et al., 2002) (Lane Change Assistant).

For mass production, a sensor array has to be cheap. Hence the number of sensor elements is limited. Nevertheless, the DOA estimates have to be as accurate as possible. For a fixed DOA estimation algorithm, the positions of the sensors have to be optimized, to get the best DOA estimates according to a suitable criterion.

A frequently used optimization criterion is the Cramr-Rao bound (Athley et al., 2004), which is a lower bound on the variance of any unbiased estimator. The Cramr-Rao bound only takes local DOA errors into account (Athley, 2005). Local errors are errors smaller than half the size of the array's main lobe. Unfortunately, with a cheap sensor and thus few snapshots and low SNR, global errors are not negligible. Athley (2005) approximated the global errors analytically and calculated an error probability per side lobe of the so called beampattern. This allows the analytical calculation of the variance of DOA estimates, which can be used as a cost function for array optimization. A typical beampattern is shown in Fig. 1.

This approach has two shortcomings: First, the maximum of each side lobe has to be computed numerically. So one needs a fast and reliable algorithm to find all local maxima of the beampattern. Second, the variance function does a weighting preferring smaller errors. Having small local errors is obviously useful, but at least in automotive applications, having a DOA error of 10° is not better than having an error of 20°. Larger errors can even be advantageous, as they are easier to classify as outliers and to reject over time by using tracking.

These considerations lead to a minimization of global errors, where only the height of the side lobes is important and not their position. Because of the huge gradient in Athley's error probability (Fig. 2) it is reasonable to assume, that minimizing the maxima of the side lobes will lead close to the

Correspondence to: P. Häcker
(patrick.haecker@lss.uni-stuttgart.de)

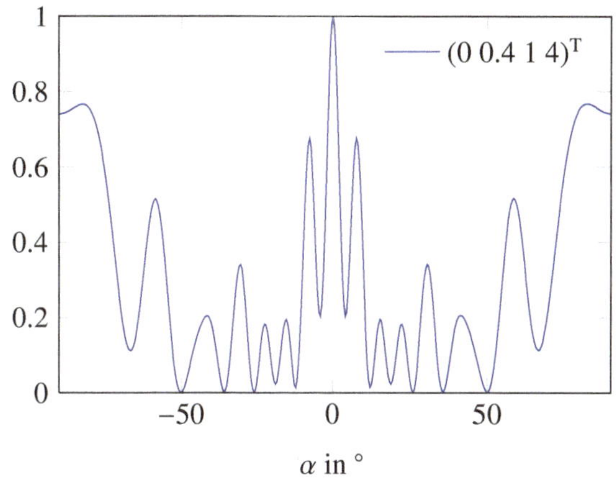

Fig. 1. Beampattern with main lobe at $\alpha = 0$ and several side lobes.

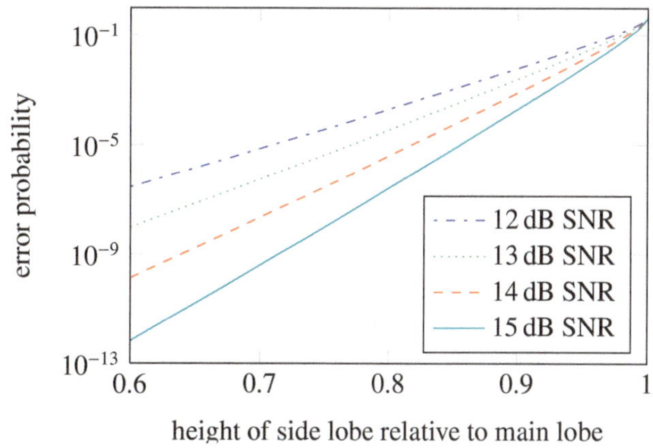

Fig. 2. Error probability for single side lobe.

optimum, as only the highest side lobe has a significant error probability. The error probability in Fig. 2 is derived for white noise. If this condition is violated by clutter or another object, the error probability is much higher.

The criterion has to be calculated for a huge number of arrays during the array optimization. In this paper, we will not investigate which global optimization algorithm to use. Instead we focus on the fast calculation of beampatterns. We introduce the beampatterns in Fig. 2 and derive needed mathematics in Sect. 3. The fast evaluation of the cost functions are shown in Sect. 4, having Eq. (5) as the most important equation. We present improvements of this basic algorithm for our specific problem in Sect. 5. We show some results in Sect. 6 and conclude our work in Sect. 7.

2 Beampatterns

We assume that all N sensors differ only in their positions having identical (e.g. omnidirectional) characteristics. If their characteristics differ only slightly, the following results will be approximately be true and are still of value. If the characteristics are completely different, the proposed algorithm cannot be used.

As we want to estimate only the azimuth angle, the sensors can all be placed along a horizontal line orthogonal to the driving direction. Without loss of generality, we choose the first sensor of the array as the reference element at position 0. Let $\boldsymbol{p} = (p_1 \ \dots \ p_N)^{\mathrm{T}}$ be a vector containing the positions of the sensors normalized by the wavelength. For the following algorithm, $\boldsymbol{p}$ may only contain rationals, possibly after factorization of a common irrational factor. The steering vector $\boldsymbol{a}(u)$ of length N can then be written as

$$\boldsymbol{a}(u) = \frac{1}{\sqrt{N}}\left(1 \ \mathrm{e}^{\mathrm{j}2\pi p_1 u} \ \dots \ \mathrm{e}^{\mathrm{j}2\pi p_N u}\right)^{\mathrm{T}}, \tag{1}$$

with $u = \sin(\alpha)$, where α is the broadside azimuth angle.

Based on those assumptions, we can derive a general beampattern $b(u; u_0, \boldsymbol{y})$ by correlating the steering vector $\boldsymbol{a}(u)$ with that for a DOA value u_0:

$$b(u; u_0, \boldsymbol{y}) = \left|\boldsymbol{a}^{\mathrm{H}}(u_0)\boldsymbol{a}(u)\right|^2 = \frac{1 + \sum_{k=1}^{N}\left(\mathrm{e}^{\mathrm{j}2\pi p_k(u-u_0)} + \mathrm{e}^{-\mathrm{j}2\pi p_k(u-u_0)} + \sum_{m=1}^{N}\mathrm{e}^{\mathrm{j}2\pi(p_k-p_m)(u-u_0)}\right)}{N^2} \tag{2}$$

u can be seen as the direction of the object and u_0 as the test direction. Using Eq. (2), the beampatterns needed for Athley (2005) as mentioned in the introduction can be calculated.

The correlation of the steering vectors corresponds to the similarity of the array responses from the appropriate directions. The Similarity of steering vectors can lead to ambiguities in the angle estimation. Note, that Eq. (2) can be written as a one-dimensional function, as all values only depend on the difference of u and u_0. Two special cases are exceptionally important.

In automotive and most other applications, the array should often uniquely differ an object in front ($u_0 = 0$) from every other angle. This is useful for applications like ACC. The beampattern becomes then

$$b(u; 0, \boldsymbol{y}) = \frac{1 + \sum_{k=1}^{N}\left(\mathrm{e}^{\mathrm{j}2\pi p_k u} + \mathrm{e}^{-\mathrm{j}2\pi p_k u} + \sum_{m=1}^{N}\mathrm{e}^{\mathrm{j}2\pi(p_k-p_m)u}\right)}{N^2} \tag{3}$$

Sometimes every array response from one angle should differ from the response from all other angles. This is especially important in applications like LCA. This condition can be formulated as

$$b(u; -u, \boldsymbol{y}) = \frac{1 + \sum_{k=1}^{N}\left(\mathrm{e}^{\mathrm{j}4\pi p_k u} + \mathrm{e}^{-\mathrm{j}4\pi p_k u} + \sum_{m=1}^{N}\mathrm{e}^{\mathrm{j}4\pi(p_k-p_m)u}\right)}{N^2}. \tag{4}$$

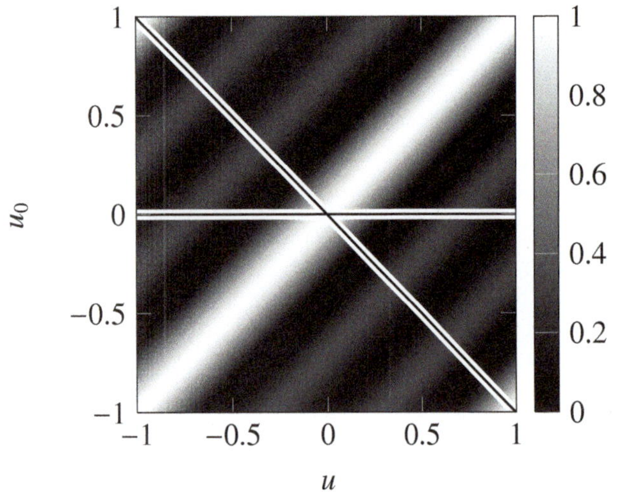

Fig. 3. Possible beampatterns and used ones.

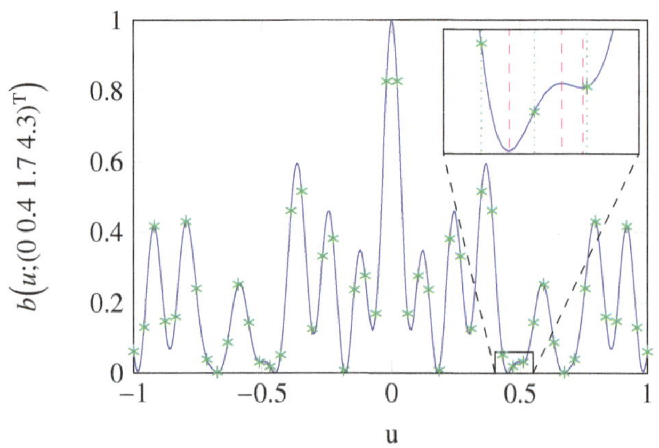

Fig. 4. Beampattern having closely spaced extrema.

The result of Eq. (4) can also be achieved by letting $u \to 2u$ in Eq. (3). In Fig. 3 all possible beampatterns according to Eq. (2) are shown. The horizontal line denotes the beampattern resulting from Eq. (3). All maxima are included in the diagonal line following Eq. (4).

We want to have the extrema of a continuous signal, i.e. their positions, their types and their values in a specific interval. A simple method would be sampling the continuous signal by calculating values in uniform distance u_Δ, determining coarse extrema by comparison of neighbor samples and performing a local optimization on each coarse extremum. The problem with this approach is that two extrema can be very tight. This happens when two extrema converge to a saddle point by changing the sensors' positions. In Fig. 4 an example of extrema (magenta, dashed) being closer than the sampling distance (green, dotted) is shown. That's why the sampling frequency must be much higher than $1/u_\Delta$ to make only few errors. Alternatively, we could do a local optimization starting from every sample, even though this is definitely slow. Although most maxima are found with that approach, there are some exceptions. This can be rapidly verified with Eq. (4) and $\boldsymbol{p} = (0\ 0.5\ 11)^T$ around $u = 0.5$. Thus a sampling algorithm cannot be fast and free of errors at the same time.

3 Generalized gcd

Before having a closer look into the algorithm we need to generalize the gcd and therefore the lcm. The classical gcd is the greatest common divisible *integer* of *two integers*, whereas the classical lcm is the least common multiple *integer* of *two integers*. The lcm can be computed efficiently with Euclid's or Stein's algorithm (Cormen et al., 2001). The gcd can be calculated using the lcm (Rosen and Michaels, 2000). We now want to generalize the emphasized terms.

3.1 gcd/lcm of multiple integers

Let us generalize the gcd of multiple integers i_1, i_2, i_3, i_4 (lcm analog):

$$\gcd(i_1, i_2, i_3, i_4) = \gcd(\gcd(i_1, i_2), \gcd(i_3, i_4))$$

This effectively transforms the problem of multiple integers into the recursive gcd calculation of two integers. The gcd of a vector $\boldsymbol{p}$ is $\gcd(\boldsymbol{p}) = \gcd(p_1, \ldots, p_{N-1})$

3.2 gcd of two rationals

We want to use gcd with rationals r_1, r_2. Note, that all fixed and floating point numbers in a computer are rationals, so this is a very general case from a practical point of view. We will define the generalized gcd by giving an algorithm to compute it. The algorithm consists of three steps:

1. Convert the floating-/fixed-point numbers into fractions e.g. by regular continued fraction expansion or generalized continued fraction expansion (Rockett and Szsz, 1992):

$$r_1 = \frac{n_1}{d_1},\ r_2 = \frac{n_2}{d_2}$$

2. Expand the fractions from step 1 so they have a common denominator, by using the classical lcm (preferably with intermediate results from continued fractions):

$$r_1 = \frac{(\mathrm{lcm}(d_1,d_2)n_1)/d_1}{\mathrm{lcm}(d_1,d_2)},\quad r_2 = \frac{(\mathrm{lcm}(d_1,d_2)n_2)/d_2}{\mathrm{lcm}(d_1,d_2)}$$

3. Calculate the classical gcd of the numerators and divide by the common denominator from step 2:

$$\gcd(r_1, r_2) = \frac{\gcd((\mathrm{lcm}(d_1,d_2)n_1)/d_1, (\mathrm{lcm}(d_1,d_2)n_2)/d_2)}{\mathrm{lcm}(d_1,d_2)}$$

As an example, it is

$$\operatorname{gcd}(0.1,0.25)=\operatorname{gcd}\left(\frac{1}{10},\frac{1}{4}\right)=\operatorname{gcd}\left(\frac{2}{20},\frac{5}{20}\right)=\frac{\operatorname{gcd}(2,5)}{20}$$
$$=\frac{1}{20}=0.05.$$

During the paper, the generalized gcd function is used with an arbitrary number of rationals.

4 Proposed algorithm

The suggested algorithm consists of the following steps:

1. Representing the problem as a polynomial
2. calculate roots subject to constraints in z-domain
3. inverse substitution
4. shifting, constraining or continuation of interval
5. classification and calculation of extrema

We will go into more detail in the following corresponding subsections.

4.1 Representation as polynomial

By calculating the gcd of all positions, we get the position step size $P=\operatorname{gcd}(\boldsymbol{p})$ to build a grid, where every position is onto. Furthermore we see, that if every position is on a grid with the calculated step size, every difference must be, too, which will be important in Sect. 2. This is because $\operatorname{lcm}(p_i,p_j)=\operatorname{lcm}(p_i,|p_i-p_j|)$ having $i\neq j$.

By substituting

$$n_k=\frac{p_k}{\operatorname{gcd}(\boldsymbol{p})}=\frac{p_k}{P}\qquad n_k\in\mathcal{N}$$

into Eq. (4), we get

$$b(u)=\frac{1}{N^2}\left(1+\sum_{k=1}^{N}\left(\mathrm{e}^{\mathrm{j}4\pi n_k Pu}+\mathrm{e}^{-\mathrm{j}4\pi n_k Pu}+\sum_{m=1}^{N}\mathrm{e}^{\mathrm{j}4\pi(n_k-n_m)Pu}\right)\right).$$

Derivating $b(u)$ leads to

$$\frac{\mathrm{d}b(u)}{\mathrm{d}u}=\frac{\mathrm{j}4\pi P}{N^2}\sum_{k=1}^{N}\left(n_k\mathrm{e}^{\mathrm{j}4\pi n_k Pu}-n_k\mathrm{e}^{-\mathrm{j}4\pi n_k Pu}\right.$$
$$\left.+\sum_{m=1}^{N}(n_k-n_m)\mathrm{e}^{\mathrm{j}4\pi(n_k-n_m)Pu}\right).$$

Doing another substitution $z=\mathrm{e}^{\mathrm{j}4\pi Pu}$ results in

$$f(z)=\frac{\mathrm{j}4\pi P}{N^2}\sum_{k=1}^{N}\left(n_k z^{n_k}-n_k z^{-n_k}+\sum_{m=1}^{N}(n_k-n_m)z^{n_k-n_m}\right).$$

Note, that $f(z)$ is a polynomial in z as all exponents are natural numbers. Doing a similar substitution without using the generalized gcd would not lead to natural numbers in the exponents and thus not to a polynomial. As $f(z)$ is a polynomial, it can be written as

$$f(z)=\sum_{l=-L}^{L}c_l z^l \tag{5}$$

using $L=\frac{\max(n_k)}{\operatorname{gcd}(\boldsymbol{p})}$ and the coefficients $\boldsymbol{c}$, which must be calculated.

4.2 Constrained rooting

The polynomial $f(z)$ has to be rooted to find the extrema. Several polynomial root finding algorithms having different properties exist, which is one of the main advantage of this new approach: The problem of global optimization is transformed into the more mature area of polynomial rooting algorithms. An advantage of such a rooting algorithm is the knowledge when the algorithm can terminate due to the fundamental theorem of algebra.

As $u\in\mathcal{R}$ it follows from the previous substitution, that $|z|=\left|\mathrm{e}^{\mathrm{j}4\pi Pu}\right|=1$. Therefore, we have to remove all roots subject to $|z|\neq 1$. This can, of course, also be done by a problem adjusted rooting algorithm. Due to the square in Eq. (2) all zeros occur in complex conjugate pairs. If there is an adjusted rooting algorithm available this knowledge should be used.

4.3 Inverse substitution

The z-substitution has to be inverted, as we are interested in the extrema positions in u and not in z. Normally one would calculate the complex logarithm to do this, but because of the constraint, calculating the arguments of the positions of the zeros in the z-plane is identical:

$$\mathrm{e}^{\mathrm{j}4\pi Pu_0}=0\Leftrightarrow u_0=\frac{\arg(u_0)}{4\pi P}$$

The denominator P scales the arguments from $[0,\ 2\pi)$ or $[-\pi,\ \pi)$ to $[0,\ \frac{1}{2P})$ or $[-\frac{1}{4P},\ \frac{1}{4P})$ so that the unit circle is mapped to the interval size $\frac{1}{P}$. Zeros at the interval limit have to be assigned to the correct end of the interval depending on the problem. They might also be duplicated, if the inclusive interval is of interest.

4.4 Interval adjustments

Depending on the calculated section of the period, there has to be a shift of the zeros. The shift is a circular shift because of the periodicity of the function and thus the extrema.

It can happen, that the periodicity is larger than the interested interval size depending on P and the limits of u. If that happens, the interval must be constrained discarding the values outside the interval.

Alternatively, if periodicity is smaller than the interval of interest, the extremas' values must be duplicated by periodic continuation. In general the number of continued periods is not an integer, so there must be special care on the last continuation or another constrain after the continuation.

4.5 Classification and calculation of extrema

After the interval adjustments, we know the position of all critical points, i.e. extrema or saddle points. Theoretically an arbitrary number of saddle points could be between two extrema. Thus we have to calculate all values and have to classify them by sign change of the differences, as a maximum must have a positive gradient to the left and a negative one to the right.

To save computational time we can neglect the existence of saddle points, so maxima and minima alternate. Thus as the functions in Sect. 2 always have a maximum at $u=0$, we know the positions of all maxima and only have to calculate these values. Saddle points are unlikely, but not entirely impossible, so the negligence can introduce errors.

5 Algorithm improvements

5.1 Polynomial long division

The order of the polynomial to be rooted can be decreased by polynomial long division. Therefore, some roots have to be known a priori. In Eqs. (2) and (4) we have always a zero-frequency component. As these equations only contain cosines, which are symmetric, there must be always an extrema at $u=0$ and thus a zero in the derivative. These transform to zeros at 1 in the z-plane. Using the generalized gcd function, we select the interval to include the highest frequency at the interval limit, leading to a zero at $z=-1$. Hence,

$$f(z)=(z-1)(z+1)g(z), \tag{6}$$

where $g(z)$ is a new polynomial with order $2L-2$.

5.2 Using symmetry to calculate coefficients

To calculate the coefficients c_l the symmetry of the beampattern $b(u;0,\mathbf{y})$ or $b(u;-u,\mathbf{y})$ can be used

$$b(u;-u,\mathbf{y})=\frac{1+2\sum\limits_{k=1}^{N}\left(\cos(4\pi p_k u)+\sum\limits_{m=1}^{k-1}\cos(4\pi(p_k-p_m)u)\right)}{N^2}. \tag{7}$$

As every cosine covers two exponentials, only about half the number of coefficients have to be calculated.

5.3 Using symmetry to reduce rooting order

The same idea can be used to reduce the order of the polynomial which has to be rooted. This is useful if no special rooting algorithm is used in Sect. 4.2. As this case is probably more likely, in the following another approach is shown to get a similar effect. We rewrite the problem as a polynomial in cosines instead of exponentials to approximately half the order of the polynomial.

Derivating Eq. (7) basically leads, because of the antisymmetric coefficients, to a weighted sum of sines

$$f(z)=-8\pi P\sum_{l=1}^{L} l c_l \sin(l4\pi P u)=-8\pi P\sum_{l=1}^{L}\gamma_l\sin(lx).$$

We use an addition theorem (Bronshtein et al., 2004) and rewrite the sines to get only terms of $\sin(x)$. We separate one sine to have only even multiplicities in the sum:

$$\begin{aligned}\sin(lx)&=\sum_{k=1}^{\left\lceil\frac{l}{2}\right\rceil}(-1)^{k+1}\binom{l}{2k-1}\sin(x)^{2k-1}\cos(x)^{l-2k+1}\\&=\sin(x)\left(\sum_{k=1}^{\left\lceil\frac{l}{2}\right\rceil}(-1)^{k+1}\binom{l}{2k-1}\sin(x)^{2k-2}\cos(x)^{l-2k+1}\right)\end{aligned} \tag{8}$$

Applying Pythagoras we can write even multiplicities of sines as cosines. Using the binomial equation (Bronshtein et al., 2004) the products of $\sin(x)$ can be written as a sum of products of $\cos(x)$ as

$$\sin(x)^{2k-2}=\left(1-\cos(x)^2\right)^{k-1}=\sum_{m=0}^{k-1}\binom{k-1}{m}(-1)^m\cos(x)^{2m}.$$

Using this result and the cosines from Eq. (8) leads to cosines only in the sums and the derivation of b can be written as

$$\begin{aligned}f(z)=-8\pi P\sin(4\pi P u)&\left(\sum_{l=1}^{L} l c_l\sum_{k=1}^{\left\lceil\frac{l}{2}\right\rceil}(-1)^{k+1}\binom{l}{2k-1}\right.\\&\left.\cdot\sum_{m=0}^{k-1}\binom{k-1}{m}(-1)^m\cos(4\pi P u)^{2m+l-2k+1}\right).\end{aligned} \tag{9}$$

Dividing Eq. (9) by $\sin(4\pi P u)$, the function can be interpreted as a polynomial using $z=\cos(4\pi P u)$. The division by the sine removes the zeros at $z=\pm1$, so this approach already includes the polynomial division from Sect. 5.1. Rooting that polynomial is less effort than rooting Eq. (5), because the order has been halved, as the order of z is $2(k-1)+L-2k+1=L-1$. The "-1" follows from the division of the sine.

The rest of the algorithm stays the same with two notable exceptions. First, the constraint is $\Im(z_0)=0\wedge|z_0|\leq 1$. Second, the inverse substitution has to be adjusted to

$$\cos(4\pi P u_0)=0\Leftrightarrow u_0=\frac{\arccos(z_0)}{4\pi P}.$$

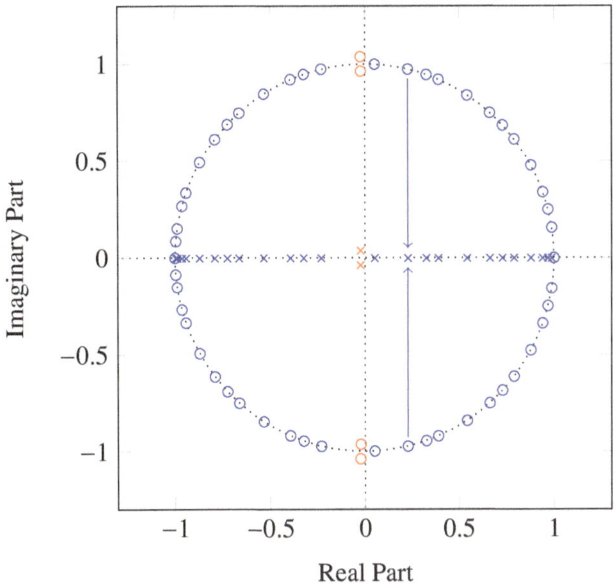

Fig. 5. Roots of a typical polynomial.

6 Results

It's interesting to note, that due to the various differences in Eqs. (3) and (4), coefficients can have components from more than one position in Eq. (5). An example would be a ULA (Uniform Linear Array). On the other hand, MRAs (Moffet, 1968) (Minimum Redundancy Array) consisting of optimum Golomb rulers (Tavares et al., 2005) have no frequency twice. So the multiplicity of frequencies is a measure of the redundancy of the array.

In Fig. 5 the zeros of the polynomial belonging to the beampattern – according to Eq. (3) – of the array with $\boldsymbol{p} = (0\ 0.1\ 0.2\ 0.3\ 1.4\ 1.5\ 2.6\ 2.7)^{\mathrm{T}}$ are shown. Both the basic algorithm described in Sect. 4 and the improved algorithm from Sect. 5 are depicted in the figure. The zeros of the basic algorithm are drawn as circles. The red roots have to be discarded due to the rooting constraint. The crosses are the roots of the optimized algorithm. Again, the roots which do not satisfy the constraint are shown in red. Every complex conjugate pair of the circular roots from the basic algorithm results in one root from the improved algorithm as a projection on the real axis.

A comparison between a sampling algorithm and a rooting algorithm strongly depends on the used algorithms. Nevertheless some general facts can be stated.

- The order of the polynomial, which must be rooted, is given by $2\frac{p_{N-1}}{\gcd(\boldsymbol{p})}$ with the general algorithm. The optimized algorithm leads to an order of $\frac{p_{N-1}}{\gcd(\boldsymbol{p})} - 1$.
- As the rooting is the dominant part, the algorithm is independent of the number of sensors for constant $\gcd(\boldsymbol{p})$.

Table 1. Polynomial orders of arrays for general and improved algorithm.

Antenna positions	Order	Reduced Order
0, 0.5, 1.5	6	2
0, 0.4, 1.5	30	14
0, 1.3, 2.05	82	40
0, 0.4, 1.7, 4.3	86	42
0, 0.1, 0.2, 0.3, 1.4, 1.5, 2.6, 2.7	54	26
0, 0.78, 2.44, 5.32	532	–

- Keeping the gcd constant, the polynomial scales linearly with the aperture.
- Scaling the whole array does not change the algorithm's efficiency.
- The algorithm is slow if the sensor positions are nearly relatively prime. As manufacturing tolerances limit the positional precision, only allowing positions on a grid with a grid size in the magnitude of the production tolerances does not add an additional constraint.
- Most of the roots of the polynomials satisfy the constraints and lead to extrema, as seen in Fig. 5.
- The calculation of the binomial coefficients in Eq. (9) can be time consuming for larger polynomial orders, reducing the rooting time benefits. Precalculating the binomials and saving them into a lookup table avoids this additional effort. Nevertheless for very large polynomial orders (> 50), some binomial coefficients get too large resulting in numerical problems. This can be seen in Fig. 5 as well, as the roots at about 1 and −1 get very close due to the negligence of the imaginary part.

Some examples of sensor positions and polynomial orders are given in Table 1. The order of the last array for the improved algorithm is not given, as only the base algorithm can be used due to the aforementioned numerical problems.

In Fig. 6 we compare the evaluation time of a sampling and the basic rooting algorithm. The sampling algorithm samples with a spatial frequency slightly higher than Nyquist's frequency and performs a local maximization starting at every sample. The sampling algorithm does not always find all maxima as shown in Fig. 4. We defined two parametrized array positions depending on o. The positions of both parameterizations get more and more "prime" with increasing o. The first parametrization uses a fixed aperture $\boldsymbol{p} = 0\ (\frac{1}{o+1}\ 1)^{\mathrm{T}}$, which is the best case (bc) for the sampling algorithm. The second one is a dynamic aperture $\boldsymbol{p} = (0\ 0.5\ 0.5(o+1))^{\mathrm{T}}$, being some sort of worst case (wc) parametrization regarding the sampling.

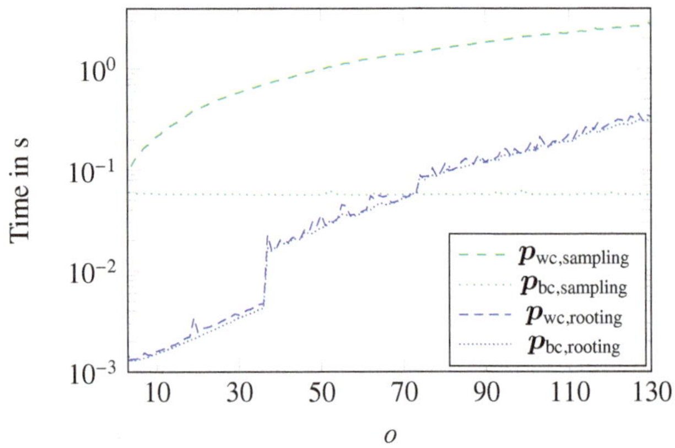

Fig. 6. Evaluation time comparison between sampling and rooting algorithm.

We see, that the sampling algorithm runs approximately in constant time having a constant aperture, i.e. constant Nyquist frequency Eq. (4). With the growing aperture realization the sampling algorithm is slow even for relatively small o. These two lines are the lower and upper limit of the sampling performance, so that most arrays will have a sampling performance in the area between both lines.

The rooting algorithm executes very fast for normal arrays. The unsteady behavior from $o = 36$ to $o = 37$ is a reproducible problem of Matlab's rooting algorithm. The rooting algorithm scales as before independent of the aperture increase. If the gcd is very small compared to the aperture, the rooting algorithm is slower than the sampling algorithm. For most practically important arrays (small o), the new algorithm is between one and two magnitudes faster than the sampling algorithm.

To show the new possibilities with the described algorithm, we performed a systematic search for arrays with a minimum maximum similarity in the whole angular range. Therefore we used an array with 5 elements and an aperture of ten wavelengths. Note, that this is nearly twice the aperture of an optimum MRA with 5 elements and violates Nyquist by a factor of five, so the array will be extremly sparse and most of the possible arrays' beampattern have very huge side lobes. The grid size of the systematic search was $0.1°$ which lead to the calculation of $\binom{99}{3} = 156849$ beampatterns. The minimum maximum similarity is 0.462 for the array positions $(0\ 1.6\ 4.8\ 6\ 10)^{\mathrm{T}}$. Only 0.063% of all calculated beampatterns have a side lobe maximum lower than 0.5. The empirical cumulative distribution function of all maximum side lobe heights is shown in Fig. 7.

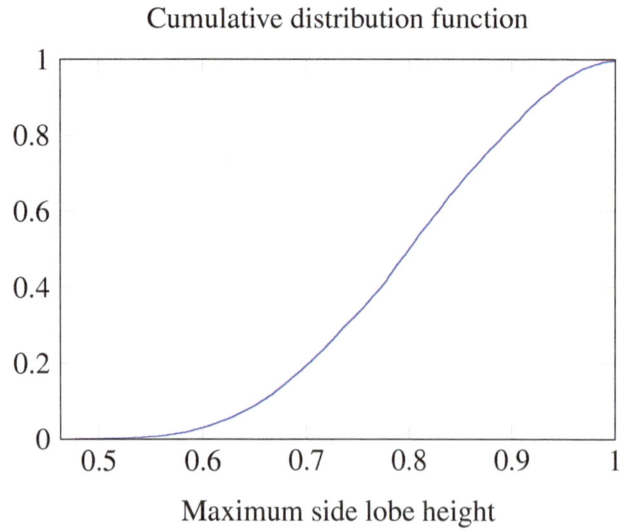

Fig. 7. Maximum side lobe height distribution of 156 849 calculated beampatterns.

7 Conclusions

We proposed an algorithm finding all maxima of a beampattern. Therefore the antenna positions have to be on a grid, which can always be approximated with arbitrary precision. The algorithm formulates the problem as a polynomial via a generalized version of gcd and locates the maxima by rooting the derivative. The algorithm accelerates sensor array optimization. We suggested an extension to the algorithm, which reduces the time for low and medium polynomial orders even more. We showed the successful application of the algorithm by finding an array with a very good aperture-to-side-lobe-height ratio. Further work should investigate if a similar algorithm can be found for combined azimuth and elevation estimation.

Acknowledgements. This research has been conducted in cooperation with the Robert Bosch Corporation, to which the authors would like to express their gratitude.

References

Athley, F.: Threshold Region Performance of Maximum Likelihood Direction of Arrival Estimators, IEEE Trans. on Signal Processing, 53, 1359–1373, 2005.

Athley, F., Engdahl, C., and Sunnergren, P.: Model-based Detection and Direction of Arrival Estimation in RADAR using Sparse Arrays, 2, 1953–1957, doi:10.1109/ACSSC.2004.1399505, 2004.

Bronshtein, I. N., Musiol, G., Mhlig, H., and Semendyayev, K. A.: Handbook of Mathematics, Springer, 4th edn., 2004.

Cormen, T. H., Leiserson, C. E., Rivest, R. L., and Stein, C.: Introduction to Algorithms, MIT Press and McGraw-Hill, 2 edn., 2001.

Jurgen, R. K.: Adaptive Cruise Control, SAE International, 2006.

Moffet, A.: Minimum-Redundancy Linear Arrays, IEEE Trans. on Antennas and Propagation, 16, 172–175, 1968.

Rockett, A. M. and Szsz, P.: Continued Fractions, World Scientific, 2 edn., 1992.

Rosen, K. H. and Michaels, J. G.: Handbook of discrete and combinatorial mathematics, CRC Press, 3 edn., 2000.

Ruder, M., Enkelmann, W., and Garnitz, R.: Highway Lane Change Assistant, in: IEEE Intelligent Vehicle Symposium, 1, 240–244, doi:10.1109/IVS.2002.1187958, 2002.

Tavares, J., Leito, T., Pereira, F. B., and Costa, E.: Evolving Segments Length in Golomb Rulers, in: Adaptive and Natural Computing Algorithms, Proc. International Converence in Coimbra, 239–242, doi:10.1007/3-211-27389-1_57, 2005.

23

A single channel speech enhancement technique exploiting human auditory masking properties

F. X. Nsabimana, V. Subbaraman, and U. Zölzer

Department of Signal Processing and Communications, Helmut Schmidt University, Hamburg, Germany

Abstract. To enhance extreme corrupted speech signals, an Improved Psychoacoustically Motivated Spectral Weighting Rule (IPMSWR) is proposed, that controls the predefined residual noise level by a time-frequency dependent parameter. Unlike conventional Psychoacoustically Motivated Spectral Weighting Rules (PMSWR), the level of the residual noise is here varied throughout the enhanced speech based on the discrimination between the regions with speech presence and speech absence by means of segmental SNR within critical bands. Controlling in such a way the level of the residual noise in the noise only region avoids the unpleasant residual noise perceived at very low SNRs. To derive the gain coefficients, the computation of the masking curve and the estimation of the corrupting noise power are required. Since the clean speech is generally not available for a single channel speech enhancement technique, the rough clean speech components needed to compute the masking curve are here obtained using advanced spectral subtraction techniques. To estimate the corrupting noise, a new technique is employed, that relies on the noise power estimation using rapid adaptation and recursive smoothing principles. The performances of the proposed approach are objectively and subjectively compared to the conventional approaches to highlight the aforementioned improvement.

1 Introduction

The enhancement of speech degraded by environmental or background noise still remains an open topic, although many significant approaches have been presented over years. The enhancement becomes more complicated especially for single channel noise reduction techniques, where no additional information about the corrupting noise and the real clean speech are available. Since the background noise is the factor that degrades the most the quality and intelligibility of the speech, it should therefore be estimated first using adequate techniques such as (Doblinger, 1995; Martin, 2001; Cohen, 2002, 2003; Rangachar, 2004; Stouten, 2006; Nsabimana, 2009). Thereafter, the estimated noise power is used in the derivation of the gain function for a desired noise reduction technique such as (Ephraim, 1985; Gustafsson, 1998; Virag, 1999; Cohen, 2002; Nsabimana, 2009).

In (Tsoukalas, 1993; Gustafsson, 1998; Virag, 1999; Yi, 2004; Hu, 2004; Nsabimana, 2009), Psychoacoustically Motivated Spectral Weighting Rules (PMSWR), which derive a gain function based on the psychoacoustical properties of the human hearing system, were proposed. Unlike the techniques like the Log Spectral Amplitude (LSA) and the Optimally Modified Log Spectral Amplitude (OMLSA) (Ephraim, 1985; Cohen, 2002), the Psychoacoustically Motivated Spectral Weighting Rules (Gustafsson, 1998; Nsabimana, 2009) do not try for a complete noise removal, they preserve instead a predefined amount of the original corrupting noise throughout the enhanced speech to account for the loss of weak speech components. Based on the error minimization of the distortions of speech and noise power components compared to the masking curve of the rough clean speech estimate, the gain function is thereafter derived (Gustafsson, 1998; Nsabimana, 2009). While the PMSWR approach (Gustafsson, 1998) generally accounts for a predefined constant amount of the original corrupting noise throughout the enhanced speech, the IPMSWR approach (Nsabimana, 2009) controls instead the level of the residual noise based on the discrimination between the regions with speech presence and speech absence by means of segmental SNR within critical bands.

This paper presents for the noise reduction technique and speech enhancement an algorithm relying on the IPMSWR approach (Nsabimana, 2009), where the estimated corrupting noise power is obtained using rapid adaptation and recursive smoothing principles (RARS) (Nsabimana, 2009) instead of OSMS approach (Martin, 2001; Nsabimana, 2009). The investigations in (Nsabimana, 2009) revealed that the RARS

Correspondence to: F. X. Nsabimana
(nsabfran@yahoo.fr)

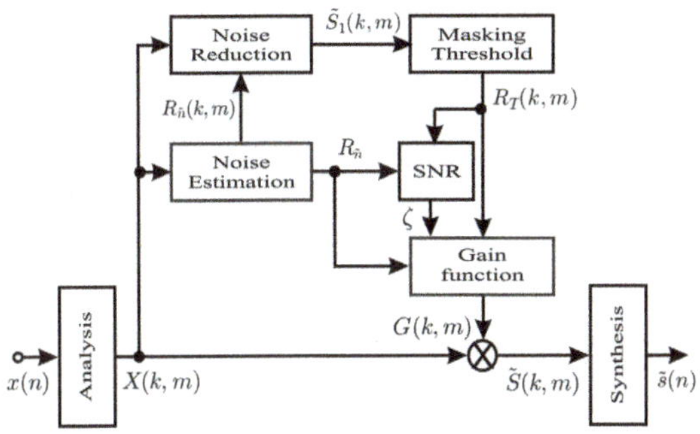

Fig. 1. Improved Psychoacoustically Motivated Spectral Weighting Rule (IPMSWR approach).

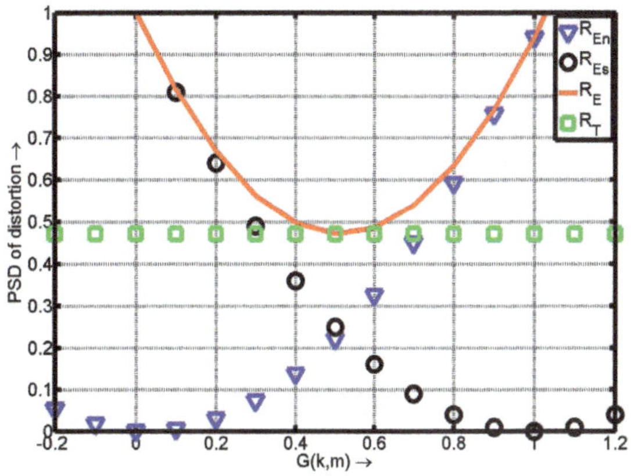

Fig. 2. Error minimization for the derivation of $G(k,m)$. PSD of residual noise distortion R_{E_n}, PSD of speech distortion R_{E_s}, PSD of estimation error R_E and masking threshold R_T.

approach adapts fast and provides accurate mean estimates than OSMS approach (Martin, 2001) especially for very low SNRs. To motivate further steps of improvements in the IPMSWR approach, the importance of the phase is also here emphasized during experimental results.

The outline of the paper is as follows: Sect. 2 presents the proposed technique, while experimental results and conclusion are presented in Sects. 3 and 4 respectively.

2 The proposed approach

Figure 1 depicts the complete system of the proposed approach. In the analysis stage, the corrupted speech is processed frame by frame with an overlapping rate of 75%. The estimated noise power $R_{\tilde{n}}(k,m)$ is computed using RARS approach (Nsabimana, 2009), while the rough clean speech estimate $\tilde{S}_1(k,m)$ needed for the computation of the masking threshold $R_T(k,m)$ is obtained using the OMLSA approach (Cohen, 2002). The masking curve $R_T(k,m)$ is computed as described in (Virag, 1999; Johnston, 1988; Zwicker, 1990; Zölzer, 2005) and summarized in (Virag, 1999). In the following, the derivation of the gain function is detailed.

Let consider the spectrum of a corrupted speech signal $X(k,m)$ to be defined as

$$X(k,m) = S(k,m) + N(k,m), \tag{1}$$

where $S(k,m)$ and $N(k,m)$ are the short-time DFT coefficients at frequency bin k and frame number m for the clean speech and additive noise respectively. $S(k,m)$ and $N(k,m)$ are also assumed to be statistically independent and zero mean. As a complete noise removal is not intended for psychoacoustically motivated spectral weighting rules, the desired spectrum of the enhanced speech is therefore defined as

$$\hat{S}(k,m) = S(k,m) + \zeta(k,m)N(k,m), \tag{2}$$

where $\zeta(k,m)N(k,m)$ represents the estimated amount of the residual noise. But the estimated magnitude spectrum of the enhanced speech is given by (s. Fig. 1)

$$\tilde{S}(k,m) = G(k,m)[S(k,m) + N(k,m)]. \tag{3}$$

The difference between Eqs. (2) and (3) yields the estimation error

$$E(k,m) = S(k,m)[G-1] + N(k,m)[G-\zeta], \tag{4}$$

with the PSD of the error expressed as

$$R_E(k,m) = R_s(k,m)[G-1]^2 + R_n(k,m)[G-\zeta]^2, \tag{5}$$

where the indexes k and m are omitted for G and ζ only for the sake of simplicity. $R_s(k,m)$ and $R_n(k,m)$ represent here the PSD of the clean speech $s(n)$ and the corrupting noise $n(n)$ respectively. Equation (5) is thus composed of the speech power distortion $R_{E_s} = R_s(k,m)[G-1]^2$ and the residual noise power distortion $R_{E_n} = R_n(k,m)[G-\zeta]^2$. The optimal $G(k,m)$ can be obtained by computing the minimum of the solid red parabola (R_E) of Fig. 2, while $G(k,m)$ for the just noticeable distortion case is derived considering the crossing point between the green curve with square (R_T) and the blue curve with triangle (R_{E_n}) of Fig. 2.

As a complete masking of both distortions $R_E < R_T$ is practically not possible, only the masking of the residual noise power distortions is taken into account. By masking the residual noise power distortions, the speech power distortions are also assumed to be minimized (Gustafsson, 1998). So equating noise power distortion R_{E_n} to masking curve of the rough clean speech R_T, the spectral weighting rule is derived as

$$G(k,m) = \min\left(\sqrt{\frac{R_T(k,m)}{R_n(k,m)}} + \zeta(\lambda,m), 1\right), \tag{6}$$

where λ represents herein a frequency bandwidth and $\zeta(\lambda,m)$ is chosen based on the corresponding subband segmental SNR:

$$SNR_i(\lambda,m) = 10\log_{10}\left(\frac{\sum_{k=k_s}^{k_e} R_T(k,m)}{\sum_{k=k_s}^{k_e} R_{\tilde{n}}(k,m)}\right), \tag{7}$$

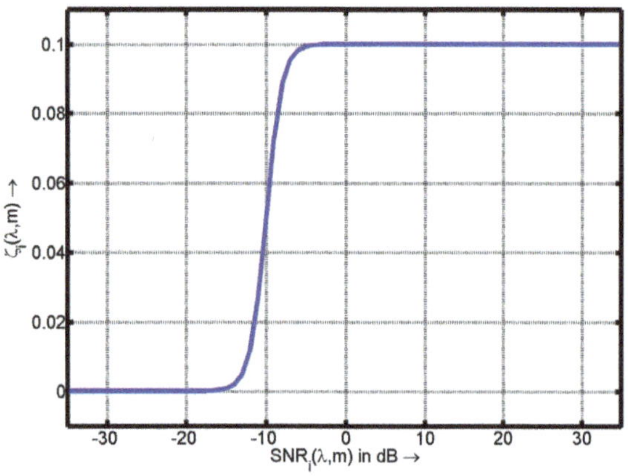

Fig. 3. $\zeta(\lambda,m)$ vs. segmental SNR in band i with bandwidth λ.

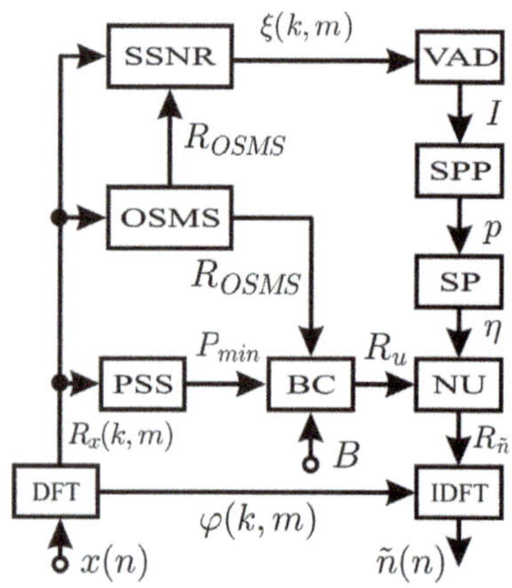

Fig. 4. RARS approach. Power Spectrum Smoothing (PSS), Bias Correction (BC), Noise Update (NU), Smoothing Parameter (SP), Speech Presence Probability (SPP), Voice Activity Detector (VAD), Smoothed SNR (SSNR).

as shown in Fig. 3, that is computed from a shifted sigmoid function. k_s and k_e represent in Eq. (7) the starting and ending bin of the ith band.

To reduce the spectral outliers in specific frequency bands, the gain function is manipulated based on the energy of the coefficients within critical bands as follows:

$$\tilde{G}(k,m) = \frac{G(k,m)}{k_e - k_s + 1} \cdot \sum_{k=k_s}^{k_e} |G(k,m)|^2. \tag{8}$$

The results with this technique are shown in Figs. 5 and 6, where it is clear that the corrupting noise has been properly controlled and sibilant sounds preserved.

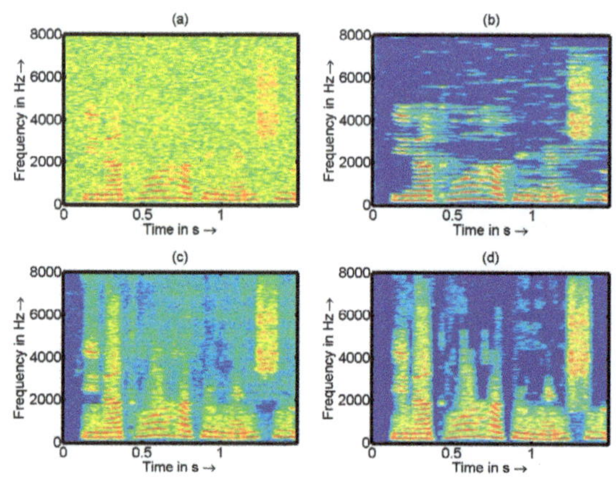

Fig. 5. Results from investigated speech enhancement techniques. Corrupted speech at 10 dB **(a)**, OMLSA aprroach **(b)**, PMSWR approach **(c)** and IPMSWR approach **(d)**.

2.1 Noise estimation

As the computation of Eq. (6) requires the knowledge of the corrupting noise power, only an estimated noise power can be used for the single channel case. Therefore, the Rapid Adaptation and Recursive Smoothing (RARS approach) (Nsabimana, 2009), that is depicted in Fig. 4, is applied here.

In the RARS approach, first the noise power is estimated first using Optimal Smoothing and Minimum Statistics (OSMS) approach (Martin, 2001) with a very short window. This yields an overestimation of the estimated noise power. Based on the smoothed posteriori SNR from the OSMS noise power a VAD index I is derived to compute the speech presence probability P and a smoothing parameter η. This smoothing parameter is finally applied to the unbiased estimated noise power R_u from OSMS approach to account for the overestimation. In order to improve the adaptation time for the estimated noise power, a condition BC is used to track quickly the fast changes in the noise power. Results from (Nsabimana, 2009) reveal that the RARS approach adapts fast and provides accurate mean estimates than OSMS approach (Martin, 2001) especially for very low SNRs.

3 Experimental results

This section presents the performance evaluation of the proposed enhancement technique using the phase of the corrupted speech on one hand (s. Figs. 5 and 6) and the phase of the clean speech on the other hand (s. Fig. 8). To get a fair comparison, tests were carried out for different SNRs using additive white gaussian noise. A window length of 512 samples with a hop size of 25% for analysis and synthesis is applied for all approaches. Figures 5 and 6 present a subjective comparison in terms of spectrogram. These results show that the IPMSWR approach preserves sibilants (s-like sounds) even for very low SNRs (5–10 dB).

Figure 7 presents again the results obtained during listening test with headphones (Nsabimana, 2009). The fifteen subjects recruited for this test were all working in our lab. For this test, subjects had first to find the hidden reference signal and assign it 100%. The results from the

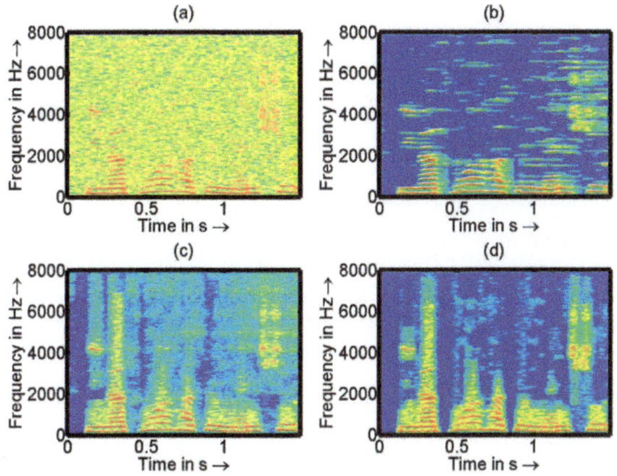

Fig. 6. Results from investigated speech enhancement techniques. Corrupted speech at 5 dB **(a)**, OMLSA aprroach **(b)**, PMSWR approach **(c)** and IPMSWR approach **(d)**.

simulated algorithms are then compared to the reference signal grade. The Mean Opinion Score (ITU-T P.862) represents the grades of the three enhancement techniques for three different kinds of noise. Figure 7 reveal that the IPMSWR approach was graded best.

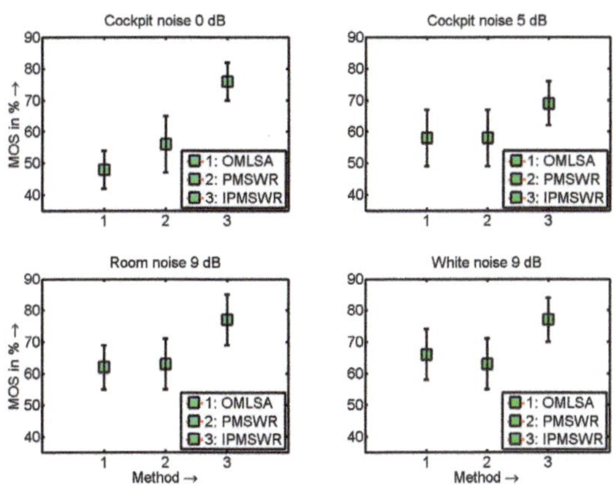

Fig. 7. Results from listening test using headphones. Bars denote 95% confidence interval.

3.1 Usefulness of phase information

The importance of the phase information in speech enhancement is currently being investigated (Shannon, 2006; Shi, 2006; Aarabi, 2006; shi, 2007). To motivate further steps of improvements in the IPMSWR approach, the role of the phase is here emphasized using clean speech degraded with artificial additive white gaussian noise at different SNRs from 0 to 35 dB (s. Fig. 8).

Figure 8 depicts the segmental SNR improvement with IPMSWR approach using the phase of the disturbed speech for the resynthesis on one hand and the phase of the clean speech for the resynthesis on the other hand. The black curve with square represents here the segmental SNR of the disturbed speech, which stands for the reference segmental SNR. The red curve with circle, which depicts the results with the IPMSWR using the phase of the disturbed speech, clearly reveals an overall segmental SNR gain of ≅5 dB. The dashed blue curve with diamond, which depicts the results with the PMSWR using the phase of the disturbed speech, remains close to the results of the IPMSWR approach only for SNRs higher than 15 dB as expected. The dashed green curve with triangle, which depicts the results with the IPMSWR using the phase of the clean speech, reveals instead an overall segmental SNR gain of ≅8 dB. This clearly outlines the usefulness of the phase information.

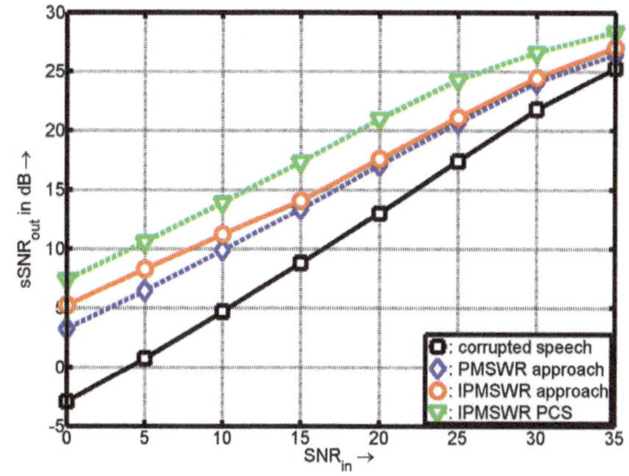

Fig. 8. Segmental SNR improvement with PMSWR and IPMSWR approaches. PCS means Phase of Clean Speech.

4 Conclusions

A speech enhancement technique based on psychoacoustics principles is proposed here. The key components of this approach are a time-frequency dependent control parameter for the residual noise within critical bands and a better estimate of the rough clean speech. As additional information on the corrupting noise is not available, a technique to estimate the corrupting noise power has been presented. Simulations results at different SNRs reveal that the proposed technique performs best and preserves sibilant sounds even at very low SNRs. To motivate further steps of improvements in the IPMSWR approach, the importance of the phase information has been emphasized here during experimental results. The obtained results show that an increase of SNR gain is achieved when the phase of the clean speech is used.

Future works should thus address the estimation of the phase to increase the speech intelligibility. To avoid abrupt jumps, the gain coefficients should also be properly controlled. The Parameter optimization remains necessary.

References

Doblinger, G.:Computationally Efficient Speech Enhancement by Spectral Minima Tracking in Subbands, in: Eurospeech 1995, vol. 2, pp. 1513–1516, September 1995.

Martin, R.: Noise power spectral density estimation based on optimal smoothing and minimum statistics, IEEE Transactions on Speech and Audio Processing, 9(5), pp. 504–512, July 2001.

Cohen, I. and Berdugo, B.:Noise estimation by minima controlled recursive averaging for robust speech enhancement, IEEE Signal Processing Letters, vol. 9(1), pp. 12–15, Januar 2002.

Cohen, I.:Noise spectrum estimation in adverse environments: improved minima controlled recursive averaging, IEEE Transactions on Speech and Audio Processing, vol. 11(5), pp. 466–475, September 2003.

Rangachari, S., Loizou, P. C., and Hu, Y.: A noise estimation algorithm with rapid adaptation for highly nonstationary environments, IEEE International Conference on Acoustics, Speech, and Signal Processing, Proceedings of ICASSP'04, vol. 1, pp. I–305–308, 17–21 May 2004.

Stouten, V., Van Hamme, H. and Wambacq, P.: Application of Minimum Statistics and Minima Controlled Recursive Averaging Methods to Estimate a Cepstral Noise Model for Robust ASR, IEEE International Conference on Acoustics, Speech and Signal Processing, ICASSP 2006, 1, pp. 765–768, May 2006.

Nsabimana, F. X.,Subbaraman, V., and Zölzer, U.: Noise power estimation using rapid adaptation and recursive smoothing principles, in: Proc. of the International Conference on Signal Processing and Multimedia Applications (SIGMAP 2009), pp. 13–18, Milan, Italy, 7–10 July 2009.

Ephraim, Y. and Malah, D.: Speech enhancement using a minimum mean-square error log-spectral amplitude estimator, IEEE Transactions on Acoustics, Speech and Signal Processing, vol. 33(2), pp. 443–445, April 1985.

Tsoukalas, D., Paraskevas, M. and Mourjopoulos, J.: Speech enhancement using psychoacoustic criteria, IEEE International Conference on Acoustics, Speech, and Signal Processing, ICASSP-93, vol. 2, pp. 359–362, April 1993.

Gustafsson, S., Jax, P., and Vary, P.: A novel psychoacoustically motivated audio enhancement algorithm preserving background noise characteristics, IEEE International Conference on Acoustics, Speech and Signal Processing, ICASSP'98, vol. 1, pp. 397–400, 12–15 May 1998.

Virag, N.: Single channel speech enhancement based on masking properties of the human auditory system, IEEE Transactions on Speech and Audio Processing, vol. 7(2), pp. 126–137, March 1999.

Cohen, I.: Optimal speech enhancement under signal presence uncertainty using log-spectral amplitude estimator, Signal Processing Letters, IEEE, vol. 9(4), pp. 113–116, April 2002.

Yi, H. and Loizou, P.C.: Incorporating a psychoacoustical model in frequency domain speech enhancement, Signal Processing Letters, IEEE, vol. 11(2), pp. 270–273, ISSN 1558-2361, Februar 2004 April 2002.

Hu, R. and Anderson, D.V.: Improved perceptually inspired speech enhancement using a psychoacoustic model, IEEE Journal on Signals, Systems and Computers, Record of the Thirty-Eighth Asilomar Conference, vol. 1, pp. 440–444, November 2004 International Conference on Acoustics, Speech, and Signal Processing, ICASSP-93, vol. 2, pp. 359-362, April 1993.

Nsabimana, F. X., Subbaraman, V., and Zölzer, U.: A single channel speech enhancement technique using psychoacoustic principles, in: Proc. of the 17th European Signal Processing Conference (EUSIPCO 2009), pp. 170–174, Glasgow, Scotland, 24–28 August 2009.

Johnston, J. D.: Transform coding of audio signals using perceptual noise criteria, IEEE Journal on Selected Areas in Communications, vol. 6(2), pp. 314–323, February 1988.

Zwicker, E. and Fastl, H.: Psychoacoustics facts and models, Springer Verlag, Berlin, 1990.

Zölzer, U.: Digitale Audiosignalverabeitung, B. G. Teubner, Wiesbaden 2005, ISBN 3-519-26180-4.

ITU-T P.862: Perceptual evaluation of speech quality (PESQ), an objective method for end-to-end speech quality assessment of narrow-band telephone networks and speech codecs, 2001.

Shannon, B. J. and Paliwal, K. K.: Role of phase estimation in speech enhancement, in: Interspeech 2006 – ICSLP, pp. 1423–1426, Pittsburgh, Pennsylvania, 17–21 September 2006.

Shi, G., Shanechi, M., Aarabi, P., On the importance of phase in human speech recognition, IEEE Transactions on Audio, Speech, and Language Processing, vol. 14(5), pp. 1867–1874, September 2006.

Aarabi, P., Shi, G., Shanechi, M. M and Rabi, S.A: Phase-Based Speech Processing, Published by, World Scientific Publishing Co. Pte Ltd, ISBN 981-256-612-0.

Shi, G., Aarabi, P., and Jiang, H.: Phase-Based Dual-Microphone Speech Enhancement Using A Prior Speech Model, IEEE Transactions on Audio, Speech, and Language Processing, vol. 15(1), pp. 109–118, Januar 2007.

24

Real time control of curved laser welding processes by cellular neural networks (CNN): first results

L. Nicolosi and R. Tetzlaff

TU Dresden, Faculty of Electrical Engineering and Information Technology, Institute of Principles of Electrical and Electronic Engineering, Germany

Abstract. In the last decades the laser beam welding (LBW) has outclassed older welding techniques in the industrial scenario. Despite the improvement in welding technology, sophisticated methods of fault detection are not commonly used in commercially available equipments yet. A recent analysis of process images have revealed the possibility to build up a real time closed loop control system. By the use of image based quality features, a feedback signal can be provided to maintain the process in the desired state. The development of the presented visual control system has been focused on the adjustment of the laser power according to the detection of the so called full penetration hole. Due to the high dynamics of the laser welding, a fast real time image processing with controlling rates in the multi kilo Hertz range is necessary to have a robust feedback control. In this paper an algorithm for the real time control of welding processes is described. It has been implemented on the Eye-RIS v1.2, a visual system which mounts a cellular structure. By applying this algorithm in real time applications, controlling rates of about 7 kHz can be reached. In the following some real time control results are also described.

1 Introduction

Nowadays image processing is used to monitor the quality of laser material processes such as laser welding, cut or ablation. In this paper laser beam welding processes are treated. The most important characteristic of the laser in welding processes is the huge power density that the beam can convey on the joining partners (Fortunato and Ascari, 2008). Such energy allows the local vaporization of the metal and the formation of a capillary, conventionally called "keyhole" (Fabbro et al., 2004). If an opportune laser power is used, due to the hydrostatic pressure of the metal vapor, all the material sheets are penetrated through the keyhole, creating a *full penetration*. Meanwhile the laser beam advances, the fluid material solidifies behind the keyhole creating a deep and slender weld seam. By the use of this strategy, moreover, high feeding rates up to 50 m/min can be reached. Figure 1 shows a longitudinal section of the welding setup and the resulting image of a coaxial process control camera. The latter represents the thermal radiation of the melt in the spectral range of 820–980 nm. The state of full penetration is visible in the coaxial camera image as a dark area directly behind the laser interaction zone, the so called "full penetration hole". It ensures that the two materials are properly connected over the whole cross section after re-solidification and, therefore, it represents an important quality feature which guarantees the strength of the connection. In the past, several authors have proposed closed-loop control systems by using, e.g. the laser power, the focal-point position, or other parameters as actuator (Bardin et al., 2004). This paper proposes a Cellular Neural Network (CNN) based visual closed loop control system for the real time adjustment of the laser power according to the detection of the full penetration hole in process images. It is important to underline the importance of controlling laser welding processes. The conventional welding strategy consists in regulating the laser power or the feeding rate until the full penetration is reached and add 10% of the laser power as a safety factor. Such uncontrolled full penetration weld is characterized by significant defects like smoke residue, spatters and craters. Furthermore, this strategy does not allow changing welding conditions during the process, such as feeding rate and material thickness. Due to the high dynamics of LBW, visual algorithms with controlling rates in the multi kilo Hertz range are needed to provide a robust real time feedback control. In order to reach such controlling rates, only a few CNN based operations for the full penetration hole detection in process images can be used to speed the algorithm up.

Correspondence to: L. Nicolosi
(leonardo.nicolosi@tu-dresden.de)

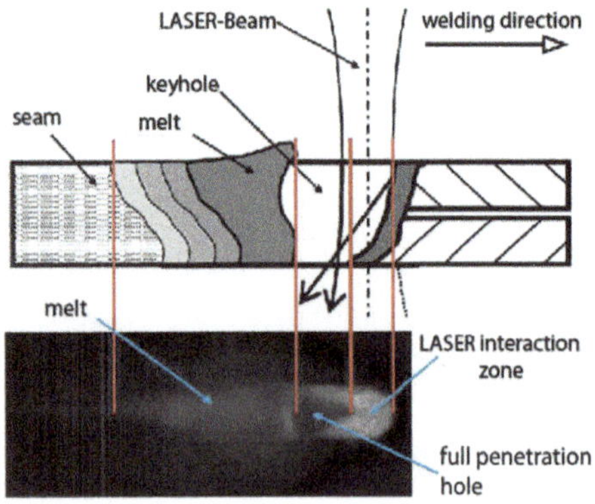

Fig. 1. Schematics of a welding process in an overlap-joint and nomenclature. The picture shows at the top the longitudinal section of the materials during the welding process and at the bottom the resulting image of a coaxial process control camera.

2 A visual system based on CNN operations

This paper is focused on the description of a new algorithm for the full penetration hole detection and for the real time control of laser welding processes. It has been implemented on the Eye-RIS v1.2 (Anafocus), a visual system essentially composed by a cellular structure, called "Q-Eye", for image sensing and early processing, an "FPGA board" (Altera NIOS II processor) for controlling the execution flow and for decision-making operations and "I/O modules" to interface external devices. In particular, the Q-Eye chip hosts the whole image sensing and evaluation. It has a quarter CIF resolution fully programmable smart image sensor array. It consists of 176×144 cells, each containing an optical sensor merged with a processor, memory circuitry and interconnections to the 8 neighboring cells. This means that the Q-Eye can both sense the corresponding spatial sample of the image and process this data in close interaction and cooperation with other neighboring pixels. This concept finds its basis in the CNN theory (Chua and Yang, 1988).

CNNs are analog dynamic processors which have some of the key features of neural networks and important potential applications in such areas as image processing. The first model of a CNN was introduced by Chua and Yang which consists of a grid of $M \times N$ interconnected cells. Each cell $C(i,j)$ can directly interact with its neighboring cells and indirectly with the other cells because of the propagation effects of the continuous-time network dynamics. The fundamental property that distinguishes a CNN from a general neural network is the local connectivity of the coupling structure. The hardware realization of large-complexity CNN chips has required to modify the original Chua-Yang model into an other one, called Full Range model (Corinto et al., 2002). It is described by:

$$\dot{x}_{ij} = -x_{ij} + \sum_{|k|\le r, |l|\le r} \hat{A}_{kl} y_{i+k,j+l} - g(x_{ij}) + \\ + \sum_{|k|\le r, |l|\le r} \hat{B}_{kl} u_{i+k,j+l} + \hat{I} \quad (1)$$

where

$$y_{ij} = \frac{1}{2}(|x_{ij}+1| - |x_{ij}-1|) \quad (2)$$

$$g(x_{ij}) = \begin{cases} m(x_{ij}+1) & if\, x_{ij} < -1 \\ 0 & if\, |x_{ij}| \le 1 \\ m(x_{ij}-1) & if\, x_{ij} > 1 \end{cases} \quad (3)$$

where m is supposed to be large enough for approximating the nonlinear characteristic shown in Linan et al. (1999).

In the previous equations x_{ij}, u_{ij} and y_{ij} represent the state, the input and the output of the cell (i,j), r denotes the neighborhood of interaction of each cell and the indexes i,j are such that $1 \le i \le M$ and $1 \le j \le N$ for a CNN arranged in M rows and N columns. Finally, the matrices $\hat{\mathbf{A}}_{kl}$ and $\hat{\mathbf{B}}_{kl}$ are linear weight functions defined, respectively as "feedback" and "feedforward" templates and allow programming the CNN in order to obtain the desired output. Defining an image as a grid of pixels, the pixel of the input image in the position (i,j) can be mapped on the CNN cell (i,j), so that each pixel is associated with the input or initial state of a particular cell. The input image dynamically evolves to a corresponding output image due to the CNN evolution, according to the state equation.

Unfortunately, the Q-Eye chip is not governed by traditional state equations like in Eq. (1) and its internal status is not accessible. Nevertheless it can be programmed by the specification of templates, thereby it is possible to implement algorithms based on traditional local neighborhood operations which can be efficiently run on any CNN architecture.

3 Algorithm for the full penetration hole detection

As described in Nicolosi et al. (2009a), algorithms based on dilating operations are allowing the detection of the full penetration hole in laser welding processes. By these strategies an evident improvement in the quality of the welding result has been obtained, especially in situations characterized by speed or material thickness changes. They have been implemented supposing the welding orientation constant during the whole process. This means that they can be applied only for welding of straight lines. Nevertheless, some real life welding processes are executed with orientation changes, such as curved weld lines. With regards to this, a new orientation-independent strategy called "omnidirectional algorithm" has been implemented (Nicolosi et al., 2009b). In the following,

Table 1. Omnidirectional algorithm: simulation results.

Sequence	05_004	05_021	05_022	05_023	05_026
Hit [$\approx$ %]	87	83	82	89	87
Images with full penetration hole	236	310	334	312	254
Images without full penetration hole	147	77	35	74	137
False positives	20	22	13	9	25
False negatives	21	45	52	33	27

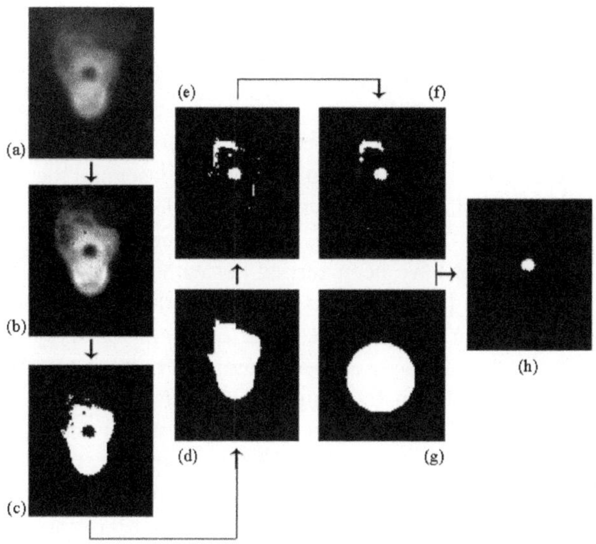

Fig. 2. Omnidirectional algorithm flow chart. The source image **(a)** is filtered **(b)** and binarized **(c)**. Afterwards, closings are executed **(d)** and logical operations (XOR and AND) are applied in order to keep white pixels in the closed area only **(e)**. Openings are performed in order to reduce the defects **(f)** and a mask **(g)** is applied to obtain the result shown in **(h)**. The mask used in this example was built offline to fit the specific source image. The images have been zoomed to improve the visibility.

besides the omnidirectional algorithm, a tool for creating automatically a mask at the beginning of the process will be described. The mask is used to reduce the defects resulting from the image elaboration.

3.1 Omnidirectional algorithm

The first step consists in enhancing the source image contrast by the application of sharpening filters and binarizing the resulting image by the use of a global threshold value within 1 and 255. The second step consists in detecting the full penetration hole by the use of closing and opening operations (Chua and Roska, 2002). The former is a dilation followed by an erosion and allows filling small black objects in a white background. The latter, instead, is an erosion followed by a dilation and it is used to delete small isolated white objects in a black background. Thereby, at first iterative morphological closings are executed in order to fill the full penetration hole. The application of a logical XOR and a logical AND comes after to keep white pixels in the full penetration hole area only. Since the closing operations generate defects due to the elaboration of the external edges of the interaction zone, the last step consists in reducing such defects by the use of opening operations and by the subsequent application of a mask. As the typical size of the full penetration hole is pretty known, its presence can be decided simply referring to the total number of white pixels in the last resulting image. Figure 2 shows the algorithm flow chart.

The omnidirectional algorithm has been simulated with several image sequences acquired by the Eye-RIS v1.2 during laser welding processes. As shown in Table 1, hit percentages greater than 82% have been obtained by analysing the simulation results through visual inspections. False detections can be explained making reference to the following: *False positives* are a consequence of the opening and closing operations. The defects can reach the typical size of the full penetration hole and the application of the opening operation could be insufficient to delete it. *False negatives* are due to the fact that the opening operation, used to reduce the defects, deletes small white objects despite they represent the full penetration hole. As explained in Nicolosi et al. (2009b), only a few false detections of the same type occur consecutively (from 1 to 4). In order to analyse the effect of such false detections, the reacting time of the physical process has been estimated in the range of 4–20 ms, by applying the Fourier transform to the laser response of different welding processes. Assuming that the controlling rate was slowed down to about 2 kHz by the image storage (not needed in real time control), the worst case of four consecutive false detections of the same type would take a total "false detection" time of about 2 ms. It is considerably less than the estimated reacting time of the physical process and made even worse by the image storage. This means that such false detections should not significantly influence the real time control of welding processes.

3.2 Mask builder

As shown in the previous section, the last elaboration step consists in applying a mask to the resulting image. The mask builder allows to fit automatically the mask to the interaction

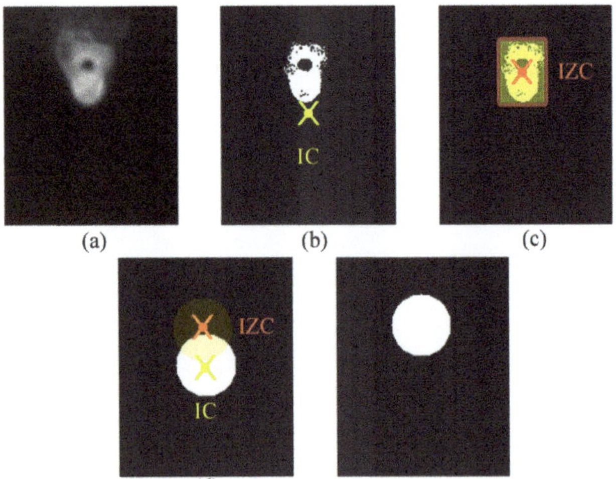

(a) (b) (c) (d) (e)

Fig. 3. Mask builder for the omnidirectional algorithm. The source image **(a)** is binarized **(b)** and the coordinates of the interaction zone center are extracted **(c)**. By the use of this information, the default mask, initially placed in the image center IC, is shifted to the interaction zone center (IZC) position, as in **(d)**, in order to obtain the final mask **(e)**.

zone position, making the algorithm independent of the specific optical setup. A default mask centered in the image is adopted. It has a roundish shape whose diameter must be chosen in order to limit the extension of the interaction zone tail which could be source of false detections. At the beginning of the process the center coordinates of the interaction zone in the first available source image are extracted. Through this information it is possible to shift the default mask from the image center to the interaction zone center. Figure 3 clarifies in details the mask builder.

4 Available hardware

In order to obtain a closed loop control of the process, the Eye-RIS system must be connected to the welding machine. The available hardware essentially consists of the Eye-RIS system v1.2, a numerically controlled (NC) machine, the laser and the interface board. The NC machine is responsible for safety interlocks, laser ON/OFF signals and movement. The laser is controlled by an analog voltage between 0 and 10 V which corresponds to 0–100% of laser power. The NC machine sends the starting 24 V signal in order to activate the laser and the algorithm flow. Consequently, the machine movement is set in action and the Eye-RIS system starts executing the control algorithm. At first the mask builder draws the mask by the use of the first available acquired image. As soon as the mask is created, while a new image is being sensed, the image acquired at the previous time step is evaluated by the omnidirectional algorithm. If the full penetration hole is detected the Eye-RIS system decreases the laser power, or increases it otherwise. Consecutively sensing, evaluation and control operations are performed until the stopping signal is received from the NC machine.

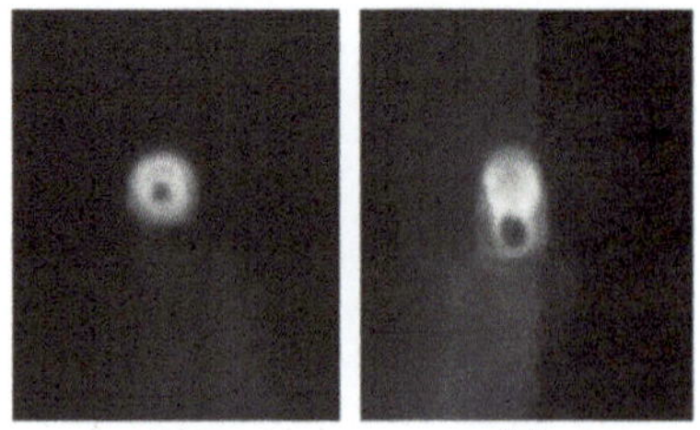

(a) (b)

Fig. 4. Images acquired during laser welding processes at 3 m/min **(a)** and at 7 m/min **(b)**.

The interface board was built to interface the laser and the NC machine with the Eye-RIS system. More details can be found in Abt et al. (2008). The welding experiments discussed in this paper have been carried out with a 2-D-laser scanner setup. The laser source is a 6 kW, 1030 nm Trumpf TruDisk 6002 Yb:YAG thin disk with a 200 μm transport fibre. The laser scanner used – a Trumpf PFO-33 – was equipped with a 450 mm focusing optic which resulted in a focal diameter of 600 μm. The Eye-RIS system is connected to the scanner optic through a 90° beam splitter. Thus the camera perspective is coaxial to the laser beam, allowing an invariant field of view regardless of the scanner position. Furthermore, it is necessary to use an aligned lens system in combination with an optical band-pass filter in order to acquire high quality camera images.

5 Experimental results

As explained in previous sections, the omnidirectional algorithm can be used to detect the full penetration hole. In addiction the mask builder represents an important tool to create the mask, which can be used to cut away the defects resulting from the image elaboration. It is important to consider for the real time control sensing and controlling operations which are two additional phases. As already mentioned, sensing and image evaluation are simultaneously performed: while the image *i* is being acquired, the image *i-1* is elaborated. Therefore, the sensing exposure time directly depends on the elaborating time. Controlling operations, instead, allow changing the laser power accordingly to the image evaluation. The application of these operations takes an additional time leading to total controlling rates up to about 7 kHz.

The experiments shown in the following have been executed to weld an overlap joint of two 0.9 mm thick steel sheets with a gap of 0.2 mm and a constant feeding rate. Despite the omnidirectional algorithm has been conceived for welding of curved seams, in a first attempt it has been used to control the weld of straight lines, to be compared with

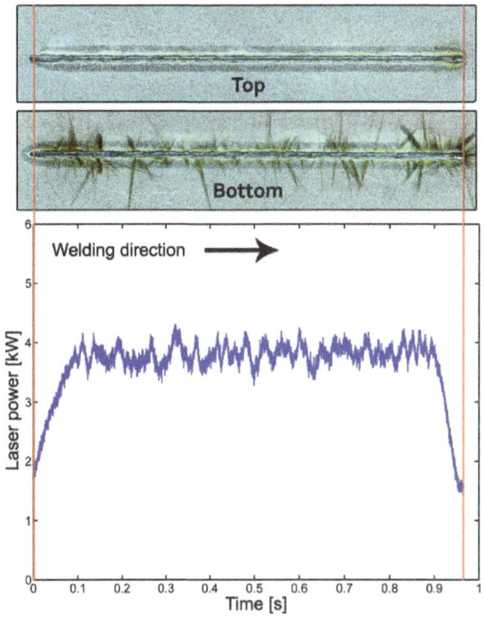

Fig. 5. Weld of an overlap joint of two sheets 0.9 mm thick, with 0.2 mm gap at 5 m/min, controlled by the omnidirectional algorithm setting the threshold value at 180. The welding result and the laser power are shown at the top and at the bottom respectively. The full penetration state was reached and kept during the process. The control was executed without image storing (Controlling rates around 7 kHz).

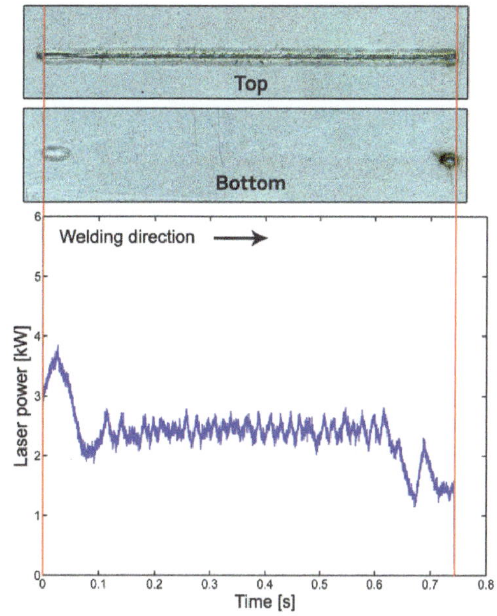

Fig. 6. Weld of an overlap joint of two sheets 0.9 mm thick, with 0.2 mm gap at 7 m/min, controlled by the omnidirectional algorithm setting the threshold value at 180. The welding result at the top presents a full penetration loss due to the low level power showed at the bottom. The control was executed without image storing (Controlling rates around 7 kHz).

the results obtained by the dilation algorithm in (Nicolosi et al., 2009a). Furthermore, some of the following experiments have been executed with image saving in order to analyse the process images offline. As already mentioned, image storing slows down the controlling rate from 7 to about 2 kHz. Nevertheless, in the considered speed range (3–9 m/min) the image storing has not negatively influenced the process control. Therefore, the omnidirectional algorithm is expected to work even more so for higher controlling rates.

The experimental results pointed out a dependence of the threshold value on the process speed. In low speed processes (1–4 m/min) the full penetration hole is placed inside the interaction zone (Fig. 4a). Because of the vapor plume, there is not a significant intensity difference between full penetration hole and interaction zone. Thereby, a high threshold value in this case is needed to distinguish the two areas. On the contrary, in higher speed processes the full penetration hole is situated directly behind the interaction zone which presents an elongated shape (Fig. 4b). In this case full penetration hole and interaction zone can be easily distinguished by the use of a low threshold value, since they show a big intensity difference. By the application of a high threshold value several areas inside the interaction zone would be discriminated because of the intensity fluctuation due to the vapor plume. Thus the false positives would increase and the laser power level would decrease, creating full penetration

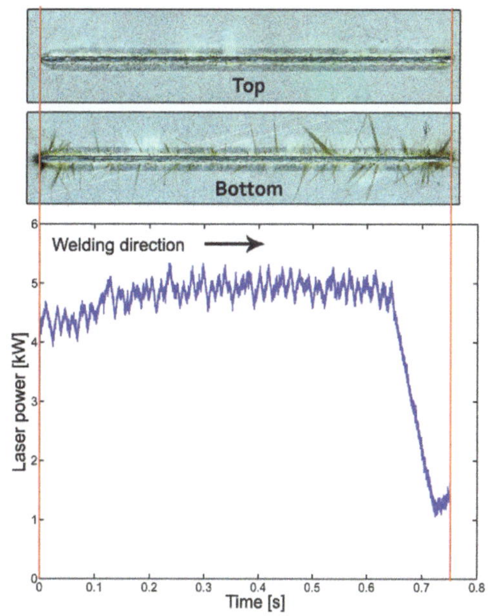

Fig. 7. Weld of an overlap joint of two sheets 0.9 mm thick, with 0.2 mm gap at 7 m/min, controlled by the omnidirectional algorithm setting the threshold value at 130. The welding result and the laser power are shown at the top and at the bottom respectively. The full penetration state was reached and kept during the process. The control was executed with image storing (Controlling rates around 2 kHz).

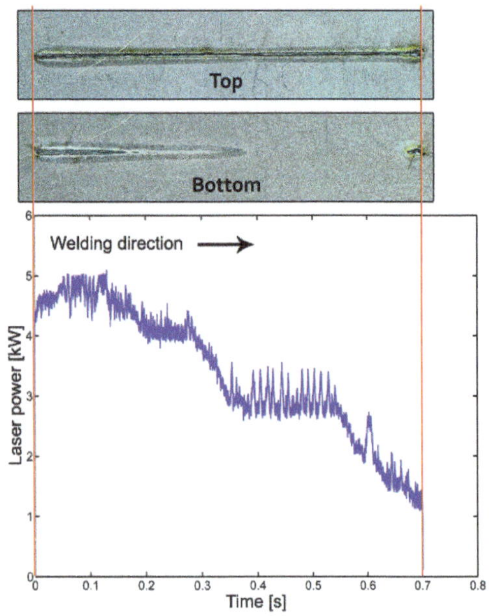

Fig. 8. Weld of an overlap joint of two sheets 0.9 mm thick, with 0.2 mm gap at 8 m/min, controlled by the omnidirectional algorithm setting the threshold value at 130. The welding result at the top presents a full penetration loss due to the low level power showed at the bottom. The control was executed with image storing (Controlling rates around 2 kHz).

losses. Nevertheless, it is possible to overcome to this problem choosing the opportune threshold value according to the feeding rate. In fact, as shown in Figs. 5 and 6, setting the threshold value to 180, the full penetration has been reached at the speed of 5 m/min, but not at the speed of 7 m/min. In order to obtain the full penetration also in 7 m/min speed processes a threshold value of 130 had to be used, as shown in Fig. 7. This behavior is confirmed by the experimental result in Fig. 8 obtained by using 130 as threshold value for a 8 m/min process.

6 Conclusions

A novel closed loop visual control system for the laser power adjustment in omnidirectional LBW processes has been presented. It is based on the detection of the full penetration hole in process images. The proposed algorithm is "omnidirectional" because no a priori knowledge of the welding direction is needed, allowing also the control of the weld of curved seams. Controlling rates up to 7 kHz can be reached. First experiments have pointed out a dependence between the process speed and the threshold value used to binarized the source images. In particular, increasing the process speed the threshold value must be decreased to reach the full penetration state. Finally, experimental results have shown that variable orientation welding processes can be also controlled by CNN based strategies. Further studies will be focused on the algorithm improvement in order to apply the same threshold value for a wide range of welding processes and feeding rates.

References

Abt, F., Nicolosi, L., Carl, D., Blug, A., Geese, M., Dausinger, F., Deininger, C., Höfler, H., and Tetzlaff, R.: Closed loop control of laser welding processes with cellular neural networks (CNN) cameras, in Proc. of the 27th International Congress on Applications of Lasers and Electro-Optics LIA, ICALEO, Temecula, CA, USA, 20–23 October, pp. 817–825, 2008.

Anafocus: www.anafocus.com, Sevilla, Spain, 14 December 2009.

Bardin, F., Cobo, A., Lopez-Higuera, J., Collin, O., Aubry, P., Dubois, T., Högström, M., Nylen, P., Jonsson, P., Jones, J., and Hand, D.: Process Control of Laser Keyhole Welding, in: Proc. of the 23rd International Congress on Applications of Lasers and Electro-Optics, ICALEO, San Francisco, USA, 2004.

Chua, L. O. and Roska, T.: Cellular neural network and visual computing: Foundations and applications, Cambridge University Press 2004, first published in printed format, 190–194, 2002.

Chua, L. O. and Yang, L.: Cellular Neural Networks: Theory, IEEE Transactions on Circuits and Systems, 35(10), 1257–1272, 1988.

Corinto, F., Gilli, M., and Civalleri, P. P.: On stability of full range and polynomial type CNNs, in: Proc. of the 7th IEEE International Workshop on Cellular Neural Networks and their Applications, CNNA, pp. 33–40, 2002.

Fabbro, R., Coste, F., Goebels, D., and Kielwasser, M.: Study of Nd-YAG Welding of Zn-Coated Steel Sheets, in Proc. of the 23rd International Congress on Applications of Lasers and Electro-Optics, ICALEO, San Francisco, USA, 2004.

Fortunato, A. and Ascari, A.: Tecnologie di giunzione mediante saldatura. Procedimenti non convenzionali e allo stato solido, Vol.2, Progetto Leonardo, Bologna, 2008.

Linan, G., Dominguez-Castro, R., Espejo, S., and Rodriguez-Vazquez, A.: Design of Large-Complexity Analog I/O CNNUC, in Proc. of the European Conference on Circuit Theory and Design, ECCTD, Design Automation Day on Cellular Visual Microprocessor, Stresa (Italy), pp. 42–57, 1999.

Nicolosi, L., Tetzlaff, R., Abt, F., Blug, A., Carl, D., and Höfler, H.: New CNN based Algorithms for the Full Penetration Hole Extraction in Laser Welding Processes: Experimental Results, in Proc. of the International Joint Conference on Neural Networks, IJCNN, Atlanta, GE, USA, pp. 2256–2263, 2009a.

Nicolosi, L., Tetzlaff, R., Höfler, H., Blug, A., Carl, D., and Abt, F.: Omnidirectional Algorithm for the Full Penetration Hole Extraction in Laser Welding Processes, in: Proc. of the 19th European Conference on Circuit Theory and Design, ECCTD, Antalya, Turkey, pp. 177–180, 2009b.

Deterministic and stochastic channel models implemented in a physical layer simulator for Car-to-X communications

J. Nuckelt, M. Schack, and T. Kürner

Institut für Nachrichtentechnik, Technische Universität Braunschweig, Germany

Abstract. This paper presents a physical (PHY) layer simulator of the IEEE 802.11p standard for Wireless Access in Vehicular Environments (WAVE). This simulator allows the emulation of data transmission via different radio channels as well as the analysis of the resulting system behavior. The PHY layer simulator is part of an integrated simulation platform including a traffic model to generate realistic mobility of vehicles and a 3D ray-optical model to calculate the multipath propagation channel between transmitter and receiver. Besides deterministic channel modeling by means of ray-optical modeling, the simulator can also be used with stochastic channel models of typical vehicular scenarios. With the aid of this PHY layer simulator and the integrated channel models, the resulting performance of the system in terms of bit and packet error rates of different receiver designs can be analyzed in order to achieve a robust data transmission.

1 Introduction

Car-to-car or car-to-infrastructure communications – summarized under the term C2X communications – have been of great research interest in recent years. This technology shall contribute to increase the safety as well as the efficiency of road traffic. The dedicated IEEE standard for vehicular communications is 802.11p, also known as WAVE (IEEE, 2010). The European equivalent is the ETSI standard Intelligent Transportation Systems (ITS)-G5 (ETSI, 2010). Both standards operate in the 5.9 GHz band and are mainly based on the well-known Wireless Local Area Network (WLAN) standard IEEE 802.11a (IEEE, 2007). Since IEEE 802.11a was primarily designed for nomadic indoor usage it has to be ensured that 802.11p will reliably operate in typical vehicular environments with high relative velocities between transmitter and receiver as well as long multipath propagation delays.

In order to investigate and validate the system performance in different C2X communication scenarios, a PHY layer simulator compliant to IEEE 802.11p has been developed. In this paper, we give a description of the simulator structure and provide an overview of the implemented channel models. Since the characteristic of the C2X radio channel has a huge impact on the system performance and each channel model has its advantages and disadvantages, we intend to use different kinds of channel models for the simulations. On the one hand, a self-developed ray-optical tool is used to model the radio channel deterministically. This approach allows us to analyze the system behavior in a specific situation, e.g. two cars are approaching a certain intersection. On the other hand, the stochastic channel models, which can be found in the literature (Acosta-Marum and Ingram, 2007), are used to investigate the system performance in scenarios that represent a certain C2X communication environment, e.g. a highly mobile highway scenario. In this way, several receiver designs like different channel estimation strategies can be evaluated and compared to each other.

This paper is structured as follows: Sect. 2 gives a detailed description of the developed PHY layer simulator. Section 3 describes the deterministic and stochastic channel models implemented in the simulator. Some performance results in terms of packet error rates are presented in Sect. 4. The paper finally ends with a summary and a conclusion in Sect. 5.

2 Description of the PHY layer simulator

The developed PHY layer simulator is part of an integrated simulation environment for C2X communications as illustrated in Fig. 1. This simulation platform includes scenario data of the city of Braunschweig, a 3D ray-optical tool to

Correspondence to: J. Nuckelt
(nuckelt@ifn.ing.tu-bs.de)

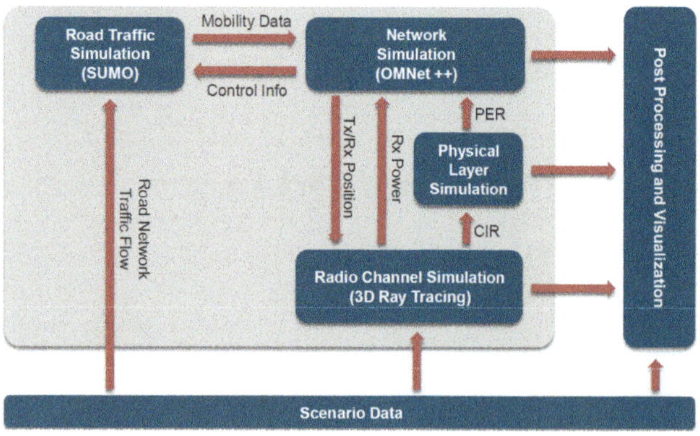

Fig. 1. Integrated simulation environment for C2X communications.

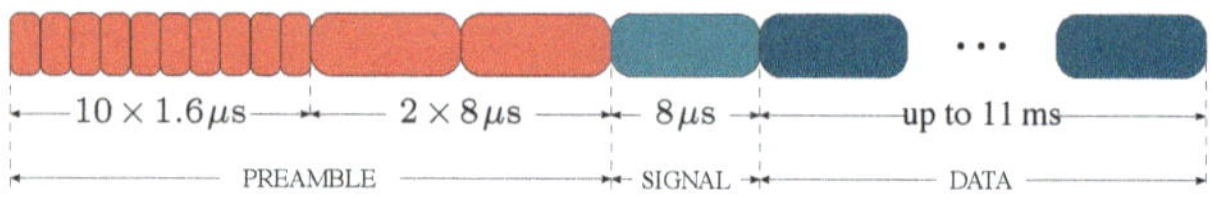

Fig. 2. IEEE 802.11p frame structure with preamble, signal and data fields.

model the radio channel, a road traffic and network simulator as well as the PHY layer simulator presented in this paper. A more detailed description of the integrated simulation environment can be found in (Schumacher et al., 2009; Kürner and Schack, 2010). Basically, the part of the PHY layer simulator can be divided into three blocks, namely transmitter, radio channel, and receiver. In the following subsections, the structure of these blocks is described in detail.

2.1 Transmitter

In IEEE 802.11p, data is transmitted as packets as illustrated in Fig. 2. First, the transmitter creates the Physical Layer Convergence Protocol (PLCP) preamble consisting of 10 short and 2 long training symbols. Depending on the chosen data rate and payload size, the subsequent signal field is generated. After that the payload data is appended. In order to generate the payload, the transmitter creates randomly distributed binary data which thereafter is scrambled using the generator polynomial $S(x) = x^7 + x^4 + 1$. The output of the scrambler passes the convolutional encoder to introduce a forward error correction (FEC). The generator polynomials of the coder are $g_0 = 133_8$ and $g_1 = 171_8$ offering a coding rate of $1/2$. Coding rates of $2/3$ and $3/4$ are generated by employing puncturing patterns. The encoded binary data is interleaved by a two-step permutation algorithm. Afterwards, the data stream is modulated to complex symbols, whereas binary phase shift keying (BPSK), quadrature phase shift keying (QPSK), 16- and 64-quadrature amplitude modulation (QAM) are defined by the standard, and is then converted from serial to parallel. The data symbols are mapped to the corresponding subcarriers and four pilot tones are inserted. A 64-IFFT block converts the data from the frequency to the time domain. The IFFT output is parallel-to-serial converted and a guard interval in the form of a cyclic prefix (CP) of a period $T_{\text{cp}} = 1.6\,\mu\text{s}$ is added to overcome the impact of multipath propagation. Several data OFDM symbols as well as the signal field and the preamble are merged to a packet that passes the radio channel block. The complete structure of the transmitter block is illustrated in Fig. 3.

2.2 Channel

The time-variant frequency-selective radio channel is modeled by two stages, namely a tapped-delay line model and an additive white Gaussian noise (AWGN) block. First, a discrete convolution of the transmitted packet and channel impulse response (CIR) is performed. The CIR is updated for each sample period, i.e. $T = 1/10\text{MHz} = 0.1\,\mu\text{s}$. The CIR is calculated either by means of 3D ray-optical modeling or derived from the stochastic channel models presented in (Acosta-Marum and Ingram, 2007). The implemented channel models are described in more detail in Sect. 3. Afterwards, AWGN with respect to the selected signal-to-noise ratio (SNR) level is added to the signal. The resulting signal is the input of the receiver block. The block diagram of the radio channel implementation is depicted in Fig. 4. Note that the receiver is also fed with a noiseless version of the signal that has only been convolved with the CIR. In this way, reference information like perfect channel knowledge can be obtained at the receiver.

2.3 Receiver

According to the frame structure of IEEE 802.11p, the whole signal processing in the receiver can be divided into two parts: processing of the preamble field and decoding of the signal and data fields. For a better understanding, the structure of the receiver is depicted in Fig. 5. First, the short preamble and the long preamble are used to perform coarse and fine time as well as frequency synchronization. The current version of the implemented simulator performs ideal time synchronization. Coarse and fine frequency synchronization is carried out according to the approach presented in (Terry and Heiskala, 2002). Afterwards, an initial estimation of the channel transfer function (CTF) based on the long preamble is performed.

In the second step, the signal field is equalized using the initial channel estimate and decoded in order to recover the transmitted information about the current data rate and payload size. After that, the OFDM symbols of the data field are decoded. Basically, the receiver performs the inverse signal processing as in the transmitter with some additional steps. First, the CP is removed. The remaining core OFDM symbol is converted from serial to parallel and is transformed to the frequency domain by applying a 64-FFT. After that, a

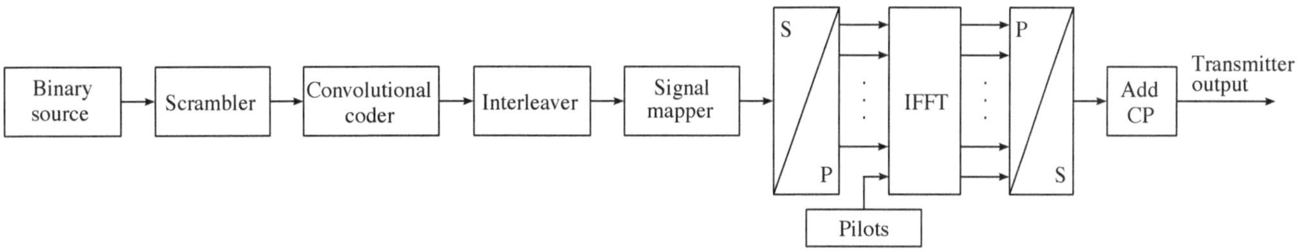

Fig. 3. Block diagram of the transmitter.

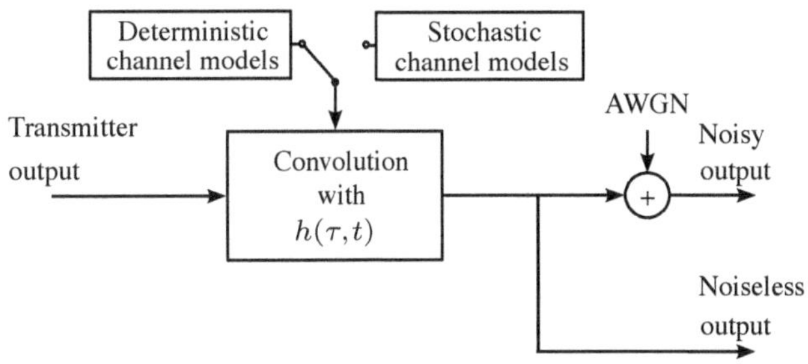

Fig. 4. Block diagram of the radio channel with deterministic and stochastic channel models.

frequency domain channel estimation is performed. On the one hand, the output signal of the channel that is not affected by AWGN can be used to perform an ideal channel estimation. On the other hand, the PHY layer simulator is equipped with an adaptive decision-directed (DD) channel estimation technique as presented in (Skupin et al., 2010). Either the ideally or the realistically estimated CTF is used to perform zero-forcing equalization before the signal is again parallel-to-serial converted. In the next step, the complex symbols are demapped to binary data, whereas hard or soft (i.e. approximate log-likelihood ratio (LLR)) decision is performed. Afterwards, the binary data is deinterleaved, decoded by the Viterbi decoder and descrambled. Finally, by a comparison of the originally transmitted data and the decoded received data, the resulting bit and packet error rates are calculated with respect to the current SNR.

In order to validate the results of the PHY layer simulator, the simulated BERs in the AWGN channel have been compared to the theoretical values, whereas both match very well.

3 Implemented channel models

Basically, two classes of channel models, namely deterministic and stochastic models, have been implemented in the PHY layer simulator. The concepts of these models are described in this section.

3.1 Deterministic channel models

As a deterministic channel model, a 3D ray-optical approach is used in our integrated simulation environment. This model considers the direct path, specular reflections as well as diffuse scattering. Specular reflections are calculated with a ray-tracing model based on the image method (McKown and Hamilton Jr., 1991) up to the n-th order, but with respect to the computation time, only first and second order reflections are considered in this investigation. Diffuse scattering is taken into account on surfaces seen by both the transmitter and the receiver. Thus, all surfaces are segmented into tiles and single scattering processes are determined.

For each ray, full-polarimetric antenna patterns are used and channel polarization matrices are computed. The output of the ray-optical model for each communication link is a time-variant CIR $h(\tau,t)$, which completely characterizes the frequency-selective channel and can be expressed in the complex baseband as

$$h(\tau,t) = \sum_{k=1}^{N(t)} a_k(t) \cdot e^{j(2\pi f \tau_k(t) + \varphi_k(t))} \cdot \delta(\tau - \tau_k(t)) \tag{1}$$

$$= \sum_{k=1}^{N(t)} \underline{a}_k(t) \cdot \delta(\tau - \tau_k(t)), \tag{2}$$

where the k-th multipath component is described by an amplitude $a_k(t)$, a delay $\tau_k(t)$ and an additional phase shift $\varphi_k(t)$ at time t. The number of multipath components, given by $N(t)$, is also time-variant. The amplitude and the phase term can be combined for each multipath component in a complex coefficient $\underline{a}_k(t)$, which can be written as

$$\begin{aligned}\underline{a}_k(t) = &\boldsymbol{e}_{\mathrm{Rx}}(\Theta_{\mathrm{Rx},k}, \Phi_{\mathrm{Rx},k})^H \cdot \mathbf{P}_k \\ &\cdot \boldsymbol{e}_{\mathrm{Tx}}(\Theta_{\mathrm{Tx},k}, \Phi_{\mathrm{Tx},k}) \cdot L_k(t),\end{aligned} \tag{3}$$

where the subscript H denotes the Hermitian transpose. The complex channel polarization matrix $\mathbf{P}_k$, which is not normalized and thus includes the reflection or scattering losses of the considered path, is multiplied by the complex polarization vectors $\boldsymbol{e}_{\mathrm{Rx}}(\Theta_{\mathrm{Rx},k}, \Phi_{\mathrm{Rx},k})$ and $\boldsymbol{e}_{\mathrm{Tx}}(\Theta_{\mathrm{Tx},k}, \Phi_{\mathrm{Tx},k})$ for the Rx and Tx antennas, respectively. These vectors comprise the antenna gains and fully describe the polarization characteristics of the antennas at the angles of arrival $(\Theta_{\mathrm{Rx},k}, \Phi_{\mathrm{Rx},k})$

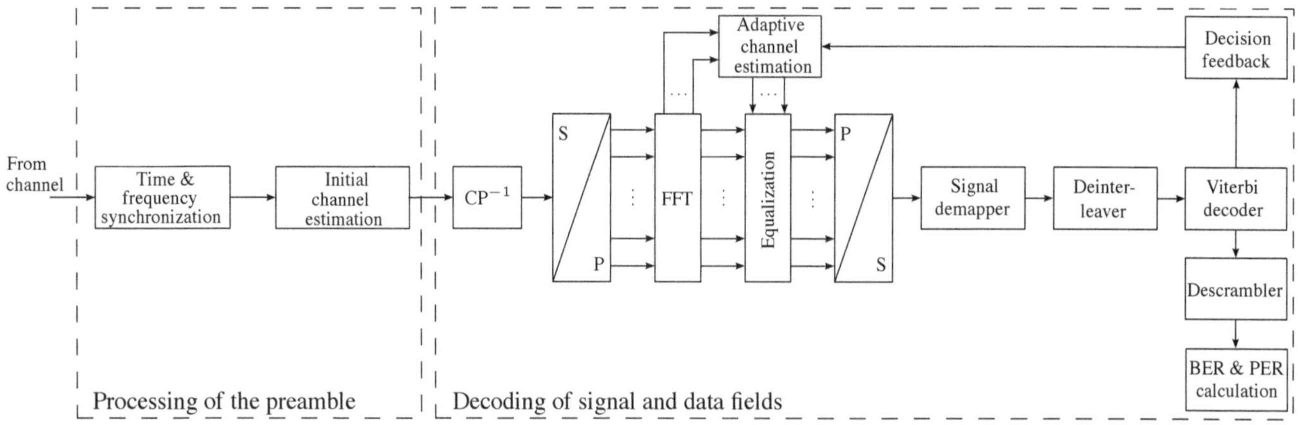

Fig. 5. Block diagram of the receiver.

and the angles of departure ($\Theta_{\mathrm{Tx},k}, \Phi_{\mathrm{Tx},k}$) for the k-th multipath component. The complex factor $L_k(t)$ comprises the propagation loss and the phase shift based on the delay τ of the considered path. For the direct path, $\mathbf{P}_k$ is the identity matrix, whereas for reflected and scattered paths of the order n, $\mathbf{P}_k$ is calculated according to (Maltsev et al., 2009) using

$$\mathbf{P}_k = \mathbf{R}(\varphi_{\mathrm{Rx},k}) \cdot \prod_{m=0}^{n-2} \left(\mathbf{R}_{n-m,k} \cdot \mathbf{R}(\varphi_{n-m,k})\right) \cdot \mathbf{R}_{1,k} \cdot \mathbf{R}(\varphi_{\mathrm{Tx},k}), \tag{4}$$

where

$$\mathbf{R}(\varphi) = \begin{bmatrix} \cos\varphi & \sin\varphi \\ -\sin\varphi & \cos\varphi \end{bmatrix} \tag{5}$$

is the rotation matrix. The value $\varphi_{\mathrm{Tx},k}$ characterizes the angle between the normal vector of the incidence plane and the field strength component E_θ at the first reflection or scattering point, whereas $\varphi_{\mathrm{Rx},k}$ is the angle between the scattering or reflecting plane and E_θ at the last reflection or scattering point for the k-th multipath component. For multiple reflections or scattering processes, the term $\varphi_{i,k}$ is the angle between E_θ and the normal vector of the incidence plane at the i-th reflection or scattering point for the k-th multipath component, respectively. The reflection or scattering matrix is given by

$$\mathbf{R}_{i,k} = \begin{bmatrix} r_{\perp_{i,k}} & \zeta_{1_{i,k}} \\ \zeta_{2_{i,k}} & r_{\parallel_{i,k}} \end{bmatrix}, \tag{6}$$

where the elements $r_{\perp_{i,k}}$ and $r_{\parallel_{i,k}}$ are the reflection coefficients of the surface belonging to the i-th reflection or scattering point of the k-th multipath component for perpendicular and parallel components of the electric field, respectively. For reflection processes, these elements are determined based on the Fresnel equations, whereas for scattering processes, the values are calculated using the model of a Lambertian emitter (Degli-Esposti et al., 2007). For the latter, both elements are equal. The elements $\zeta_{1_{i,k}}$ and $\zeta_{2_{i,k}}$ are cross polarization coupling coefficients and are set to zero in this investigation. An extension of this approach for multiple input multiple output (MIMO) systems and an investigation of different antenna configurations can be found in (Schack et al., 2010).

Based on 3D building data, which is available for the city of Braunschweig, and a microscopic road traffic simulator, several traffic scenarios in a dense urban area have been created. These scenarios represent typical traffic situations that are interesting for C2X applications like intersection collision warning or traffic jam assistants. They can be applied in order to investigate how the underlaying system will operate in specific situations and under different radio channel conditions, e.g. the impact of non-line-of-sight (NLOS) and line-of-sight (LOS). An example scenario with two oncoming cars that cross each other in an urban canyon is depicted in Fig. 6.

The speed of each vehicle is 120 km/h and there are LOS conditions at any time.

One great challenge of C2X communications are the high Doppler shifts due to the relative motion between communicating cars. In this scenario, for example, the relative velocity of transmitter and receiver is $v = 240\,\mathrm{km/h}$ corresponding to a maximum Doppler shift of

$$f_{\mathrm{D,max}} = \frac{v}{c} f_{\mathrm{c}} = 1311\,\mathrm{Hz}, \tag{7}$$

whereas c and f_{c} denote the speed of light and the carrier frequency of 5.9 GHz, respectively. In Fig. 7, the Doppler spectra with respect to the scenario time are depicted. The Doppler shift of the LOS components at $\pm 1311\,\mathrm{Hz}$ can clearly be observed. Besides the dominant LOS component, lower Doppler shifts caused by reflected and scattered waves are also present. The calculated spectra fit very well with the expected ones.

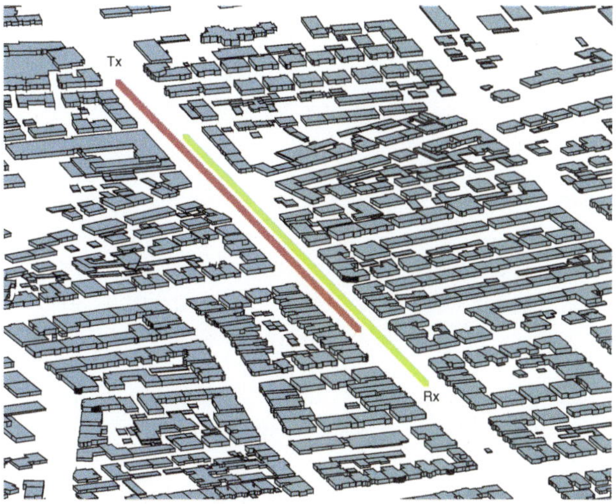

Fig. 6. Exemplary scenario of two oncoming cars in the city of Braunschweig.

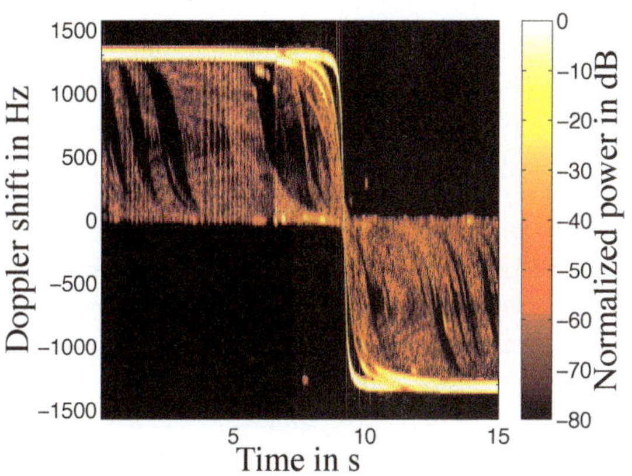

Fig. 7. Calculated Doppler spectra versus scenario time of two oncoming cars.

Furthermore, the corresponding power delay profiles (PDPs) are shown in Fig. 8. As expected, a lot of delayed multipath components due to reflected or scattered waves on surrounding buildings can be observed.

3.2 Stochastic channel models

Besides deterministic channel models also six doubly selective stochastic channel models published in (Acosta-Marum and Ingram, 2007) have been implemented. The authors in (Acosta-Marum and Ingram, 2007) present three vehicle-to-vehicle (VTV) and three roadside-to-vehicle (RTV) models:

- VTV-Expressway oncoming (EO),
- VTV-Urban canyon oncoming (UCO),
- RTV-Suburban street (SS),

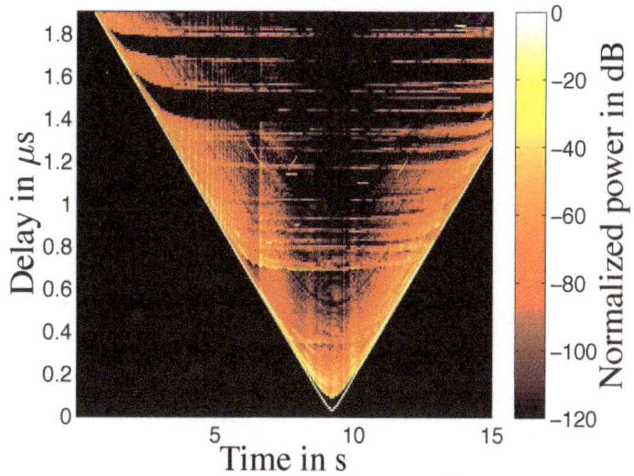

Fig. 8. Power delay profiles of the scenario with oncoming cars.

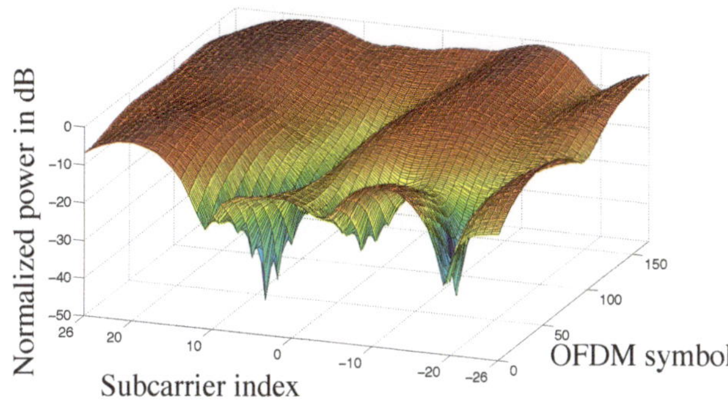

Fig. 9. Realization of a doubly selective channel transfer function of the RTV-SS scenario.

- RTV-Expressway (E)[1],
- VTV-Expressway same direction with wall (ESDWW) and
- RTV-Urban canyon (UC).

Each scenario is characterized by a more or less dominant LOS component and several superimposed multipath components. All paths are described by a delay value, the relative path loss, a K factor as well as Doppler-related parameters like maximum Doppler shift, LOS Doppler component and the shape of the Doppler spectra. As all Doppler frequencies are scaled to 120 or 140 km/h, these models can be employed to carry out PHY layer simulations of highly mobile propagation scenarios. A limitation of these channel models is the fact that nonstationarity, which is an important issue for the modeling of C2X radio channels, is not considered at all. In Fig. 9, a realization of a CTF corresponding to the RTV-SS scenario is exemplarily shown.

The time-variant and frequency-selective character of the typical C2X propagation channel can clearly be observed.

[1] According to the note in (Ivan et al., 2010), the K factor of the first path of this channel model has to be changed to 5.3 dB (instead of −5.3 dB as given).

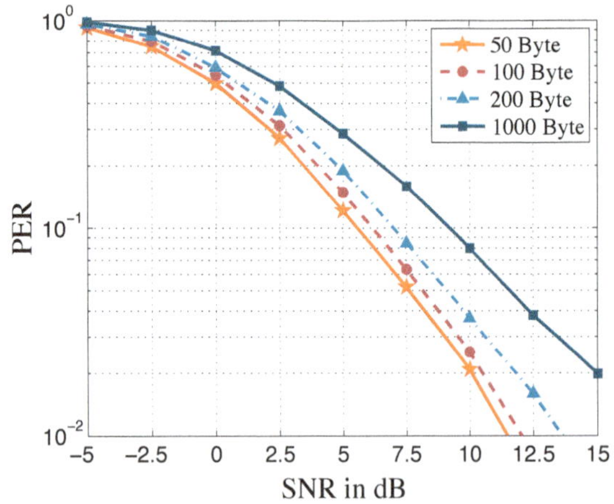

Fig. 10. Impact of the payload size on PER performance in the VTV-EO scenario.

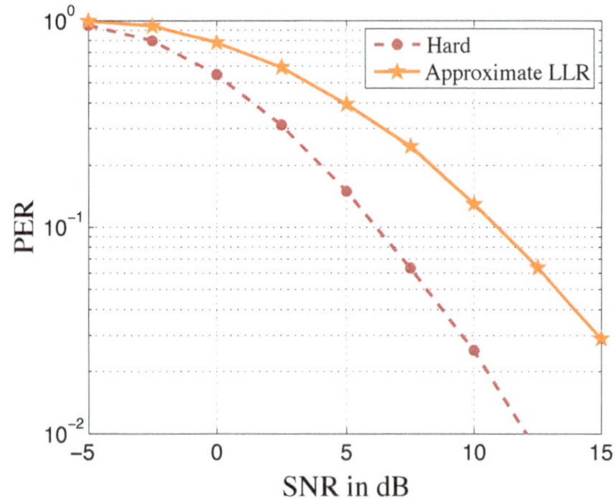

Fig. 11. Performance evaluation of hard and soft (approximate LLR) decision in the VTV-EO scenario.

The variation of the radio channel with respect to the time is one of the most challenging issues when designing robust C2X receivers.

4 Exemplary simulation results

This section presents some exemplary simulation results in terms of packet error rates (PERs). The PHY layer of IEEE 802.11p offers a lot of system parameters that can be modified. For example, the payload size can be chosen up to 4095 byte leading to different packet periods. The length of the transmitted packets has a significant impact on the system performance. In Fig. 10, the PERs of several packet lengths are shown for a data rate of 3 Mbit/s. Perfect channel knowledge was assumed at the receiver and soft-input Viterbi decoding has been performed. The applied channel model is the VTV-EO scenario. It can be observed that the length of a packet has a considerable impact on the resulting performance and, hence, the choice of the payload size has to be carefully decided with respect to the performance requirements of the considered application.

Furthermore, a sophisticated receiver design can also enhance the performance. Different concepts of time and frequency synchronization, channel estimation or diversity techniques have to be investigated on their suitability in C2X communication systems. For instance, a soft-decision demapper in combination with a preamble-based SNR estimation as it is proposed in (Ren et al., 2009) applied in time-variant and frequency-selective C2X channel has been compared to simple hard decision. The results in terms of PERs are depicted in Fig. 11. This chart clearly demonstrates that advanced receiver techniques offer a great potential to improve the system performance.

5 Conclusions

This paper presents a PHY layer simulator compliant to the IEEE 802.11p WAVE standard. This simulator facilitates the analysis of future C2X communication systems operating in realistic environments. Deterministic channel models by means of 3D ray-optical modeling as well as stochastic channel models derived from an extensive measurement campaign have been implemented in the simulator. In this way, the influence of the fast-fading characteristic of typical C2X channels on the system performance can be analyzed. Since the radio channel is one of the most challenging issues with respect to PHY layer implementations, different receiver designs can be evaluated in order to develop a robust receiver design with the aid of this simulator.

Future work will focus on PHY layer simulations of the implemented determinstic and stochastic channel models. The results will be evaluated with respect to the advantages and disadvantages of both channel models.

Acknowledgements. The authors are grateful to the Ministry of Science and Culture of Lower Saxony for funding this work within the Connected Cars in a Connected World (C3World) project. Furthermore, they would like to thank Hendrik Hoffmann and Alexander Tschiene for their valuable support.

References

Acosta-Marum, G. and Ingram, M.: Six time- and frequency-selective empirical channel models for vehicular wireless LANs, Vehicular Technology Magazine, IEEE, 2, 4–11, 2007.

Degli-Esposti, V., Fuschini, F., Vitucci, E. M., and Falciasecca, G.: Measurement and Modelling of Scattering From Buildings, Antennas and Propagation, IEEE Transactions on, 55, 143–153, 2007.

ETSI: Intelligent Transport Systems (ITS); European profile standard for the physical and medium access control layer of Intelligent Transport Systems operating in the 5 GHz frequency band, ETSI ES 202 663 V1.1.0 (2010-01), 2010.

IEEE: IEEE Standard for Information Technology – Telecommunications and information exchange between systems – Local and metropolitan area networks – Specific requirements – Part 11: Wireless LAN Medium Access Control (MAC) and Physical Layer (PHY) Specifications, IEEE Std 802.11-2007 (Revision of IEEE Std 802.11-1999), 2007.

IEEE: IEEE Standard for Information technology – Telecommunications and information exchange between systems – Local and metropolitan area networks – Specific requirements Part 11: Wireless LAN Medium Access Control (MAC) and Physical Layer (PHY) Specifications Amendment 6: Wireless Access in Vehicular Environments, IEEE Std 802.11p-2010, 1–51, 2010.

Ivan, I., Besnier, P., Bunlon, X., Le Danvic, L., Crussiere, M., and Drissi, M.: Influence of propagation channel modeling on V2X physical layer performance, Antennas and Propagation (EuCAP), Proceedings of the Fourth European Conference on, 1–5, 2010.

Kürner, T. and Schack, M.: 3-D Ray-Tracing Embedded Into an Integrated Simulator for Car-to-X Communications, in: URSI Commission B International Symposium on Electromagnetic Theory (EMT-S), Berlin, 2010.

Maltsev, A., Maslennikov, R., Lomayev, A., Sevastyanov, A., and Khoryaev, A.: IEEE 802.11-09/0431r0, 2009.

McKown, J. and Hamilton Jr., R.: Ray tracing as a design tool for radio networks, Network, IEEE, 5, 27–30, 1991.

Ren, G., Zhang, H., and Chang, Y.: SNR estimation algorithm based on the preamble for OFDM systems in frequency selective channels, Communications, IEEE Transactions on, 57, 2230–2234, 2009.

Schack, M., Kornek, D., Slottke, E., and Kürner, T.: Analysis of Channel Parameters for Different Antenna Configurations in Vehicular Environments, Vehicular Technology Conference Fall (VTC 2010-Fall), IEEE 72nd, 1–5, 2010.

Schumacher, H., Schack, M., and Kürner, T.: Coupling of simulators for the investigation of car-to-x communication aspects, in: IEEE Asia-Pacific Services Computing Conference, APSCC, 58–63, 2009.

Skupin, C., Hofmann, F., Garcia, A., and Nuckelt, J.: Adaptive Channel Estimation for VANETs, in: 7th International Workshop on Intelligent Transportation (WIT 2010), 91–96, 2010.

Terry, J. and Heiskala, J.: OFDM Wireless LANs: A Theoretical and Practical Guide, Sams Publishing, Indianapolis, 2002.

A multi channel coupling based approach for the prediction of the channel capacity of MIMO-systems

F. Hagebölling and U. Zölzer

Department of Signal Processing and Communications, Helmut Schmidt University – University of the Federal Armed Forces Hamburg, Holstenhofweg 85, 22043 Hamburg, Germany

Abstract. When installing Multiple Input Multiple Output (MIMO)-systems, the antenna positioning has a major influence upon the achievable transmission quality. To determine those antenna positions, which maximize the transmission quality, in adequate time, a computer based prediction of the channel capacity is imperative.

In this paper, we will show that Ray Tracing, which is a very popular prediction method and well suited for the prediction of transfer functions or power delay profiles, produces unacceptable errors when predicting the channel capacity of MIMO-systems. Furthermore we identify the source of the prediction errors and present a new algorithm, based on an approach known as Multi Channel Coupling (MCC), which avoids this error source.

Finally a comparison of the prediction results of our algorithm with prediction results gained with an Image Ray Tracer as well as with measured results is used to show the formidable increasement of prediction accuracy which can be gained by using our algorithm.

1 Introduction

In Foschini and Gans (1998) it was shown, that MIMO-systems using spatial multiplexing increase the channel capacity to such an extent, that an immensely higher spectral efficiency becomes possible.

When installing such MIMO-systems the count of positions, where antennas can be placed, is usually higher than the number of antennas, which are to be placed. Being aware of the fact, that the antenna positioning has a big impact on the spatial multiplexing capabilities of the MIMO-system and thereby the achievable transmission quality, a lot of effort has already been taken to find antenna positionings which maximize transmission quality.

A lot publications concentrate on measuring channel capacities for different positionings in certain scenarios, Tang and Mohan (2005) and Fügen et al. (2002) shall be mentioned exemplarily. But measuring based approaches are always limited by the fact, that the number of possible MIMO-systems increases very fast with the number of possible antenna positions.

To be able to compare several thousand or more possible systems, computer based simulation is needed. In Talbi (2001) and Ziri-Castro et al. (2005) and a lot of other publications the received power was predicted and compared with measurements.

Though the results presented in those publications accord very well with measurements, the received power is, in our opinion, not a sufficient quantity to judge the transmission quality of MIMO-systems, because it doesn't contain any information about the linear independence of the Multiple Input Single Output (MISO)-systems contained in each MIMO-system. The latter is important, because in a spatial multiplexing MIMO-system the received signals can only be decoded properly, if the MISO-systems are linearly independent.

Though the channel capacity contains this important information only a comparatively small amount of publications deals with the prediction of this quantity. And most of the publications, which do so, don't present measurements as comparison for the prediction results. Elnaggar et al. (2004) may be named as example.

In the following we will briefly summarize some fundamentals in Sect. 2, compare results of a ray tracing based prediction algorithm with measurements and point out the disadvantages of this approach in Sect. 3, present our new algorithm in Sect. 4 and verify the better performace of our algorithm with more measurements in Sect. 5. A summary and conclusion marks can be found in Sect. 6.

Correspondence to: F. Hagebölling
(falko.hageboelling@hsu-hh.de)

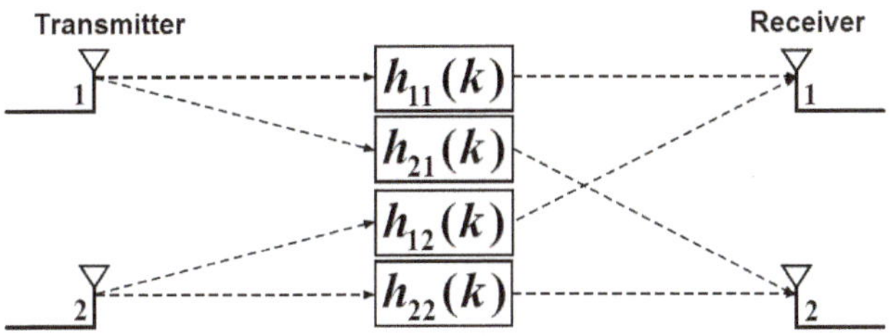

Fig. 1. SISO impulse responses of a MIMO Channel, $n_\mathrm{T} = n_\mathrm{R} = 2$

2 Fundamentals

2.1 Channel description

The MIMO Channel with n_T transmit antennas and n_R receive antennas consists of one complex Single Input Single Output (SISO) impulse response between every transmit antenna m and every receive antenna n (see Fig. 1).

Assuming the same order L for each of these impulse responses $\boldsymbol{h}_{n,m} = \left[h_{n,m}(0)\, h_{n,m}(1) \ldots h_{n,m}(L_{n,m})\right]^T$ the frequency selective MIMO channel can be described by $L+1$ complex channel matrices $\mathbf{H}(k)$ with $k = 0, ..., L$:

$$\mathbf{H}(k) = \begin{bmatrix} h_{1,1}(k) & \cdots & h_{1,n_T}(k) \\ \vdots & \ddots & \vdots \\ h_{n_R,1}(k) & \cdots & h_{n_R,n_T}(k) \end{bmatrix}. \tag{1}$$

2.2 Channel capacity

Assuming that no power allocation strategy is used, the transmitted power P_T is uniformly distributed over the bandwidth B and the Channel Capacity of a frequency selective MIMO system can be written as (Palomar et al., 2000)

$$C = \int_{-B/2}^{+B/2} \log_2 \left(\det \left[\mathbf{I}_m + \frac{P_\mathrm{T}/(n_\mathrm{T} \cdot B)}{\mathbf{S}_{nn}(f)} \mathbf{Q}(f) \right] \right) df, \tag{2}$$

with $S_{nn}(f)$ being the power spectral density matrix of the noise, $m = \min(n_\mathrm{T}, n_\mathrm{R})$ and

$$\mathbf{Q}(f) = \begin{cases} \mathbf{H}_\mathbf{F}(f)\mathbf{H}_\mathbf{F}(f)^H, & n_\mathrm{R} < n_\mathrm{T} \\ \mathbf{H}_\mathbf{F}(f)^H\mathbf{H}_\mathbf{F}(f), & n_\mathrm{R} \geq n_\mathrm{T} \end{cases}. \tag{3}$$

Discretizing this equation, assuming additive white noise and normalizing channel energy according to Bauch and Al-Dhahir (2002)

$$\sum_{k=0}^{L} E\left\{ \left| h_{n,m}(k) \right|^2 \right\} = 1 \qquad \forall\, 1 \leq m \leq n_\mathrm{T},\ 1 \leq n \leq n_\mathrm{R} \tag{4}$$

leads to the formula

$$C = \frac{1}{N_\mathrm{F}} \sum_{f=1}^{N_\mathrm{F}} \log_2 \left(det \left[\mathbf{I}_m + \frac{\rho}{n_\mathrm{T}} \mathbf{Q}(f) \right] \right), \tag{5}$$

with N_F being the number of discrete frequencies and ρ being the mean Signal to Noise Ratio (SNR).

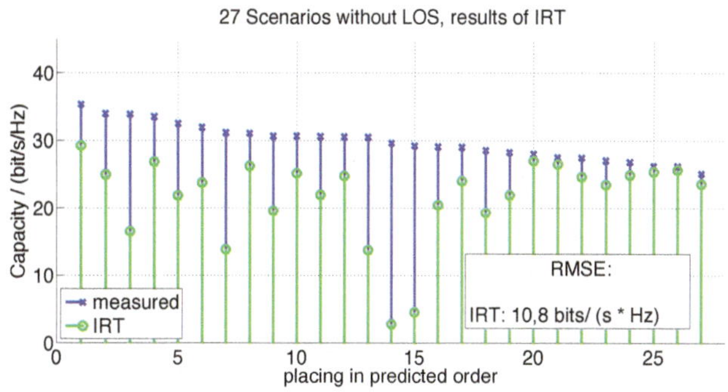

Fig. 2. Prediction results gained with image ray tracing.

The normalization according to Eq. (4) is needed to replace the term $\frac{P_\mathrm{T}/(n_\mathrm{T} \cdot B)}{\mathbf{S}_{nn}(f)}$ by ρ, but it abolishes differences in the path loss of the SISO-channels and reduces the comparability of different MIMO systems. To maintain the differences and to enable a comparison of N_M MIMO systems the condition has to be relaxed to be (Hagebölling et al., 2006)

$$\sum_{i=1}^{N_M} \sum_{m=1}^{n_\mathrm{T}} \sum_{n=1}^{n_\mathrm{R}} \sum_{k=0}^{L} E\left\{ \left| h_{(n,m)_i}(k) \right|^2 \right\} = N_M \cdot n_\mathrm{T} \cdot n_\mathrm{R}. \tag{6}$$

3 Image Ray Tracing

In Hagebölling et al. (2006) we presented a prediction algorithm based on the very popular method of Image Ray Tracing (IRT). The algorithm determines all possible paths between each pair of transmit and receive antennas in a given scenario and with a given number of maximal reflections per path. It then identifies the points, where rays following these paths interact with the surrounding and calculates the impact of these interactions upon the electromagnetic field. We showed that the prediction of the channel capacity using this algorithm is in general very accurate but produces certain discrepancies for some scenarios.

Figure 2 shows some actual predictions of this algorithm for 27 indoor scenarios. The measurement of the channel capacity of this scenarios has been done using our laboritory MIMO system presented in Weikert and Zölzer (2005). In all scenarios there are $n_\mathrm{T} = 4$ transmit antennas and $n_\mathrm{R} = 4$ receive antennas, the carrier frequency is 2.49 GHz and the SNR amounts 30 dB.

Especially for the systems number 4, 5, 13, 14 and 15 the prediction error is very high.

In order to find the reason for the erroneous prediction of this scenarios, we prospected the propagation paths, which were used during the Ray Tracing process and those which were not used due to a too high number of reflections. It turned out that in all of this scenarios there are paths with a very low damping but with a very high number of reflections, mostly at metallic materials.

Due to their low damping those paths imperatively have to be considered when calculating the impulse response, but because of their high number of reflections they were not considered by the algorithm. This error source is inherent for ray tracing based algorithms and can only be bated by increasing the maximum number of reflections per path.

However this measure does not solve the problem generally and is very expensive in terms of computation time and needed memory, because the complexity of Ray Tracing is of the order w^R for w walls and a maximal number of reflections per path R.

4 Multi Channel Coupling

Multi Channel Coupling (MCC) is a prediction algorithm, which takes an infinite number of interactions per path into account. It thereby avoids the mentioned error source, which is inherent for Ray Tracing based algorithms. MCC was first presented as algorithm to predict the transmission of power in 2-D-scenarios in Liebendorfer and Dersch (1997). This algorithm is based upon the following considerations:

- For every pair of walls there are two channels, one in each direction.
- Transmit antennas couple a certain percentage of their transmitted power into each channel.
- Each channel couples a certain percentage of the present power into each other channel.
- Each channel couples a certain percentage of the present power into each receive antenna.

Each coupling of transmit antennas into channels includes one reflection at or transmission through the start wall of the channel and each coupling from a channel into another channel is a reflection or transmission. Every coupling is described by a coupling factor between 0 and 1 and the coupling factors are organized in matrices: a $(n_C \times n_T)$ matrix $\mathbf{T}$ for the coupling factors of the n_T transmit antennas into the n_C channels, a $(n_C \times n_C)$ matrix $\mathbf{C}$ for the coupling factors of channels into each other and a $(n_R \times n_C)$ matrix $\mathbf{R}$ for the coupling factors of the channels into the n_R receive antennas.

Assuming no direct component and considering an infinite number of couplings from channels into channels, i.e. an infinite count of reflections and transmissions, the power transfer matrix from the transmit antennas to the receive antennas can then be calulated by the equation

$$\begin{aligned}\mathbf{P} &= \mathbf{R}\cdot(\mathbf{T}+\mathbf{C}\mathbf{T}+\mathbf{C}^2\mathbf{T}+\cdots+\mathbf{C}^\infty\mathbf{T}) \\ &= \mathbf{R}\cdot\textstyle\sum_{i=0}^{\infty}\mathbf{C}^i\mathbf{T}=\mathbf{R}\cdot(\mathbf{I}-\mathbf{C})^{-1}\mathbf{T}\end{aligned} \qquad (7)$$

Thus MCC considers an infinite number of reflections and transmissions with the computation time being finite. Moreover the most computation time is needed for the calculation of the matrix $(\mathbf{I}-\mathbf{C})^{-1}$, which only depends on the location and the material parameters of walls in the surrounding. Thus MCC is very effective, when calculating the same scenario several times with different antenna positions.

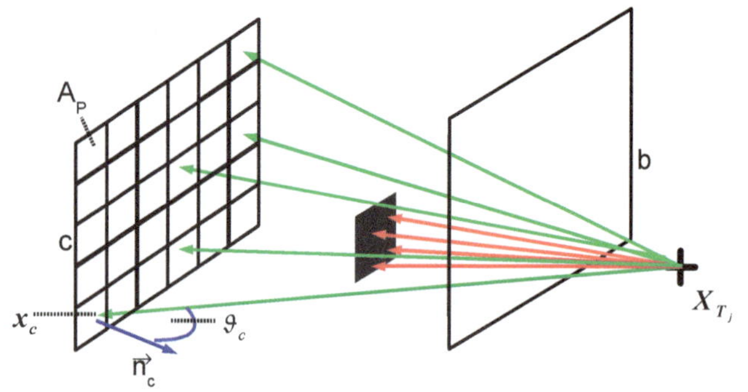

Fig. 3. Coupling from antenna j into channel i; green rays contribute to the coupling factor.

It was already shown, that the MCC method can also be used to predict other values than power. In Karthaus (2001) it was used to predict power delay profiles in 3-D-scenarios.

To be able to predict the channel capacity of frequency selective MIMO systems with and without line of sight using the MCC method, we write Eq. (7) as

$$\mathbf{H}(f)=\mathbf{R}(f)\cdot(\mathbf{I}-\mathbf{C}(f))^{-1}\mathbf{T}(f)+\mathbf{D}(f) \qquad (8)$$

with f being the frequency, $\mathbf{H}(f)$ being the complex transfer function in the frequency domain and the $(n_R \times n_T)$ matrix $\mathbf{D}(f)$ containing the coupling factors of the direct paths between each pair of transmit and receive antennas.

In the following we will present our formulas for the prediction of the channel capacity with Eq. 8.

4.1 The elements of the matrix $\mathbf{T}(f)$

The element T_{ij} describes, how an impulse at the transmit antenna j is coupled into channel i.

To calculate T_{ij}, the end wall of channel i is discretized. Then for each ray p from the transmit antenna T_j through one of the discrete wallelements with the area A_c it is calculated, how the amplitude and the phase of an impulse transmitted by the antenna is altered by the coupling.

Only rays, which intersect with the start wall of the channel and are not disturbed by other walls can contribute to the coupling factor. The impact on the amplitude is represented by τ_p and calculated using the portion of the solid angle, under which the wallelement is seen from the position of the transmit antenna, the antenna gain G_j and the absolute directional characteristic $|C_{j,p}|$ of the antenna at path p and the transmission / reflection coefficient $r_p(f)$ at the start wall of the channel. The superposition of all rays yields the absolute value of the coupling coefficient.

The impact on the phase is the sum of the phase of the transmission / reflection coefficient $r_p(f)$ and the phase shift

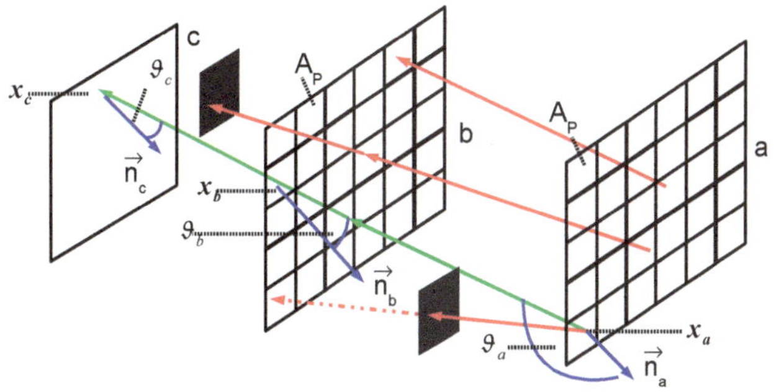

Fig. 4. Coupling from channel j into channel i; green rays contribute to the coupling factor.

φ_0 caused by the free space propagation from the antenna to the wallelement. The phase of each ray p is weighted with the according value τ_p and the mean value of all weighted phases is the phase of the coupling coefficient:

$$T_{ij}(f) = \sqrt{\sum_p \tau_p(f)} \cdot e^{j\varphi(f)} \tag{9}$$

with

$$\begin{aligned}
\tau_p(f) &= \Delta_1 \frac{A_c \cdot \cos(\vartheta_c)\left|x'_{T_j} - x_C\right|^{-2}}{4\pi} \cdot G_j \left|C_{j,p}\right| \left|r_p(f)\right| \\
\varphi(f) &= \frac{\sum_p (\arg(r_p(f)) + \varphi_{0_p}(f)) \cdot \tau_p(f)}{\sum_p \tau_p} \\
\Delta_1 &= \begin{cases} 1 \text{ if ray from } T_j \text{ through } A_c \text{ and } b \\ \quad \text{is not interrupted} \\ 0 \text{ else} \end{cases}
\end{aligned} \tag{10}$$

4.2 The elements of the matrix C(f)

The element C_{ij} describes the coupling from channel j into channel i and is the expectation value of the ratio between the transfer function in channel i and that one in channel j.

For the calculation, the start and the end wall of channel j are discretized. Of all rays p_α from a discrete wall element of the start wall a of channel j through a discrete wall element α of the end wall b of channel j, only those, which hit the end wall of channel i and are not disturbed by other walls, contribute to the coupling. Again the absolute value of the coupling factor is obtained by the superposition of all contributing rays and its phase is the average of the weighted phases of the rays.

The wall element at wall a has the area A_a and is thought of as full radiator, which transmits in every direction proportional to $A_a \cos(\vartheta_a)$. The remaining, the coupling coeficcient defining quantities are the portion of the solid angle, under which the wallelement of wall b is seen from the position x_a and the transmission / reflection coefficient $r_p(f)$ at wall b. For the calculation of all transmission or reflexion coefficients the wave matrix method (Layer, 2001) is used, which allows the definition of walls with layers of different materials.

$$C_{ij}(f) = \sqrt{\frac{\sum_\alpha \sum_{p_\alpha} X_{p_\alpha}}{\sum_\alpha \sum_{p_\alpha} \chi_{p_\alpha}(f)}} \cdot e^{j\varphi(f)} \tag{11}$$

with

$$\begin{aligned}
\chi_p(f) &= \Delta_1 A_a cos(\vartheta_a) A_b \cos(\vartheta_b) \left|\boldsymbol{x}_a - \boldsymbol{x}_b\right|^{-2} \\
X_p &= \Delta_2 \chi_p(f) \left|r_p\right| \\
\varphi(f) &= \frac{\sum_\alpha \sum_{p_\alpha} (\arg(r_{p_\alpha}(f)) + \varphi_{0_{p_\alpha}}(f)) \cdot \Delta_2 \chi_{p_\alpha}(f) \left|r_{p_\alpha}(f)\right|}{\sum_\alpha \sum_{p_\alpha} \Delta_2 \chi_{p_\alpha}(f) r_{p_\alpha}(f)} \\
\Delta_1 &= \begin{cases} 1 \text{ if line from } A_a \text{ to } A_b \\ \quad \text{is not interrupted} \\ 0 \text{ else} \end{cases} \\
\Delta_2 &= \begin{cases} 1 \text{ if ray from } A_a \text{ through } c \\ \quad \text{is not interrupted} \\ 0 \text{ else} \end{cases}
\end{aligned} \tag{12}$$

If the points $\boldsymbol{x}_b$ and $\boldsymbol{x}_a$ are equal, which is possible, if the walls a and b do intersect each other, only the portion of the solid angle, under which the wallelement of wall c is seen from the position x_a and the number N_c of rays starting at $\boldsymbol{x}_a$, which hit wall c are used to calculate the coupling factor. χ_p then is $\chi_p = 1$ and

$$X_p = \frac{1}{N_c} \sum_p \Delta_2 \frac{A_c \cos(\vartheta_c) \left|\boldsymbol{x}_a - \boldsymbol{x}_c\right|^{-2}}{4\pi}. \tag{13}$$

4.3 The elements of the matrix R(f)

The element R_{ij} describes how the signal is coupled from channel j into the receive antenna i. The coupling factor is unequal to zero for all non disturbed rays P_α from a discrete wall element of the start wall a of channel j through the position of the receive antenna and a discrete wall element α of the end wall b of channel j. Every ray is thought of as representing a subchannel of channel j, having the cross section area A_a at wall a, A_R at the receive antenna and A_b at wall b. The discrete wall element at wall a is again thought of as full radiator. The coupling factor is further determined by the portion of the solid angle, under which the wallelement of wall b is seen from the position x_a and the ratio between the cross section area A_R and the aperture of receive antenna $\frac{G_i |C_{i,p}| \lambda^2}{4\pi}$, where G_i ist the antenna gain and $C_{i,p}$ is the directional characteristic of antenna i at path p.

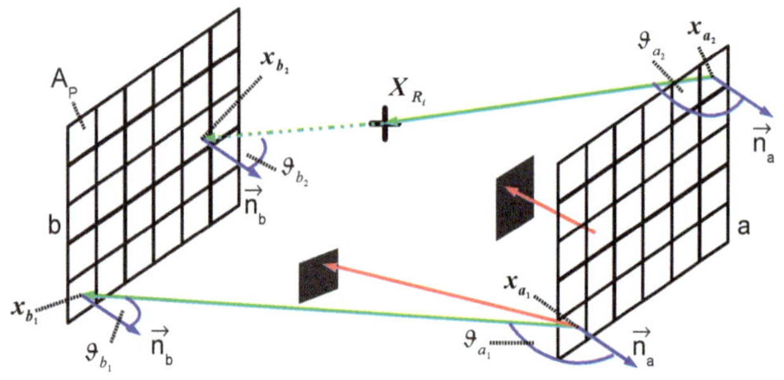

Fig. 5. Coupling from channel j into antenna i; green rays contribute to the coupling factor.

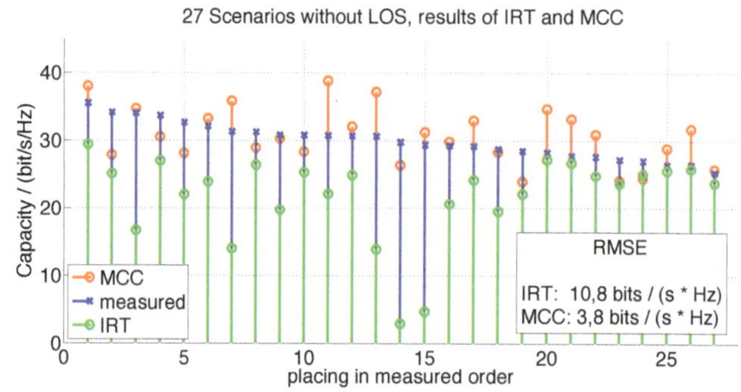

Fig. 6. Comparison of MCC and IRT.

$$R_{ij}(f) = \sqrt{\frac{\sum_\alpha \sum_{p_\alpha} \Delta_1 G_i \left|C_{i,p}\right| \frac{\lambda^2}{4\pi A_R} \rho_\alpha}{\sum_\alpha \sum_{p_\alpha} \Delta_2 \rho_\alpha}} \cdot e^{j\varphi(f)} \tag{14}$$

with

$$\rho(f) = A_a \cos(\vartheta_a) A_b \cos(\vartheta_b) |\boldsymbol{x}_a - \boldsymbol{x}_b|^{-2}$$

$$\varphi(f) = \sum_\alpha \frac{\varphi_{0_{p_\alpha}}(f) \rho_{p_\alpha}(f)}{\sum_\alpha \rho_{p_\alpha}(f)}$$

$$\Delta_1 = \begin{cases} 1 \text{ if ray from } A_a \text{ through } x_{R_i} \text{ hits } A_b \\ \quad \text{without interruption} \\ 0 \text{ else} \end{cases} \tag{15}$$

$$\Delta_2 = \begin{cases} 1 \text{ if ray from } A_a \text{ hits } A_b \\ \quad \text{without interruption} \\ 0 \text{ else} \end{cases}$$

Using the intercept theorem one can show that

$$A_R = 4\left(\frac{|\boldsymbol{x}_b - \boldsymbol{x}_R|}{|\boldsymbol{x}_a - \boldsymbol{x}_b|}\right)^2 A_a \cos(\vartheta_a) \tag{16}$$

and that leads to a formula for $R_i j$ which needs only one summation over all rays r from the discrete wall elements of wall a to the receive antenna:

$$R_{ij}(f) = \sqrt{\frac{\sum_r \kappa_r}{\sum_\alpha \sum_{p_\alpha} \Delta_2 \rho_\alpha}} \cdot e^{j\varphi(f)} \tag{17}$$

with

$$\kappa(f) = \Delta_1 \frac{\lambda^2}{16\pi} G_i \left|C_{i,p}\right| \cos(\vartheta_a) |\boldsymbol{x}_a - \boldsymbol{x}_{BS}|^{-2}$$

$$\varphi(f) = \sum_r \frac{\varphi_{0_r}(f) \kappa_r(f)}{\sum_r \kappa_r(f)} \tag{18}$$

and ρ, Δ_1 and Δ_2 remaining the same as in Eq. (15).

4.4 The elements of the matrix $\mathbf{D}(f)$

The element D_{ij} describes the free space distribution between transmit antenna j and receive antenna i. It is calculated as

$$D_{ij}(f) = \frac{r_0}{r} \cdot e^{-j\frac{2\pi}{\lambda}(r - r_0)} \tag{19}$$

with r being the distance between the antennas and $r_0 = \lambda(4\pi)^{-1}$. If the first fresnel zone is not free of obstacles, D_{ij} is proportional to r^{-2} instead of r^{-1}, which is known as double regression model.

5 Appraisal of prediction results

Figure 6 shows the results of the prediction of the 27 scenarios mentioned in Sect. 3 using the new algorithm. For comparison with the results of the Ray Tracing based algorithm and the measurements, this data is also shown in the figure. One can see, that the prediction using MCC does not produce the same big errors as that using IRT.

By avoiding the errors, which are inherent to measures, which can only handle a finite count of interactions, Multi Channel Coupling outperforms Image Ray Tracing in terms of the root mean square error (RMSE). While the RMSE of the Ray Tracing based prediction amounts up to 10.8 bits $(\text{Hz s})^{-1}$, MCC has a RMSE of only 3.8 bits $(\text{Hz s})^{-1}$, which is approximately a third of the first one. To further support the thesis above, Fig. 5 shows the absolute deviations of MCC and IRT from measurements for 108 indoor scenarios, 54 with line of sight (LOS) and 54 without one. In a scenario with a line of sight, usually the LOS is the path, where the most power is transfered to the receiver. Concomitantly, the LOS is always taken into recognition while the Ray Tracing based prediction, because the number of reflections on that path is zero.

Because of that and as the Ray Tracing based approach fails only when paths with a lot of interactions transfer a big part of the received power, one would expect the Ray Tracing based approach, to perform as good as MCC when the predicted scenario contains a line of sight. The results of the 54 systems with LOS confirm this expectation and that of the 54 systems without LOS show again, that MCC avoids the big errors, which cannot be avoided using Ray Tracing.

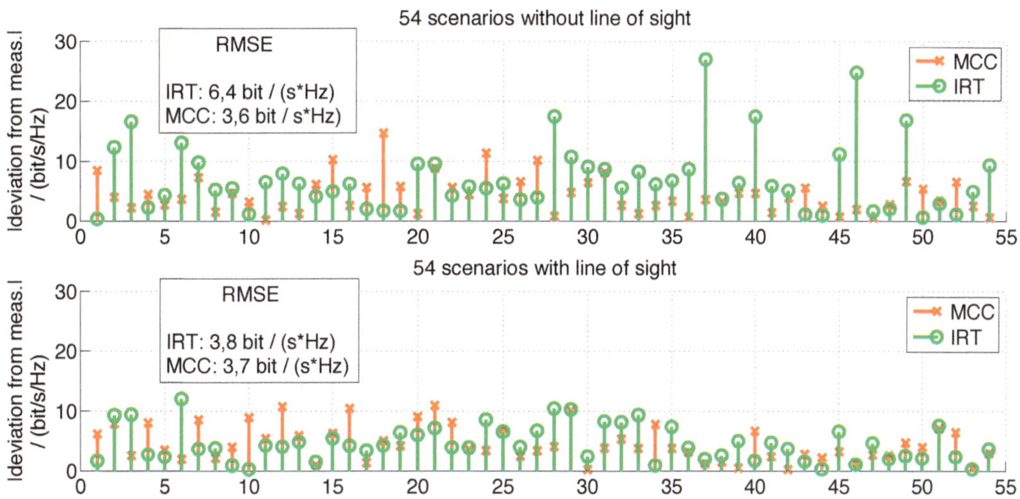

Fig. 7. Comparison of MCC and IRT: 108 scenarios.

6 Conclusions

We explained, why in our opinion only the channel capacity is an adequate quantity to judge the transmission quality of MIMO systems and showed, that Ray Tracing under certain circumstances produces big prediction errors when predicting the channel capacity of scenarios without a line of sight. Those errors can not be excluded in general as long as one uses Ray Tracing as prediction measure.

We then presented a new algorithm based upon the concept of Multi Channel Coupling, which accounts for an infinite number of interactions of the rays with the surrounding and thereby avoids the identified error source. The new algorithm was validated against a Ray Tracing based algorithm and against measurements.

For scenarios with line of sight, the prediction results using our approach were as good as with Image Ray Tracing, for scenarios without a line of sight, our approach outperformed the Ray Tracing based algorithm remarkably.

References

Bauch, G. and Al-Dhahir, N.: Reduced-Complexity Space-Time Turbo-Equalization for Frequency-Selective MIMO Channels, IEEE Transactions on Wireless Communications, 1, 819–829, 2002.

Elnaggar, M. S., Safavi-Naeini, S., and Chaudhuri, S. K.: Simulation of the Achievable MIMO Capacity by Using Adaptive Phased-Array, in: Proc. of the IEEE Radio and Wireless Conference (RAWCON), pp. 155–158, Atlanta (GA), 2004.

Foschini, G. J. and Gans, M. J.: On Limits of Wireless Comunication in a Fading Environment when Using Multiple Antennas, Wireless Personal Communications, 6, 311–335, 1998.

Fügen, T., Sommerkorn, G., Maurer, J., Hampicke, D., Wiesbeck, W., and Thomä, R.: MIMO Capacities for Different Antenna Array Structures Based on Double Directional Wide-Band Channel Measurements, in: Proc. of the 13th IEEE International Symposium on Personal, Indoor and Mobile Radio Communications (PIMRC), 3, 1777–1781, Lissabon, 2002.

Hagebölling, F., Weikert, O., and Zölzer, U.: Deterministic Prediction of the Channel Capacity of Frequency Selectice MIMO Systems, in: Proc. 11th International OFDM-Workshop 2006, pp. 208–212, Hamburg, 2006.

Karthaus, U.: Vermessung und Simulation breitbandiger zeitvarianter Funkkanäle im Bereich 30–60 GHz, Dissertation, Universität Gesamthochschule Paderborn, Shaker Verlag, Aachen, ISBN 3-8265-8699-9, 2001.

Layer, F.: Synthese und Analyse zeitvarianter Indoor-Mobilfunkkanäle mittels strahlenoptischer Verfahren, Dissertation, Universität Gesamthochschule Kassel, Shaker Verlag, Aachen, ISBN 3-8265-8455-4, 2001.

Liebendorfer, M. and Dersch, U.: Multi Channel Coupling, a new method to predict indoor propagation, in: Proc. of the 47th IEEE Vehicular Technology Conference, Phoenix, AZ, USA, 3, 1405–1409, 1997.

Palomar, D., Fonollosa, J., and Lagunas, M.: Capacity Results on Frequency-Selective Rayleigh MIMO Channels, in: IST Mobile Communications Summit, Galway, Irland, pp. 491–496, 2000.

Talbi, L.: Simulation of Indoor UHF Propagation Using Numerical Technique, in: Proc. of the Canadian Conference on Electrical and Computer Engineering 2001, 2, 1357–1362, Toronto, 2001.

Tang, Z. and Mohan, A. S.: Experimenal Investigation of Indoor MIMO Ricean Channel Capacity, IEEE Antennas and Wireless Propagation Letters, 4, 55–58, 2005.

Weikert, O. and Zölzer, U.: A Flexible Laboratory MIMO System Using Four Transmit and Four Receive Antennas, in: Proc. of the 10th International OFDM-Workshop 2005, Hamburg, Germany, pp. 298–302, 2005.

Ziri-Castro, K. I., Scanlon, W. G., and Evans, N. E.: Prediction of Variation in MIMO Channel Capacity for the Populated Indoor Environment Using a Radar Cross-Section-Based Pedestrian Model, IEEE Transactions on Wireless Communications, 4, 1186–1194, 2005.

Forward and backward RLS-DDCE processing in MIMO-OFDM spatial multiplexing receivers

M. Muxfeldt, P. Beinschob, and U. Zölzer

Department of Signal Processing and Communications, Helmut-Schmidt-Universität/University of the Federal Armed Forces Hamburg, Holstenhofweg 85, 22043 Hamburg, Germany

Abstract. In this paper we present a novel approach in frequency domain channel estimation technique. Our proposal is based on the recursive least squares (RLS) algorithm combined with the decision making process called decision directed channel estimation (RLS-DDCE). The novelty and key concept of this technique is the block-wise causal and anti-causal RLS processing that yields two independent processing of RLS along with the associated decisions. Due to the implemented low density parity check (LDPC) code the receiver operates with soft information, which enables us to introduce a new modification of the Turbo principle as well as simple addition of the a-posteriori log-likelihood ratios (LLRs). Although the computational complexity is increased by both of our approaches, the latter is relatively less complex than the earlier. Simulation results show that these implementations outperform the simple RLS-DDCE algorithm and yield lower bit error rates (BER) and more accurate channel information.

1 Introduction

A widespread modulation technique used in today's communication systems is orthogonal frequency-division multiplexing (OFDM), which combines high spectral efficiency, robustness against inter-symbol interference and an easy implementation using the fast Fourier transform (FFT). Combining the OFDM system with a multiple-input-multiple-output (MIMO) system, a MIMO-OFDM system is created, which results in a higher spectral efficiency and link reliability (Bölcskei et al., 2002; Uysal et al., 2001).

Although the development of OFDM is already very advanced, there still exists enough research potential in MIMO-OFDM. Today MIMO systems can be found in wireless local area network access points, WiMAX or some 3GPP specifications.

Especially under bad transmission conditions with small signal-to-noise ratios (SNR) or high mobility, there are broad possibilities to improve the performance of MIMO-OFDM systems. High mobility involves a highly time-variant channel, which causes the spectral efficiency to decrease. A solution to countervail this degradation is the enhancement of detailed knowledge of the channel state information (CSI).

An exceedingly important application area for improvement is in the domain of vehicle to vehicle (V2V) and vehicle to infrastructure (V2I) communication, where high relative velocities create a time-variant channel combined with bad SNR scenarios due to obstacle obstructed communication channels.

Receiver designs for MIMO-OFDM which make acceptable use of diversities are rare. There are few researches focusing on iterative receiver architechture (Akhtman and Hanzo, 2007b; Zhang et al., 2006; Liu et al., 2003; Sun et al., 2004), which exploit the Turbo principle with its iterative decoding structure. Even though they result in higher computational complexity these receivers seem to be promising in relation to BER performance.

Since the LDPC codes possess similar performance when compared to Turbo Codes, they are implemented in MIMO-OFDM systems as well (Lu et al., 2004). Also the combination of the Turbo Principle and LDPC codes is under research (Salari et al., 2007).

In this paper we propose a novel Turbo processing based on the LDPC codes, where the information gain is obtained from a causal RLS-DDCE processing and an independent anti-causal processing. These two possible strategies are presented here. Simple information combined by summing up of a-posteriori information and Turbo processing by exchange of extrinsic information between the forward RLS-DDCE process and the backward fork.

Correspondence to: P. Beinschob
(patric.beinschob@hsu-hamburg.de)

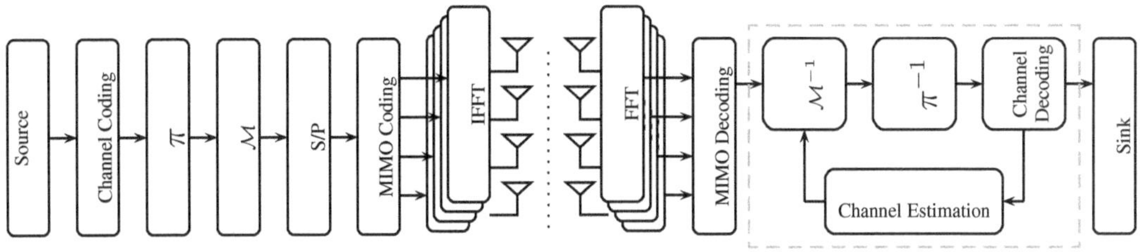

Fig. 1. General structure of the underlying system, represented for a 4×4 system.

The rest of the paper is organized as follows. The underlying system model and structure are presented in Sect. 2, followed by the description of the RLS-DDCE algorithm in Sect. 3. Our novel approach with detailed information about the modified Turbo principle and summation of a-posteriori LLRs is presented in Sect. 4, respectively in Sects. 4.1 and 4.2. The paper is concluded by illustrating our simulation results in Sect. 5 and a conclusion in Sect. 6.

2 System model and structure

The vector of received values $\boldsymbol{r}$ at the time sample m of a MIMO system is the superposition of $L \cdot n_{\mathrm{T}}$ previously sent samples and the current n_{T} samples, where $L+1$ is the length of the sampled channel impulse response and n_{T} is the number of transmit antennas. It is given by

$$\boldsymbol{r}[m] = \sum_{l=0}^{L} \mathbf{h}[l,m] \cdot \mathbf{s}[m-l] + \mathbf{w}[m], \tag{1}$$

where $\mathbf{s}[m]$ denotes the current vector of symbols of each of the transmit antenna, $\mathbf{w}$ is an identically, independently distributed (iid) additive white Gaussian noise term and $\mathbf{h}[l,m]$ is the MIMO channel matrix in delay and time domain, indexed with l respectively m. The past sent samples are denoted by $\mathbf{s}[m-l]$, for $l \neq 0, l \leq L$. For simulations the data symbols of the K subcarriers are modulated by an inverse fast Fourier transform (IFFT). In simulations every value corresponding to a transmit antenna of the resulting vectors is transmitted using the formula above.

In frequency domain the system model in Eq. (1) can be described as

$$\boldsymbol{r}[n,k] = \mathbf{H}[n,k] \cdot \mathbf{s}[n,k] + \mathbf{w}[n,k], \tag{2}$$

where n denotes the time index of an OFDM symbol and k its subcarrier index. The vectors $\boldsymbol{r}[n,k]$ and $\mathbf{w}[n,k]$ are of dimension $n_{\mathrm{R}} \times 1$, $\mathbf{s}[n,k]$ of $n_{\mathrm{T}} \times 1$ and the matrix $\mathbf{H}[n,k]$ of $n_{\mathrm{R}} \times n_{\mathrm{T}}$, at which n_{R} is the number of receive antennas. In simulations the MIMO channel coefficients $H_{\mathrm{r,t}}[n,k]$, $r = 1,\ldots,n_{\mathrm{R}}$, $t = 1,\ldots,n_{\mathrm{T}}$ are modeled using the 3GPP spatial model which is developed to evaluate receiver algorithms in MIMO scenarios (3GP, 2008).

An overview of the implemented general system structure can be seen in Fig. 1. At first the bits for one OFDM transmission frame, arranged in vectors $\boldsymbol{u}_{\mathrm{v}}$, $v = 1,\ldots,n_{\mathrm{c}}$, with n_{c} being the number of codewords, are LDPC (MacKay, 1999) encoded and implemented according to Richardson et al. (2001), using the coding matrix $\mathbf{G}$

$$\boldsymbol{u}_{\mathrm{v,c}} = \mathbf{G} \cdot \boldsymbol{u}_{\mathrm{v}}. \tag{3}$$

The vectors $\boldsymbol{u}_{\mathrm{v,c}}$ contain the regular information bits u_{h} and the parity bits u_{f}, $h = 1,\ldots,n_{\mathrm{s}}$, $f = 1,\ldots,n_{\mathrm{p}}$, where n_{s} denotes the information length and n_{p} denotes the number of parity bits. Following the transmitter scheme the systematic and parity bits are then interleaved and quadrature amplitude modulated as shown by π respectively $\mathcal{M}$ in Fig. 1. At this stage the generated symbols are serial to parallel converted and MIMO encoded, which is performed by spatial multiplexing by multiplying with the unity matrix in the underlying system, resulting in vectors $\boldsymbol{s}[n,k]$, which are symbol wise fed into the IFFT, one for each transmit antenna.

In the receiver, the superposed received signals are transferred back into the frequency domain with the help of a FFT, resulting in the vectors $\boldsymbol{r}[n,k]$ of Eq. (2). Subsequently the RLS-DDCE algorithm is performed on each of the OFDM symbols. The demodulator produces soft information in form of a-priori LLRs, $L(u_{\mathrm{h}})$, which are defined as the logarithm of the ratio of the probabilities of a bit u_{h} being 0 or 1 . These LLRs are LDPC decoded, producing the estimated a-posteriori LLRs at the output of the LDPC decoder:

$$L(u_{\mathrm{h}}|\mathbf{y}) = \ln\left(\frac{P(u_{\mathrm{h}}=1|\mathbf{y})}{P(u_{\mathrm{h}}=0|\mathbf{y})}\right), \tag{4}$$

where $\mathbf{y}$ is the sequence of received bits for one codeword. Together with the estimated a-posteriori LLRs the LDPC decoder produces the parity check sum (PCS):

$$\mathbf{A} \cdot \boldsymbol{u}_{\mathrm{v,c}} \equiv \boldsymbol{\gamma}, \tag{5}$$

where $\boldsymbol{\gamma}$ is the syndrome vector and $\mathbf{A}$ the parity check matrix. The PCS is the Hamming distance of the syndrome vector from the null vector. In case of a codeword with PCS of zero, the codeword is assumed to be correct and to produce the transmitted symbols, the a-posteriori LLRs are decided on a hard basis, followed by the encoding, interleaving and modulating of the decided bits.

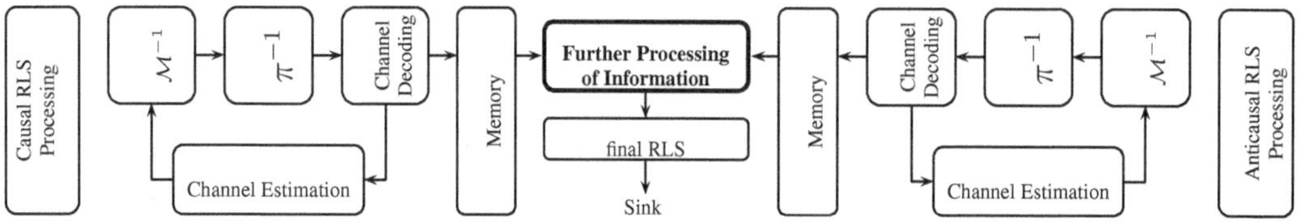

Fig. 2. Novel approach in forward and backward RLS-DDCE processing implemented on the receiver side of the system depicted in Fig. 1.

When the PCS is not zero, the soft information of the a-posteriori LLRs is used to transform the received symbols into the transmitted symbols. The symbols gained by either of the strategies are then used to calculate an estimate of the channel transfer function vectors $\tilde{\mathbf{H}}[n,k]$ via the RLS algorithm, which is further explained in Sect. 3.

The BER is calculated on the basis of the equalized received symbols. These symbols are therefore decoded,de-interleaved and decided on a hard basis.

Figure 2 shows the structure of our novel approach to the channel estimation and is explained in Sect. 4.

3 RLS-DDCE

The RLS algorithm as described in Akhtman and Hanzo (2007a) is suitable for tracking a communication channel as it computes an estimate of the current channel matrix $\tilde{\mathbf{H}}[n,k]$ upon arrival of new received data $\boldsymbol{r}[n,k]$ and converges within just a few OFDM symbols.

The introduced forgetting factor ξ associates an exponential weighting of the past transmitted signals onto the current channel factor. Therefore it can be used to adapt to the time-variant channel conditions.

For calculating the channel transfer function the autocorrelation matrix (6) of the transmitted signals and the cross-correlation matrix (7) of the transmitted and received signals are needed. Initialisation of both matrices is performed by using the identity matrix $\mathbf{I}$.

$$\begin{aligned}\boldsymbol{\Phi}[n,k] &= \sum_{m=1}^{n} \xi^{n-m} \cdot \mathbf{s}[m,k] \cdot \mathbf{s}^H[m,k] \\ &= \xi \cdot \boldsymbol{\Phi}[n-1,k] + \mathbf{s}[n,k] \cdot \mathbf{s}^H[n,k] \end{aligned} \tag{6}$$

$$\begin{aligned}\boldsymbol{\theta}[n,k] &= \sum_{m=1}^{n} \xi^{n-m} \cdot \mathbf{s}[m,k] \cdot \boldsymbol{r}^H[m,k] \\ &= \xi \cdot \boldsymbol{\theta}[n-1,k] + \mathbf{s}[n,k] \cdot \boldsymbol{r}^H[n,k] \end{aligned} \tag{7}$$

Following the development of the RLS algorithm the solution for calculating the channel factors ends up in the matrix form of the normal equation, that can be formed to calculate the channel factors by inverting the autocorrelation matrix, as can be seen in the following equation:

$$\tilde{\mathbf{H}}[n,k] = \left(\boldsymbol{\Phi}^{-1}[n,k] \cdot \boldsymbol{\theta}[n,k]\right)^H . \tag{8}$$

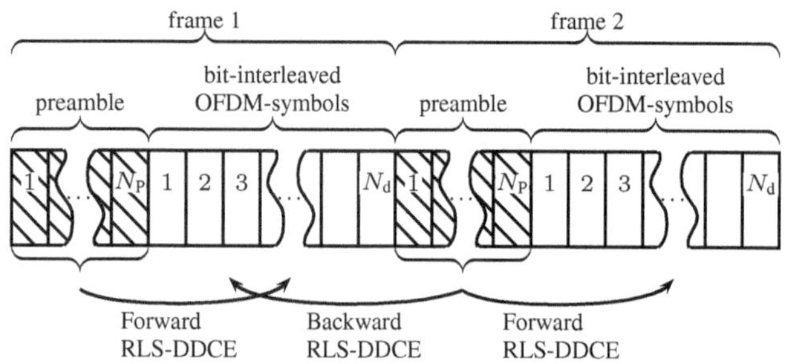

Fig. 3. Frame structure for proposed MIMO-OFDM RLS-DDCE Forward and Backward Filtering with re-use of next and previous frame preambles.

The inversion of the autocorrelation matrix can be avoided by using the matrix inversion lemma, the description of which is omitted here.

Summation of rank-1 matrices in Eqs. (6) and (7) is avoided by starting each transmission frame with a training sequence of known pilot symbols, as the matrices have full condition after a few summations. The transmitted symbols are known in the receiver and the channel transfer function estimates $\tilde{\mathbf{H}}[n,k]$ can instantly be calculated. The symbols following the pilot symbols have to be decided in the receiver, for that reason the received symbols are equalized, decoded and detected before calculation of the channel factors.

4 Forward and backward RLS processing

The RLS-DDCE algorithm provides the receiver with information on the channel and also results in the transmitted signals. Due to the uncertainty in the decision making process it is prone to error propagation which might lead to a large amount of errors and entirely destroyed subcarriers.

Our novel proposal for this is to perform the RLS algorithm twice, in causal and anti-causal direction, which uses block-wise processing of received data, as shown in Fig. 3. This is done independently of each other since the additional information gained from that technique can be evaluated to eliminate errors and correct wrong decisions.

Figure 2 illustrates our proposed approach. On the left side the causal RLS processing is shown, which yields the a-posteriori LLRs $L_1(u_h|\mathbf{y})$ out of the received symbols $\boldsymbol{r}[n,k]$

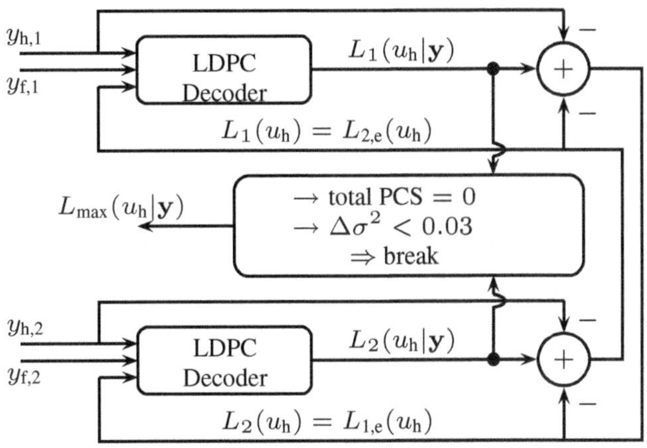

Fig. 4. Modified Turbo principle.

by soft demodulation and decoding, depicted by $\mathcal{C}^{-1}$ in Eq. (9). The right part of the figure processes the received data anti-causal wise, Eq. (10), where the length of the pilot symbols is denoted by N_P and the data length by N_d. This routine can be described as:

$$L_1(u_h|\mathbf{y}) = \mathcal{C}^{-1}(\mathcal{M}^{-1}\{\mathbf{H}^{\dagger}[n,k]\mathbf{r}[n,k]\}), \quad (9)$$

$$L_2(u_h|\mathbf{y}) = \mathcal{C}^{-1}(\mathcal{M}^{-1}\{\mathbf{H}^{\dagger}[\nu,k]\mathbf{r}[\nu,k]\}), \quad (10)$$
$$\nu = (N_P + N_d) - n,$$

where $\mathbf{H}^{\dagger}$ denotes the pseudo-inverse of the channel matrix and ν the anti-causal time index.

The gained information in the form of a-posteriori LLRs (4) is passed on for further evaluation. Due to this post-evaluation of the first two RLS processing, depending on the SNR and actual channel conditions, some of the codewords in the receiver can be considered as being correct. These codewords provide additional reliable information for the final RLS processing and will not be decided again. In case the post-evaluation does not result in any correct codewords, the final RLS processing is performed using the information of either the causal or anticausal RLS processing.

An important part of the proposed approach is the application of soft information, which is used to calculate the transmitted symbols in case of incorrect decoding. This additional usage of the soft information is exploited to calculate the BER, contrary to the normal RLS-DDCE algorithm.

The underlying frame structure only provides a training sequence at the beginning of each frame, and therefore the training sequence of the subsequent transmission frame can be used for the anti-causal RLS processing. Exploiting the incremental overhead twice comes at no additional overhead cost, but results in an additional channel information gain. Figure 3 gives an idea of the double usage of the pilot symbols.

4.1 Modified Turbo principle for forward and backward RLS processing

For our purpose the original Turbo principle of Berrou et al. (1993) and Berrou and Glavieux (1996), nicely explained by Hanzo et al. (2002), is changed, as presented in Fig. 4. We totally ignore the encoding part of the original Turbo Coding and solely perform the normal LDPC coding as presented in Sect. 2, which will lead to the combination of LDPC and Turbo decoding in the receiver.

On the receiver side we retain the Turbo decoding layout, though we change the inputs to the component decoders. As the RLS algorithm is processed in causal and anti-causal manner the soft information of the received bits y_h is available twice, $y_{h,1}$ and $y_{h,2}$. The availability of two different inputs, which are supposed to be the same under perfect conditions, replaces the usage of two different codes. The extrinsic information is created and exchanged in the same way as in the original Turbo principle:

$$L_e(u_h) = L(u_h|\mathbf{y}) - L(u_h) - L_c \cdot y_h, \quad (11)$$

where L_c denotes the channel reliability.

Our proposed Turbo decoding is performed twice, starting with the soft information from the causal RLS first, followed by the anti-causal information with the adequate extrinsic information (12) and vice versa (13).

$$L_1(u_h|\mathbf{y}) \Rightarrow L_{1,e}(u_h) = L_2(u_h) \Rightarrow L_2(u_h|\mathbf{y}) \quad (12)$$

$$L_2(u_h|\mathbf{y}) \Rightarrow L_{2,e}(u_h) = L_1(u_h) \Rightarrow L_1(u_h|\mathbf{y}) \quad (13)$$

In our proposal we use the break criterion provided by Robertson (1994), which evaluates the variance of the a-posteriori LLRs. This is sensible as this criterion describes the alteration of the a-posteriori LLRs and the iteration breaks when subsequent a-posteriori LLRs do not change more than 0.03.

Nevertheless we also include an additional break criterion in order to avoid unnecessary iterations. In case the PCS for all codewords of either the causal or anti-causal RLS are zero, which is mostly the case for high SNR, the iteration of the Turbo decoding will not even start. The two break criteria are also shown in Fig. 4.

The comparably large codeword distance of LDPC codes (MacKay, 1999) can be used to evaluate the performance of certain iteration steps. A PCS of zero is likely to be equivalent to an error free codeword. And in addition, simulations have shown that a small total PCS over all codewords is connected to a lower BER.

The total PCS is then used to decide on the output of the two different iteration directions, Eqs. (14) and (15). The lower total PCS of the second component decoder determines

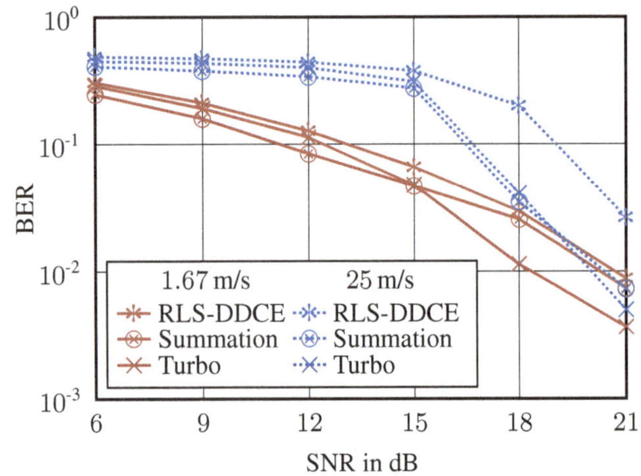

Fig. 5. BER for different SNR at velocities of 1.67 m/s and 25 m/s.

the a-posteriori LLRs $L_{\max}(u_\mathrm{h}|\mathbf{y})$ that are to be further processed:

$$\sum_{m=1}^{n_c} p_{\mathrm{v,fw}} \leq \sum_{m=1}^{n_c} p_{\mathrm{v,bw}} \Rightarrow L_{\max}(u_\mathrm{h}|\mathbf{y}) = L_2(u_\mathrm{h}|\mathbf{y}), \tag{14}$$

$$\sum_{m=1}^{n_c} p_{\mathrm{v,bw}} \leq \sum_{m=1}^{n_c} p_{\mathrm{v,fw}} \Rightarrow L_{\max}(u_\mathrm{h}|\mathbf{y}) = L_1(u_\mathrm{h}|\mathbf{y}), \tag{15}$$

where $p_{\mathrm{v,fw}}$ respectively $p_{\mathrm{v,bw}}$ describe the PCS of the codeword v after the second LDPC decoder operating on the anti-causal respectively causal data.

After the iteration the RLS algorithm is again processed in causal or anti-causal direction, depending on the origin of $L_{\max}(u_\mathrm{h}|\mathbf{y})$. The correct codewords are determined by codewords with a PCS of zero.

4.2 Summation of a-posteriori LLRs

Our second approach to improve the channel estimation is to simply sum up the a-posteriori LLRs from the forward and backward RLS processing:

$$L_{\max}(u_\mathrm{h}|\mathbf{y}) = L_1(u_\mathrm{h}|\mathbf{y}) + L_2(u_\mathrm{h}|\mathbf{y}). \tag{16}$$

This, in comparison to the modified Turbo principle, has less computational complexity and still corrects a large amount of errors.

The key issue behind this idea is the availability of the soft information, which is presented by the sign and magnitude of the LLRs. Out of the LLRs, the probability of a bit being a 1 or 0 can easily be computed:

$$P(u_\mathrm{k}=1) = \frac{1}{1+\exp(-L(u_\mathrm{h}))}, \tag{17}$$

$$P(u_\mathrm{k}=0) = \frac{1}{1+\exp(+L(u_\mathrm{h}))}. \tag{18}$$

Table 1. MIMO-OFDM system parameters.

Parameter	Value
carrier frequency	2.412 GHz
channel bandwidth	20 MHz
number of subcarriers K	128
number of OFDM symbols p. frame	576
number of pilot symbols p. frame	32
channel model	3GPP SCM
channel order L	6
norm. Doppler freq. $f_{\mathrm{d,n}}$	$0,\ldots,1.35\times10^{-3}$
equiv. velocity v in m/s	$0,\ldots,25$
LDPC design code rate	1/2

In the case when both of the RLS processing results in the same decided bits, the summation of the LLRs will not change the decisions, but enhances the reliability of the right decision. When the RLS processing decide on different bits, the larger magnitude of the soft information will determine the final bits.

The reliable codewords to conduct the final RLS processing are determined by comparing the hard decided bits, based on the a-posteriori LLRs after the summation, with the hard decided bits of the a-posteriori LLRs before the summation. A codeword is considered to be reliable when all a-posteriori LLRs of a certain codeword do not change due to the summation. To ensure the correctness of that codeword the total PCS for the codeword has to be zero, regardless of whether it is based on the causal or anti-causal RLS.

5 Simulation results

The simulation was performed on a 4×4 MIMO-OFDM system with the simulation parameter given in Table 1. For the modulation, a 4-QAM was taken so that an OFDM symbol consisted of 1024 bits. The calculated frame duration based on the paramters resulted in 3.86 ms.

The forward, backward and final RLS processing used the simple zero forcing equalizer due to computational complexity and retention of soft information. Thereby the resulting BER was not as small as possible, but the operating principle and the improvements due to our new approach were demonstrated.

Simulations were done for several velocities, ranging from 0 m/s, for comparison, to 25 m/s, suitable for a micro urban cell as simulated by the channel model. The range of interesting SNR values lied between 6 dB and 21 dB.

The forgetting factor ξ was chosen according to (Akhtman and Hanzo, 2007a) with a value of 0.7, so that the algorithm worked fine over a large range of velocities.

Figure 5 shows the comparison of the implemented algorithms for a velocity of 1.67 m/s and 25 m/s. Over the entire SNR range the simple RLS-DDCE performed worst

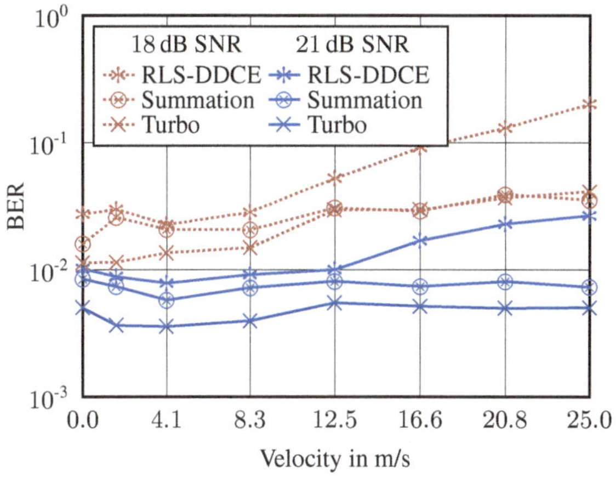

Fig. 6. BER for different velocities at 18 dB SNR respectively 21 dB SNR.

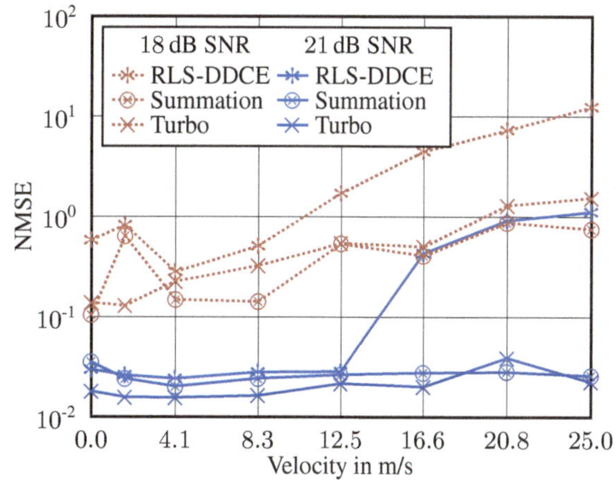

Fig. 7. NMSE for different velocities at 18 dB SNR respectively 21 dB SNR.

for both velocities. For smaller SNR, up to about 15 resp. 19 dB, the summation of a-posteriori LLRs dominated the BER performance. The Turbo principle was slightly worse, but increased in performance for higher SNR values. For small SNR values the receiver did not yield correct codewords, so the performance increase in comparison to the simple RLS-DDCE algorithm was solely due to the extended soft information evaluation of the final RLS processing. In addition the summation used the added a-posteriori LLRs for the BER evaluation in low SNR regions, which explains the better performance for small SNR values. For the upper SNR range the Turbo principle worked better due to the larger amount of correct codewords, which resulted in a smaller BER.

The simulation results in relation to the velocities are shown in Fig. 6. The RLS-DDCE algorithm again performed worst. For the SNR of 18dB the summation was comparable to the modified Turbo principle for velocities above 12.5 m/s. In case of lower velocities and for the entire velocity range at 21 dB the modified Turbo principle performed better, because of the available number of correct codewords.

In order to evaluate the channel estimation the normalized mean squared error (NMSE) is applied:

$$\mathrm{NMSE} = \frac{\sum_{n=0}^{N-1}\sum_{k=0}^{K-1}\sum_{t=1}^{n_\mathrm{T}}\sum_{r=1}^{n_\mathrm{R}} \left| H_{\mathrm{r,t}}[n,k] - \tilde{H}_{\mathrm{r,t}}[n,k] \right|^2}{\sum_{n=0}^{N-1}\sum_{k=0}^{K-1}\sum_{t=1}^{n_\mathrm{T}}\sum_{r=1}^{n_\mathrm{R}} \left| H_{\mathrm{r,t}}[n,k] \right|^2} \quad (19)$$

Figure 7 shows the NMSE in relation to the velocities for a SNR of 18 and 21 dB. For 21 dB SNR the behaviour was the same as for the BER, the modified Turbo principle resulted in the lowest NMSE values. The results were different for 18 dB SNR, where the summation of a-posteriori LLRs performed the better channel estimation. The values at 1.67 m/s and 20.8 m/s for 18 resp. 21 dB for the summation and Turbo principle presented runaway values due to insufficient channel realisations.

Figure 8 shows the iterative behavior of the implemented Turbo principle for a SNR of 18 dB and a velocity of 8.33 m/s. The figure nicely depicts how the number of codewords with a total PCS of zero increased with increasing number of iterations, for the forward and backward iteration. Along with it the number of wrong codewords, codewords with PCS of zero and biterrors, also rose. This was due to the LDPC decoder, which ran into wrong codewords due to the exchange of extrinsic information. As one can see, the curve's slope for the total PCS was flattening with increasing iterations, so that the variance break criterion became active at one point and stopped the iteration. In case the final RLS processing is able to make better decisions with more reliable channel information, then the number of iterations should not be too large in order to avoid incorrect codewords. In addition Fig. 8 presents the difference between the causal and anticausal iteration direction, as can be seen at the starting values of the curves.

6 Conclusion

In this paper we have presented a novel approach to the channel estimation process for challenging time-variant channels and the performance with respect to BER and NMSE has been evaluated over a large range of velocities. The modified Turbo principle, based on different input data for the component decoder, shows increased performance over the entire velocity range for larger SNR values, especially at the upper limit of the velocity range the performance compared to the simple RLS-DDCE is superior. At lower SNR values the performance is still better than the simple RLS-DDCE, though the applied linear equalizer prohibits better performance. The summation of a-posteriori LLRs in contrast performs better for smaller SNR values as the summation

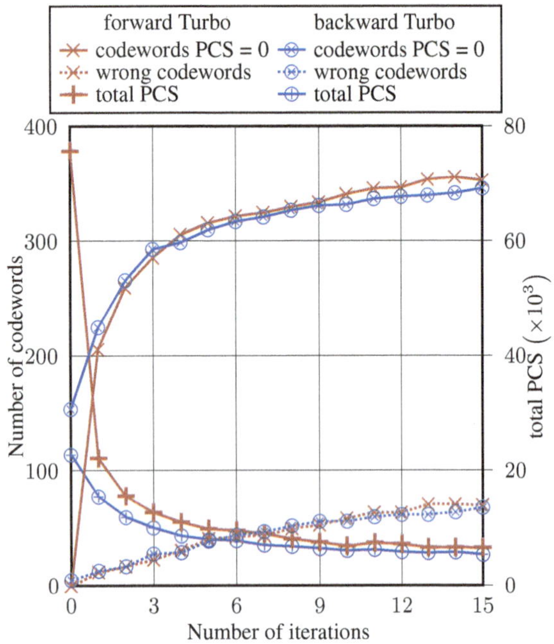

Fig. 8. Iterative behavior of the Turbo principle showing the number of correct codewords, wrong codewords and total PCS for the forward and backward Turbo iteration, at a SNR of 18 dB and 8.33 m/s.

corrects a certain amount of wrong decided symbols. The performance increase for the NMSE is comparable to the BER performance and results in a better channel estimation for the proposed approaches.

References

3GPP: Spatial channel model (SCM) for Multiple Input Multiple Output (MIMO) simulations, TR 25.996, 2008.

Akhtman, J. and Hanzo, L.: Advanced Channel Estimation for MIMO-OFDM in Realistic Channel Conditions, IEEE International Conference on Communications, 2007, ICC '07, pp. 2528–2533, doi:10.1109/ICC.2007.418, 2007a.

Akhtman, J. and Hanzo, L.: Iterative Receiver Architectures for MIMO-OFDM, in: Wireless Communications and Networking Conference, 2007.WCNC 2007, IEEE, pp. 825–829, doi: 10.1109/WCNC.2007.157, 2007b.

Berrou, C. and Glavieux, A.: Near optimum Error Correcting Coding and Decoding: Turbo-Codes, IEEE Transactions on Communications, 44, 1261–1271, 1996.

Berrou, C., Glavieux, A., and Thitimajshima, P.: Near Shannon limit error-correcting coding and decoding: Turbo-codes (1), IEEE International Conference on Communications, 2, 1064–1070, 1993.

Bölcskei, H., Gesbert, D., and Paulraj, A.: On the Capacity of OFDM-Based Spatial Multiplexing Systems, IEEE Transactions on Communications, 50, 225–234, 2002.

Hanzo, L., Liew, T. H., and Yeap, B. L.: Turbo coding, turbo equalisation, and space-time coding for transmission over fading channels, Wiley, Chichester, 2002.

Liu, H., Wang, X., and Xiong, Z.: Iterative receivers for OFDM coded broadband MIMO fading channels, IEEE Workshop on Statistical Signal Processing, pp. 355–358, 2003.

Lu, B., Yue, G., and Wang, X.: Performance Analysis and Design Optimization of LDPC-Coded MIMO OFDM Systems, IEEE Transactions on Signal Processing, 52, 348–361, 2004.

MacKay, D.: Good error-correcting codes based on very sparse matrices, IEEE Transactions on Information Theory, 45, 399–431, doi:10.1109/18.748992, Mar 1999.

Richardson, T. J., Shokrollahi, M. A., and Urbanke, R. L.: Design of capacity-approaching irregular low-density parity-check codes, IEEE Transactions on Information Theory, 47, 619–637, doi:http://dx.doi.org/10.1109/18.910578, 2001.

Robertson, P.: Illuminating the structure of code and decoder of parallelconcatenated recursive systematic (turbo) codes, IEEE Global Telecommunications Conference, 3, 1298–1303, 1994.

Salari, S., Ardebilipour, M., Ahmadian, M., Cances, J., and Meghdadi, V.: Turbo Receiver Design with Carrier-Frequency Offset Estimation for LDPC-Coded MIMO OFDM Systems, International Conference on Advanced Communication Technology, 3, 1911–1915, 2007.

Sun, S., Wu, Y., Li, Y., and Tjhung, T.: A novel iterative receiver for coded MIMO OFDM systems, IEEE International Conference on Communications, 4, 2473–2477, 2004.

Uysal, M., Al-Dhahir, N., and Georghiades, C.: A SpaceTime Block-Coded OFDM Scheme for Unknown Frequency-Selective Fading Channels, IEEE Communications Letters, 5, 393–395, 2001.

Zhang, L., Sellathurai, M., and Chambers, J.: A Space-Time Coded MIMO-OFDM Multiuser Application With Iterative MMSE-Decision Feedback Algorithm, 8th International Conference on Signal Processing, 3, 2006.

Cyber-physical networking for wireless mesh infrastructures

C. Mannweiler, C. Lottermann, A. Klein, J. Schneider, and H. D. Schotten

Chair for Wireless Communications and Navigation, University of Kaiserslautern, Germany

Correspondence to: C. Mannweiler (mannweiler@eit.uni-kl.de)

Abstract. This paper presents a novel approach for cyber-physical network control. "Cyber-physical" refers to the inclusion of different parameters and information sources, ranging from physical sensors (e.g. energy, temperature, light) to conventional network information (bandwidth, delay, jitter, etc.) to logical data providers (inference systems, user profiles, spectrum usage databases). For a consistent processing, collected data is represented in a uniform way, analyzed, and provided to dedicated network management functions and network services, both internally and, through an according API, to third party services. Specifically, in this work, we outline the design of sophisticated energy management functionalities for a hybrid wireless mesh network (WLAN for both backhaul traffic and access, GSM for access only), disposing of autonomous energy supply, in this case solar power. Energy consumption is optimized under the presumption of fluctuating power availability and considerable storage constraints, thus influencing, among others, handover and routing decisions. Moreover, advanced situation-aware auto-configuration and self-adaptation mechanisms are introduced for an autonomous operation of the network. The overall objective is to deploy a robust wireless access and backbone infrastructure with minimal operational cost and effective, cyber-physical control mechanisms, especially dedicated for rural or developing regions.

1 Introduction

Information and communication technology (ICT) infrastructure (both operator equipment and consumer electronics), besides being very costly, is generally developed according to needs and requirements of urban areas in developed countries. In rural areas and even more so in developing countries (except for the big cities), we encounter completely different conditions and deployment environments and significant financial restrictions. Therefore, the most feasible approach is to conceptually and technically adapt, extend, and improve existing equipment so that it better reflects the needs of sparsely populated regions in developing countries. The SolarMesh project, a consortium funded by the German Ministry of Education and Research, consisting of two research institutions and three companies, develops such an autonomous, resource-efficient, easily deployable and maintainable system, as presented in this work. One of the key challenges is to model the infrastructure as a multi-domain system, thus reflecting the dependencies between different parameters and facilitating a network management across different scales (Horstemeyer, 2010).

The remainder of the paper is structured as follows: the next section briefly describes the overall objectives of the SolarMesh project. Section 3 then provides an overview of related work, in particular multi-domain systems and how according concepts can be used to intelligently manage the SolarMesh infrastructure. Section 4 outlines the architectural concept of the considered wireless mesh network (WMN). Key innovations the project addresses are presented in Sect. 5, whereas Sect. 6 discusses expected results. The paper concludes with a summary and description of future work in Sect. 7.

2 Technical objectives

The SolarMesh wireless mesh network has been designed for deployment in rural areas of both industrialized and developing countries as a means to provide cost-efficient backhaul connectivity. The key objectives guiding the development, as listed below, have been derived from that major motivation:

1. *High-performing, adaptive and extensible system architecture*
   The overall architecture will be developed in a way that technologies beyond those considered in the project

(WLAN IEEE 802.11 and GSM) can be integrated easily. This holds particularly true for those alternative radio technologies reducing costs, energy consumption, or access barriers. Therefore, the integration of GSM as an additional access technology is motivated by three reasons: GSM terminals are relatively affordable, their energy consumption is comparably low, and, finally, they are very common in (rural) Africa. Moreover, many services, including governmental services, merely rely on SMS signaling. However, the termination of the GSM connection already takes place in the access points in order to avoid expensive GERAN infrastructure.

2. *Development of a mesh network based on autonomous energy supply*
   Energy supply of the mesh nodes is based on regenerative sources (solar power). Hence, all components (hardware, software, backbone, and air interface) and mechanisms (routing, scheduling, handover, radio transmission) have to be optimized taking into account fluctuating power availability and limited energy storage capacities. The partners will develop according advanced energy-aware methods for network control.

3. *Advanced auto-configuration capabilities*
   Deployment and start up of the mesh network will not necessarily require technically trained personnel. Individual nodes as well as the network as a whole will be mostly configured and adapted autonomously to their respective deployment area.

4. *Advanced self-adaptation capabilities*
   Besides considering energy efficiency as a main factor, network management and load balancing have to support a large-area mesh network with time-variant topology and dynamically changing traffic load. Availability of individual nodes will be significantly reduced due to limited energy resources, extreme weather conditions (high temperatures, strong rain, sand storms), and maintenance work. The network will have to cope with these characteristics in an autonomous and signaling-efficient manner.

5. *Implementation of efficient handover und routing mechanisms*
   Handover and routing processes will be optimized and executed in a situation-aware manner to facilitate mobility support for the user and multi-KPI (key performance indicator) network management for the operator (Mihailovic et al., 2009). For the latter, this allows for both control and prediction of power consumption, radio link availability, load distribution in the network, and cognitive use of available spectrum. The user will benefit from seamless connectivity, independent of service and radio technology.

3 Related work

Carrier-grade wireless mesh networks have been subject of former research projects. In EU FP7 CARMEN (Kretschmer and Robitzsch, 2010), for example, a wireless mesh infrastructure has been developed that abstracts from different link layer technologies by the usage of IEEE 802.21. The design of the infrastructure and the routing mechanisms provides support for different QoS classes.

3.1 Energy models

Generally, system-level models optimize the overall system performance with respect to a chosen set of parameters (Chan et al., 2006; Schrijver, 1986). When including energy consumption, transmission power is an important factor. Accordingly, current energy models particularly optimize network topology (i.e. the location of access points or base stations). Energy consumption is modeled based on path loss ($\frac{1}{d^n}$, where d denotes the distance between transmitter and receiver and n represents the path loss exponent that depends on the environments of the considered system), with additional consideration of large-scale fading using log-normal distributions as well as small-scale fading using Rayleigh distributions, e.g. Rodoplu and Meng (1999) and Wieselthier et al. (2000).

3.2 Energy efficiency metrics

For deployed nodes of wireless ad-hoc networks running with regenerative energy sources (e.g. solar power for countries of sub-Saharan Africa), routing protocols also play an important role in energy consumption (Chang and Tassiulas, 2000). Several energy-aware routing algorithms have been developed that take the characteristics of regenerative energy supplies and the referring outage situations into consideration. In Badawy et al. (2008), an energy-aware routing within the scope of wireless mesh networks based on a WLAN architecture is mapped to an assignment problem and solved by a generic algorithm. In Serrano et al. (2010), the general relation between energy consumption and throughput in the scope of WLAN infrastructures is given. Frequently, total energy consumption is modeled in a linear way, using metrics for consumed energy per routed information unit, such as "energy per bit" or "energy per packet". Obviously, these linear models do not properly reflect energy consumed due to unsuccessful attempts to access the channel (media contention), messages lost due to collision, bit errors, or loss of wireless connectivity (Feeney and Nilsson, 2001).

3.3 Multi-domain modeling

The notion of cyber-physical systems (CPS) has been introduced, among others, by Lee (2008). Cyber-physical systems are characterized by the integration of physical systems and processes with networked computing (Krogh et al., 2008).

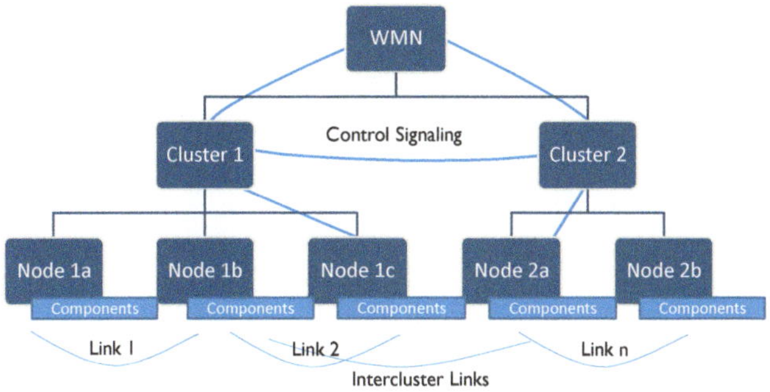

Fig. 1. Hierarchical Energy-Aware Network Management.

The "integration" refers to a tight interaction and mutual dependency of physical resources, such as ICT infrastructures and computational resources, e.g. algorithms to intelligently control these infrastructures. Insofar, numerous systems are categorized as "cyber-physical" or "multi-domain", among them robotic systems, automotive and avionic systems, infrastructures (utilities, communication networks), defense systems, or medical systems. While models describing either domain separately frequently exist, there is a need to design *cross-domain* models for analyzing and optimizing multi-domain systems. These models better reflect the close interdependency of the physical, computational, and network domain. Within the scope of this paper, the authors present an approach for describing wireless mesh networks as multi-domain systems. In particular, energy consumption and efficiency (physical domain) can be modeled depending on, among others, network load and topology (network domain) and user mobility (physical domain). Energy consumption will then be optimized using sophisticated analytical tools (computational domain) that can trigger network management decisions, such as shutting down entire nodes or selected radio interfaces of individual nodes or adapting transmission power levels. Figure 1 depicts the different hierarchical levels on which an energy-aware network management can be executed, ranging from a global view, i.e. the entire network, to components of individual nodes and selected wireless links.

4 System concept

With a special focus on the multi-domain modeling challenges, the following section describes the system concept of SolarMesh. It depicts an introduction to the overall system architecture, the auto-configuration and self-adaptation mechanisms, and finally to the handover and routing mechanisms that are implemented.

4.1 Network architecture

Several SolarMesh nodes create a wireless, autonomous meshed backhaul network based on IEEE 802.11 wireless technology, as depicted in Fig. 2. The mesh network has three tasks: providing connectivity to the external core network, offering heterogeneous wireless access to user terminals and, most importantly, serving as a wireless backhaul network within a particular area, thus representing an alternative to wired infrastructure. The first one is realized by several gateway connections to the core network and the Internet, whereas the second one is provided by utilizing commonly used Wide Area Network (WAN) technologies, e.g. GSM (in order to take advantage of the cheap off-the-shelf user devices) or Wireless LAN (thus exploiting its affordable network components). Backhaul connectivity is solely based on IEEE 802.11, incorporating mechanisms for carrier-grade QoS, i.e. different services and the accordingly required QoS classes are supported by the overall SolarMesh architecture, namely best-effort and delay-sensitive VoIP services.

The overall system architecture consists of several nodes, disposing of different functionalities:

- *SMN* (SolarMesh Node) holds the SolarMesh capabilities and embodies a basic interface for other nodes within the architecture. The main task of the SMNs is to build up the mesh backhaul network and to forward the user- and control-plane traffic with respect to the QoS requirements among the SMNs (to the target sink).

- *SMGW* (SolarMesh Gateway) constitutes the gateway of the wireless backhaul network to the external backbone networks via Global Area Networks (GANs), e.g. a satellite link.

- *SMAP* (SolarMesh Access Point) provides the wireless access interface to the user terminals. It is equipped with at least one radio interface that does not carry backhaul traffic but is dedicated to the user access. Supported access technologies are IEEE 802.11 and GSM.

- *SMC* (SolarMesh Coordinator) performs cluster-wide, centralized decisions within a cluster of SMNs. The task of the SMCs is to consider both cluster- and network-wide optimization goals for radio planning, energy-aware node and link configuration, capacity management, or handover decisions.

Besides a comprehensive multi-domain modeling of the system for simulation purpose, a future testbed will be set up with mesh nodes (disposing of at least three 802.11 interfaces) which will be used to deploy the SolarMesh backhaul network. To realize the GSM-based access network, SMAPs will be extended with additional *USRP 2* from Ettus LLC controlled by *openBTS* and an *Asterisk PBX* software.

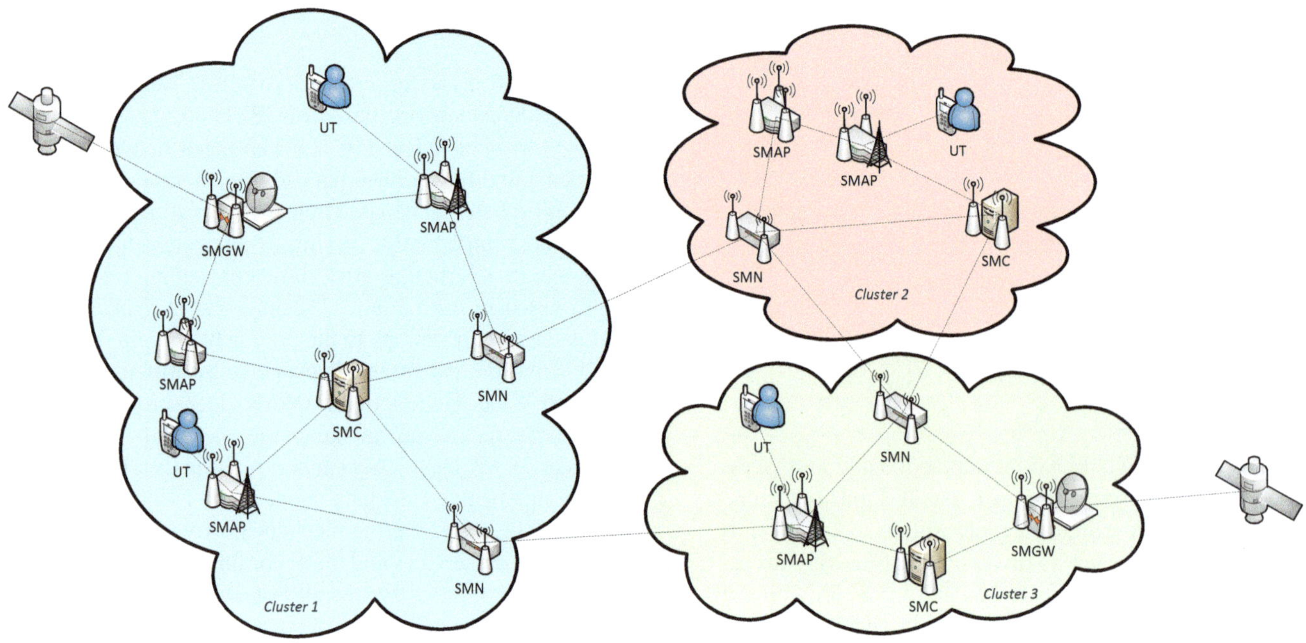

Fig. 2. Overview of System Concept.

4.2 Multi-domain self-optimization

One of the main characteristics of SolarMesh is the ability of auto-configuration and self-adaptation, as shown in Fig. 3. This is realized by an autonomous topology detection and configuration of the deployed SMNs. They are grouped into clusters in order to decrease the overall signaling effort and to perform local, i.e. cluster-wide, optimization concerning routing and handover decisions. For these optimization decisions, aspects from the physical domain (e.g. user movements), the computational domain (e.g. energy efficiency statistics), and the network domain (e.g. traffic load) are taken into account. Within a cluster, a SMC is determined that is responsible for the cluster-wide decisions and optimizations. Hence, a SMC can decide to switch off a SMN due to low utilization of the according SMN or its low energy status, e.g. caused by an almost empty buffer battery and, simultaneously, no energy supply by solar panels. As a result, the network topology is reconfigured and the traffic of the considered SMN is dynamically re-routed via other routes. SMNs are able to modify transmission power, modulation and coding schemes, respectively, as well as provided QoS in an autonomous way. Depending on the actual objective function, the emphasis of the optimization can be put on different domains of the system.

For example, target access points for the user terminals are selected depending on the load situation and energy level of the target SMAPs, the capabilities of the user terminals and the SMAPs, and the QoS that has been requested by the user terminal.

4.3 User mobility and energy-aware routing

As outlined in previous sections, handover and routing decisions are primarily based on mobility, load situation, and energy situation of the SMNs as well as the provided and requested QoS parameters. Mobility-based handovers are triggered by the local SMNs, whereas the remaining handover decisions are performed by the SMCs in order to allow for a cluster-wide or global optimization of traffic load and energy consumption. Both horizontal, i.e. within the same radio technology, and vertical, i.e. across different radio networks, handovers can be performed. For each end-to-end connection, the most efficient path with respect to energy status or load situation is selected.

5 Key innovations

Within SolarMesh, several research challenges need to be tackled which will be described in the following section.

5.1 Energy awareness and autonomous power supply

Individual components of the SolarMesh network will be designed to work in regions with specific characteristics: harsh environmental and climatic conditions and unreliable infrastructures, e.g. with respect to energy supply or availability of maintenance resources. These constraints are explicitly considered in the system model as they are typical characteristics of a cyber-physical system. In order to guarantee a continuous operation mode, nodes are equipped with their own energy supply. Power consumption of the SMNs needs

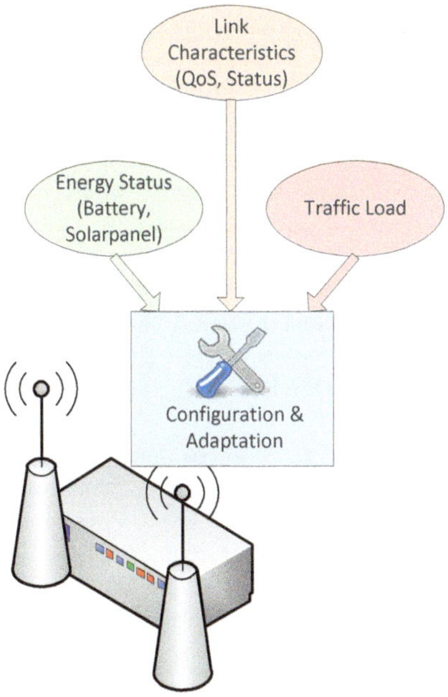

Fig. 3. Auto-Configuration and Self-Adaptation Functionalities of a SMC Node.

to be kept at a minimum aiming at achieving a long operation time. Therefore, an energy management function (EMF) is required that is capable of reducing power consumption by shutting down certain capabilities. This EMF employs a detailed multi-domain energy model of the SMNs. The model uses the power consumption of the supported wireless access networks (e.g. IEEE 802.11 and GSM), depending on different modulation modes, supported states of operation as well as state changes, as input parameters. Additionally, the influence of traffic load on the SMNs is considered, which in turn is dependent on the type of the node: SMCs need to handle more control plane traffic than simple SMNs which in turn leads to a higher power consumption. In general, decisions concerning energy management are taken at different hierarchy levels, as depicted in Fig. 4. As an example, modulation schemes for the terminal link can generally be decided at the local node. In contrast, the decision of shutting down an entire node can only be taken at higher levels (cluster or global), since the individual node is not foreseen to have enough information to take that decisions and to assess the consequences.

5.2 Network management

Within the meshed network infrastructure, an advanced network management entity is required in order to perform the required reconfigurations of the constantly changing network topology. This entity has instances at node, cluster, and global level and implements, among others, routing and re-routing mechanisms, handover algorithms, as well as network topology management functionalities. Besides network-side parameters, such as dynamic characteristics of

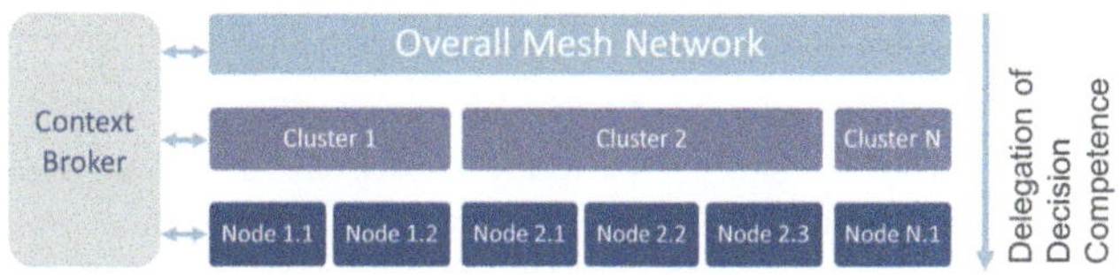

Fig. 4. Network Energy Management Concept.

load or traffic situation with respect to QoS classes, it takes into consideration mobility and capabilities of registered user terminals (such as supported air interfaces, display size, or available codecs), thus also reflecting relevant input parameters from the physical domain.

6 Expected results

Within SolarMesh, a wireless mesh infrastructure will be realized that keeps the overall deployment and operational costs at a very low level, as required for operation in rural areas of sub-Saharan Africa. From a theoretical perspective, multi-domain modeling concepts (such as cyber-physical systems) are expected to facilitate the according design of the mesh network and to fulfill these objectives. The overall system design will be evaluated by a mixed simulation and emulation and will finally be implemented and validated in a testbed environment. Routing and handover decisions will be optimized and accelerated by the usage of context-aware decision functions. Seamless connectivity will be realized by the support of mobility (horizontal and vertical handover) and dynamic routing and re-routing. Due to the usage of advanced energy management functions, a significant reduction of the overall power consumption will be realized. Most importantly, the mesh network will rely on and being able to cope with autonomous energy supply solutions, thus gaining independence from locally deployed, potentially unreliable energy grids.

7 Conclusion and outlook

The authors have presented how analytical concepts of cyber-physical systems can be used for an advanced management of a wireless mesh network. In particular, the exploitation of information that is available from different domains (physical, computational, network) supports the design of a multi-domain system model. Moreover, motivation, objectives, research challenges, and the initial system concept of the SolarMesh project have been outlined. Deployment scenarios, such as rural areas in both developed and developing countries, have been described.

Applying multi-domain models to wireless mesh networks is particularly useful for improving energy efficiency that depends on factors from different domains, such as, among others, traffic load in the network (utilization), amount and mobility of users, network and environment topology, transmission power levels, and employed radio technologies. Analyzing and mathematically describing the dependencies between these quantities facilitates the optimization of the network with regards to selected parameters, e.g. energy consumption.

In a first step, future work therefore includes the detailed specification and development of a system architecture disposing of hardware and software components that meet the defined requirements. In a second step, the system will be validated and evaluated based on simulations and testbed experiments. Here, the developed multi-domain models will be used to generate quantitative results with respect to the applicability of the selected parameters for optimized management of the wireless mesh network.

Acknowledgements. This work has been funded by the Federal Ministry of Education and Research of the Federal Republic of Germany (Förderkennzeichen 01 BU 1116, SolarMesh-Energieeffizientes, autonomes großflächiges Sprach- und Datenfunknetz mit flacher IP-Architektur). The authors alone are responsible for the content of the paper.

References

Badawy, G. H., Sayegh, A. A., and Todd, T. D.: Energy Aware Provisioning in Solar Powered WLAN Mesh Networks. Proceedings of 17th International Conference on Computer Communications and Networks (ICCCN), St. Thomas, U.S. Virgin Islands, 2008.

Chan, T. F., Cong, J., Shinnerl, J. R., Sze, K., Xie, M., and Zhang, Y.: Multiscale Optimization in VLSI Physical Design Automation, in: Multiscale optimization methods and applications, edited by: Hager, W. W., Huang, S.-J., Pardalos, P. M., and Prokopyev, O. A., Springer, 2006.

Chang, J.-H. and Tassiulas, R.: Energy Conserving Routing in Wireless Ad-hoc Networks. Proceedings of the IEEE International Conference on Computer Communications (INFOCOM), Tel Aviv, Israel, 2000.

Feeney, L. M. and Nilsson, M.: Investigating the Energy Consumption of a Wireless Network Interface in an Ad Hoc Networking Environment. Proceedings of the IEEE International Conference on Computer Communications (INFOCOM), Anchorage, USA, 2001.

Horstemeyer, M. F.: Multiscale Modeling: A Review, in: Practical Aspects of Computational Chemistry, edited by: Leszczynski, J. and Shukla, M. K., Springer, 2010.

Kretschmer, M. and Robitzsch, S.: Wireless Mesh Network Coverage with QoS Differentiation for Rural Areas. Proceedings of the First International Workshop on Wireless Broadband Access for Communities and Rural Developing Regions (WIRELESS4D), Karlstad, Sweden, 2010.

Krogh, B. H., Lee, E., Lee, I., Mok, A., Pappas, G., Rajkumar, R., Sha, L. R., Vincentelli, A. S., Shin, K., Stankovic, J., Sztipanovits, J., Wolf, W., and Zhao, W.: Cyber-Physical Systems – Executive Summary, CPS Steering Group, 2008.

Lee, E. A.: Cyber Physical Systems: Design Challenges. Proceedings of the IEEE International Symposium on Object-Oriented Real-Time Distributed Computing, Orlando, USA, 2008.

Mihailovic, A., Chochliouros, I. P., Georgiadou, E., Spiliopoulou, A. S., Sfakianakis, E., Belesioti, M., Nguengang, G., Borgel, J., and Alonistioti, N.: Situation awareness mechanisms for cognitive networks. Proceedings of the International Conference on Ultra-Modern Telecommunications (ICUMT), St. Petersburg, Russia, 2009.

Rodoplu, V. and Meng, T. H.: Minimum energy mobile wireless networks, IEEE J. Sel. Area. Comm., 17, 1333–1344, 1999.

Schrijver, A.: Theory of Linear and Integer Programming, John Wiley & Sons, 1986.

Serrano, P., Hollick, M., and Banchs, A.: On the Trade-Off between Throughput Maximization and Energy Consumption Minimization in IEEE 802.11 WLANs, J. Commun. Netw., 12, 1976–5541, 1229–2370, 150–157, 2010.

Wieselthier, J. E., Nguyen, G. D., and Ephremides, A.: On the Construction of Energy-Efficient Broadcast and Multicast Trees in Wireless Networks. Proceedings of the IEEE International Conference on Computer Communications (INFOCOM), Tel Aviv, Israel, 2000.

29

Reliability analysis of buffer stage in mixed signal application

S. More[1], M. Fulde[2], F. Chouard[1], and D. Schmitt-Landsiedel[1]

[1]Lehrstuhl für Technische Elektronik,Technische Universität München, Germany
[2]Infineon Technologies AG, Villach, Austria

Abstract. This paper discusses reliability analysis of a buffer circuit targeted for an analog to digital converter application. The circuit designed in a 32 nm high-κ metal gate CMOS technology was investigated by circuit simulation and sensitivity analysis. This analysis was conducted for realistic time varying (AC) stress. As aging effects, negative and positive bias temperature instability, conducting and non-conducting hot carrier injection are taken into consideration. The aging contributions of these effects on the different transistors in the buffer circuit and on different buffer performance figures are evaluated. Using these results, the impact of an aged buffer circuit on the performance of a successive approximation ADC circuit is evaluated. The most severely affected performance due to aging is amplifier offset, which leads to time varying gain error in the ADC circuit.

1 Introduction

Non-constant field scaling in nanoscale CMOS technology has led to undesirable reliability issues due to effects such as Negative Bias Temperature Instability (NBTI), Positive Bias Temperature Instability (PBTI), Conducting and Non-Conducting Hot-Carrier Injection (HCI,NCHCI). Significant device level research is carried out on these effects (Grasser et al., 2007; Kaczer et al., 2009). However, since the superposition of these effects in a circuit is quite complex, the study of their impact on circuit level is still in a preliminary stage (Jha et al., 2005; Martin-Martinez et al., 2009; Chouard et al., 2010a).

This paper targets to investigate the impact of aging on a buffer stage designed for Analog to Digital Converter (ADC) application. The circuit was implemented and simulated using 32 nm high-κ metal gate regular V_{th} nMOS and pMOS devices (Chen et al., 2008), so the effect of PBTI which appears due to high-κ is also present. The effects of buffer circuit degradation on ADC performance is demonstrated and the need for countermeasures is highlighted. The degradation effects are evaluated using a combination of sensitivity analysis and circuit simulation. This methodology as described in (More et al., 2010) was developed, as it needs considerably less computing effort than sole use of circuit simulation, and it provides more intuitive insight into the various degradation contributions.

Section 2 presents the models used for aging simulation. The investigated buffer circuit topology is shown in Sect. 3. Section 4 discusses results of aging simulation of buffer circuits. The analytical evaluation of these degradation effects using sensitivity analysis, its comparison with simulation results and the observed impact on the circuit performance are presented in Sect. 5. Effects of buffer circuit aging on ADC performance and need of countermeasures are discussed in Sect. 6. A conclusion is given in Sect. 7.

2 Modeling of device degradation

Device degradation due to aging resulting from NBTI or PBTI mainly shifts the threshold voltage (V_{th}) of MOS transistors, whereas HCI and NCHCI reduce the transistor current (I_d) and also give a smaller contribution to the threshold voltage shift. This can be modeled by replacing the MOSFET with an equivalent circuit shown in Fig. 1. The V_{th} shifts are modeled by an equivalent voltage source vt_{shift} in series to the gate terminal, the hot carrier effects by a current controlled current source (CCCS) Id_{shift} between the drain and source terminal. The threshold voltage shift by each of the BTI effects can be modeled by $(\Delta V_{th})_{\text{BTI}}$ as in Eq. (1). BTI stands for NBTI in pMOS transistors and for PBTI in nMOS transistors. The current degradation is modeled by

Correspondence to: S. More
(shailesh.more@tum.de)

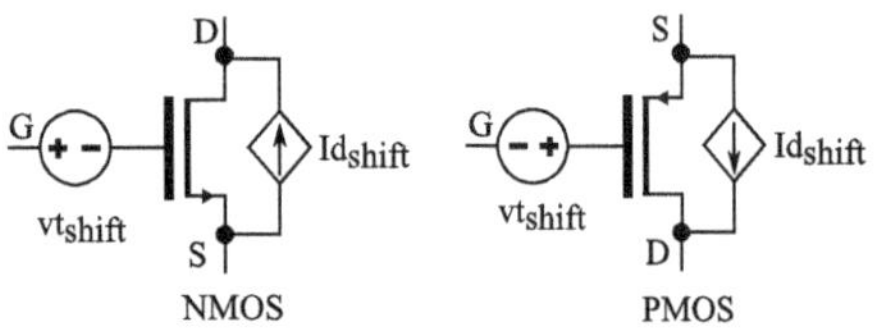

Fig. 1. Models for MOSFETs with degradation.

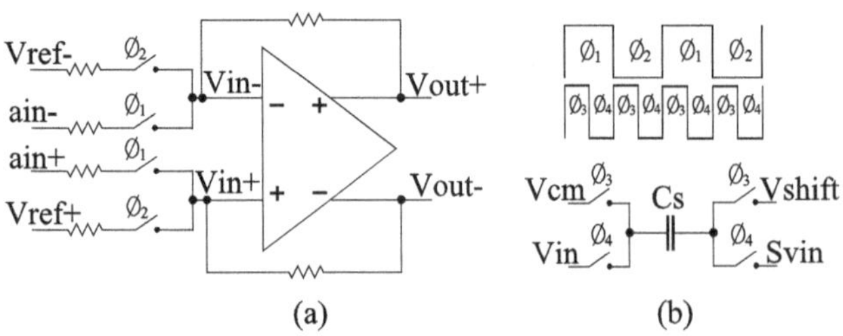

Fig. 2. Schematic of buffer and DC level shifting circuit.

Eq. (2), separately for HCI and NCHCI. The values of the equivalent sources are determined using Eqs. (1)–(2), which describe the relation of different contributing factors to the overall degradation, similar to (Martin-Martinez et al., 2009; Huard et al., 2009). The total vt_{shift} is the sum of two contributions, see Eq. (3). The total current degradation is modeled by Eq. (4). The model parameters were fitted to single device stress measurements.

$$(\Delta V_{th})_{\text{BTI}} = A \cdot \left(\frac{V_{gs}}{T_{inv}}\right)^m \cdot e^{\left(\frac{\Delta E}{kT}\right)} \cdot L^{\alpha} \cdot W^{\beta} \cdot t^n \tag{1}$$

$$(\Delta I_d)_{\text{HCI/NCHCI}} = I_d \cdot B \cdot V_{ds}^p \cdot e^{\left(\frac{\Delta E}{kT}\right)} \cdot L^{\delta} \cdot t^q \tag{2}$$

$$vt_{\text{shift}} = a \cdot (\Delta V_{th})_{\text{BTI}} + b \cdot (\frac{\Delta I_d}{I_d})_{\text{HCI}} \tag{3}$$

$$Id_{\text{shift}} = c \cdot (\frac{\Delta I_d}{I_d})_{\text{HCI}} + d \cdot (\frac{\Delta I_d}{I_d})_{NCHCI} \tag{4}$$

For the evaluations based on circuit simulation and to account for time varying (AC) stress, the aging simulation tool RelXpert^{TM} (RelXpert, 2010) was used, which employs the above discussed models.

3 Buffer circuit for ADC application

The schematic for the fully differential buffer circuit is illustrated in Fig. 2a. It basically consists of an Operational Transconductance Amplifier (OTA), resistors and switches. Capacitors of 2 pF are used to emulate the load of a Digital to Analog Converter (DAC) connected at the output of the buffer circuit. The primary function of this buffer circuit in the charge distribution based successive approximation (SAR) ADC as depicted in Fig. 3 (Fulde et al., 2009), is to drive the DAC implemented using capacitors, within a short settling time (≈4 ns) with reference voltage ($V_{\text{ref}}\pm$) and analog input ($a_{\text{in}}\pm$) during different phases of the clock signals that control the switches. The clock signals are shown only schematically, not displaying the more complex clocking during the ADC operation.

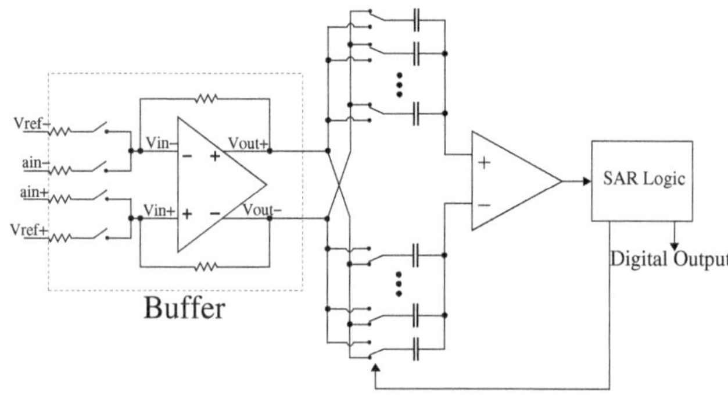

Fig. 3. Schematic of successive approximation ADC.

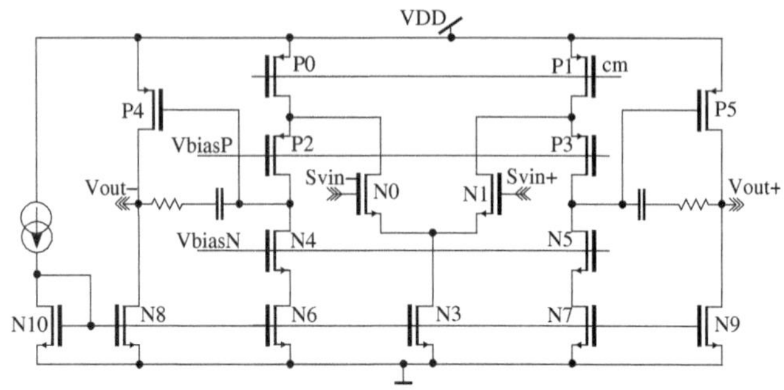

Fig. 4. Schematic of the OTA circuit.

The used OTA is illustrated in Fig. 4. It is a two stage folded cascode amplifier with Miller compensation. Implementing this circuit using regular V_{th} pMOS and nMOS transistors with $V_{th} \approx 0.45$ V and supply $V_{DD} = 1$ V leads to significant challenges in using the common mode voltage $V_{\text{cm}} = 0.5$ V at the gate of input differential pair (N_0 and N_1) and maintaining the tail transistor (N_3) in saturation. To overcome this problem a simple level shifting circuit as depicted in Fig. 2b was used at the input of the OTA. This circuit shifts each input signal (V_{in}) of the OTA with $V_{\text{cm}} = 0.5$ V to a new shifted signal (S_{vin}) depending on the difference between the common mode and the shift voltage (V_{shift}).

Table 1. Simulated Gain(dB)/Phase margin(°)/Gain Bandwidth(MHz) resulting for a stress time of 10 Yrs.

	V_{DD} (V)					
Temp (°C)	1			1.1		
25	59.81	88.98	736.52	59.79	88.96	736.27
85	59.79	89	736.56	59.72	88.84	736.52

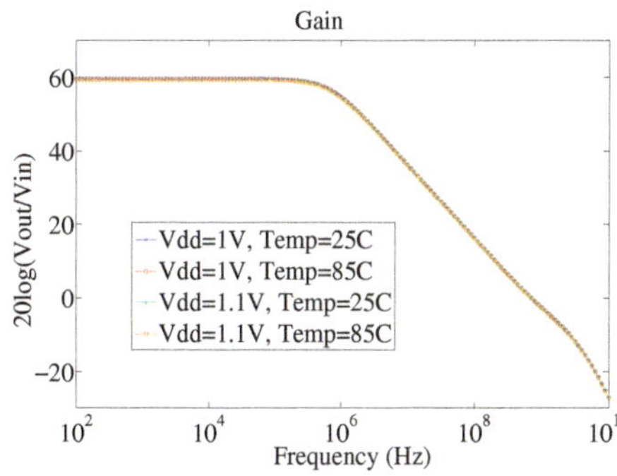

Fig. 5. Simulated Gain (dB) resulting in the buffer circuit for a stress time of 10 Yrs.

4 Aging simulation results

OTA performances at nominal condition (before stress, Temp = 25°C and $V_{DD} = 1$ V) are listed below. Here V_{offset} refers to output referred offset.

DC Gain = 59.71 dB   Gain Bandwidth = 736.8 MHz
Phase Margin = 89° $V_{\text{offset}} = 0$ V

The aging simulations were performed using AC input signals of $a_{\text{inp}} = 0.6V_{p-p}$, 1.1MHz sine wave, and $a_{\text{inn}} = 0.6V_{p-p}$, 1.1 MHz, 180° phase shifted sine wave. The bias voltages are $V_{\text{refp}} = 0.8$ V, $V_{\text{refn}} = 0.2$ V, $V_{\text{cm}} = 0.5$ V, $V_{\text{shift}} = 0.8$ V, Clk = 20 MHz). Figure 5 illustrates the gain Bode plot for the OTA at different V_{DD} and Temp stress conditions, simulated for stress time (Age) of 10 years. In the OTA circuit, mainly the bias current through transistors N_3 and N_{10} decides the gain, phase margin and bandwidth. These transistors see only limited gate and drain voltages, so no remarkable degradation occurs. Consequently, it can be observed from Fig. 5 and the results given in Table 1 that no significant performance degradation of these parameters (less than 1%) was observed after aging since this bias current is not affected. This shows that the current mirror structures are robust towards aging effects.

The spectral behavior of the buffer circuit was studied to see if any significant non-linearity is introduced due to circuit aging. The simulation results are depicted in Figs. 6 and 7. There is no significant difference in the spectral performances before and after aging.

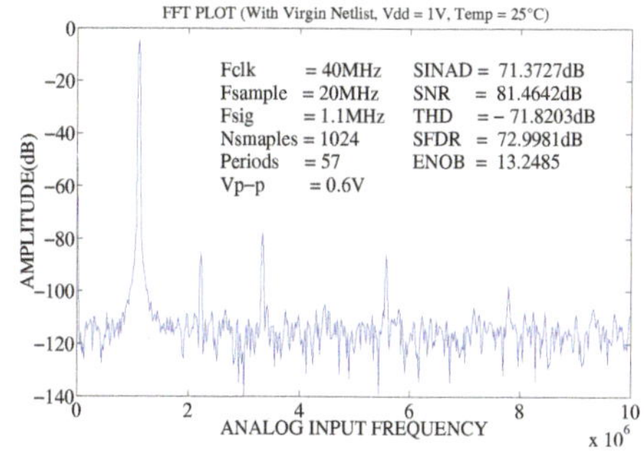

Fig. 6. FFT Spectrum results from fresh netlist.

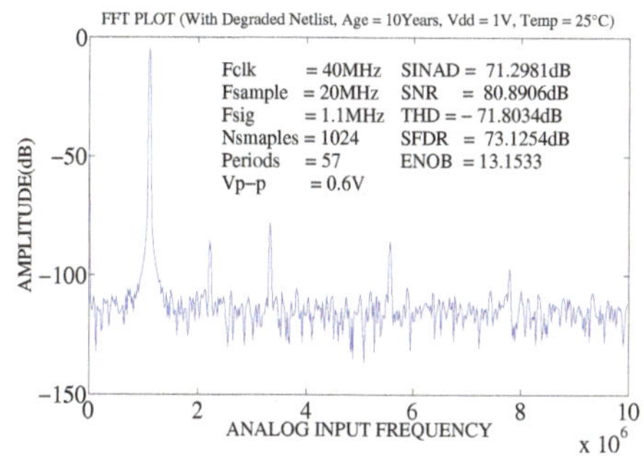

Fig. 7. FFT Spectrum results from degraded netlist.

However in closed loop configuration aging degradation leads to considerable V_{offset} in the differential output pair due to the asymmetry of the stress conditions on these two transistors: P5 is connected to $V_{\text{out}}+$ and sees high positive voltages, whereas P4 sees much lower voltages at $V_{\text{out}}-$. Table 2 summarizes the degradation-induced output referred V_{offset} simulated at different V_{DD} and Temp stress conditions for an age of 10 years. The corresponding input referred offset is obtained by dividing these values with the gain of the OTA, and it is found to be quite small due to a large gain.

5 Impact of aging on circuit performance

To study more closely the particular contribution of each transistor in the buffer circuit to V_{offset}, we evaluate its sensitivity towards V_{th} and I_d shift resulting from circuit aging. V_{offset} is calculated in this sensitivity analysis based analytical evaluation (More et al., 2010) using Eq. (5). For each transistor (T_n), the sensitivity ($S_{V_{T_n}}$) of V_{offset} towards V_{th} shift as well the sensitivity ($S_{I_{T_n}}$) towards I_d shift is determined by circuit simulation and is then multiplied with the respective V_{th} and I_d shifts. Finally, the total offset voltage shift is summed up. The results of this analysis are matching closely with results obtained by circuit simulation using RelXpert, as illustrated in Table. 2.

$$V_{\text{offset}} = S_{V_{T1}} \cdot Vt_{\text{shift}_{T1}} + S_{V_{T2}} \cdot Vt_{\text{shift}_{T2}} + \ldots + S_{I_{T1}} \cdot Id_{\text{shift}_{T1}} + S_{I_{T2}} \cdot Id_{\text{shift}_{T2}} + \ldots \quad (5)$$

Table 2. Output referred V_{offset} in (mV) resulting in the buffer circuit for a stress time of 10 Yrs.

Temp (°C)↓	Simulated	Sens. Analysis
$V_{DD} \rightarrow$	1/1.1	1/1.1
25	−1.357/8.446	−1.33/8.48
85	6.722/18.086	6.735/18.185

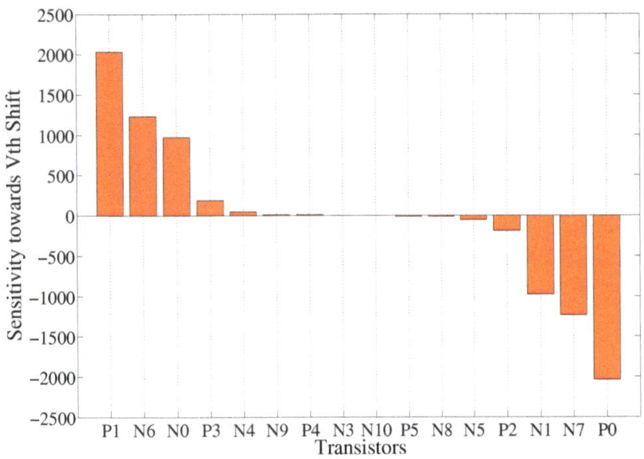

Fig. 8. Sensitivity of V_{offset} towards V_{th} shift of different transistors in the OTA circuit.

The sensitivity of V_{offset} towards V_{th} and I_d shift of all transistors in the buffer circuit is illustrated in Figs. 8 and 9.

The data in Figs. 10 and 11 represents V_{th} and I_d shift respectively, in different transistors of the OTA circuit resulting due to aging. The V_{th} shift is a combined effect of BTI and HCI and I_d shift is a combined effect of HCI and NCHCI degradation.

In Figs. 12 and 13 the contributions to V_{offset} due to aging of the individual transistors are given. These are the theoretical intermediate results obtained by multiplying the V_{th} and I_d shift for each transistor with respective sensitivity of V_{offset}. The NBTI degradations of the pMOS transistors P_0 and P_1 in Fig. 12 are of equal magnitude and compensate each other in their effect on the offset, since they see the same gate to source voltage coming from the common mode (cm) feedback circuit. Thus their NBTI degradation causes negligible V_{offset} change. The same holds true for transistors P_2 and P_3. As mentioned before, the dominant contribution to V_{offset} due to V_{th} shift comes from the pMOS transistors of the output stage, P_4 resulting from HCI (50.56%) and NBTI (49.44%) followed by P_5 resulting from HCI (38.87%) and NBTI (61.13%). Although the sensitivity of V_{offset} towards V_{th} shift of transistors P_4 and P_5 is low (±12.67) the difference between their V_{th} shifts leads to a significant impact on V_{offset}.

Figure 13 shows that the dominant contribution to V_{offset} due to I_d shift comes from pMOS transistors P_4 and P_5 re-

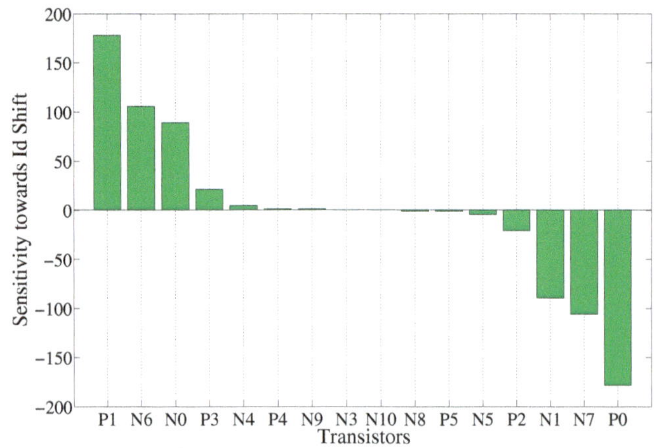

Fig. 9. Sensitivity of V_{offset} towards I_d shift of different transistors in the OTA circuit.

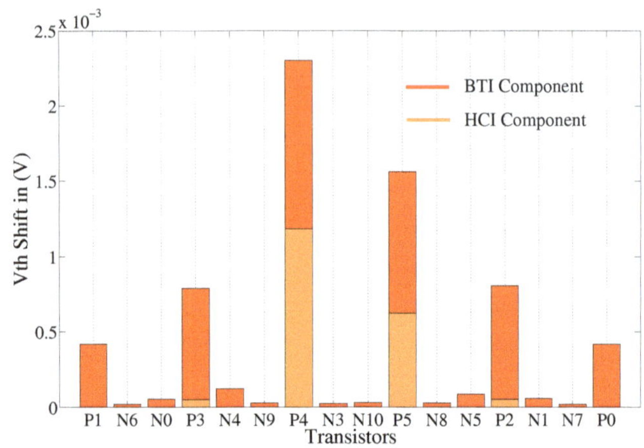

Fig. 10. NBTI, PBTI and HCI degradation component in V_{th} shift simulated for stress time of 10Yrs at $V_{DD} = 1$ V and Temp = 85 °C.

sulting from HCI degradation. Although the sensitivity of V_{offset} towards I_d shift of transistors P_4 and P_5 is low (±1.09) the difference between their I_d shift leads to a significant impact on V_{offset}. The effects of transistors P_2 and P_3 again cancel each other. Comparison of Figs. 12 and 13 shows that V_{th} shift is the dominant contribution (85.9%) to V_{offset} compared to I_d shift (14.1%).

In summary, the degradation due to circuit aging affects almost all transistors in the buffer circuit. The most degraded circuit performance is V_{offset}. Since the OTA in the buffer circuit always operates in closed loop configuration, its input transistors see smaller stress compared to the output stage transistors. Simulation results show that these output transistors are the main contributors to V_{offset}. The results presented here are in-line with experimental findings from (Chouard et al., 2010b).

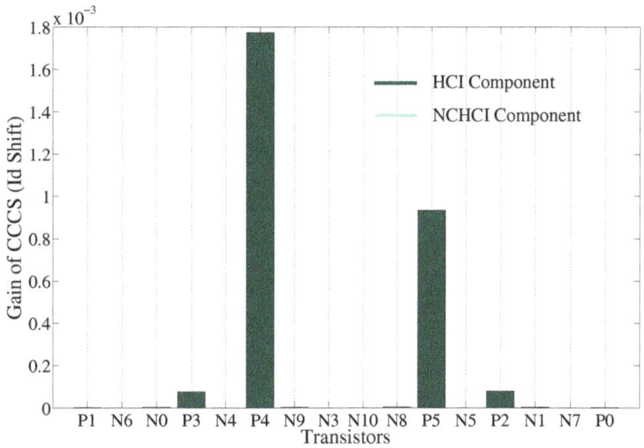

Fig. 11. HCI and NCHCI degradation component in I_d shift simulated for stress time of 1 Yrs at $V_{DD} = 1$ V and Temp = 85 °C.

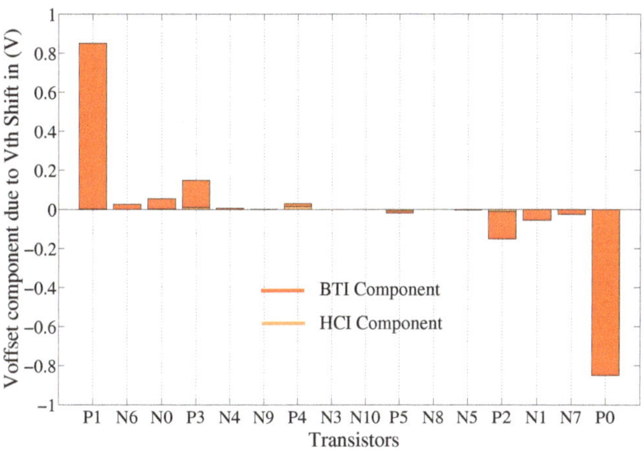

Fig. 12. Contribution to V_{offset} due to V_{th} shift resulting from NBTI, PBTI and HCI degradation, simulated for stress time of 10 Yrs at $V_{DD} = 1$ V and Temp = 85 °C.

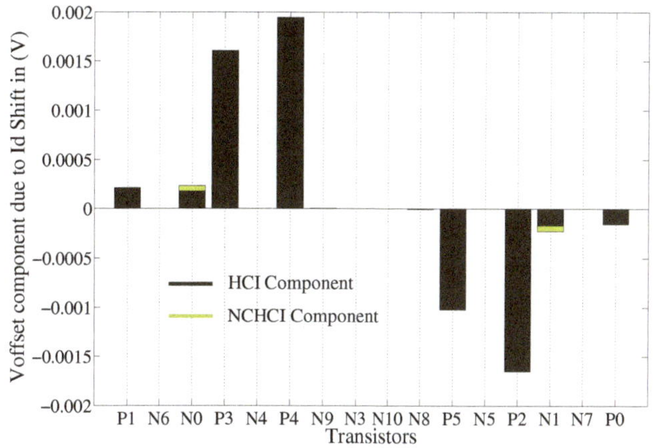

Fig. 13. Contribution to V_{offset} due to I_d shift resulting from HCI and NCHCI degradation, simulated for stress time of 10Yrs at $V_{DD} = 1$ V and Temp = 85 °C.

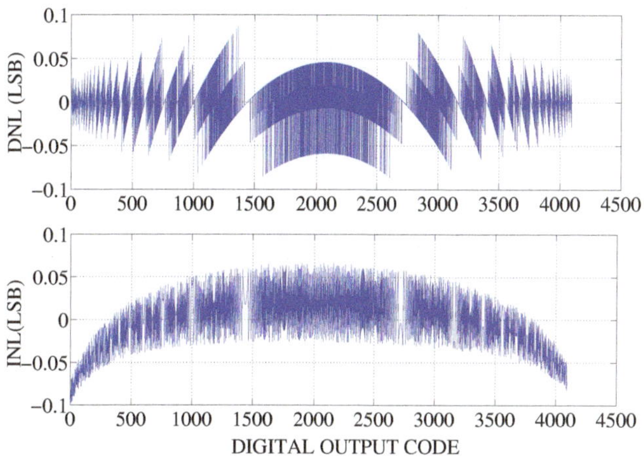

Fig. 14. Simulated DNL and INL of ADC with 10mV input referred offset in the buffer circuit.

6 Impact on ADC performance

To study the impact of an aged buffer circuit on the performance of a successive approximation ADC circuit, a 12 bit SAR ADC with charge redistribution as illustrated in Fig. 3 was modeled. All the circuit blocks were modeled as ideal elements using Verilog-A to reduce the complexity and the simulation time. A stress induced input referred offset as high as 5mV has been reported for an OTA circuit operated in closed loop configuration (Chouard et al., 2010b). So an input referred offset of 10 mV was assumed as worst case condition for the simulations. The converter was simulated with a clock frequency of 33.33 MHz, and a conversion rate of 1.19 MS/s was used with $V_{\text{refp}} = 0.8$ V, $V_{\text{refn}} = 0.2$ V and $V_{DD} = 1$ V.

The simulated differential non-linearity (DNL) and integral non-linearity (INL) are shown in Fig. 14. The resultant ENOB is 11.98 bits, which shows that the linearity of this SAR ADC is not affected by the aged buffer circuit. It was confirmed by simulations that also the spectral characteristics of the ADC were not affected. However, the transfer characteristics depicted in Fig. 15 shows that a gain error arises in the converter. All SAR ADC's having resolution (1 LSB) value smaller than the worst case input referred offset will be affected by aging degradation in this manner, with magnitude depending on the offset. There are proven methods to correct a gain error. But it is important to note that this error will vary over time. Hence special countermeasures need to be implemented to guarantee a stable and correct circuit function for the whole lifetime of the circuit.

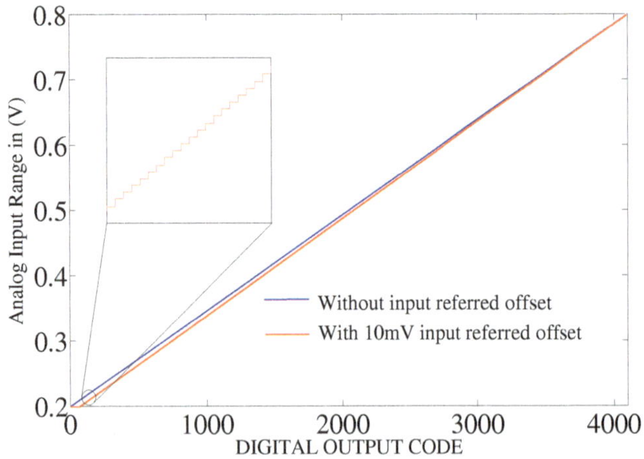

Fig. 15. Simulated input vs. output characteristics of ADC with (10 mV) and without input referred offset in the buffer circuit.

7 Conclusions

In this paper a buffer circuit designed in a 32 nm high-k, metal gate technology was analyzed by simulation for reliability with device aging. It was found that the most severely affected performance due to aging is amplifier offset, which leads to time varying gain error in an ADC circuit. Comparing Bode plots and spectal characteristics of the buffer circuit before and after aging, it could be observed that the amplifier gain, bandwidth, phase margin and spectral performance are not degraded due to aging. In a successive approximation ADC, the main effect due to aging is a time varying gain error. This leads to the conclusion that there is a need to implement special countermeasures in ADCs which correct for time varying errors resulting from stress induced degradation.

Acknowledgements. This work was supported by the European Commission and the Austrian government in the MODERN project (ENIAC-120003). The authors would like to thank the entire design and reliability team at Infineon for support and discussions.

References

Chouard, F., Fulde, M., and Schmitt-Landsiedel, D.: Reliability Assessment of Voltage Controlled Oscillators in 32nm High-k Metal Gate Technology, in: European Solid-State Circuits Conference, ESSCIRC Proceedings, 410–413, 2010a.

Chouard, F., Fulde, M., Werner, C., and Schmitt-Landsiedel, D.: A test concept for circuit level aging demonstrated by a differential amplifier, in: IEEE International Reliability Physics Symposium, IRPS, 826–829, 2010b.

Fulde, M., Schmitt-Landsiedel, D., and Knoblinger, G.: Analog and RF design issues in high-k and multi-gate CMOS technologies, in: IEEE International Electron Devices Meeting, IEDM, p. 447, 2009.

Grasser, T., Kaczer, B., Hehenberger, P., Gos, W., O'Connor, R., Reisinger, H., Gustin, W., and Schlunder, C.: Simultaneous Extraction of Recoverable and Permanent Components Contributing to Bias-Temperature Instability, in: IEEE International Electron Devices Meeting, IEDM, 801–804, 2007.

Huard, V., Parthasarathy, C., Bravaix, A., Guerin, C., and Pion, E.: CMOS device design-in reliability approach in advanced nodes, in: IEEE International Reliability Physics Symposium, 624–633, 2009.

Jha, N., Reddy, P., Sharma, D., and Rao, V.: NBTI degradation and its impact for analog circuit reliability, IEEE Transactions on Electron Devices, 52, 2609–2615, 2005.

Kaczer, B., Grasser, T., Martin-Martinez, J., Simoen, E., Aoulaiche, M., Roussel, P., and Groeseneken, G.: NBTI from the perspective of defect states with widely distributed time scales, in: IEEE Internationa lReliability Physics Symposium, 55 –60, 2009.

Martin-Martinez, J., Rodriguez, R., Nafria, M., and Aymerich, X.: Time-Dependent Variability Related to BTI Effects in MOSFETs: Impact on CMOS Differential Amplifiers, IEEE Transactions on Device and Materials Reliability, 9, 305–310, 2009.

More, S., Fulde, M., Chouard, F., and Schmitt-Landsiedel, D.: Sensitivity Analysis Based Analytical Evaluation of Aging Degradation in Linear Circuits, in: European Solid-State Circuits Conference Fringe Session, ESSCIRC, 2010.

RelXpert, C.: Users Manuals BSIMPro+/RelXpert/UltraSim, Cadence Design Systems Inc, 2010.

X. Chen et al.: A cost effective 32nm high-K metal gate CMOS technology for low power applications with single-metal gate-first process, in: Symposium on VLSI Technology, 88–89, 2008.

Photo detector IC for Blu-ray-Disc applications: a realization applying efficient design methodologies

S. Lange, T. Reich, J. Nowak, B. Dimov, M. Meister, and E. Hennig

Institut für Mikroelektronik- und Mechatronik- Systeme gemeinnützige GmbH, Erfurt, Germany

Abstract. A high-speed photo detector IC for application in Blu-ray/DVD/CD drives is presented. Bandwidths for the highest gain of 254 MHz and 221 MHz for 405 nm (Blu-ray) and 635 nm (DVD) wavelengths, respectively, were achieved by applying novel design methodologies. The combination of this outstanding speed performance with its low power dissipation of 192 mW at 5V supply and the low noise power of −72 dBm at 300 MHz makes it the best in literature reported optical transceiver IC for Blu-ray and Blu-ray/DVD/CD multi drives. Beside the excellent performance results, the usage of the novel design methodologies gave us an increased design efficiency with 25% compared to earlier similar design processes.

1 Introduction

In 2008 the Blu-ray Disc won the competition for the distribution media of high-definition video data. Since the data storage density of Blu-ray Discs is about one order higher compared to DVDs and in consequence much higher data rates have to be transferred, high-speed electronics is needed to read, write and rewrite the discs.

In this work a photo detector IC (PDIC) with the capability of converting high-frequency blue and red laser light into electrical signals for data reading and writing is presented. The PDIC development had to face extreme requirements of signal amplifier bandwidth with a large dynamic range at the same time. The maximum possible rotation speed of recent Blu-ray Disc drives corresponds to a 12x readout speed and maximum data rates of 432 Mbit/sec, which requires 3 dB bandwidths larger than 200 MHz for the whole amplifier chain at all gains. In previous works maximum 3 dB bandwidths for either, DVD (260 MHz in Sturm, 2005) or Blu-ray applications (204 MHz in Ko, 2009) are reported. But especially for BD/DVD multi drives a PDIC with high bandwidths for multiple wavelengths is indispensable. We combined three novel design and optimization methodologies within the design process of the presented 12x-PDIC to achieve this sophisticated goal.

The PDIC was fabricated in a 0.6 μm BiCMOS process (XFAB, 2010) using integrated PIN-photodiodes with sensitivities of 0.26, 0.42 and 0.40 A/W for wavelengths of 405 nm, 635 nm and 780 nm (CD), respectively. With the implemented 8 gains one can adjust the high dynamic sensitivity between 28.0 and 0.2 mV/μW. Furthermore the device is featured by an overall power consumption of only 192 mW for a 10-channel system as depicted in Fig. 1.

2 Architecture of the photo detector IC

The schematic diagram in Fig. 1 gives an overall architecture overview of the PDIC. The PIN photodiode array is formed by 12 single diodes, divided into 4 center diodes (A-D) for beam focusing and data transfer and 8 satellite diodes (E-L) for tracking control of the laser beam. The diode array is followed by a 10-channel programmable (8 gains) 4-stage amplifier chain. The 6 amplifier chains A-D, RFp and RFn need to be high-speed for transmitting data up to 500 Mbit/sec. The 4 channels S1–S4 operate at lower speed and the connection to the input diodes E-L is reconfigurable via a mode decoder.

Each amplifier chain consists of a current amplifier (CAp, CAn) followed by a constant transimpedance amplifier (TIA) and a buffer amplifier (VA). Applying a 2-stage current amplifier with distributed gain control ensured high-speed operation and stability for all gains within the large dynamic range. For the differential channel RFp–RFn the diode currents A-D are summed up. Gain and mode decoder can be

Correspondence to: S. Lange
(steffen.lange@imms.de)

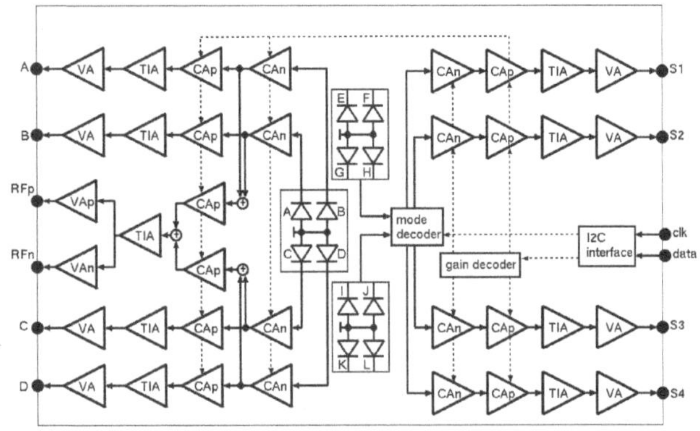

Fig. 1. Architecture overview of the complete 12x-PDIC.

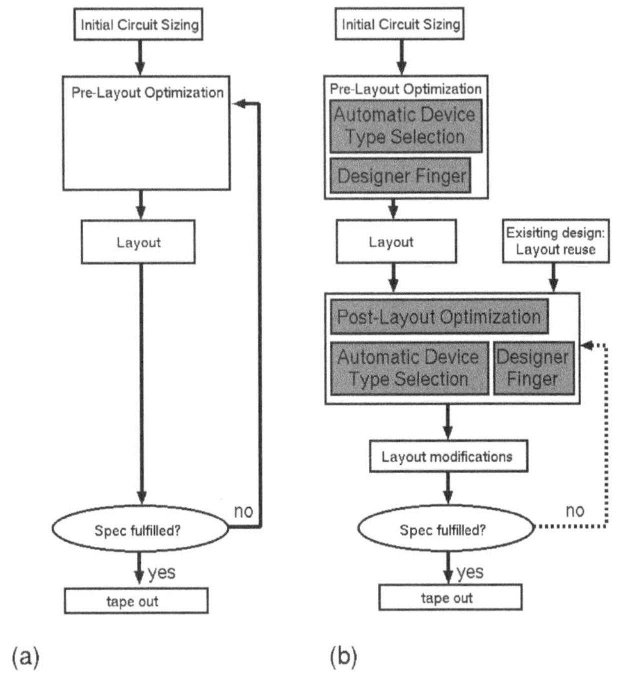

Fig. 2. Design flow within the 12x-PDIC development: **(a)** conventional optimization loop, **(b)** novel design flow applying post-layout optimization, automatic device type selection and Designer Finger.

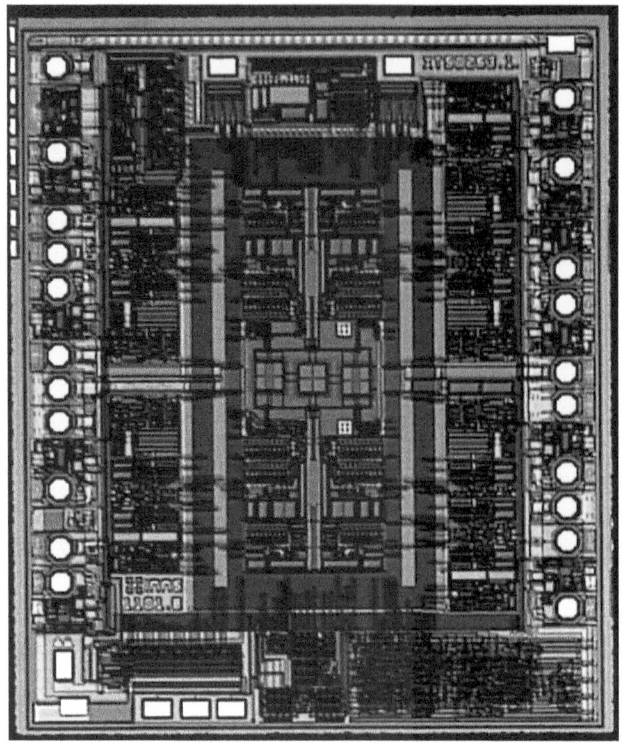

Fig. 3. Chip photograph of the PDIC.

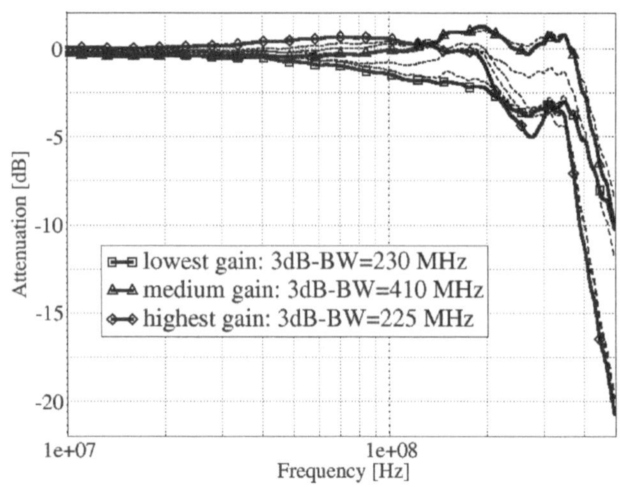

Fig. 4. Frequency response of an example high-speed channel for all 8 gains: curves for lowest, medium and highest gain are highlighted as bold lines.

configured via an I^2C interface. The presented PDIC is optimized with respect to signal bandwidth, sensitivity and noise for the high-speed channels to meet the challenging requirements of Blu-ray applications.

3 Novel methodologies for circuit optimization

Since almost ten years our group develops successfully optical transceiver ICs starting from CD and DVD applications up to today's Blu-ray drives. Raised requirements on amplifier chain bandwidths due to higher data rates made new design methodologies indispensable. We implemented a combination of three novel methods within the overall design process. In Fig. 2 the difference between the conventional design and optimization flow and the novel flow including all three methodologies for efficient optimization is shown. The proposed flow in Fig. 2b is suitable for complete design processes beginning from the initial topology selection and circuit sizing as well as for cost saving reuse of existing

Table 1. Simulated and measured performances of the presented 12x-PDIC.

Parameter	Simulated value	Measured value
sensitivity [mV/μW]		
· Blu-ray, highest/lowest gain	28.02/0.22	27.89/0.22
· DVD, highest/lowest gain	–	43.66/0.34
3 dB bandwidth [MHz]		
· Blu-ray, medium/highest gain	454/276	369/254
· DVD, medium/highest gain	454/276	355/221
Peaking [dB]	0.27	0.73
Slew rate rise/fall [V/μs]	133/−204	145/−135
Offset voltage [mV]		
· Mean value (108 channels)	1.39	1.0
· Standard deviation	–	3.3
Power dissipation @ Vdd=5V [mW]	190	191.7
Noise power (single-ended RF channel) @ 300 MHz [dBm]	−78	−77

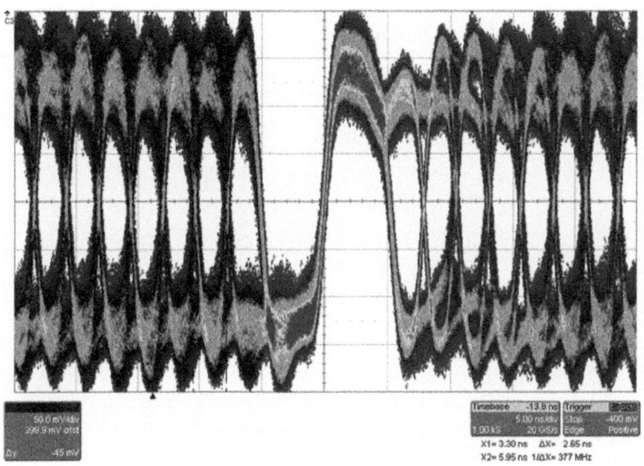

Fig. 5. Eye diagram in 12x-mode with an eye width of 2.5 ns (data rate = 400 Mbit/sec).

designs and layouts up to fulfilled specifications. The base for the design within this work ranged from existing low-speed sub designs including layouts (e.g. VA blocks) to completely new designed blocks (e.g. CAp, CAn). Combining layout reuse with the methodologies automatic device type selection (ADTS) and post-layout optimization (PLO) for the buffer amplifier VA for the fast channels A-D, RFp and RFn was very successful in terms of circuit performance and design efficiency. The post-layout 1 dB bandwidth of the single block VA could be increased from 220 MHz to 630 MHz. Furthermore, the discrepancy between pre- and post-layout performance was significantly reduced due to the intrinsic parasitics estimation of the PLO method. The layout reuse allowed a cost saving of 90% in the layout process for this particular block. Both, the PLO and ADTS methodology, are described in detail in (Reich, 2010) and (Dimov, 2008), respectively. The third novel methodology Designer Finger (Krausse, 2008) was successfully applied to the TIA. Its 1dB bandwidth was increased from 260 MHz to 385 MHz with a remaining peaking < 1 dB.

The presented methodologies are easy to implement into common design and optimization environments – within the presented design process Cadence 5.1 (Cadence Virtuoso, 2008) and WiCkeD analog circuit optimizer (MunEDA, 2008) were used.

4 Experimental results

Before PDIC characterization the laser spot characteristics was measured using a beam profiler. The full width at half maximum (FWHM) parameter was determined to 3.4 μm and 6.6 μm at wavelength of 405 nm and 635 nm. A very good focused laser beam was used for determining the photo diode sensitivity and homogeneity of this sensitivity. The measured sensitivity of the integrated PIN diode was 0.26, 0.42 and 0.40 A/W for wavelengths of 405 nm, 635 nm and 780 nm, respectively, which fit well with the simulated values.

A chip photograph of the fabricated 12x-PDIC is shown in Fig. 3. In Table 1 the simulated and measured key performances of the PDIC are listed. Intending high frequency operation, the buffer amplifier outputs have been matched with an integrated output resistance, since the downstream low-capacitive flexible PCB is specified with a load of 2 pF || 110 Ω. Within the measurement setup the load resistance was 110 Ω, whereas the load capacitance was slightly higher than the one for simulations. This fact helps the reader to understand the significant discrepancy in simulated and measured performances. However, considering the strong influence of

Table 2. Comparison of published PDIC realizations for DVD and Blu-ray drives.

Parameter		(Sturm, 2005)	(Seidl, 2005)	(Ko, 2009)	This work
Highest Sensitivity [mV/μW]	Blu-ray	–	–	12.3	27.9
	DVD	100	186	–	43.7
3 dB bandwidth [MHz] (high gain)	Blu-ray	–	–	204	254
	DVD	260	145	–	221
Offset voltage [mV]		n.a.	6.0	n.a.	3.3
Noise power	[dBm]	n.a.	–66.0	–89.8	–77.0
	@Freq. [MHz]	n.a.	200	120	295
Power dissipation [mW]		300	163	140	192
Number of channels		10	10	6	10

every small parasitics due to the measurement setup at high operation frequency, the achieved agreement between simulated and measured performances is superb.

In Fig. 4 the small-signal response of one high-speed channel for all gains with blue light illumination (wavelength = 405 nm) is shown. A mean 3 dB bandwidth of 254 MHz with 0.73 dB peaking was determined by measuring 25 high-speed channels on 5 PDIC devices. Furthermore, a medium gain mean 3 dB bandwidth of 369 MHz with 1.28 dB peaking was measured. For red light (wavelength = 635 nm) laser source the highest gain mean 3 dB bandwidth was determined to 221 MHz.

The eye diagram in Fig. 5 shows the highest gain measurement with 400 Mbit/sec pseudorandom data input, which is sufficient for the focused Blu-ray application. The limit for the realized 12x-PDIC is bound to 500 Mbit/sec data rates, since rise and fall times of about 2 ns were determined within experiments.

The statistical distribution of the measured offset voltage with a mean value of 1 mV and 5.8 mV standard deviation was achieved for measuring 108 channels. By applying the internal offset compensation the standard deviation improves to 3.3 mV. A power consumption of 192 mW is state-of-the-art among PDICs for DVD and Blu-ray drives.

The comparison with earlier publications of PDIC realizations in Table 2 substantiates the outstanding performance of the presented PDIC. It is state-of-the-art for optical transceiver ICs for fastest DVD drives with respect to bandwidth, sensitivity, noise and power dissipation. It is the best known PDIC for Blu-ray drives with the maximum 12x speed, since we achieved a 3 dB bandwidth larger than 250 MHz for the first time. All other known publications suffer from the fact, that they are characterized for either, DVD or Blu-ray. In contrast our PDIC for Blu-ray/DVD/CD multi drives shows excellent performance for both, the fastest DVD and Blu-Ray applications.

5 Conclusions

A high-speed 10-channel photo detector IC for Blu-ray, DVD and CD storage drives has been presented. The implementation of novel design methodologies within the overall design flow was the base for a successful IC realization in a 0.6 μm-BiCMOS technology meeting the requirements of Blu-ray applications with up to 500 Mbit/sec data transfer rates. The 12x PDIC with 8 programmable gains features a power consumption of 192 mW and a maximum transimpedance of 214 kΩ resulting in 28 mV/μW sensitivity. For the highest gain 3 dB bandwidths of 254 MHz and 221 MHz were achieved for wavelengths of 405 nm (Blu-ray) and 635 nm (DVD), respectively. The presented PDIC is the best known device for 12x speed Blu-ray drives and Blu-ray/DVD/CD multi drives. Furthermore its performance is state-of-the-art for DVD applications.

Acknowledgements. This work was mainly supported by the Thüringer Aufbaubank (TAB) within the research project no. 2006 VF 0046. Parts of this work were funded by the German Federal Ministry of Eduction (BMBF) within research projects HONEY (grant no. 01 M 3184 C) and SYENA (grant no. 01 M 3086).

References

Sturm, J., Leifhelm, M., Schatzmayr, H., Groiß, S., and Zimmermann, H.: Optical Receiver IC for CD/DVD/Blue-Laser Application, IEEE Journal of Solid-State Circuits, 40(7), 1406–1413, 2005.

Ko, J., Baek, K., and Kim, S.: High speed OEIC with offset drift compensation for bluray disc system, Proc. SPIE, 7219, 2009.

X-FAB Semiconductor Foundries AG: datasheet on 0.6 μm-BiCMOS process, http://www.xfab.com, 2010.

Cadence, Virtuoso Front To Back Design Environment 5.1. www.cadence.com, 2008.

MunEDA GmbH: WiCkeD Tool Suite, http://www.muneda.com, 2008.

Cadence Design Systems: Virtuoso NeoCircuit Circuit Sizing and Optimization, http://www.cadence.com/products/custom_ic/neocircuit/index.aspx, 2008.

Reich, T., Dimov, B., Lang, Ch., Boos, V., and Hennig, E.: A Post-Layout Optimization Method With Automatic Device Type Selection for BiCMOS Analog Circuits, IEEE 16th International Conference on Electronics, Circuits and Systems, 803–806, 2010.

Dimov, B., Boos, V., Reich, T., Lang, C., Hennig, E., and Sommer, R.: A novel technique for CAD-optimization of analog circuits with bipolar transistors, Adv. Radio Sci., 7, 219–223, doi:10.5194/ars-7-219-2009, 2009.

Krausse, D., Nowak, J., Schaefer, E., Sommer, R., and Hennig, E.: Frequency Compensation by Automated Topology Modification Using Mixed Analytical and Numerical Methods for Design of Fast TIA for HDTV Applications, Proc. SM^{2}ACD'08, Erfurt, Deutschland, 118–126, 2008.

Seidl, C., Schatzmayr, H., Sturm, J., Groiss, S., Leifhelm, M., Spitzer, D., Schaunig, H., and Zimmermann, H.: „A Programmable OEIC for Laser Applications in the Range from 405 nm to 780 nm, Proceedings of ESSCIRC, 439–442, 2005.

31

Real-time delay statistics in wireless IP networks

D. Huo

Bell Laboratories, 67 Whippany Road, Whippany, NJ 07981, USA

Correspondence to: D. Huo (dhuo@ieee.org)

Abstract. In the wireless communication, the variation of the transmission delay is called jitter and is one of the variables responsible for the degradation of the service quality. Jitter is present in every section of the transmission system. Its stochastic behavior depends on the technology implemented in the system and the service provided by the system. This paper focuses on mathematical modeling and phenomenological analysis of the jitter encountered by the real-time services in a wireless network. Using the data made available to the public by the wireless industry, we explore the stochastic characterizations of the jitter in a wireless IP networks. Within the scope of real-time service, we studied the relation between delay, jitter and the inter-packet time. Evaluation of the sample data indicates a long range dependence of the inter-packet time of the received packets in a real-time connection. The result helps understanding the transmission delay encountered by the real-time service over wireless IP networks.

1 Introduction

To provide IP (Internet Protocol) based packet transmission service a wireless network relies on an infrastructure that consists of multiple sections. Each of these sections introduces a time delay to the packet transmission. The temporary variation of the time delay is referred to as jitter and it depends not only on the underlying transmission technology, but also on the service the packet transmission is supposed to support. The real-time service requires a fixed transmit and receive rate and, as such, its quality is impaired by the jitter. In wireless technologies such as HSPA (High Speed Packet Data) or LTE (Long Term Evolution), jitter is primarily caused by two techniques introduced to improve the radio interface performance: the packet scheduling at the RNC (radio network controller) for UMTS or at the eNodeB for LTE, and the HARQ (hybrid automatic request-response) protocol on the radio link. The scheduling in the air-interface enables efficient resource sharing among users, and the HARQ enhances the utilization of the radio spectrum assigned to each user. In order to study the jitter caused by the air-interface in a wireless network, we consider a mobile-to-mobile connection as a chain of transmission sections. In specifics, the transmission system consists of the following sub-systems in sequence: source, radio up-link, core network, radio down-link, destination. Providing real-time services such as voice requires transmission of a continuous stream of speech frames with fixed packet rate. While the packet stream has a constant packet rate at the input of the transmission system, the received packet stream at the output of this transmission system may not have the same constant packet rate, because the time between speech frames may change as result of the delay jitter. It is thus the task of system architects to find ways, e.g. via buffer design, to cope with the jitter. Finding jitter mitigation strategies while not introducing excessive delay requires a better understanding of the jitter statistics. Here lies the motivation to the work reported in this paper. To confine our scope, we investigate jitter phenomena on the radio access network specified by 3GPP (2012). Instead of analyzing the protocols, we take a phenomenological approach by regarding each sub-system in the transmission chain as a black box and evaluate the experimental results available for the black boxes. In a project by 3GPP-RAN1 (2006), the VoIP provided by an HSPA was studied through simulations, where both up-link and down-link are implemented according to the technical specifications 3GPP-RAN1 (2011). The simulation is based on a cellular network layout of 59 sectors, using 4 channel models and 4 network loading scenarios. The detailed configuration and parameters can be found by 3GPP-SA4 (2006); ITU-R (2012). The packet delay and error event observed in the simulation can be further processed to produce sample paths of jitter. Thus, 54 sample paths can be obtained as raw material for our study.

In Sect. 2 we will discuss the definition of the random variables delay and inter-arrival time, and characterize the packet streams of constant rate in a rather general way. Section 3 is dedicated to the characterization of the compound transmission system, consisting of concatenated sub-systems. The mining of the aforementioned data for the purpose of jitter study is performed in Sect. 4, making use of the theory developed in the previous sections. In Sect. 5 we will construct and analyze the mobile-to-mobile connection using existing sample paths, before giving a conclusion in Sect. 6.

2 Transmission system

A packet stream is defined by indexing each packet by a unique integer $i \in \mathbf{Z}$, where $\mathbf{Z}$ is the set of integer numbers, so that consecutive packets are packets that should have adjacent indices. While being transmitted through the transmission system, packet experience delay, which can be determined by measurement at two points: One at the input port of transmission system and the other at the output port of the transmission system. The arrival time, as well as the departure time, is a mapping of the packet index to a real number $\mathbf{Z} \mapsto \mathbf{R}$, where $\mathbf{R}$ is the set of real numbers. Note that the interpretation of the term "arrival" and "departure" depends on the context, i.e. depending on the system we observe. As far as a pair of transmitter and receiver is concerned, the arrival time at the receiver is the time when the packet departs the system of transmission medium that connects the transmitter and receiver. For a packet stream with constant data rate, the time between consecutive packets at the data source should be the same. Then,without transmission impairments the arrival time at the receiver is a monotone increasing step function with a fixed increment. Let $T(i)$ denote the time when the packet with index i departs the transmitter. Then $T(i) = C + i/r$ for $i \in \mathbf{Z}$, where the real number $r > 0$ is the packet rate and C is the reference time. It is reasonable to assume that all packets in a packet stream are generated in order at the transmitter, i.e. $T(i) < T(j)$ for $i < j$. Let $t(i)$ be the time when the same packet arrives at the receiver. Then, $t(i) > T(i)$ for all $i \in \mathbf{Z}$. Obviously, the transmission delay of a packet $d(i) = t(i) - T(i)$ is positive for any non-trivial transmission system. Moreover it in fact a random number. For the further study, we make the assumption that the delay is a random variable defined in $(\Omega, \mathbf{B}, \mathbf{P})$, where Ω is the space of random event, $\mathbf{B}$ the Borel algebra on real numbers and $\mathbf{P}$ the probability measure. Since $t(i) = C_1 + d(i) + i/r$ with C_1 accounting for the fixed delay of the system and the reference time at the transmitter, it is readily seen that $t(i) - E\{t(i)\} = d(i) - E\{d(i)\}$, where $E\{\}$ refers to the statistical expectation.

Another related random number is the inter-packet time. As mentioned before, with the term of inter-packet time we could be talking about the inter-arrival time or the inter-departure time, depending on the context. For both cases the inter-packet time is a mapping of index $x : \mathbf{Z} \mapsto \mathbf{R}$. In particular, the inter-arrival time at the receiver is well defined by $x(i) = t(i) - t(i-1)$ for $\lim_{i \to -\infty} t(i) = 0$, and we have

Table 1. Example of re-ordering.

i	0	1	2	3	4	5	...
$t(i)$	10	11	14	12	13	15	...
$\pi \circ t(i)$	10	11	14	14	14	15	...

Proposition 1 *Between the inter-packet time and the delay there are following relations*

$$x(i) = d(i) - d(i-1) + \frac{1}{r} \tag{1}$$

$$d(i) = d(0) + \sum_{k=0}^{i} x(k) \tag{2}$$

For a stable transmission system, we can assume the stationarity. Then $E\{d(i)\} = E\{d(i-1)\}$ and $E\{x(i)\} = 1/r$, meaning the transmission system does not change the average inter-packet time, which is a necessary requirement of the real time service. Assuming the packet loss is handled by the upper layer protocol, we can ignore the packet loss in our study. Normally a packet with smaller index will arrive earlier that one with a larger index. But, due to the randomness of the delay, a packet with smaller index may be received later than a packet with larger index, resulting in packet disorder in the packet stream at the receiver. The requirement of order preservation for the delivery of real-time services necessitates a posterior re-ordering operation at the receiver: It is a mapping of a set of the packet indices into itself, so that the sequential order of the packets is restored as given by the source. While the re-ordering action re-shuffles the received packets for the purpose of re-establishing the original order, it modifies the receive time, resulting in jitter.

Definition 1 *Let* $\mathbf{Z}$ *be the set of all packet indices and* $\mathbf{R}$ *the set of the departure time. A re-ordering operation is a mapping* $\pi : \mathbf{Z} \times \mathbf{R} \mapsto \mathbf{Z} \times \mathbf{R}$ *and for* $i < j$

$$\pi(i,t) := (\pi \circ t)(i) = \begin{cases} t(j), & \text{for } t(i) > t(j) \\ t(i), & \text{for } t(i) \leq t(j) \end{cases} \tag{3}$$

The operation π permutes the packets with respect to the indices. Thus, for a given positive $\tau \in \mathbf{R}$,

$$\pi^{-1}(\tau) = \{i \in \mathbf{N} | (\pi \circ t)(i) = \tau\}, \tag{4}$$

the set of indices associated with τ after the re-ordering, is non-trivial and can have more than one elements, causing the phenomena of bursty arrival time, i.e. accumulation of packets in the receive buffer at a given time. Thus, the action of reordering changes the delay statistics. An example of disorder and re-odering operation is shown in Table 1.

Generally speaking the re-ordering introduces delay:

Proposition 2 *Let $t(i)$ be the departure time of packet i and $Var\{t(i)\} < \infty$. Then, for the re-ordering function $\pi : \mathbf{Z} \times \mathbf{R} \mapsto \mathbf{Z} \times \mathbf{R}$, there is $(\pi \circ t)(i) \geq t(i)$*

Proof: The re-ordering operation works index after index in ascending order, starting from a point $i = 0$. If $t(i) \geq t(i-1)$, then the order between packet i and packet $i-1$ is not changed by the transmission system, hence there is no need of re-ordering and $(\pi \circ t)(i) = t(i)$. But, if $t(i) < t(i-1)$, then a disorder occurs during the transmission, there is a need of re-ordering which maps $t(i)$ to a number no less than $t(i-1)$, resulting in $(\pi \circ t)(i) > t(i)$. QED.

How does the re-ordering affect the inter-arrival time and jitter ? Using the knowledge obtained so far, we can draw the following conclusion for the inter-arrival time: For $x(i) \geq 0$, no re-ordering is needed, as there is no disorder. A re-ordering operation due when $x(i) < 0$ is . Starting from i, one checks through $j = i, i+1, .., i+m-1$ until $i+m$ with $x(i+m) \geq 0$. Then, all those packets need to be buffered because they are out of order. Hence, the re-ordering operation is equivalent to setting $(\pi \circ x)(j) := 0$ for $j \in [i, i+m-1]$ and $(\pi \circ x)(i+m) := x(i+m) - \sum_{j=i}^{i+m-1} x(j)$. Then, it follows

$$\sum_{j=i}^{i+m} (\pi \circ x)(j) = x(i+m) - \sum_{j=i}^{i+m-1} x(j) \tag{5}$$

Now, assume that the arrival after $i+m-1$ is ordered for $n > 0$ consecutive packets, i.e. $x(j) \geq 0$ with $j \in [i+m, i+m+n]$ until again $x(i+m+n+1) < 0$. Then

$$\sum_{j=i+m}^{i+m+n} (\pi \circ x)(j) = \sum_{j=i+m}^{i+m+n} x(j) \tag{6}$$

Thus, we obtain finally

$$\sum_{j=i}^{i+m+n} (\pi \circ x)(j) = \sum_{j=i+m}^{i+m+n} x(j) - \sum_{j=i}^{i+m-1} x(j) \geq \sum_{j=i}^{i+m+n} x(j) \tag{7}$$

meaning that the inter-arrival time tends to increase after the re-ordering operation at the receiver. The conclusion can be formalized as the following

Corollary 1 $\lim_{n \to \infty} Pr\{\frac{1}{n} \sum_{i=1}^{n} [(\pi \circ x)(i) - x(i)] \geq 0\} = 1$

Proof:: Consider the set $N = \{1, 2, ..., n\}$ of n elements and $N = N^+ \cup N^-$ with $N^+ \cap N^- = \emptyset$, where $N^+ = \{i \in N | x(i) \geq 0\}$ and $N^- = \{i \in N | x(i) < 0\}$. Then

$$\sum_{j \in N^+} [(\pi \circ x)(j) - x(j)] = 0 \tag{8}$$

and

$$\sum_{j \in N^-} [(\pi \circ x)(j) - x(j)] > - \sum_{j \in N^-} x(j) \geq 0 \tag{9}$$

where the equal sign applies when $N^- = \emptyset$. Hence

$$E\{\frac{1}{n} \sum_{j \in N} [(\pi \circ x)(j) - x(j)]\} \geq 0 \tag{10}$$

Since the process is stationary, the relation holds for $n \to \infty$ and the weak convergence follows. QED

3 Compound transmission system

A heterogeneous transmission path can be decomposed into mutual independent sections. We call such a transmission system a compound system, consisting of concatenated sub-systems. In the chain of concatenated sub-systems the packet departure time at one sub-system is at the same time the packet arrival-time for the following sub-system. Thus, the delay observed at the output of the very last sub-system in the chain is the accumulation of delays from all sub-systems. Let $d_k(i)$ denote the packet delay caused by the k-th sub-system and $x_k(i)$ the corresponding inter-packet time. Denoting the random delay of the i-th packet of the compound system of n sub-systems with $D_n(i)$, we have $D_n(i) = \sum_{k=1}^{n} d_k(i)$. Accordingly, the inter-packet time $X_n(i)$ of the i-th packet of the compound system can be determined by

$$X_n(i) = D_n(i) - D_n(i-1) + \frac{1}{R_n} \tag{11}$$

following the equation (1). Then,

$$E\{X_n - \frac{1}{R_n}\} = \sum_{k=1}^{n} E\{x_k - \frac{1}{r_k}\} \tag{12}$$

Therefore, we have the following result:

Proposition 3 *For a compound system of n independent concatenated sub-systems,*

$$E\{X_n\} = \frac{1}{R_n} \tag{13}$$

when $E\{x_k\} = 1/r_k$ for $k = 1, 2, ..., n$. Then, the variances are subject to

$$Var\{X_n\} = \sum_{k=1}^{n} Var\{x_k\}. \tag{14}$$

The second statement follows from the independence between the sub-systems, i.e. $E\{(x_k - 1/r_k)(x_i - 1/r_i)\} = 0$ for $k \neq i$, and it indicates a spreading of the variance as result of concatenation. Moreover, if $E\{d_k(i) - d_k(i-1)\} = 0$ for all $i \in \mathbf{Z}$ and $k = 1, .., n$, we have $R_n = r_k = r$.

Corollary 2 *If a compound system is the concatenation of n independent sub-systems, then the compound system is rate preserving only when all n sub-systems are rate preserving.*

At each inter-face between the sub-systems, the inter-packet time can be positive or negative. Negative inter-packet time means packet index disorder in the packet stream. The inter-packet time after the re-ordering operation is always non-negative.

4 Data mining

4.1 Radio interfaces

All modern wireless networks utilize the techniques of resource sharing over the air-interface. Users are multiplexed onto the frequency and time resources, using radio technologies such as CDMA (Code Division Multiple Access), TDMA (Time Division Multiple Access), OFDMA (Orthogonal Frequency Division Multiple Access) or any combination of them, to achieve spectrum efficiency and thus the network capacity. The radio-interface HSPA is based on CDMA with dynamic resource allocation. Similar to LTE, the protocol architecture of HSPA consists of, from bottom to top, PHY(Physical Layer) –MAC(Medium Access Control Protocol) – RLC(Radio Link Control Protocol) –PDCP(Packet Data Convergence Protocol) – IP(Internet Protocol), from bottom to top, with services carried in the payload of IP stack. A critical entity in the MAC layer is the HARQ that combines the request-response protocol and the adaptive channel coding to enhance the link performance. While improving the throughput of the radio channel, HARQ also introduces packet delay that varies with the changing radio channel condition. The merit is similar to that of a scheduler implemented at the radio network controller to assign radio resource among users and services, so that the air-interface can be fairly shared and the resource utilization be optimized. While scheduler and HARQ are essential for the performance of the radio interface, both introduce delay variation. Thus, through HARQ and the scheduler, the delay performance of a wireless network depends on both the radio channel condition and the cell traffic demand. As result, the jitter occurs. A project of 3GPP-RAN1 (2011) has provided ample simulation data for HSPA air interface performance in 3GPP-RAN1 (2006), in which traces of the packet delay and packet loss are recorded. Following 3GPP-SA4 (2006), the data were produced for the purpose of development of voice codec. As such, they are ideal for the study of the jitter behavior in an wireless cellular environment, because the underlying network configuration, channel conditions and traffic loads are predetermined and normalized by the standard development organizations, and as such are representative and authentic. We will make use of those data in the sequel.

4.2 Stochastic sample paths

In 3GPP-RAN1 (2006), an input stream of identical packets was sent every 20 ms, and the packet delay and packet loss at the receiver were recorded for each packet. The samples for the down-link (DL) are collected on the RLC layer at the mobiles with a fixed geometry, while the samples for the up-link (UL) are collected on the RLC layer at RNC. The cellular network consists of 19 cells, each with 3 sectors. The 4 channel models used are based on ITU-R (2012):

- Pedestrian A with 3 km h^{-1} (PA3),
- Pedestrian B with 3 km h^{-1} (PB3),
- Vehicular A with 30 km h^{-1} (VA30),
- Vehicular A with 120 km h^{-1} (VA120),

representing the different speed of mobility with respective channel model. In addition, 4 intensities of traffic load are considered:

- 40 users per sector (40U),
- 45 users per sector (45U),
- 60 users per sector (60U),
- 100 users per sector (100U).

The simulation was performed by two independent teams, using different simulators. As result, 54 traces were produced, among which 36 are for the down-link and 18 are for the up-link. Sequence of delay values in units of millisecond, together with the error flag are recorded in each trace, where the error flags indicate whether the packet is lost or in error. The latter is ignored because we assumed the proper error handling by the upper layer protocols. In order to make those data useful for our study, we have to transform these sequences into traces of inter-packet time after an re-ordering operation. The result of this preprocessing is referred to as sample path of inter-arrival time: A sample path is a sequence of real numbers representing the random inter-packet time, from which we can observe the jitter. Before evaluating the 54 sample paths, an introduction to fractional Brownian motion is due. The process of long range dependence has been extensively studied in the past and has found widely application Mandelbrot (1998); Beran (1994). Among the processes of long range dependence, the self-similar process is perhaps the mostly quoted in the information and telecommunication industry, e.g. by Paxson (1995); Geist (2000); Crovella et al. (1997); Taqqu et al. (1997); Willinger et al. (1997, 1994); Park et al. (2000).

Table 2. Slope of the regression line as a two dimensional distribution.

j	=	ACF	PER	VAR
$a(j,i)$	=	$-\alpha(\text{ACF},i)$	$-\beta(\text{PER},i)$	$-\gamma(\text{VAR},i)$

Table 3. Slope statistic for the DL (df=35).

Statistic	a(ACF)	a(PER)	a(VAR)
Mean	−0.39318	−0.3601396	−0.4713244
Stdev	0.223543	0.1697654	0.03629753
90 %-EM	0.048695	0.036981	0.007907
H	0.803409	0.8199302	0.76433781

Table 4. Slope statistic for the UL (df=17).

Statistic	a(ACF)	a(PER)	a(VAR)
Mean	−0.36799	−0.2607483	−0.4728221
Stdev	0.162307	0.2737584	0.02107574
90 %-EM	0.050881	0.085819	0.006607
H	0.816006	0.6303742	0.76358897

Definition 2 *A stochastic process $X(t)$ is self-similar when it has a distribution identical to that of $w^{-H}X(wt)$ for $t > 0$, $w > 0$ and $0.5 < H < 1$, where H is called the Hurst parameter and w the scaling factor. A stationary stochastic process is called a fractional Brownian motion when*

1. $X(0) = 0$
2. $X(t_2) - X(t_1)$ has a Gaussian distribution for $t_1 < t_2$
3. $E\{X(t_2) - X(t_1)\} = 0$
4. $\mathrm{Var}\{X(t_2) - X(t_1)\} \propto |t_2 - t_1|^H$ for $1/2 < H < 1$

Based on this definition, long range dependence of stochastic process can be quantified by the Hurst parameter H, which can be estimated by means of proper asymptotes, as we will do in the following.

4.3 Asymptotes

For the packet inter-arrival time, the following second-order statistics can be computed using the sample paths:

1. Autocorrelation(ACF) :

$$\rho(k) = \frac{\mathrm{Cov}\{x(\cdot + k)x(\cdot)\}}{\mathrm{Var}\{x(\cdot)\}} \tag{15}$$

for $k \in \mathbf{Z}$ with $\rho(k) = O(k^{-\alpha})$ for $k \to \infty$.

2. Periodogram(PER):

$$p(\omega) = \frac{1}{2\pi n} |\sum_{k=1}^{n} [x(k) - E\{x\}] e^{k\omega\sqrt{-1}}|^2 \tag{16}$$

for sufficiently large $n > 1$ with $p(\omega) = O(\omega^{-\beta})$ for $\omega \to 0$.

3. Variance (VAR):

$$\sigma^2(k) = \mathrm{Var}\{\frac{1}{k}\sum_{i=1}^{k} x(i)\} \tag{17}$$

for large $k > 1$ with $\sigma^2(k) = O(k^{-\gamma})$ for $k \to \infty$.

In the above asymptotes, the speed of the decay characterizes the dependence between entries in the sample path. A stochastic process is self-similar, when the sample estimate of $\rho(k)$ and $\sigma^2(k)$ tends to $k^{-\alpha}$ for large sample size k and $\alpha > 1/2$, or equivalently the sample estimate of $p(\omega)$ tends to $\omega^{-\beta}$ for $1 < \beta < 2$ for small ω. The Hurst parameter H used to characterize the self-similar process is related to the power of the asymptotes through $H = 1 + \alpha/2 = (1 + \beta)/2$. The stochastic process has the long range dependence when the corresponding Hurst parameter is in the range of $0.5 < H < 1$. In order to obtain an estimation of this parameter from our samples, we observe the slope of logarithm of the respective second order statistics, i.e. the parameter $a(\mathrm{ACR}, i) = \alpha$, $a(\mathrm{PER}, i) = \beta$, $a(\mathrm{VAR}, i) = \gamma$, respectively. In more general term, as shown in Table 2, we observe the two dimensional variables $a(j, i)$, where j is the category index with $j = \mathrm{ACR}, \mathrm{PER}, \mathrm{VAR}$, and i is the index of the sample path used.

Using all 54 samples, the intercept and the slope of the logarithm of the autocorrelation, the periodogram and the variance are estimated, respectively: All in total, 36 slopes are estimated for the down-link, while 18 slopes are estimated for the up-link. The sample mean of the slope $a(\mathrm{ACF}, i)$, $a(\mathrm{PER}, i)$ and $a(\mathrm{VAR}, i)$ are computed by averaging over i, resulting in

$$a(\mathrm{ACF}) = \frac{1}{n_s}\sum_{i=1}^{n_s} a(\mathrm{ACF}, i),$$

$$a(\mathrm{PER}) = \frac{1}{n_s}\sum_{i=1}^{n_s} a(\mathrm{PER}, i),$$

$$a(\mathrm{VAR}) = \frac{1}{n_s}\sum_{i=1}^{n_s} a(\mathrm{VAR}, i),$$

where n_s is the number of samples. The results are shown in Tables 3 and 4 for the down-link and the up-link, respectively. The 90 percent confidence interval is also computed for Student-t distribution and shown as the error margin, i.e. 90 %-EM in the tables.

As Tables 3 and 4 show, a long range dependence of the inter-packet time is visible. The Hurst parameter is estimated over the marginal distribution of three categories ACF, PER and VAR, and shown in Table 5.

Table 5. Estimated H.

Link	Hurst Parameter	Stdev.	#Samples
DL	0.796	5.26	$86 \times 3 = 258$
UL	0.747	1.20	$18 \times 3 = 54$

5 Mobile-to-mobile connection

A mobile-to-mobile connection requires at least an up-link, a down-link and at least an intermediate node representing the core network. However, since we only have sample paths for the radio interface, it is reasonable to focus our attention to the jitters caused by the air-interface on the jitter. For this purpose, the model can be simplified by assuming that the core network processes the incoming traffic according to a deterministic round-robin scheduler. As result, the observed statistics of the mobile traffic will not be affected by the other traffics that join the node in the core network, which adds only a fixed delay to the packet stream of the refrence mobile traffic. Thus, the inter-arrival time, hence the jitter, becomes independent of the core network under this assumption, and we are left with a concatenation of only two sub-systems: One is the up-link radio interface and the other is the down-link radio interface. Using data available in 3GPP-RAN1 (2006) for up-link and down-link, one can construct new sample paths to study the jitter statistics of an end-to-end connection. Now that the compound system consists of two concatenated sub-systems, 4 groups of new sample paths can be generated, by combining four different types of sub-systems, i.e. low mobility (lm), high mobility (hm), low traffic (lt), high traffic (ht). The existing sample paths can, according to their respective physical configuration, be associated with these 4 groups as the following,

1. lm-lt: low mobility and low traffic (PA3,PB3,VA30; 40U)
2. hm-lt: high mobility and low traffic (VA120; 40U)
3. lm-ht: low mobility and high traffic (PA3, PB3,VA30; 100U)
4. hm-ht: high mobility and high traffic (VA120; 100U)

for the down-link and as

1. up-lm: low mobility (PA3,PB3,VA30)
2. up-hm: high mobility(VA120)

for the up-link. Since the inter-packet time of the sub-systems demonstrates long range dependence, we are interested in the same long range dependence of the inter-packet time in a compound system. To find more about this, let $\rho(k)$ denote the autocorrelation function of the inter-packet time of the compound system, and $\rho_i(k)$ be the autocorrelation function of the inter-packet time of the sub-systems with $i = 1, 2, ...m$. When independence among the sub-systems is assumed, we have

Table 6. Estimated α_c of the mobile-to-mobile connections.

Mean	Stdev	90 % EM
0.078454	0.369239	0.027325

$$\rho(k) = \sum_{i=1}^{m} \rho_i(k) w_i \tag{18}$$

for any $k \geq 0$, where

$$w_i = \frac{\text{Var}(x_i)}{\sum_{i=1}^{m} \text{Var}(x_i)} \tag{19}$$

By arranging the terms on the right hand in ascending order of magnitude α_i,

$$\lim_{k\to\infty} \rho(k) = \sum_{i=1}^{m} \lim_{k\to\infty} \rho_i(k) w_i = k^{-\alpha_c}(w_1 c_1 + w_2 c_2 k^{-\alpha_2+\alpha_c} + ... + w_m c_m k^{-\alpha_m+\alpha_c})$$

for $\alpha_c = \min_{1\leq i\leq m} \alpha_i$ and $c_i > 0$ with $i = 1, 2, ..m$. This indicates that the asymptote of the auto-correlation is dominated by the sub-system with the slowest decay. As an example we evaluate the parameter of the concatenated system consisting of 18 up-link samples and 18 down-link samples, randomly selected from the available data base. Combining them to make an end-to-end connection results in $18 \times 18 = 324$ samples. Now parameter α_c is estimated using these 324 samples, the result is shown in Table 6. Comparing to the numbers in Table 3 and 4, the estimated α_c in Table 6 is less than both α_0 and α_1. Thus, the example is consistent with the analysis. On the other hand, when the constants c_i are also taken into account, we obtain the following more accurate result:

Proposition 4 *For a compound system of m concatenated independent sub-systems, assume the processes of inter-departure time x_i are pair-wise independent for $i = 1, 2, .., m$, and the autocorrelation function $\rho_i(k)$ for the i-th subsystem has the asymptotic $\lim_{k\to\infty} \rho_i(k) = c_i k^{-\alpha_i}$ with $c_i > 0$ and $\alpha_i > 0$. Then, the autocorrelation function of the compound system $\rho(k)$ has the asymptotic behavior*

$$\lim_{k\to\infty} \rho(k) = O(k^{-\alpha_c}) \tag{20}$$

with

$$\alpha_c \leq \frac{\sum_{i=1}^{m} \alpha_i c_i w_i}{\sum_{i=1}^{m} c_i w_i}. \tag{21}$$

Proof: Introducing $c_w = \sum_{i=1}^{m} c_i w_i$ in (18) we obtain

$$\lim_{k\to\infty} \rho(k) = c_w \sum_{i=1}^{m} k^{-\alpha_i} c_i w_i / c_w \tag{22}$$

The sum on the right hand of the equation can be viewed as the expectation of the convex function k^{-u}. By Jensen's inequality,

$$k^{-\min_{i\leq i\leq m}(\alpha_i)} \geq \sum_{i=1}^{m} k^{-\alpha_i} \frac{c_i w_i}{c_w} \geq k^{-\sum_{i=1}^{m} \alpha_i \frac{c_i w_i}{c_w}} \tag{23}$$

for $c_w > 0$ and $\alpha_i > 0$. Since $\ln(k^{-u}) \geq \ln(k^{-v})$ for $u \leq v$,

$$\lim_{k\to\infty} \frac{-\ln[\rho(k)]}{\ln[k]} \leq \sum_{i=1}^{m} \alpha_i \frac{c_i w_i}{c_w} \tag{24}$$

Obviously, the term on the left hand is α_c and the equal sign holds when $\alpha_c = \alpha_i$ for $i = 1, 2, ..., m$. QED.

6 Conclusions

So far we have studied the jitter statistics experienced by the real time service like voice, when packet streams are transmitted over HSPA wireless systems, using analytical as well as empirical methods. By analysis we established the relationship between delay and inter-packet time, and investigated the stochastic behavior of those quantities. With help of simulation data available for HSPA, we could estimate the second order statistics of the jitter and evaluate their asymptotes. Combining the existing samples to construct new samples representing new physical scenarios allows us further to investigate the jitter statistics of a mobile-to-mobile connection.

An important assumption made here was that the service provided by the packet stream requires rate preservation, i.e. the data are generated with constant rate and also expected to be received with the same rate. This requirement makes it necessary to deploy buffer to enable re-ordering at the receiving end. Only the operation of re-ordering can mitigate the delay jitter introduced by the deployment of the scheduler and the HARQ in a wireless transmission system. Evaluation of the sample paths with respect to the asymptotic decay of the autocorrelation, the periodogram and the variance reveals a statistical long range dependence of the inter-packet time, where the power law of the asymptote of the order $k^{-\alpha}$ has typically $1/2 < \alpha < 1$. For a compound system consisting of concatenated sub-systems, the long range dependence behavior of the compound system is dominated by the sub-system with the slowest decay, a finding confirmed by the sample evaluation.

References

3GPP(Third Generation Partnership Project): "TS 36.300 V8.8.0", available at: http://www.3gpp.org, Aug. 2012.

3GPP: "LS on RAB and the error-delay-profile for the performance characterization of VoIMS over HS-DPA/EULLS", RAN1-44, 13–17 Febuary 2006, Denver, avilable at: http://www.quintillion.co.jp/3GPP/TSG_RAN/TSG_RAN2006/TSG_RAN_WG1_RL1_2.html, Aug. 2012.

3GPP, "Specification 25 Series", RAN1, available at: http://www.3gpp.org July, 2011.

3GPP: "Working Document for the Performance Characterization of VoIMS over HSDPA/EDCH, v0.0.4", 3GPP-SA4-40, 28 August–2 September 2006, Sophia-Antipolis, available at: http://www.3gpp.org, Dec. 2006.

International Telecommunications Union, "Guidelines for Evaluation of Radio Transmission", Recommendation ITU-R M.1225, available at: http://www.itu.int/rec/R-REC-M.1225/en, April, 2012.

Paxson, V.: "Fast Approximation of Self-Similar Network Traffic", LBL-36750, Lawrence Berkeley Laboratory, April, 1995.

Geist, R. and Westall, J.: "Practical Aspects of Simulating Systems Having Arrival Processes with Long-range Dependence", IEEE Proc. of the 2000 Winter Simulation Conference, 2000.

Mandelbrot, B. B.: Mutifractals and 1/f Noise, Springer-Verlag, 1998.

Beran, J.: Statistics for Long-Memory Processes, Chapman and Hall, 1994.

Crovella, M. E. and Bestavros, A.: Self-Similarity in World Wide Web Traffic: Evidence and Possible Causes, IEEE/ACM, Trans. on Networking, 5(6), December, 1997.

Taqqu, M., Willinger, W., and Sherman, R.: Proof of a Fundamental Results in Self-Similar Traffic Modeling, ACM SiGCOMM Compiuter Communication Review, 27(2), 5–23, April, 1997.

Willinger, W., Taqqu, M. S., Sherman, R., and Wilson, D. V.: Self-Similar Through High-Variability: Statistical Analysis of Ethernet LAN Traffic at the Source Level, IEEE/ACM, Trans. on Networking, 5(1), February, 1997.

Willinger, W., Taqqu, M. S., Sherman, R., and Wilson, D. V.: On the Self-Similar Nature of Ethernet Traffic (Extended Version), IEEE/ACM Trans. on Networking. 2(1), 1–15, February, 1994.

Park, K. and Willinger, W. (Ed.): Self-Similar Network Traffic and Performance Evaluation, John Wiley and Sons, Inc., 2000.

Quantitative comparison of performance analysis techniques for modular and generic network-on-chip

M. C. Neuenhahn, J. Schleifer, H. Blume, and T. G. Noll

Chair for Electrical Engineering and Computer Systems, RWTH Aachen Univ., Schinkelstraße 2, 52062 Aachen, Germany

Abstract. NoC-specific parameters feature a huge impact on performance and implementation costs of NoC. Hence, performance and cost evaluation of these parameter-dependent NoC is crucial in different design-stages but the requirements on performance analysis differ from stage to stage. In an early design-stage an analysis technique featuring reduced complexity and limited accuracy can be applied, whereas in subsequent design-stages more accurate techniques are required.

In this work several performance analysis techniques at different levels of abstraction are presented and quantitatively compared. These techniques include a static performance analysis using timing-models, a Colored Petri Net-based approach, VHDL- and SystemC-based simulators and an FPGA-based emulator. Conducting NoC-experiments with NoC-sizes from 9 to 36 functional units and various traffic patterns, characteristics of these experiments concerning accuracy, complexity and effort are derived.

The performance analysis techniques discussed here are quantitatively evaluated and finally assigned to the appropriate design-stages in an automated NoC-design-flow.

1 Introduction

Due to advances in VLSI-technology chip complexity increases, leading to the implementation of Systems-on-Chip (SoC) that include several functional units (e.g. processor cores, memories and embedded FPGAs) on a single chip. With further technology improvements the number of components in a SoC grows continuously (Blume et al., 2005).

One key problem concerning SoC is the communication among the functional units of a SoC. Applications mapped to a SoC often come with complex communication demands: communication should feature low latency as well as a high throughput rate with a guaranteed quality of service (Angiolini et al., 2006). It should be flexible to map different applications to a SoC. Furthermore, the communication should cause minimal costs in terms of area and power dissipation. For compatibility to future SoC with even more functional units the communication architecture should be scalable (Wielage and Goossens, 2002).

Classic communication architectures like busses or point-to-point connections can not fulfill these requirements (Bainbridge and Furber, 2002). An often mentioned communication architecture which might solve these problems are so-called Networks-on-Chip (NoC) (Benini and De Micheli, 2002). These NoC are derived from computer networks and adapted to the needs of SoC. First prototypes and even products featuring NoC have been recently published (Vangal et al., 2007; Tilera Corporation, 2007).

The performance and costs caused by these NoC strongly depend on NoC-specific parameters like topology, data word length, routing algorithm and many more (Bjerregaard and Mahadevan, 2006). This multitude of parameters creates a large design space associated with NoC. Finding a suitable setup for each application thus requires tools for efficient performance analysis to evaluate the NoC in different stages of the design flow.

In this work several static and dynamic performance analysis techniques are quantitatively compared concerning

- simulation/emulation speed,
- monitoring options
- related modeling/analysis effort and
- accuracy

in various experimental setups differing for example in NoC size or traffic pattern.

Correspondence to: J. Schleifer
(schleifer@eecs.rwth-aachen.de)

The static performance analysis is based on timing models, the NoC-parameters chosen and the traffic-patterns (see Sect. 3). The dynamic performance analysis techniques evaluated here are

- a VHDL-Simulator,
- a SystemC-Simulator,
- a Colored Petri Net-based simulation and
- a FPGA-based emulation.

Each of these techniques was first evaluated and then assigned to an appropriate design stage of a design flow for NoC.

2 Modular and generic NoC

The modular and generic NoC framework which was introduced in Neuenhahn et al. (2006) consists of three basic building blocks: Network Interface (NI), Routing Switch (RS) and Link. The Network Interface is the gateway between NoC and a functional unit, the Routing Switch routes data through the NoC and the link represents the connections between these building blocks (see Fig. 1). The arrangement of these blocks is defined by the NoC-parameter *topology*. The options of possible topologies range from classic topologies like mesh, torus or star topologies to application specific hierarchical topologies.

Each NoC component (NI, RS, Link) consists of different modules which can be implemented to provide the desired functionality. The modules for error-correction and – detection in NI and RS for example can be implemented using no coding, parity code or Huffman-codes (Neuenhahn et al., 2007).

3 Traffic-patterns

A reliable performance analysis of a NoC can only be performed for given communication demands of an application. These communication demands are usually given as traffic-patterns that can be distinguished into synthetic and application specific ones.

3.1 Synthetic traffic-patterns

Using synthetic traffic-patterns, the performance of NoC can be compared, as these traffic-patterns are representative for a certain application class.

These traffic-patterns are defined by the requested load of the NoC, the size of data blocks transferred over the network and the way traffic-source and traffic destination are chosen. For the last parameter different patterns are proposed like e.g. random traffic (Duato et al., 2003).

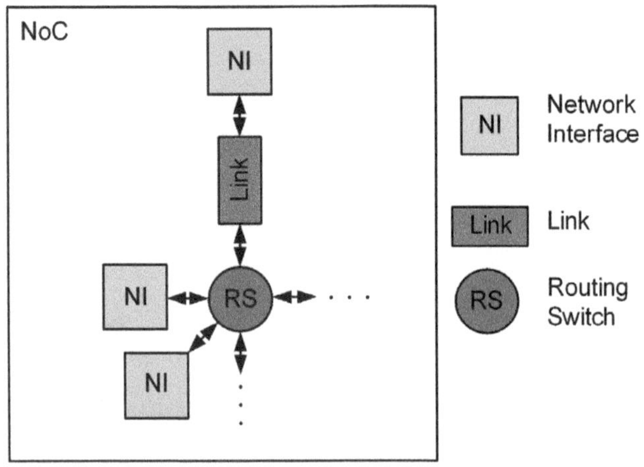

Fig. 1. Network-on-Chip.

3.2 Application specific traffic-patterns

The communication demands of an application can be modeled using a task graph (Dick et al., 1998). This graph consists of nodes representing certain tasks and directed edges, connecting these nodes. The weights of the edges represent an amount of data, which has to be transferred from one node to another. The weights of the nodes represent computation time. The task graph is restarted after a defined period of time, called hyper period.

4 Static performance analysis

During a static performance analysis traffic-patterns are mapped onto a NoC. Using this mapping, the expected loads on certain NoC-components are calculated. Dynamic effects, like e.g. already used links and/or repeated connection establishment due to blocked links can not be considered.

5 Dynamic performance analysis

5.1 FPGA-based NoC-emulator

The FPGA-based NoC-Emulator emulates a NoC on an FPGA (Neuenhahn, 2007). As even modern FPGAs lack the capacity to emulate a complex large SoC the functional units are replaced with abstract data sources and sinks.

The FPGA-based generic emulation environment consists of an FPGA and a PC. The PC is used to synthesize the FPGA-configuration, initiate experiments and evaluate the experiment results. The FPGA contains the NoC to be emulated (see Sect. 2), traffic-sources and -sinks which emulate functional units of a SoC and a soft-core processor (Nios II, Altera Corporation, 2007) which is used to control the data-sources and evaluate the raw results generated by the data-sinks.

5.2 SystemC-based NoC-simulator

Using a SystemC based simulation represents one possible alternative for performance analysis. SystemC is a C++ class library, which combines hardware description languages like VHDL or Verilog-HDL and object-oriented programming languages like C++ (IEEE, 2006). In addition, the open source SystemC library provides simulation capabilities.

Many SystemC-based simulators have been presented at different levels of abstraction. In this work a cycle-accurate NoC simulator was used. The structure of this simulator resembles that of the FPGA-based emulator presented in Sect. 5.1. It consists of a SystemC description of the NoC and a SystemC-model of data sources and sinks, which are controlled by the same C++-code as used in the FPGA-based emulator.

5.3 Simulator using colored petri nets

Another possibility to model and analyze NoC performance is simulation based on a Colored Petri Net (CPN) model.

Petri Nets (Petri, 1962; Girault and Valk, 2004) provide a graphical modeling formalism based on automata theory and find widespread use in different fields of engineering (Dotolia and Fanti, 2006; Shang et al., 2004). Petri Nets feature two basic components: places and transitions. Places are related to states in an automata graph and can be marked with an arbitrary number of tokens. Transitions are connected to places as input and output and can remove and generate tokens in these places. By allowing several places to be marked at the same time, Petri Nets facilitate modeling of concurrent processes.

Modeling with Colored Petri Nets (CPN) (Jensen, 1980) is a further expansion of these possibilities, especially by inclusion of data structures into tokens.

For design space exploration purposes a library of CPN models of NoC components was compiled (Schleifer, 2008). As only traffic generation is relevant for NoC performance modeling, clients are replaced by simple data sources and sinks that generate traffic either according to a random distribution or a task graph. Messages passing through the network are modeled as tokens containing the appropriate control data. Each component of the NoC is modeled in a modular fashion. It is thus possible to exchange modules such as the router or traffic generator without further changes to the model.

5.4 VHDL simulator

The VHDL-code which is used for VLSI-implementation and the FPGA-based emulator can be used for simulation, too. The control-signals for data sinks have to be derived from the NoC-description.

For the evaluation of this analysis technique the simulation software ModelSim (Mentor Graphics, 2007) was used.

6 Experimental results

Several experiments have been performed to determine the simulation speed and the simulation accuracy of the presented analysis techniques. Exemplarily the following NoC-specific parameters were chosen:

- Mesh-topology,
- static XY-routing,
- 16-bit data word length,
- circuit switching and
- no error protection.

The NoC-Parameter NoC-size, i.e. the number of functional units attached to the NoC, was varied.

Furthermore, traffic-patterns with synthetic random data traffic were used. The traffic parameters

- requested load and
- data block size

were varied during these experiments. The experiments were executed on a PC using a Pentium Core Duo Processor (3 GHz) and 1 GB DDR RAM. The FPGA-based emulator was implemented on an Altera Stratix II FPGA (EP2S60, Altera Corporation, 2007).

6.1 Accuracy

The performance of NoC can be determined using different measures. In this case we used the average latency of the data transmitted over the NoC. The relative error of this measure was below two percent for all analysis techniques but the static analysis technique. Using static performance analysis, an accurate performance analysis can only be performed for traffic-patterns with a low usage of the NoC.

The results produced by the FPGA-based emulator have been taken as reference values, as the VHDL-code of the NoC-emulator is identical to the code used for the VLSI-implementation. Moreover, the high simulation speed enables experiments with huge amounts of data.

6.2 Simulation speed

To analyze the performance of traffic-patterns on NoC most of the analysis techniques need some preparation, which consumes a certain lead time. This preparation has do be performed once for each NoC, e.g. the VHDL-Code used in a VHDL-simulator has to be compiled and the FPGA-based emulator has to be synthesized.

In the case of CPN modeling, at least with the software tool employed here, a syntax check of the complete model has to be performed before simulation. In case of a five by five NoC this takes approximately 10 min. This syntax

Table 1. Lead time for different analysis techniques using a 5×5 NoC with 16 Bit data word length and 1 GBit data.

	Lead time
Static Analysis	–
VHDL simulation (compilation of the VHDL-code)	<1 min
SystemC simulation	<1 s
CPN simulation (syntax check)	<10 min
FPGA-based emulation (synthesis of NoC-emulator)	<1 h

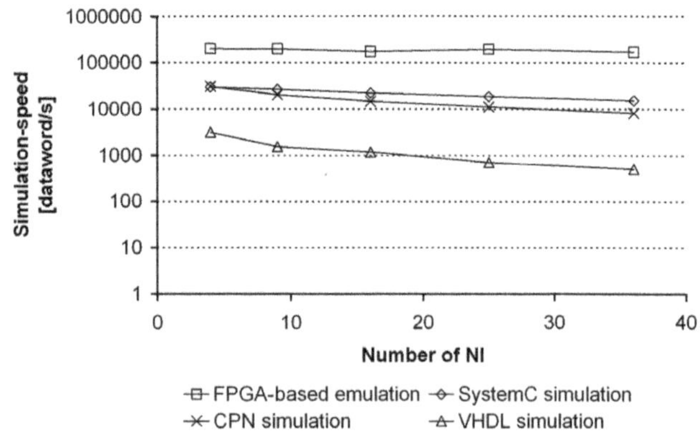

Fig. 2. Simulation speed of NoC using different NoC-sizes at 50% requested load and a data block size of 100 data words.

check is needed only after any major modification (like e.g. changing the topology of the NoC). The duration of this syntax check increases with NoC size and complexity. Small changes in the experimental scenario, such as longer messages or redistribution of traffic load amongst the different data sources, only require a partial syntax check, which is done in less than one minute.

The other techniques presented have no or a negligible lead time. In Table 1 lead times for different analysis techniques using NoC with 25 functional units and 16-bit data word length are compared.

The simulation speed in this work is measured in data words transferred over the NoC per second. Other metrics such as clock-cycles per second have been analyzed as well, which lead to similar results.

As no simulation is conducted for the static performance analysis, this technique is not regarded in the following section.

Simulation speed is affected by NoC-specific parameters as well as traffic patterns. The impact of each parameter is different for each analysis technique. As an example of NoC-specific parameter dependency the results of experiments varying NoC-size in a range from 2×2 to 6×6 (4 to 36 functional units) are shown in Fig. 2.

The number of FUs has a negligible impact on the simulation speed of the emulator due to the inherent parallelism of the FPGA. The simulation speed of the other techniques decreases with increasing number of FUs because the complexity of the NoC to be simulated increases. This effect is small for the SystemC-based simulator and stronger for the VHDL-simulator and the simulator using CPNs.

Furthermore the speed of both the FPGA-based emulator and the CPN-simulator depend linearly of the data block size. This dependency of the emulator is based on the fact that most of the time is spent generating traffic requests and evaluating received data traffic. As with an increased data block size the number of data blocks is reduced, simulation speed improves.

Due to the event-driven nature of the CPN-simulator its speed increases depending on the data block size. Increasing data word size reduces the number of events per transmitted data word and thus increases the simulation speed. The other techniques are not affected by the parameter data word size.

All these experiments show that using an FPGA-based emulator is at least one order of magnitude faster than the simulation using CPNs or the SystemC-based simulation which in turn are about one order of magnitude faster than the VHDL-simulation. CPN-based and SystemC-based simulations yield a comparable simulation speed. Depending on system-specific parameters (e.g. data block size) one or the other technique has its advantages.

7 Evaluation of performance analysis techniques

7.1 Static performance analysis

The advantages of a static timing analysis are short analysis times and good monitoring options, e.g. the utilization of a single link. As no dynamic effects can be considered, the accuracy of this technique is reduced if utilization of the NoC increases. Certain NoC-parameters like adaptive routing techniques are difficult to include into this analysis technique.

Due to the very short analysis time this technique can be advantageously used in early design stages, where various NoC-parameter settings (e.g. topology, data word length, clock frequency) can be compared to optimize the mapping of a task-graph onto the NoC. The reduced accuracy is negligible in this early design stage.

7.2 VHDL-simulator

The VHDL-simulation performs a cycle accurate simulation of the VHDL-code, which is used during the VLSI-implementation of the NoC. As every signal can be monitored an optimal tracing of the NoC is possible. Because the simulation speed of the VHDL-simulation is very slow compared to the other analysis techniques it is not applicable for huge NoC and extensive performance analysis.

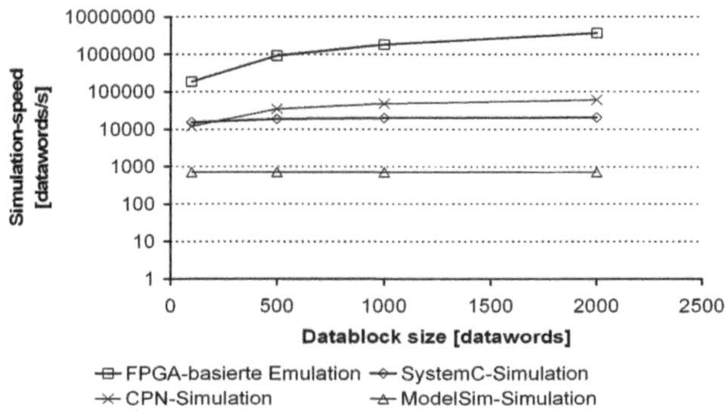

Fig. 3. Simulation speed for a NoC with 25 NI and varied data block size.

The VHDL-simulation is applied during the implementation of new NoC-components for the NoC-library. It is used for verifying the correct operation and for debugging purposes.

7.3 Simulator using Colored Petri-Nets

The model used for CPN/based simulation of a NoC in contrast to the other methods presented does not include modeling of the complete data transmission. Instead only the necessary control sequences are modeled as actual data is not relevant for network performance. The result of this approach is shown in Fig. 3: The less control overhead is needed for the transmission of a given amount of data the faster CPN-based simulation is. Furthermore, the CPN-based NoC model is very abstract and only includes parameters influencing network performance. Due to this small number of required parameters CPN modeling is well suited for performance analysis in early stages of the design flow. Since transition throughput and occupation of places in the CPN are easily accessible, all relevant data can be gained without major modification of the model.

This method of dynamic performance analysis allows identification of hotspots as well as characterization of NoC performance with parameters like average latency and load. Due to the interdependence of simulation speed and message size CPN based simulation is especially well suited to evaluation of traffic scenarios with large message sizes.

7.4 SystemC-based NoC-simulator

Using a cycle accurate SystemC-based simulator has advantages in performance analysis of NoC as no lead time is required, dynamic effects are considered while obtaining a quite high simulation speed. Other advantages are good monitoring options and the theoretical unrestricted NoC-size.

The disadvantages of this analysis technique are the increased modeling effort, as each NoC-component has to be modeled for simulation (SystemC) and VLSI-implementation (e.g. VHDL). Another important disadvantage is the fact that simulations with huge, real world data amounts last very long time and are therefore impractical.

Table 2. Qualitative comparison of performance analysis techniques.

	static performance analysis	VHDL simulator	SystemC simulator	CPN simulator	FPGA-based emulator
simulation speed	+ +	− −	+	+	+ +
lead time	+ +	o	+ +	−	− −
consideration of dynamic effects	− −	+ +	+ +	+ +	+ +
monitoring options	+	+ +	+ +	+	−
simulation-accuracy	− −	+ +	o	o	+ +
arbitrary NoC size	+ +	o	+ +	− −	− −
modeling effort	+ +	o	+	−	o

This analysis technique should be applied in the design-stage "Dynamic performance analysis" of the NoC-design flow, using traffic-patterns with small to medium data amounts. Bottlenecks can be identified during this simulation, too.

7.5 FPGA-based NoC-emulator

The fastest simulation speed is achieved by FPGA-based NoC-emulation. It is about an order of magnitude faster than the simulation speed of simulations featuring CPNs and SystemC-based simulations. Additionally, the same source-code used for VLSI-implementation is used during the emulation, too. Hence, dynamic effects are perfectly considered at cycle-accuracy. The emulations speed of the emulator is about a magnitude slower than a NoC implemented on a SoC.

As the synthesis of the NoC-emulator results in a high lead time an application in early design-stages where many NoC-parameter settings have to be compared is not advisable. The size of NoC is limited due to the capacity of the available FPGAs.

Monitoring options in FPGA-based emulators are limited. The results of performance analysis are general measures like average or maximal latency of transmitted data.

The high simulation speed enables a final functional verification of the NoC and a performance analysis using real world data amounts and different traffic-patterns. The appropriate state where the NoC-emulator should be used in the design-flow is the dynamic performance analysis, too.

All the features for the different analysis techniques are summarized in Table 2.

8 Conclusions

In this work different performance analysis techniques for a generic and modular NoC architecture were presented, quantitatively evaluated and compared. For this evaluation traffic-patterns with real world data amounts for NoC featuring up to 36 functional units have been used. The comparison of the different modeling methods discussed shows significant differences in precision, simulation speed, lead time and modeling/analysis effort.

Based on these results the analysis techniques were evaluated and assigned to the appropriate design stage: in an early design stage a static performance analysis is used, as it is very fast at the cost of reduced accuracy. In the succeeding design stages other analysis techniques (SystemC-based simulation, Colored Petri-Nets (CPN) based simulation or FPGA-based emulation) are used. Which analysis technique is best suited in each case depends on the chosen NoC-parameters and parameters describing the applied traffic-patterns. For example, the SystemC-based simulation and the simulator using CPNs are applicable for small to medium data amounts, whereas the FPGA-based emulator can be advantageously used for real world data amounts.

References

Altera Corporation: Nios II Processor Reference Handbook, Altera Corporation, 2007.

Altera Corporation: Stratix II Device Handbook, Altera Corporation, 2007.

Angiolini, F., Meloni, P., Carta, S., Benini, L., and Raffo, L.: Contrasting a NoC and a Traditional Interconnect Fabric with Layout Awareness, in: Design, Automation and Test in Europe, DATE '06, 1–6, 2006.

Bainbridge, J. and Furber, S.: Chain: a delay-insensitive chip area interconnect, IEEE Micro, 22, 16–23, 2002.

Benini, L. and De Micheli, G.: Networks on chips: a new SoC paradigm, Computer, 35, 70–78, 2002.

Bjerregaard, T. and Mahadevan, S.: A survey of research and practices of Network-on-chip, ACM Comput. Surv., 38, p. 1, 2006.

Blume, H., Feldkaemper, H. T., and Noll, T. G.: Model-Based Exploration of the Design Space for Heterogeneous Systems on Chip, J. VLSI Signal Proc., 40, 19–34, 2005.

Dick, R. P., Rhodes, D. L., and Wolf, W.: TGFF: task graphs for free, The Sixth International Workshop on Hardware/Software Codesign (CODES/CASHE '98), 97–101, 1998.

Dotolia, M. and Fanti, M. P.: An Urban Traffic Network Model via Coloured Petri Nets, Control Eng. Pract., 14, 17 pp., 2006.

Duato, J., Yalamanchili, S., and Ni, L.: Interconnection Networks – An Engineering Approach, San Francisco, USA, Morgan Kaufmann Publishers, 479-482, 2003.

Girault, C. and Valk, R.: Petri Nets for Systems Engineering, Berlin, Springer, 9–68, 2004.

IEEE: IEEE Std 1666–2005 IEEE Standard SystemC Language Reference Manual, IEEE Std 1666–2005, 0_1-423 pp., 2006.

Jensen, K.: Net Models in System Development, Ph.D. thesis, Aarhus University, 1980.

Mentor Graphics: ModelSim, Mentor Graphics, 2007.

Neuenhahn, M. C., Blume, H., and Noll, T. G.: Quantitative analysis of network topologies for NoC-architectures on an FPGA-based emulator, Kleinheubacher Tagung, 2006.

Neuenhahn, M. C., Blume, H., and Noll, T. G.: Quantitative Design Space Exploration of Routing-Switches for Network-on-Chip, Kleinheubacher Tagung Miltenberg, 145–150, 2007.

Neuenhahn, M. C., Lemmer, D., Blume, H., and Noll, T. G.: Quantitative Cost Modeling of Error Protection for Network-on-Chip, ProRISC Workshop, Veldhoven, 331–337, 2007.

Petri, C. A.: Kommunikation mit Automaten, Ph.D. thesis, Institut für instrumentelle Mathematik, Friedrich-Wilhelm-University Bonn, 1962.

Schleifer, J., Blume, H., and Noll, T. G.: Performance Analysis of Networks on Chip Using Coloured Petri Nets, in: ProRISC 2008, Veldhoven, The Netherlands, 94–99, 2008.

Shang, D. B., Koelmans, F., Yakovlev, A., and Xia, F.: Asynchronous system synthesis based on direct mapping using VHDL and Petri nets, IEE P.-Comput. Dig. T., 151, 12 pp., 2004.

Tilera Corporation: TILE64™ Processor Product Brief, Tilera Corporation, 2007.

Vangal, S., Howard, J., Ruhl, G., Dighe, S., Wilson, H., Tschanz, J., Finan, D., Iyer, P., Singh, A., Jacob, T., Jain, S., Venkataraman, S., Hoskote, Y., and Borkar, N.: An 80-Tile 1.28TFLOPS Network-on-Chip in 65 nm CMOS, in: IEEE International Solid-State Circuits Conference (ISSCC), Digest of Technical Papers, 98–589, 2007.

Wielage, P. and Goossens, K.: Networks on silicon: blessing or nightmare?, in: Digital System Design, 196–200, 2002.

Efficient modelling of IC conducted emission for power integrity analysis

R. Kazemzadeh[1,2], **S. Ludwig**[1,2], **Lj. Radić-Weissenfeld**[1,2], **and W. Mathis**[1]

[1]Leibniz University of Hannover Institute of Electromagnetic Theory, Hannover, Germany
[2]Fraunhofer Research Institution for Electronic Nano Systems, Advanced System Engineering, Paderborn, Germany

Abstract. In this paper two methodologies to reduce the complexity of IC conducted emission models for Power Integrity analysis in ICs are presented. The methodologies differ concerning the applicability in simulation tools, complexity and accuracy of the generated models. The first methodology uses a complex model and reduces its order to generate a model with a fewer number of elements. This methodology therefore involves a model order reduction approach. A second minimum complexity, module based modelling approach is introduced for rough estimations, as the order reduced model is still too complex for some applications. The two methodologies are applied to an IC conducted emission model of two digital modules of a 32 Bit microcontroller. The results of the three models are compared and discussed. Fields of application for the introduced modelling approaches are the estimation of the magnitude and time behaviour of the supply current as well as the determination of the number and position of the IC's supply pins.

1 Introduction

The ongoing development in the area of electronic circuit integration is accompanied by new, outstanding challenges in the field of low power design. This leads to more severe constraints in the field of power integrity, especially in regard to the ever decreasing supply voltages in modern electronic devices, circuits and systems. For a reduction of time-to-market by first-time right designs these low power and power integrity constraints need to be considered in an early stage of the design flow. To achieve this, the designer is reliant on modern, power-conscious EDA tools and accurate models that give insight to the power consumption and integrity behaviour at a high abstraction level of the design (Landman et al., 1994). For the estimation of the power integrity behaviour of integrated circuits (IC), IC Conducted Emission Models are created. Depending on the precision and simulation speed demands of the designer, different complexity levels for these models exist. Section 2 shows the necessity of reducing the complexity of certain IC conducted emission models and describes the structure of such models. The modelling methodologies of the two models with lower complexity, gained by reducing the size of the complex model, will be described in Sects. 3 and 4. In Sect. 5 simulation results of the three models are presented and compared. Section 6 gives a summary of the paper.

Correspondence to: R. Kazemzadeh
(reza.kazemzadeh@pb.izm.fraunhofer.de)

2 IC conducted emission models

The three IC conducted emission models employed in this section describe the conducted emission behaviour of the supply system of an IC. The model with the highest level of complexity is generated with the so-called EXPO tool (Hesidenz and Steinecke, 2005).

It consists of several modules, each describing a specific functionality on the chip (e.g. digital, analogue, memory). Each module consists of up to hundreds of identical supply line models (SLM) forming a mesh (Fig. 1). The SLM therefore represents the basic element of the EXPO model. Each SLM contains a power distribution network (PDN) consisting of passive R, L, C elements, and an independent current source (Fig. 2) (Dhia et al., 2006).

While the PDN describes the parasitic electromagnetic effects of the on-chip power supply, the current source characterises the internal activity (IA), i.e. the simultaneous switching noise (SSN) of the transistors. The parameters of the passive elements of the PDN as well as the prescribed waveforms of the current sources vary from module to module, but are the same for all SLMs within one functional module. The parameters of the model are determined either during the IC design phase or by taking measurements on an existing chip.

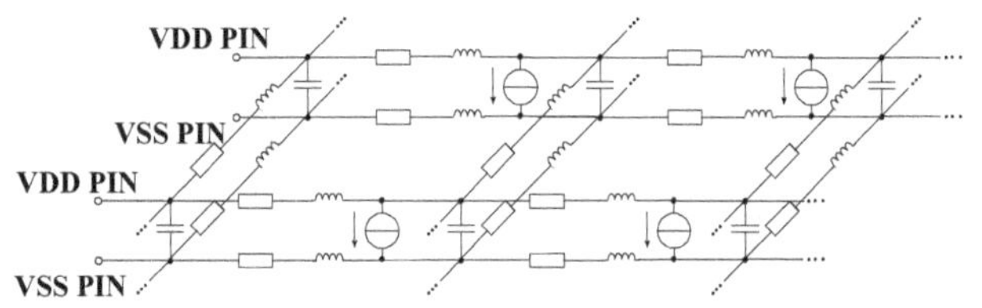

Fig. 1. Section of a functional module of the EXPO model consisting of a large number of SLMs.

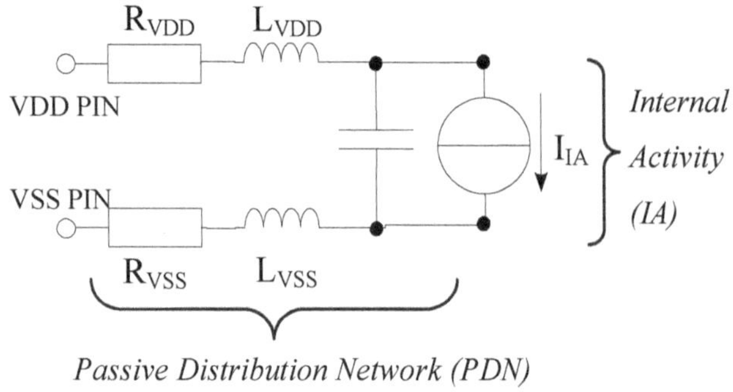

Fig. 2. Supply line model (SLM) consisting of PDN and IA.

Due to the relatively high complexity of the model gained with this approach, applicability in power integrity analysis tools is limited. IC conducted emission models with a smaller number of elements are required to lower the computational effort. Two methodologies to gain models adequate for the implementation in EDA tools by reducing the complexity of the EXPO model are presented here. While the computational effort is reduced and the ability to obtain information about the power integrity behaviour is preserved with these models, the accuracy is diminished to a certain degree compared to the complex model.

3 Modelling of an order reduced IC conducted emission model

The first modelling methodology involves a model order reduction (MOR) approach. The objective of the MOR modelling methodology is to reduce the size of the considered model and thereby its complexity while approximating the properties at the pins of the original network (Ludwig, 2008b). For MOR the electrical network of the IC conducted emission model has to be described with the help of the modified nodal analysis as a system of differential algebraic equations. Here the number of equations specifies the order of the system. To use MOR in IC conducted emission models efficiently, certain preparations have to be made in the network's description, as we have shown in Ludwig (2008a). The network description of the IC conducted emission model can now be reduced by the use of MOR algorithms. Order reduction of system descriptions of networks

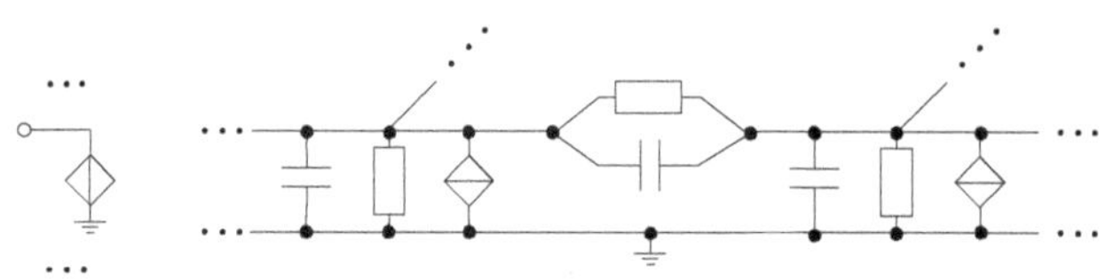

Fig. 3. MOR reduced IC conducted emission model.

has to preserve their main properties. This is for example achieved by using moment matching, as we investigated in Radić-Weissenfeld (2009). Here we use the block Arnoldi algorithm, where every system matrix is directly reduced, resulting in a drastically reduced order of the system. It can be shown that the obtained transfer function of the considered network matches the original transfer function within the investigated frequency band. In a last step, the order reduced circuit equations have to be synthesised as an electrical network for simulations with power analysis tools. The network synthesis method we proposed in Ludwig (2008b) generates a network with a low number of nodes and is exported as a SPICE netlist. By reducing the complexity of the EXPO model by means of MOR, the EXPO model's SLM structure is not maintained in the final synthesis step, resulting in a model consisting of G, C components and controlled current sources (Fig. 3).

4 Module-related IC conducted emission model

For a rough estimation of the power integrity behaviour where low simulation time takes priority over preciseness, the MOR modelling approach still leads to models that are too complex. For these applications a simplified IC conducted emission model with a very low number of elements is required. Such a model can be generated either by measurements or in a simulative approach and contains only one SLM for each functional module, i.e. the elements of the PDN and one current source (UTE 47A EMC Task Force, 2002) (Fig. 4). To meet the low complexity demands of the model, a curve fitting approach is used to obtain the R, L, C parameters of the PDN.

The fitting equation for an impedance Z_{mn} defined as the voltage V_m between the V_{DD} and V_{SS} Pins at module m divided by the input current I_n at the V_{DD} pin at module n is

$$Z_{mn}(s) = \frac{V_m}{I_n} = a_{mn} + s \cdot b_{mn} + \frac{c_{mn}}{s} \tag{1}$$

where the coefficients a_{mn}, b_{mn} and c_{mn} are obtained from the simulation results of the of the EXPO model. The relation between the coefficients and the R_{mn}, L_{mn}, C_{mn} parameters are as follows

$$\begin{aligned} R_{mn} &= a_{mn} \\ L_{mn} &= b_{mn} \\ C_{mn} &= c_{mn}^{-1} \end{aligned} \tag{2}$$

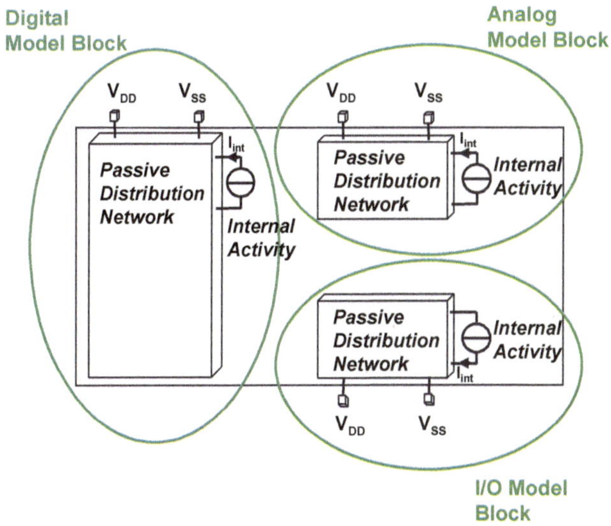

Fig. 4. Module-related model with one SLM (encircled) for each functional module.

For every module, Z_{mn}, m=n, is determined. The coefficients for $m \neq n$ are not considered, as coupling effects between modules are neglected for this low complexity approach.

For the determination of the internal activity of the SLMs, the external currents at the pins of the model are obtained through simulation. Following this, the external currents are converted into the frequency domain and multiplied with the transfer function of the previously determined PDN of the SLM, resulting in the internal current of the SLM of the model.

$$I_{\text{int}}(j\omega) = I_{\text{ext}}(j\omega) \cdot \left(1 - \omega^2 LC + j\omega RC\right) \tag{3}$$

5 Simulation results

For the analysis, the three proposed methodologies were applied to an IC conducted emission model describing two digital modules of a 32-Bit IC. The model consists of 12 SLMs with 120 elements and 51 nodes and is generated with the EXPO tool (Hesidenz and Steinecke, 2005). This model was reduced with the MOR approach to a model containing 64 elements and 14 nodes.

Since the netlist of the module-related model for rough power integrity estimation contains only 8 elements and 6 nodes, complexity is even further reduced with this approach compared to the MOR approach.

All three models were simulated in HSpice. Figure 5 shows the impedance curves Z_{11} of one digital module for all three models. The Z_{11} curve of the MOR model as well as the curve of the module-related model show very good matching with the curve of the model generated with the EXPO Tool.

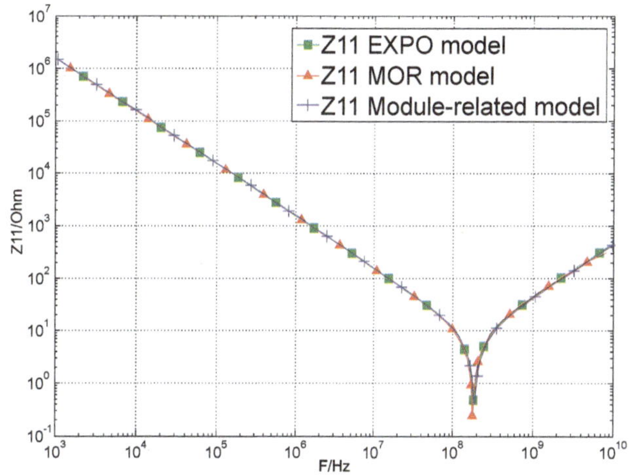

Fig. 5. Impedance curves Z_{11} of the EXPO model, MOR model, and module-related model.

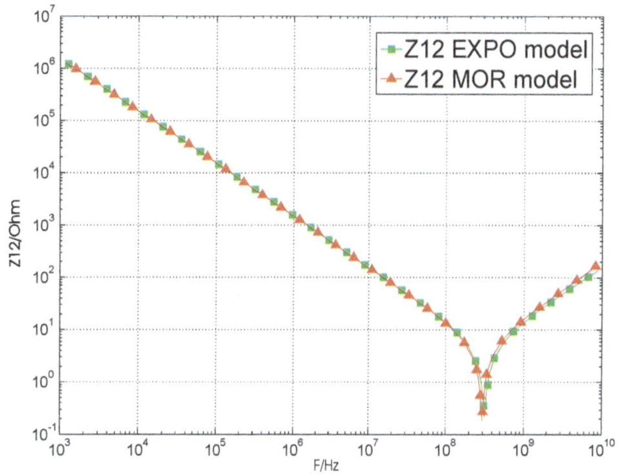

Fig. 6. Impedance curves Z_{12} of the EXPO model and MOR model.

For the Z_{12} parameter, describing the coupling between the two modules, Fig. 6 depicts good correlation between the EXPO model and the MOR model. A Z_{12} curve is not provided for the module-related model, as coupling effects between modules can not be considered for this type of model.

Figure 7 shows the transient response of the supply voltage $V_{DD1,\text{ref}}$=1.5 V for the three models at the V_{DD1}-Pin of the first digital module.

Again, the MOR model as well as the module-related model show a very good matching with the EXPO model.

The main properties of the three models are summarised in Table 1. The simulation time is reduced significantly for the MOR and the module-related approach.

For the module-related model in this work, two digital modules of the complex EXPO model were reduced to one SLM for each module. Since the SLM possesses only two pins, the model is merely suitable for module-related

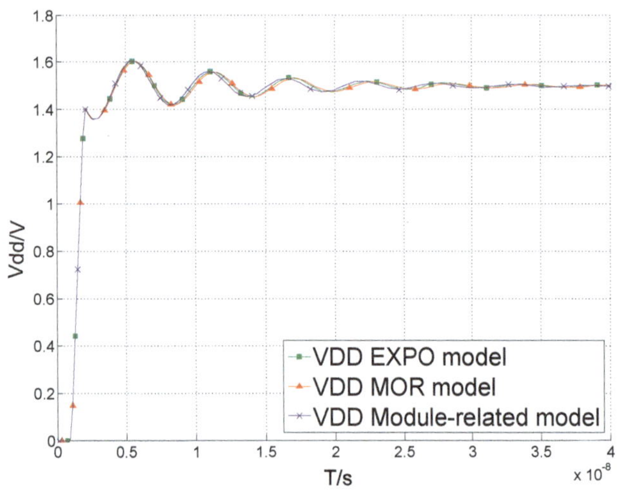

Fig. 7. Transient response of the EXPO model, MOR model, and module-related model.

Table 1. Complexity of the EXPO model, MOR model and module-related model.

	EXPO	MOR	Module-related
		Model complexity	
El. compon.	110	64	8
nodes	51	14	6
		HSpice simulation time	
AC-Analysis	100%	42%	19%
Transient	100%	74%	13%
		Accuracy	
	– Pin-related	– Pin-related	– Module-related
	– Crosstalk considered	– Crosstalk considered	– Crosstalk not considered

estimation of conducted emission, whereas for the EXPO and the MOR model, having all IC pins modelled, pin-related estimation of emission is possible.

6 Conclusion

This paper proposes two methodologies to reduce the complexity of large IC conducted emission models. The first methodology uses MOR algorithms, while the second method uses a fitting algorithm to obtain the parameters of a minimum complexity module-related model from simulation results of the complex model.

While it is possible to make a point about the disturbances at the IC pins with both modelling approaches, only the MOR model provides conclusive information about pin-coupling effects. This information is lost for the module-related model.

The MOR and module-related model were compared with the EXPO model and close alignments were observed in the time and frequency domain. A significant speed-up in simulation time was achieved for the MOR model in comparison to the EXPO model, and even more so for the module-related model.

Acknowledgements. Represented research and development work is carried out in the frame of the MEDEA+ project A701 PARACHUTE project (Parasitic Extraction and Optimization for Efficient Microelectronic System Design and Application).

This particular research was supported by the BMBF (Bundesministerium fuer Bildung und Forschung) Federal Republic of Germany under 01M 3169 A, 01M 3169 D and 01M 3169 E. The responsibility for this publication is held by the authors only.

References

Dhia, S. B., Ramdani, M., and Sicard, E.: Electromagnetic Compatibility in Intergrated Circuits, Springer-Verlag, New York, 2006.

French committee UTE 47A EMC Task Force: Cookbook for Integrated Circuit model ICEM, project number 62014-3, 2002, available at: http://www.ute-fr.com/FR/, last accessed 2009.

Hesidenz, D. and Steinecke, T.: Chip-Package EMI Modeling and Simulation Tool "EXPO", presented at the 2005 EMC Compo, Munich, Germany, 2005.

Landman, P. E.: Low Power Architectural Design, PhD dissertation, Electrical Engineering and Computer Sciences, University of California, Berkeley, CA, 1994.

Ludwig, S., Radić-Weissenfeld, Lj., Mathis, W., and John, W.: Efficient Model Reduction of Passive Electrical Networks with a Large Number of Independent Sources, presented at the International Symposium on Circuits and Systems (ISCAS), Seattle, Washington, USA, 2008.

Ludwig, S., Radić-Weissenfeld, Lj., Mathis, W., and John, W.: Efficient passive network description of IC conducted emission models for model reduction, Adv. Radio Sci., 6, 133–137, 2008, http://www.adv-radio-sci.net/6/133/2008/.

Radić-Weissenfeld, Lj., Ludwig, S., Mathis, W., and John, W.: Two-step Order Reduction of IC Conducted Emission Models, presented at the 2008 Asia-Pacific EMC Week, Singapore, to be published, 2009.

Analysis of iteration control for turbo decoders in turbo synchronization applications

T. Lehnigk-Emden, U. Wasenmüller, C. Gimmler, and N. Wehn

Microelectronic Systems Design Research Group, University of Kaiserslautern, Erwin-Schroedinger Str., 67663 Kaiserslautern, Germany

Abstract. Wireless data transmission results in frequency and phase offsets of the signal in the receiver. In addition, the received symbols are corrupted by noise. Therefore, synchronization and channel coding are vital parts of each receiver in digital communication systems. By combining the phase and frequency synchronization with an advanced iterative channel decoder (inner loop) e.g. turbo codes in an iterative way (outer loop), the communications performance can be further increased. This principle is referred to as turbo synchronization. The energy consumption and the peak throughput of the system depend on the number of iterations for both loops. An advanced iteration control can decrease the mean number of needed iterations by detecting correctly decoded blocks. This leads to a dramatic energy saving or to an increase of throughput. In this paper we present a new stopping criterion for decodable blocks for turbo decoding in interrelation with turbo synchronization. Furthermore the implementation complexity of the turbo decoder is shown on a Xilinx FPGA.

1 Introduction

Synchronization and channel decoding are essential parts of every digital receiver for wireless communication. Task of the synchronization is to present data bits to the channel decoder, where the negative influences of timing, frequency and phase offset are eliminated. A well known advanced scheme for channel coding are turbo codes (Berrou et al., 1993). Turbo code decoding is done in an iterative algorithm based on the maximum a posteriori principle. The turbo decoder produces extrinsic information for the systematic bits, which are used in the next iteration. Communications performance, throughput and power consumption of a turbo code decoder depend on the number of iterations. Thus it is desirable to find criteria for the iteration control without negative impact on the communications performance.

Synchronization is typically performed only once before channel decoding. The method of joint iterative turbo code decoding and synchronization is called turbo synchronization (Godtmann et al., 2006). Turbo synchronization allows to decrease the number of pilot symbols for synchronization purposes and thus increases the user data rate for a given bandwidth. In turbo synchronization the turbo decoder must additionally calculate new a posteriori information on the parity bits for the next synchronization refinement. The essential calculation of the extrinsic information of the parity bits in turbo synchronization provides the possibility to use this additional information for iteration control. State-of-the art iteration controls for turbo decoder cannot check for a valid codeword without adding additional redundancy bits. Therefor they can stop the decoder for invalid codewords too. Our novel iteration control is based on a check, whether the results of the actual turbo decoding iteration represent a valid codeword without extra redundancy bits.

In this paper we apply turbo synchronization for the iterative synchronization of phase and frequency offset as well as iteration control. In Sect. 2 we summarize turbo encoding as well as the the principle of turbo code decoding. In the following Sect. 3 we explain the iterative synchronization and the interrelation with the decoding algorithm. In Sect. 4 we categorize the state-of-the-art methods for iteration control and present our new iteration control method. In Sect. 5 we analyse the communications performance of turbo synchronization. We thoroughly compare our advanced iteration control to state-of-the-art taking communications performance and hardware performance into account. Communications performance is determined by bit true models. Implementation complexities and results are given based on Xilinx devices.

Correspondence to: T. Lehnigk-Emden
(lehnigk@eit.uni-kl.de)

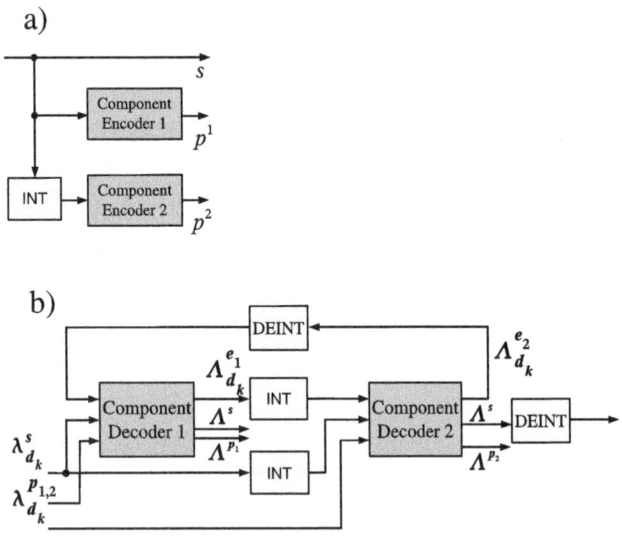

Fig. 1. **(a)** Turbo encoder, **(b)** Turbo decoder.

2 Turbo codes

With the introduction of binary turbo codes by Berrou in 1993 (Berrou et al., 1993) near optimum error correction became possible. Due to these error correction capabilities, binary and duo-binary turbo codes allow for low frame error rates (FER) at a low signal-to-noise ratio (SNR), outperforming the widely used convolutional codes. Because of this advantage turbo codes are now part of a large number of communication standards.

Turbo codes in general consist of a serial or parallel concatenation of two codes, so called component codes, and an interleaver. While the first component code encodes the information in the original order, the second one gets the information in a permuted order, see Fig. 1a). In all standards convolutional codes are used as component codes.

Decoding of turbo codes is an iterative process where probabilistic information is exchanged between component decoders (Berrou, 2003). Iterative decoding implies a big challenge with respect to low latency and high throughput decoders.

A possible realization of a decoder of turbo codes is given in Fig. 1b). The two component decoders that decode the two component codes are connected via interleaver and deinterleaver. They use log likelihood ratios (LLR) $\lambda^s_{d_k}$, $\lambda^{p^1}_{d_k}$ and $\lambda^{p^2}_{d_k}$ of the systematic and parity information to compute the extrinsic information $\Lambda^{e1}_{d_k}$ and $\Lambda^{e2}_{d_k}$ on the information bits. The iterative exchange of $\Lambda^{e1}_{d_k}$ and $\Lambda^{e2}_{d_k}$ between these component decoders is referred to as turbo principle. One (full) iteration is done if decoder one and decoder two have run once. If only one decoder has calculated new information, we call this one half iteration.

Both component decoders perform a maximum a posteriori probability (MAP) decoding. However the suboptimal Max-Log MAP algorithm with extrinsic scaling factor (ESF) is more suitable for implementations. In comparison to the optimal algorithm the Max-Log MAP results in a performance loss below 0.2 dB (Robertson et al., 1995, 1997). Moreover it was shown in Worm et al. (2000) that, when employed in turbo decoding, one does not require knowledge of the SNR.

The Max-Log MAP algorithm consists of a forward and a backward recursion. It computes for each possible information or parity bit d_k an a posteriori probability (APP) LLR Λ^s, Λ^{p1}, Λ^{p2}.

3 Turbo synchronization

The synchronization consists of the estimation of the unknown parameters of timing, frequency and phase offset, and the elimination of all possible negative influences introduced by these parameters. We focus on the frequency and phase synchronization of bursts with linear modulation (e.g. QPSK,16-QAM) in conjunction with turbo decoding. We assume, that the steps of gain control, timing and burst detection are properly carried out before. The received sample sequence r is given in the complex baseband according to Eq. (1):

$$r(l) = s(l) \cdot e^{j(2\pi f_o l + \Phi)} + n(l) \qquad l = 0, 1, \ldots, L-1 \quad (1)$$

The sample sequence r with L elements is based on modulation symbols $s(l)$ with one sample per symbol and symbol duration T, and is disturbed by a noise sequence n. The frequency offset f_o and phase offset Φ have to be estimated and corrected. They are considered fixed during an estimation interval. The synchronization is done in two main steps. Initially, a coarse synchronization is carried out. Afterwards, fine synchronization is done iteratively with the additional use of tentative decoder decisions after each decoder iteration. The methods for the coarse synchronization step depend on the existing communications system. The choice of method is influenced by the maximum frequency offset, the number of symbols per burst as well as the number and placement of known symbols.

We concentrate on the step of fine synchronization, which will be used for turbo synchronization. With the assumption, that all symbols of the burst are known the effect of the modulation by $s(l)$ can be removed by:

$$\tilde{r}(l) := r(l) \cdot s^*(l) \qquad l = 0, 1, \ldots, L-1 \quad (2)$$

A lot of algorithms exist to estimate frequency and phase offset for an unmodulated carrier as described in Eq. (2) (Meyr et al., 1998; Mengali and D'Andrea, 1997). However it must be considered, that usually the symbols of the burst are unknown or only some symbols, used for supporting the burst detection or supporting the coarse synchronization are known. Thus we replace the transmitted symbol sequence

s by an estimated symbol sequence s_e. The estimation of the transmitted symbol sequence is provided by the turbo decoder. Furthermore, the fine synchronization runs in parallel to the decoder iterations to avoid throughput degradation by turbo synchronization; thus a fast and simple algorithm is required. The fine estimation of frequency and phase offset is based on the average phase of the front part and rear part of the burst with a modulation removal by the estimated symbol sequence s_e. This is formally given by

$$Z(k) := \sum_{l=0+k\cdot L/2}^{L/2-1+k\cdot L/2} r(l) \cdot s_e(l) \qquad k = 0, 1 \tag{3}$$

With the two phase values of Eq. (3) the estimate of the frequency offset can be calculated with

$$\tilde{f_0} = \frac{arg(Z(1) \cdot Z^*(0))}{2\pi \cdot L} \tag{4}$$

The estimate of the phase offset is calculated with the help of Eq. (4)

$$\tilde{\phi} = arg(Z(1) + Z(0)) - L \cdot \tilde{f_0} \cdot \pi \tag{5}$$

The first decoder iteration is based on the LLR values $\lambda^s_{d_k}$, $\lambda^{p^1}_{d_k}$ and $\lambda^{p^2}_{d_k}$ calculated with the symbols of the coarse synchronized received sequence. For the iterative fine synchronization an estimate of the transmitted symbols is used, which is produced by the turbo code decoder after each iteration. The estimate of the transmitted symbols is gathered by the APP LLR of the decoder. A turbo code decoder computes APP LLR values Λ^s of the systematic bits by default. In turbo synchronization applications the decoder must additionally calculate the APP LLR values $\Lambda^{p1,2}$ for the parity bits.

To reduce the effect of using some erroneous reference symbols in the fine synchronization, soft values are used for the reference symbols. The values for the quadrature components of the estimated symbols are calculated by a tanh operation on the APP LLR. With the tentative soft values of systematic bits and parity bits the sequence s_e is generated and the described fine synchronization process can be carried out.

The received sequence r is corrected with the new estimates of frequency and phase offset. A new synchronized received sequence r is calculated after each full decoder iteration. Turbo decoding and fine synchronization run in parallel in our architecture. For the n-th iteration of the decoder the LLR values Λ^s, Λ^{p1}, and Λ^{p2} of the transmitted bits are calculated on base of the fine synchronized sequence, which used the APP LLR values of the (n-2)-th iteration. A discussion of the effects of the schedule regarding the update of LLR values can be found in (Alles et al., 2007).

4 Iterations control

Latency and energy consumption strongly depend on the number of iterations in the decoding process. Thus, reducing the number of iterations in the decoding process is an urgent issue. Instead of using a fixed number of iterations, the number of iterations has to be controlled by intelligent iteration control mechanisms. An efficient iteration control allows to detect correctly decoded blocks while decoding. Thus, a good iteration control is key to optimize energy consumption and reduce the mean latency. In Thul et al. (2002) it was shown that iteration control is the most efficient technique for energy saving in a turbo decoder system without sacrificing communications performance.

4.1 State-of-the-art

Cyclic Redundancy Check (CRC) rule. A separate error-detection code, such as a CRC, can be concatenated as an outer code with an inner turbo code in order to flag erroneously decoded sequences. The decoding process is stopped after iteration i whenever the syndrome of the CRC is zero. The false alarm rate in sense of recognizing a erroneous block as correct is a delta to zero. One disadvantage of the CRC is an additional code rate loss. However, in many systems like UMTS LTE the CRC code is available.

Hard decision rules. The idea behind these algorithms presented in Wu et al. (2000) and Shao et al. (1999) is to follow the changes of the hard decisions of either the extrinsic information, the APP information or the decoded bits within one or two iterations. The disadvantage of these rules are that they are not able to decide if the decoded codeword is a valid codeword.

Soft decision rules. Other approaches observe the soft LLR or APP values in the MAP decoders. With these values, several metrics can be calculated, e.g. the mean LLR value. For more details the reader is referred to Wu et al. (2000); Hagenauer et al. (1996) and Matache et al. (2000). All soft decision rules have in common, that thresholds have to be defined for aborting the decoding process. In addition, there is no guarantee that the decision is correct. The choice of the thresholds is always a trade-off between false alarm rate and detection rate of correctly decoded blocks.

4.2 Re-encoding approach

Our approach is to check whether the actually decoded information represent a valid codeword or not. One possibility to realize this, is to re-encode the hard decoded information bits and to compare the parity bits from the decoder and encoder. If a valid codeword is detected the decoder can be stopped.

The decoder calculates LLR values of the information and parity bits. The hard decision information bits are given to a turbo encoder which is identical the the encoder at the transmitter side, for re-encoding. The encoder produces new

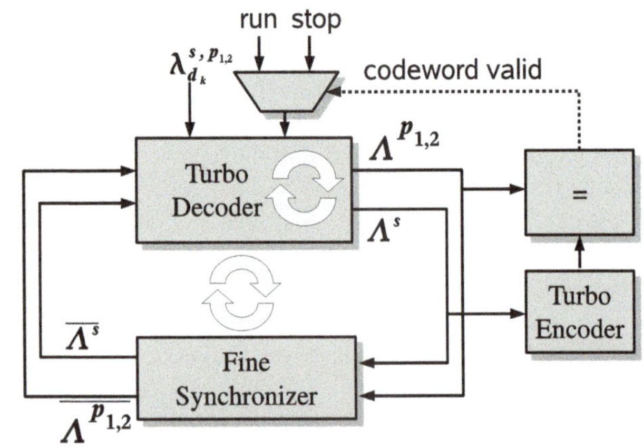

Fig. 2. Turbo decoder with fine synchronization and iteration control.

parity bits which are compared to the hard decision parity bits calculated by the two MAP decoders. If there is no difference the decoder has found a valid codeword. Further iterations make no sense. The decoder can be stopped immediately, see Fig. 2.

For implementation the check can be done in parallel to the MAP decoding which leads to no throughput degradation of the MAP decoder. In parallel to the MAP processing the encoder encodes the systematic information and compares the parity of the current MAP decoder. A valid codeword is detected if the systematic bits do not change within two successive half iterations and if the two parity bit sets do not differ. In this case the decoder and synchronization loops can be stopped and the system is ready for a new codeword.

5 Results

The new approach for iteration control is validated per software simulations in sense of communications performance. In addition the implementation complexity for the turbo synchronization with iteration control on a Xilinx FPGA is presented and compared to the non iteration control solution.

5.1 Communications performance

The new iteration control detects correctly decoded blocks and stops the decoder. This approach has no impact on the communications performance, which is shown by simulations in Fig. 3. The figure shows the results for a block length of 1056 and 1136 information bits for the code rates 0.443 and 0.8, respectively. For the lower code rate code 16-QAM gray modulation is used for the higher rate QPSK respectively. Down to a frame error rate (FER) of 10^{-3} the iteration control has no effect on the communications performance, the curves are congruent. In addition, the turbo synchronization gains up to 0.25 dB compared to the initial synchronization only.

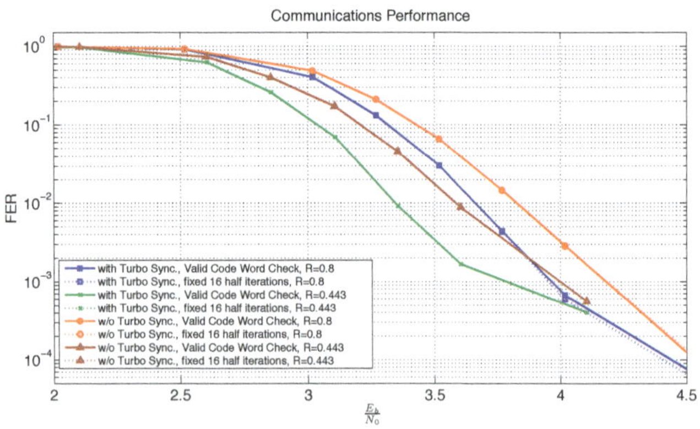

Fig. 3. Comparison communications performance - with and w/o iteration control; 16 decoding half iterations.

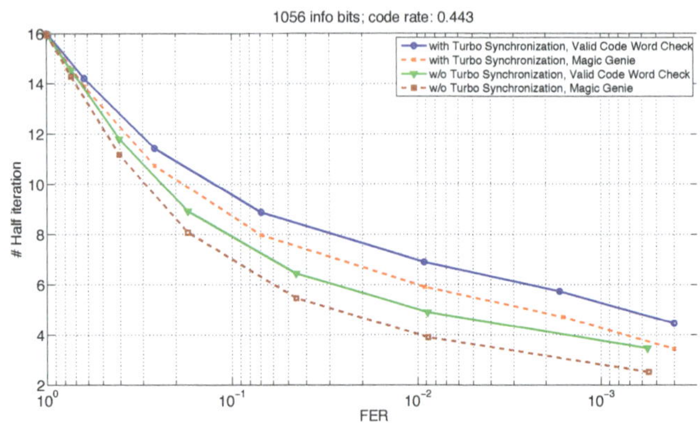

Fig. 4. Iteration Control: magic genie vs. re-encoding approach – with and w/o turbo synchronization; max 16 decoding half iterations; code rate 0.443.

An optimal stopping criterion for decodable blocks, we call this "magic-genie", can stop the decoder at once if the decoded information bits are correct. A comparison of the needed iterations between the "magic-genie" criterion and the re-encoding approach is shown in Fig. 4 for the code with rate 0.443 and in Fig. 5 for the code with the rate 0.8, respectiveley.

The re-encoding approach is close to the magic-genie criterion. The maximum difference is one half iteration. For lower FER the criterion can save up to 80% of the iterations. That means the energy consumption of the whole system can be decreased dramatically or the peak throughput can be increased. In addition it can be seen, that 50% of iterations can be saved at a frame error rate below 10^{-1} for high rate code. The turbo decoder with turbo synchronization operates at a lower SNR for the same FER than without turbo synchronization, see Fig. 3. Therefore it needs up to two more half iterations in average. In addition the new criterion can be used as quality indicator of the channel. A statistic over several blocks can be seen as a rough channel estimation.

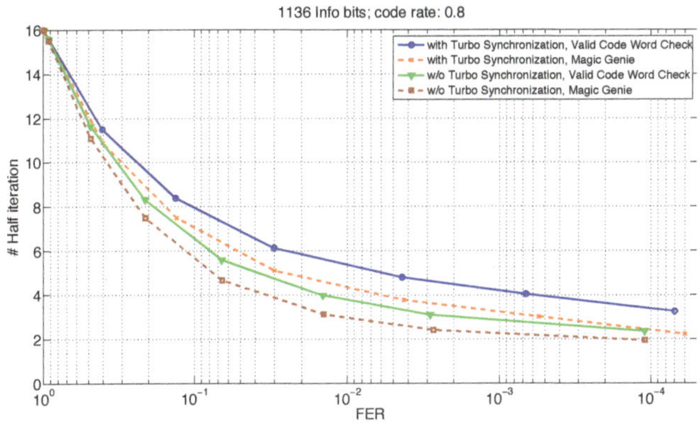

Fig. 5. Iteration Control: magic genie vs. Re-encoding approach – with and w/o turbo synchronization; max 16 decoding half iterations; code rate 0.8.

Table 1. Key parameter of the turbo decoder.

	Binary turbo decoder
Architecture	Serial MAP, 3 ACS units
Trellis states	8
Infoword size	128–5124 bits
Codeword size	128–15 372 bits
Code rates	$\geq 1/3$
Parallelism	1
Input quantization	6 bit
Algorithm	MaxLog-MAP
ESF	0.75
Max. iterations	8
Technology	FPGA
Clock frequency	233 MHz
Payload Thrpt. [Mbit/s]	4.67–14.02
Latency [μs]	10.36–414.7
Payload Bit/Cycle	0.02–0.06

5.2 Hardware implementation

The implementation complexity is analysed on a Xilinx Virtex-5 FPGA. Three versions have been implemented. First the turbo decoder without the extra parity LLR calculation unit which is needed for the turbo synchronization. The second one with the parity LLR calculation unit. The third version extends the second one with the new iteration control. The architecture of the turbo decoder is a state-of-the-art SMAP architecture with three recursion units which run in parallel. The key parameters are summarized in Table 1. For more details the reader is referred to (May et al., 2007).

Table 2 shows a comparison of implementation complexity of the three cases. The extra parity calculation and the encoder for the iteration control utilize additional 57 slices, i.e. 4%. That means that the hardware overhead is negligible in this case.

Table 2. Comparison implementation complexity.

	Turbo decoder	Turbo decoder	Turbo decoder
calc. parity LLR	no	yes	yes
iteration control	no	no	yes
Xilinx Virtex-5 XC5VSX95T @ 233 MHz			
LUT	4432	4495	4513
FF	2255	2283	2297
Slices	1427	1477	1484
BRAMs	4	4	4

6 Conclusions

In this paper we presented a new iteration control for turbo decoding in combination with turbo synchronization for turbo decoders. With turbo synchronization a communications performance gain of 0.25 dB for the low and high rate code can be achieved. The stopping criterion is based on a re-encoding approach to detect a valid codeword during the decoding process. Stopping the decoding loop with this criterion saves up to 80% of iterations and energy without losing any communications performance. The hardware overhead for the stopping criterion is below 5% of the whole turbo decoder.

References

Alles, M., Lehnigk-Emden, T., Wasenmüller, U., and Wehn, N.: Implementation Issues of Turbo Synchronization with Duo-Binary Trubo Decoding, in: Proc. 19th Annual IEEE International Symposium on Personal, Indoor and Mobile Radio Communications (PIMRC) 2007, Athens, Greece, 2007.

Berrou, C.: The Ten-Year-Old Turbo Codes are Entering into Service, IEEE Communications Magazine, 41, 110–116, 2003.

Berrou, C., Glavieux, A., and Thitimajshima, P.: Near Shannon Limit Error-Correcting Coding and Decoding: Turbo-Codes, in: Proc. 1993 International Conference on Communications (ICC '93), pp. 1064–1070, Geneva, Switzerland, 1993.

Godtmann, S., Pollok, A., Hadaschik, N., Steinert, W., Ascheid, G., and Meyr, H.: Joint Iterative Synchronization and Decoding Assisted by Pilot Symbols, in: IST Mobile & Wireless Communications Summit, Myconos, Greece, 2006.

Hagenauer, J., Offer, E., and Papke, L.: Iterative Decoding of Binary Block and Convolutional Codes, IEEE Transactions on Information Theory, 42, 429–445, 1996.

Matache, A., Dolinar, S., and Pollara, F.: Stopping Rules for Turbo Decoders, TMO Progress Report 42–142, http://tda.jpl.nasa.gov/progress_report/, Jet Propulsion Laboratory, Pasadena, California, 2000.

May, M., Neeb, C., and Wehn, N.: Evaluation of High Throughput Turbo-Decoder Architectures, in: IEEE International Symposium on Circuits and Systems (ISCAS), New Orleans, USA, 2007.

Mengali, U. and D'Andrea, A.: Synchronization Techniques for Digital Receivers, Plenum Publishing Corporation, New York, 1997.

Meyr, H., Moeneclaey, M., and Fechtel, S. A.: Digital Communication Receivers, John Wiley & Sons Inc, 1998.

Robertson, P., Villebrun, E., and Hoeher, P.: A Comparison of Optimal and Sub-Optimal MAP decoding Algorithms Operating in the Log-Domain, in: Proc. 1995 International Conference on Communications (ICC '95), pp. 1009–1013, Seattle, Washington, USA, 1995.

Robertson, P., Hoeher, P., and Villebrun, E.: Optimal and Sub-Optimal Maximum a Posteriori Algorithms Suitable for Turbo Decoding, European Transactions on Telecommunications (ETT), 8, 119–125, 1997.

Shao, R. Y., Lin, S., and Fossorier, M. C. P.: Two Simple Stopping Criteria for Turbo Decoding, IEEE Transactions on Communications, 47, 1117–1120, 1999.

Thul, M. J., Vogt, T., Gilbert, F., and Wehn, N.: Evaluation of Algorithm Optimizations for Low-Power Turbo-Decoder Implementations, in: Proc. 2002 IEEE International Conference on Acoustics, Speech, and Signal Processing (ICASSP '02), pp. 3101–3104, Orlando, Florida, USA, 2002.

Worm, A., Hoeher, P., and Wehn, N.: Turbo-Decoding without SNR Estimation, IEEE Communications Letters, 4, 193–195, 2000.

Wu, Y., Woerner, B. D., and Ebel, W. J.: A Simple Stopping Criterion for Turbo Decoding, IEEE Communications Letters, 4, 258–260, 2000.

Noise of short-time integrators for readout of uncooled infrared bolometer arrays

D. Würfel, D. Weiler, B. J. Hosticka, and H. Vogt
Fraunhofer Institute for Microelectronic Circuits and Systems, Duisburg, Germany

Abstract. As state-of-the-art readout circuits short-time integrators in Far Infrared (FIR) uncooled bolometer arrays are commonly used. This paper compares the transfer functions of an ideal continuous-time integrator with that of a real integrator with focus an OTA parameters and noise analysis. Beside the noise sources at the non-inverting input of the OTA special care has to be taken to account for capacitances at the inverting input. The Noise Equivalent Temperature Difference (NETD) as the key parameter for bolometer applications for a real short-time integrator will be derived. As the result it will be shown, that the NETD is $1/f$-noise limited.

1 Introduction

The need for uncooled infrared focal plane arrays (IRFPA) for imaging systems has increased since the beginning of the nineties. Examples for the application of IRFPAs are thermography, pedestrian detection for automotive, firefighting and infrared spectroscopy. According to Plank's law, each body with a temperature above absolute zero point emits electro-magnetic radiation. The wavelength and intensity depends on the temperature of the body. For example, a body at a temperature of 300 K has its maximum emittance at a wavelength of approx. $10\,\mu$m. FIR imager uses the IR-radiation of a warm body in the wavelength range between $8..14\,\mu$m. The FIR-sensitive element is the bolometer which changes its resistance by absorbing the radiation of the warm body. A pulsed bias with a constant voltage is applied to the bolometer and the resulting current is integrated for a fixed time. An offset current is subtracted before integration. Conventional readout circuits use a column-wise architecture.

Correspondence to: D. Würfel
(daniel.wuerfel@ims.fraunhofer.de)

2 Bolometer

The bolometer is fabricated by post-processing on CMOS wafers in Fraunhofer IMS Microsystem Lab. The realization of a microbolometer is shown in Fig. 1 as a cross section. A micromachined membrane consisting of amorphous silicon is suspended by two via stacks of metal from the CMOS substrate. The membrane forms with two other layers a good interferometric structure for radiation absorption. On top of the membrane an antireflection layer with a sheet resistance of 377 Ω/sq is deposited. The bottom structure consists of a nearly perfect reflecting metal layer (Ruß et al., 2007). To increase the thermal resistance the membrane is fixed by two small legs (Fig. 2). The optical distance between membrane and reflection metal reaches an optimum for one quarter of the radiation wavelength. To reduce thermal losses by gas conduction a vacuum package is required. The bolometer converts the infrared radiation into heat energy and this induces a temperature rise resulting in a change of the electrical resistance. Typical bolometers have pixel pitch values of $35\,\mu$m or $25\,\mu$m. Figure 2 shows a bolometer array with 4×4 sensor elements.

3 Short-time integrator

In Sect. 1, the ideal integrator model is analysed. In Sect. 2, the real integrator circuit is presented, analysed and compared to the ideal model.

3.1 Ideal model

Figure 3 shows the ideal short-time integrator model in the time-domain. By applying a bias voltage pulse $v_{\mathrm{Bias}}(t) = V_{\mathrm{Bias}}$ to the bolometer resistor $R_{\mathrm{Bolo}}(T_0)$ the bolometer current $i_{\mathrm{Bolo}}(t)$ is generated and integrated at the integration capacitor C_{int}. Only a small part of the bolometer current depends on the IR-radiation, so a bias current $i_{\mathrm{Bias}}(t) = I_{\mathrm{Bias}}$ is

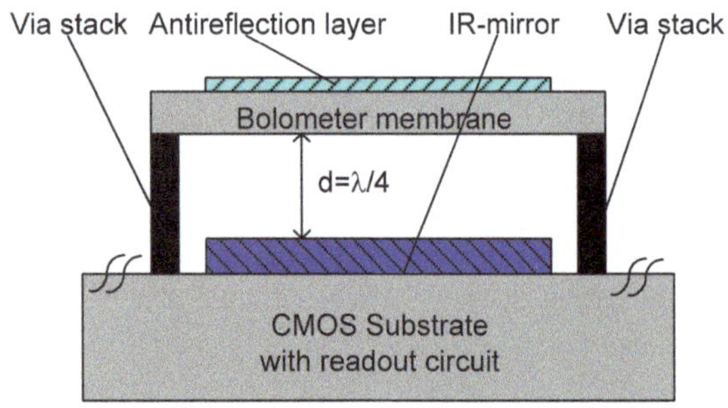

Fig. 1. Realization of a bolometer (cross section).

Fig. 2. Layout of a bolometer array.

subtracted at the first summation node in order to cancel a corresponding offset at the integrator output. The integration stops after readout time t_{Pixel}. This is modelled in the following way: the second summation node subtracts the replica of the bolometer current delayed by the readout time t_{Pixel} and integrates it. The integration constant is $-1/C_{\text{int}}$. In this way we obtain a "time-windowed" integration. Otherwise the ideal integration would never end. The output voltage $V_{\text{out}}(s)$ is calculated in the Laplace-domain in Eq. (1).

$$V_{\text{out}}(s) = \left(1 - e^{-st_{\text{Pixel}}}\right)\frac{-1}{sC_{\text{int}}}\left(\frac{U_{\text{Bias}}(s)}{R_{\text{Bolo}}(T_0)} - I_{\text{Bias}}(s)\right) \tag{1}$$

Thus the transfer functions for output voltage $V_{\text{out}}(s)$ over input voltage $V_{\text{Bias}}(s)$ and input current $I_{\text{Bias}}(s)$ are given by Eqs. (2) and (3).

$$H_{\text{in,ideal}}(s) = \frac{V_{\text{out}}(s)}{V_{\text{Bias}}(s)} = \left(1 - e^{-st_{\text{Pixel}}}\right)\frac{-1}{sR_{\text{Bolo}}(T_0)C_{\text{int}}} \tag{2}$$

$$H_{\text{in2,ideal}}(s) = \frac{V_{\text{out}}(s)}{I_{\text{Bias}}(s)} = \left(1 - e^{-st_{\text{Pixel}}}\right)\frac{-1}{sC_{\text{int}}} \tag{3}$$

3.2 Real integrator

Figure 4 shows the real readout integrated circuit (ROIC) consisting of an OTA with an integration capacitor C_{int} in the feedback and all relevant noise sources. The output resistance is assumed to be infinity. The single-stage OTA is compensated by a load capacitance C_L. The integrator is reseted by the switch "reset". A parasitic capacitor C_p caused e.g. by a column line is placed at the inverting input of the OTA. After the reset phase bolometer and a current source are connected to the inverting OTA input by the switch "select" for the readout time t_{Pixel}. The current from the current source is a constant during the integration, but there is no current if the switch "select" is off. Then the output voltage is held and sampled by a following stage not shown here. The switch resistance R_s in series with the bolometer resistor $R_{\text{Bolo}}(T_0)$ can be neglected, because $R_{\text{Bolo}}(T_0) \gg R_s$. If the transconductance of the OTA is sufficiently high, so that $g_{\text{mOTA}}R_{\text{Bolo}}(T_0) \gg 1$ and $C_{\text{int}}g_{\text{mOTA}}R_{\text{Bolo}}(T_0) \gg C_L$, the output voltage $V_{\text{out}}(s)$ is given by

$$V_{\text{out}}(s) = \left(1 - e^{-st_{\text{Pixel}}}\right)\cdot \frac{V_{n,A}(s)\left(1+s\left(C_{\text{int}}+C_p\right)R_{\text{Bolo}}(T_0)\right) + \left(I_{\text{Bias}}(s)R_{\text{Bolo}}(T_0) - V_{\text{Bias}}(s)\right)\left(1-\frac{sC_{\text{int}}}{g_{\text{mOTA}}}\right)}{sC_{\text{int}}R_{\text{Bolo}}(T_0)\left(1+\frac{sC_e}{g_{\text{mOTA}}}\right)} \tag{4}$$

with

$$C_e = C_L + C_p + \frac{C_L C_p}{C_{\text{int}}}. \tag{5}$$

From Eq. (4) 3 transfer functions can be derived:

$$H_{\text{in1}}(s) = \frac{V_{\text{out}}(s)}{V_{\text{Bias}}(s)} = \frac{-\left(1 - \frac{sC_{\text{int}}}{g_{\text{mOTA}}}\right)}{sC_{\text{int}}R_{\text{Bolo}}(T_0)\left(1 + \frac{sC_e}{g_{\text{mOTA}}}\right)}\left(1 - e^{-st_{\text{Pixel}}}\right) \tag{6}$$

$$H_{\text{in2}}(s) = \frac{V_{\text{out}}(s)}{I_{\text{Bias}}(s)} = \frac{\left(1 - \frac{sC_{\text{int}}}{g_{\text{mOTA}}}\right)}{sC_{\text{int}}\left(1 + \frac{sC_e}{g_{\text{mOTA}}}\right)}\left(1 - e^{-st_{\text{Pixel}}}\right) \tag{7}$$

$$H_{\text{in3}}(s) = \frac{V_{\text{out}}(s)}{V_{n,A}(s)} = \left(\frac{1}{sC_{\text{int}}R_{\text{Bolo}}(T_0)} + 1 + \frac{C_p}{C_{\text{int}}}\right)\frac{1}{1 + \frac{sC_e}{g_{\text{mOTA}}}}\left(1 - e^{-st_{\text{Pixel}}}\right) \tag{8}$$

$H_{\text{in1}}(s)$ and $H_{\text{in2}}(s)$ compared to $H_{\text{in1,ideal}}(s)$ and $H_{\text{in2,deal}}(s)$, respectively, shows that there is now a zero in the transfer function and an additional non-dominant pole, which is given by the gain bandwith product (GBW) of the OTA. If the zero is not dominant, its effect can be

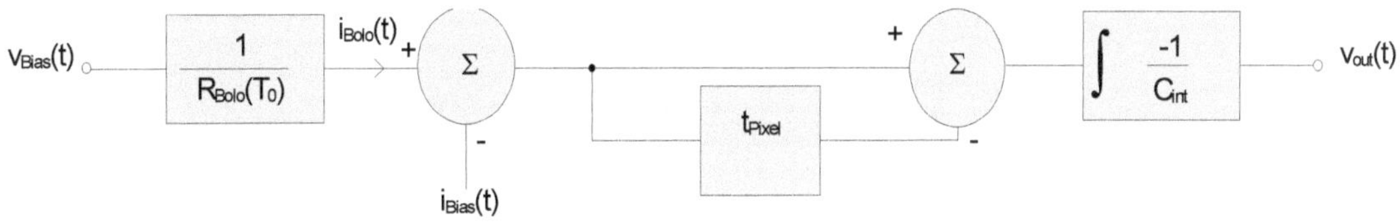

Fig. 3. Ideal integrator-model.

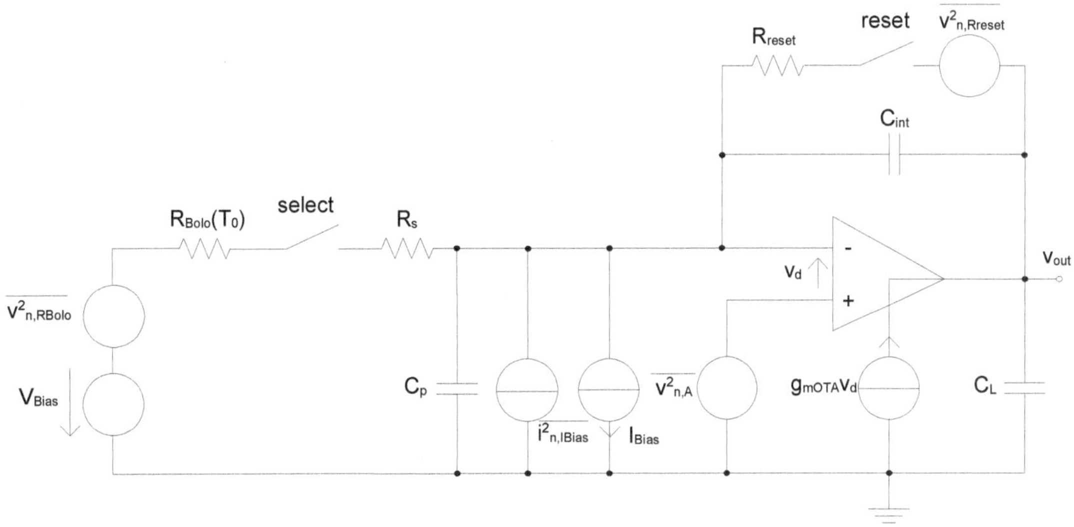

Fig. 4. Real ROIC with an OTA (noise sources included).

neglected. But $H_{\text{in3}}(s)$ is of special interest. Noise at the noninverting input of the OTA is integrated but additionally amplified by $(1+C_p/C_{\text{int}})$ up to the GBW. There could be a high amplification if there is a large parasitic capacitance and a small integration capacitance with a large bandwith dependent on OTA's GBW.

4 Noise calculation and NETD

The power noise density S_{RBolo} including white and $1/f$-noise for the bolometer resistor is given by

$$S_{\text{RBolo}}(f) = \frac{\overline{v^2_{n,\text{RBolo},w}}}{\Delta f} + \frac{\overline{v^2_{n,\text{RBolo},1/f}}}{\Delta f} = 4kT_0 R_{\text{Bolo}}(T_0) + V^2_{\text{Bias}} \frac{k_{\text{fBolo}}}{f} \tag{9}$$

with k as Boltzman's constant, k_{fBolo} the $1/f$-noise parameter and T_0 as the reference temperature. The power noise density S_A including white and $1/f$-noise for the OTA is given by

$$S_A(f) = \frac{\overline{v^2_{n,A,w}}}{\Delta f} + \frac{\overline{v^2_{n,A,1/f}}}{\Delta f} = 2\frac{4K_a kT_0 n\gamma}{g_{\text{mOTA}}} + 2K_b \frac{k_{\text{fOTA}}}{WLC^2_{ox} f} \tag{10}$$

with W as the width of the input transistor, L as the length of the input transistor, C_{ox} as the capacitance per unit of the transistor, n the weak inversion slope parameter, γ as the noise parameter and k_{fOTA} as the $1/f$-noise parameter of the input transistor (Anelli, 2000). K_a and K_b account for the effect of OTA's remaining transistors on total OTA's white noise and $1/f$-noise respectively. The power noise density of the current source depends on the realization. In the simplest case it could be a resistor with the same resistance as the bolometer resistance. To make it easy it is assumed that the current source is ideal, but has a power noise density given by

$$S_{\text{IBias}} = \frac{\overline{i^2_{n,\text{IBias},w}}}{\Delta f} = \frac{4kT_0}{R_{\text{Bolo}}(T_0)}. \tag{11}$$

At first the noise power at the integrator output caused by the bolometer for white noise is calculated.

$$\overline{v_{n,\text{out,RBolo},w}^2} = \int_{f=0Hz}^{\infty} \frac{\overline{v_{n,\text{Bolo},w}^2}}{\Delta f} |H_{\text{in1}}(f)|^2 df = \int_{f=0Hz}^{\infty} 4kT_0 R_{\text{Bolo}}(T_0) \left(\frac{2\sin(\pi f t_{\text{Pixel}})}{2\pi f R_{\text{Bolo}}(T_0) C_{\text{int}}} \right)^2 \cdot \frac{1}{1+\left(2\pi f \frac{C_e}{g_{\text{mOTA}}}\right)^2} df +$$

$$\int_{f=0Hz}^{\infty} 4kT_0 R_{\text{Bolo}}(T_0) \left(\frac{2\sin(\pi f t_{\text{Pixel}})}{2\pi f R_{\text{Bolo}}(T_0) C_{\text{int}}} \right)^2 \cdot \frac{\left(2\pi f \frac{C_{\text{int}}}{g_{\text{mOTA}}}\right)^2}{1+\left(2\pi f \frac{C_e}{g_{\text{mOTA}}}\right)^2} df$$

$$\approx \int_{f=0Hz}^{\infty} 4kT_0 R_{\text{Bolo}}(T_0) \left(\frac{2\sin(\pi f t_{\text{Pixel}})}{2\pi f R_{\text{Bolo}}(T_0) C_{\text{int}}} \right)^2 df + \int_{f=0Hz}^{\infty} 4kT_0 R_{\text{Bolo}}(T_0) \left(\frac{2}{2\pi f R_{\text{Bolo}}(T_0) C_{\text{int}}} \right)^2 \cdot \frac{\left(2\pi f \frac{C_{\text{int}}}{g_{\text{mOTA}}}\right)^2}{1+\left(2\pi f \frac{C_e}{g_{\text{mOTA}}}\right)^2} df$$

$$\approx 2kT_0 \left(\frac{t_{\text{Pixel}}}{R_{\text{Bolo}}(T_0) C_{\text{int}}^2} + \frac{2}{g_{\text{mOTA}} R_{\text{Bolo}}(T_0) C_e} \right) \tag{12}$$

This result has been obtained using several approximations. The noise power at the output caused by the current source is calculated similarly as:

$$\overline{v_{n,\text{out,IBias},w}^2} = \int_{f=0Hz}^{\infty} \frac{\overline{i_{n,\text{IBias},w}^2}}{\Delta f} |H_{\text{in2}}(f)|^2 df \approx 2kT_0 \left(\frac{t_{\text{Pixel}}}{R_{\text{Bolo}}(T_0) C_{\text{int}}^2} + \frac{2}{g_{\text{mOTA}} R_{\text{Bolo}}(T_0) C_e} \right). \tag{13}$$

The noise power caused by OTA's white noise is determined as

$$\overline{v_{n,\text{out},A,w}^2} = \int_{f=0Hz}^{\infty} \frac{\overline{v_{n,A,w}^2}}{\Delta f} |H_{\text{in3}}(f)|^2 df \approx \int_{f=0Hz}^{\infty} 2\frac{4K_a kT_0 n\gamma}{g_{\text{mOTA}}} \left[\left(\frac{1}{R_{\text{Bolo}}(T_0) C_{\text{int}}} \cdot \frac{2\sin(\pi f t_{\text{Pixel}})}{2\pi f} \right)^2 + 2\frac{\left(1+\frac{C_P}{C_{\text{int}}}\right)^2}{1+\left(2\pi f \frac{C_e}{g_{\text{mOTA}}}\right)^2} \right] df$$

$$\approx \frac{4K_a kT_0 n\gamma}{g_{\text{mOTA}}} \frac{t_{\text{Pixel}}}{(R_{\text{Bolo}}(T_0) C_{\text{int}})^2} + 2\left(1+\frac{C_P}{C_{\text{int}}}\right)^2 \frac{2K_a kT_0 n\gamma}{C_e}. \tag{14}$$

The noise power caused by the bolometer resistor's $1/f$-noise is described as

$$\overline{v_{n,\text{out,RBolo},1/f}^2} = \int_{f_1}^{\infty} \frac{\overline{v_{n,\text{Bolo},1/f}^2}}{\Delta f} |H_{\text{in1}}(f)|^2 df = \int_{f_1}^{\infty} V_{\text{Bias}}^2 \frac{k_{\text{fBolo}}}{f} \left(\frac{1}{R_{\text{Bolo}}(T_0) C_{\text{int}}} \cdot \frac{2\sin(\pi f t_{\text{Pixel}})}{2\pi f} \right)^2 \cdot \frac{1}{1+\left(2\pi f \frac{C_e}{g_{\text{mOTA}}}\right)^2} df +$$

$$\int_{f_1}^{\infty} V_{\text{Bias}}^2 \frac{k_{\text{fBolo}}}{f} \left(\frac{1}{R_{\text{Bolo}}(T_0) C_{\text{int}}} \cdot \frac{2\sin(\pi f t_{\text{Pixel}})}{2\pi f} \right)^2 \cdot \frac{\left(2\pi f \frac{C_{\text{int}}}{g_{\text{mOTA}}}\right)^2}{1+\left(2\pi f \frac{C_e}{g_{\text{mOTA}}}\right)^2} df$$

$$\approx V_{\text{Bias}}^2 k_{\text{fBolo}} (-1) \frac{-1+\cos(\xi) - \xi\sin(\xi) + \xi^2 Ci(\xi)}{\xi^2 R_{\text{Bolo}}^2 (T_0) C_{\text{int}}^2} t_{\text{pixel}}^2 + \frac{V_{\text{Bias}}^2 k_{\text{fBolo}}}{(g_{\text{mOTA}} R_{\text{Bolo}}(T_0))^2} \ln\left(\frac{\sqrt{1+\left(2\pi f_1 \frac{C_e}{g_{\text{mOTA}}}\right)^2}}{\left(2\pi f_1 \frac{C_e}{g_{\text{mOTA}}}\right)} \right)$$

$$\approx V_{\text{Bias}}^2 k_{\text{fBolo}} \left(\frac{t_{\text{Pixel}}}{R_{\text{Bolo}}(T_0) C_{\text{int}}} \right)^2 \ln\left(\frac{1}{2\pi f_1 t_{\text{Pixel}}} \right) + \frac{1}{(g_{\text{mOTA}} R_{\text{Bolo}}(T_0))^2} \ln\left(\frac{1}{\left(2\pi f_1 \frac{C_e}{g_{\text{mOTA}}}\right)} \right). \tag{15}$$

where $\xi = 2\pi t_{\text{Pixel}} f_1$, $\cos(\xi) \approx 1$, $\sin(\xi) \approx \xi$, the Cosine Integral $\text{Ci}(\xi) \approx \ln(\xi)$ with $|\ln(\xi)| \gg 1$ and $2\pi f_1 C_e / g_{\text{mOTA}} \ll 1$. A mechanical shutter in front of the IRFPA is periodically closed. This can be modelled as a high pass filter. The lower frequency f_1 is assumed to be (Wood, 1997)

$$f_1 \approx \frac{1}{4t_{\text{shutter}}}. \tag{16}$$

t_{shutter} is the shutter time, where $t_{\text{shutter}} \gg t_{\text{Pixel}}$. The contribution of the $1/f$-noise of the OTA with same approximations as above is given by

$$\begin{aligned}
\overline{v^2_{n,\text{out},A,1/f}} &= \int_{f_1}^{\infty} \frac{\overline{v^2_{n,A,1/f}}}{\Delta f} |H_{\text{in3}}(f)|^2 df \\
&\approx \int_{f_1}^{\infty} 2K_b \frac{k_{\text{fOTA}}}{WLC^2_{ox} f} \left(\begin{array}{c} \frac{(2\sin(\pi f t_{\text{Pixel}}))^2}{(2\pi f R_{\text{Bolo}}(T_0) C_{\text{int}})^2} + \\ + \left(1 + \frac{C_P}{C_{\text{int}}}\right)^2 \frac{4}{1+\left(2\pi f \frac{C_e}{g_{\text{mOTA}}}\right)^2} \end{array} \right) df \\
&\approx 2K_b \frac{k_{\text{fOTA}}}{WLC^2_{ox}} \left(\begin{array}{c} \left(\frac{t_{\text{Pixel}}}{R_{\text{Bolo}}(T_0) C_{\text{int}}}\right)^2 \ln\left(\frac{1}{2\pi t_{\text{Pixel}} f_1}\right) + \\ 4\left(1 + \frac{C_P}{C_{\text{int}}}\right)^2 \ln\left(\frac{1}{\left(2\pi f_1 \frac{C_e}{g_{\text{mOTA}}}\right)}\right) \end{array} \right).
\end{aligned} \tag{17}$$

During the reset of the integrator the kTC-noise is stored on the integration capacitor. Thus

$$\overline{v^2_{n,\text{reset}}} = \frac{kT_0}{C_{\text{int}}}. \tag{18}$$

The current responsitivity of the bolometer at the input is given for a constant voltage across the bolometer by

$$R_I \approx \frac{-V_{\text{Bias}}\alpha}{R_{\text{Bolo}}(T_0) g_{\text{Bolo}}}. \tag{19}$$

where α is the temperature coefficient (TCR) and g_{Bolo} is the heat conductance of the bolometer (derived from Wood, 1997) with $g_{\text{Bolo}} \approx g_{\text{eff}}, I_{\text{Bolo}} = V_{\text{Bias}}/R_{\text{Bolo}}(T_0), \alpha < 0, I_{\text{Bolo}}$ as the constant bolometer current, g_{eff} as the effective heat conductance). The TCR α is assumed to be constant. The total output noise power is refered to the input by multiplying by the square of the DC value of the transfer-function $H_{\text{in1}}(f)$ and converted into a current by division by $R^2_{\text{Bolo}}(T_0)$. The NETD is defined as

$$\text{NETD} = \frac{4 \cdot F^2_{no}}{\pi \cdot A_{\text{Bolo}} \cdot \varepsilon_{\text{Bolo}} \cdot \frac{dL}{dT_{target}}} \frac{\sqrt{\overline{i^2_{n,\text{total}}}}}{R_I} \tag{20}$$

where $i_{n,\text{total}}$ is the total noise current, F_{no} is the f-number of the optic, A_{Bolo} the area of the bolometer, $\varepsilon_{\text{Bolo}}$ is the emission coefficient which is equal to the absorption coefficient and dL/dT_{target} is the differential change of radiance over the change of the target temperature (Wood, 1997). Finally, total NETD is calculated to

$$\begin{aligned}
\text{NETD} = &\frac{4 \cdot F^2_{no}}{\pi \cdot A_{\text{Bolo}} \cdot \varepsilon_{\text{Bolo}} \cdot \frac{dL}{dT_{target}}} \cdot \left(\frac{1}{R_{\text{Bolo}(T_0)}}\right) \cdot \left(\frac{R_{\text{Bolo}(T_0)} C_{\text{int}}}{t_{\text{Pixel}}}\right) \cdot \\
&\cdot \frac{\sqrt{\left(\begin{array}{c} \overline{v^2_{n,\text{out},R\text{Bolo},w}} + \overline{v^2_{n,\text{out},R\text{Bolo},1/f}} + \overline{v^2_{n,\text{reset}}} \\ + \overline{v^2_{n,\text{out},I\text{Bias},w}} + \overline{v^2_{n,\text{out},A,w}} + \overline{v^2_{n,\text{out},A,1/f}} \end{array} \right)}}{R_I}
\end{aligned} \tag{21}$$

Assuming that there is only noise from the bolometer resistor (white and $1/f$-noise) and current source (white noise) with $g_{\text{mOTA}} \rightarrow \infty, C_L \rightarrow \infty$, WL$\rightarrow \infty$ the NETD reaches

$$\text{NETD} = \frac{4 \cdot F^2_{no} g_{\text{Bolo}}}{\pi \cdot A_{\text{Bolo}} \cdot \varepsilon_{\text{Bolo}} \cdot \frac{dL}{dT_{target}}} \cdot \frac{\sqrt{\frac{4kT_0 R_{\text{Bolo}}(T_0)}{t_{\text{Pixel}}} + V^2_{\text{Bias}} k_{\text{fBolo}} \ln\left(\frac{2t_{\text{shutter}}}{\pi t_{\text{Pixel}}}\right)}}{|-V_{\text{Bias}}\alpha|}. \tag{22}$$

NETD caused by white noise can be decreased by increasing the readout time and bias voltage and by decreasing the bolometer resistance assuming that the bolometer parameters and f-number are fixed. NETD caused by $1/f$-noise is independent of the bias voltage, because it cancels out. Because of the logarithm there won't be much change when varying t_{Pixel} for a fixed shutter time. Thus white noise can be reduced, but $1/f$-noise cannot. So $1/f$-noise will limit NETD.

5 Conclusions

The ideal and real short-time integrator have been modelled. The effect of the finite integration time has been included. The total noise power and the resulting NETD have been calculated. For long integration the limiting factor as far as NETD is concerned is $1/f$-noise.

References

Ruß, M., Bauer, J., and Vogt, H.: The geometric design of microbolometer elements for uncooled focal plane arrays, Proc. of SPIE Conference Infrared Technology and Applications XXXIII, Volume 6542, 654223-1–10, 2007.

Anelli, G.: Design and Characterization of Radiation Tolerant Integrated Circuits in Deep Submicron CMOS Technologies for the LHC Experiments, Ph.D. Thesis, Institut National Polytechnique de Grenoble, France, 2000.

Wood, R. A.: Monolithic Silicon Microbolometer Arrays, in Uncooled Infrared Imaging Arrays and Systems, edited by: Kruse, P. and Skatrud, D., Semiconductors and Semimetals, 47, 45–121, Academic Press, 1997.

Resolving longitudinal amplitude and phase information of two continuous data streams for high-speed and real-time processing

A. Guntoro and M. Glesner

Institute of Microelectronic Systems, Department of Electrical Engineering and Information Technology, Technische Universität Darmstadt, Germany

Abstract. Although there is an increase of performance in DSPs, due to its nature of execution a DSP could not perform high-speed data processing on a continuous data stream. In this paper we discuss the hardware implementation of the amplitude and phase detector and the validation block on a FPGA. Contrary to the software implementation which can only process data stream as high as 1.5 MHz, the hardware approach is 225 times faster and introduces much less latency.

1 Introduction

Digital signal processing which replaces the analog control system in our heavy ion acceleration environment is depicted in Fig. 1. Two analog input signals which correspond to the ion states at a specific time are received from the cavity sensors and go to the A/D block (with 14-bit resolution at 28.5 MHz sampling rate). Longitudinal amplitude and phase information of both signals are extracted by the detector and go to the validation block. The validation block outputs the longitudinal amplitude and phase differences of both signals. Special phase-corrector filter tunes up the phase of the signal. The corrected signal is used at the end by the controller to control the DDS (Direct Digital Synthesis). In this paper we focus on the amplitude and phase detector and the validation block which should be able to process two high-speed continuous data streams.

Correspondence to: A. Guntoro
(guntoro@mes.tu-darmstadt.de)

2 Longitudinal amplitude and phase

Given a discrete signal $f[n]$, the longitudinal amplitude information of the signal is defined as Iverson (2004); Hind et al. (2003):

$$\chi_X = \sqrt{x^2 + y^2}$$

and the corresponding longitudinal phase information is

$$\phi_X = arctan\frac{y}{x}$$

with $x{=}Q_1{-}Q_2$ and $y{=}I_1{-}I_2$ where I_1, I_2, Q_1, and Q_2 are the in-phase and quadrature components of the signal.

To resolve the magnitude information, one *square root*, two *square*, and one *addition* operations are involved. Accordingly one *division* operation and one *arctan* function are involved to compute the phase.

The magnitude and phase difference of two signals are easily computed with:

$$\chi_\Delta = \chi_1 - \chi_2 \qquad \phi_\Delta = \phi_1 - \chi_2$$

Our original implementation of the amplitude and phase detector and the validation block which are performed on DSP C6713 is only capable to receive two input data streams as high as 1.5 MHz on each channel Klingbeil (2004). Thus the signals are decimated by 19 which introduces friction and information loss.

2.1 CORDIC

The CORDIC (COordinate Rotation DIgital Computer) algorithm is a method to compute elementary functions by means of planar rotation and vectoring. The basic of CORDIC principle utilizes only shift and add/sub operations which are simple and cheap from the resource allocation point of view Andraka (1998). The coordinate system of the input

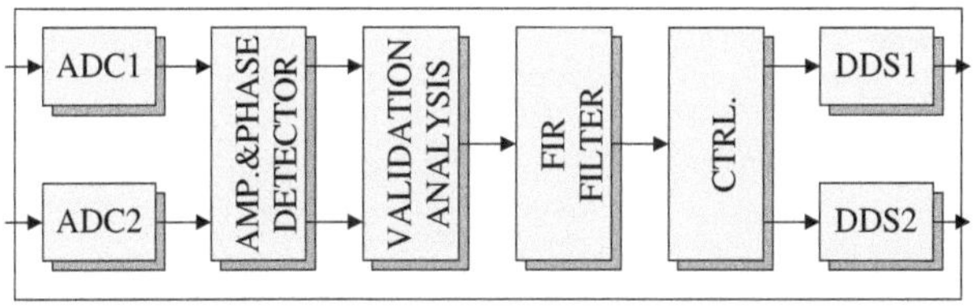

Fig. 1. Digital signal processing chain.

parameter is rotated through constant angles until the angle is reduced to zero.

Planar rotation for a vector A of (X_j, Y_j) can be defined in matrix form as:

$$\begin{bmatrix} X_j \\ Y_j \end{bmatrix} = \begin{bmatrix} \cos\theta & -\sin\theta \\ \sin\theta & \cos\theta \end{bmatrix} \begin{bmatrix} X_i \\ Y_i \end{bmatrix}$$

To calculate the rotation, trigonometric functions (cos and sin) are involved. This rotation angle θ in fact can be executed in steps by means of iterative process. With a slight rearrangement, Eq. (1) defines this process which performs the rotation stepwise.

$$\begin{bmatrix} X_{n+1} \\ Y_{n+1} \end{bmatrix} = \cos\theta_n \begin{bmatrix} 1 & -\tan\theta_n \\ \tan\theta_n & 1 \end{bmatrix} \begin{bmatrix} X_n \\ Y_n \end{bmatrix} \tag{1}$$

By selecting the angle steps θ_n such that the tangent of a step ($\tan\theta_n$) is a power of 2, the multiplication operation can be eliminated and substituted by shift operation. Equation (2) gives the angle parameter for each step.

$$\theta_n = arctan\left(\frac{1}{2^n}\right) \tag{2}$$

Their sums must equal to the rotation angle θ:

$$\sum_{n=0}^{\infty} S_n\theta_n = \theta$$

where $S_n=\{-1; +1\}$ in respect to addition or substraction is determined later. Rewriting Eq. (1), we have:

$$\begin{bmatrix} X_{n+1} \\ Y_{n+1} \end{bmatrix} = \cos\theta_n \begin{bmatrix} 1 & -S_n 2^{-n} \\ S_n 2^{-n} & 1 \end{bmatrix} \begin{bmatrix} X_n \\ Y_n \end{bmatrix}$$

Except for the common coefficient $\cos\theta_n$ which can be treated as a constant K and computed at the end, multiplications are replaced with shift operations.

Residue Z which defines then angle difference between the expected rotation and the iterative rotations is:

$$Z_{n+1} = \theta - \sum_{i=0}^{n} \theta_i = \theta - \sum_{i=0}^{n} arctan\frac{1}{2^i}$$

The rotation parameter S_n is determined by:

$$S_n = \begin{cases} -1 & \text{if } Z_n < 0 \\ +1 & \text{if } Z_n \geq 0 \end{cases}$$

Practically, the iteration does not go up to infinity. In the hardware implementation, for each iteration the resolution of the result is increased by one bit (16 iterations deliver 16 bits result). Thus, it is not necesarry to compute the *arctan* on the fly. Look-up table of 16 entries is enough to store the precomputed *arctan* values in this case.

Driving Z to zero, the CORDIC performs:

$$\begin{bmatrix} X_j \\ Y_j \\ Z_j \end{bmatrix} = \begin{bmatrix} P(X_i \cos Z_i - Y_i \sin Z_i) \\ P(Y_i \cos Z_i + X_i \sin Z_i) \\ 0 \end{bmatrix}$$

With initial values $X_i=1/P$, $Y_i=0$, and $Z_i=\theta$ we will have at the end of the iteration:

$$\begin{bmatrix} X_j \\ Y_j \\ Z_j \end{bmatrix} = \begin{bmatrix} \cos\theta \\ \sin\theta \\ 0 \end{bmatrix}$$

To suit our demand, we need to drive Y to zero instead of Z. The obtained result is:

$$\begin{bmatrix} X_j \\ Y_j \\ Z_j \end{bmatrix} = \begin{bmatrix} P\sqrt{X_i^2 + Y_i^2} \\ 0 \\ Z_i + arctan\frac{Y_i}{X_i} \end{bmatrix}$$

Again with $X_i=X$, $Y_i=Y$, and $Z_i=0$ we obtain

$$\begin{bmatrix} X_j \\ Y_j \\ Z_j \end{bmatrix} = \begin{bmatrix} P\sqrt{X^2 + Y^2} \\ 0 \\ arctan\frac{Y_i}{X_i} \end{bmatrix} \tag{3}$$

Neglecting the amplification factor P, X_j and Z_j correspond to the magnitude and phase information we need.

3 Hardware Realization

The critical constraint is to process two continuous data streams at a minimum of 28.5 MHz each without introducing any decimation. We exploit a pipeline structure within our design to guarantee this real-time demand. Figure 2 concisely depicts the tasks for each pipeline stage.

3.1 2-to-1 controller

This stage receives two input data streams. It collects four samples from each input and feeds them to the subtraction block. To save resources, two input data streams are processed alternately (i.e. the first clock cycle for the first input, the next clock cycle for the second input and so on).

3.2 Substraction

This stage handles the extraction of incoming quadrature and in-phase signals into corresponding X and Y which will be fed to the CORDIC core. These variables represent directly the vector of which the core operates.

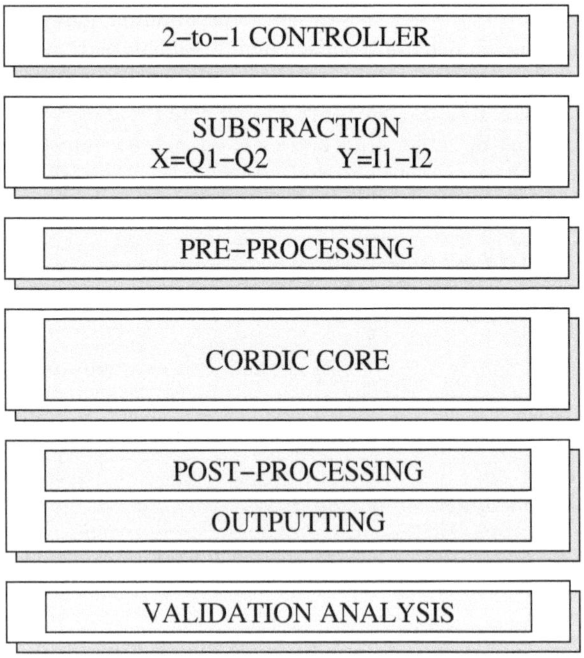

Fig. 2. Pipeline stages of the amplitude and phase detector and validation analysis block.

3.3 Preprocessing

The CORDIC vectoring is only capable of processing the signals arranged on quadrant I and IV. Since the combination of inputs X and Y may also lay on quadrant II or III, preprocessing and rearrangement of the inputs are necessary. The quadrant preprocessing conversion is formulated as

$$f'(x, y) = \begin{cases} f(x, y) & x \geq 0 \\ f(-x, -y) + \pi & x < 0, y \geq 0 \\ f(-x, -y) - \pi & x < 0, y < 0 \end{cases}$$

which tells that quadrant II is mapped to quadrant IV, thus the result will be corrected by adding π, and quadrant III is mapped to quadrant I and will be corrected by subtracting π from the result.

3.4 CORDIC core

The core stage expands the CORDIC iteration into micro pipelines to fulfill our high-speed requirement. 16 pipeline stages are needed to provide 16-bits resolution at the output. To have a full resolution swing, 16-bits two's complement binary format is used to represent the value between $-\pi$ and $+\pi$ for the angle component. LUT-based precalculated arctan 2^{-n} is used to store 16 arctan entries. Similar to the angle representation, two's complement is also used to represent the magnitude component. This is done because the iterative process inside of the core employs also addition and subtraction in estimating the magnitude.

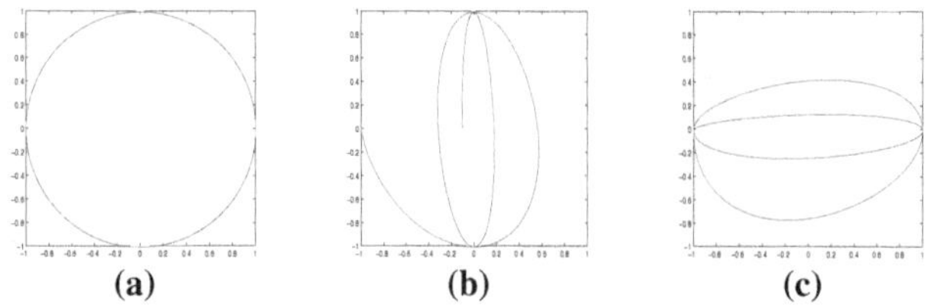

Fig. 3. Different test input sources.

Table 1. Maximum error, absolute mean, and standard deviation of the resulting error for the magnitude component (normalized to 1) for each input source.

Source	Max Error	$\lvert\bar{x}\rvert$	σ
Const. Mag.	0.000338	0.000111	0.000069
Decaying X	0.000316	0.000118	0.000072
Decaying Y	0.000359	0.000120	0.000072

3.5 Postprocessing and outputting

The Postprocessing stage corrects the output by adding or subtracting the result with π based on the information collected at the preprocessing stage. Additionally, the amplification factor which is introduced by CORDIC as defined in Eq. (3) might also be eliminated at this stage if it is demanded. Although it depends on the application, while most applications demand no unity gain correction for the magnitude output.

3.6 Validation analysis

This stage calculates the difference between two longitudinal magnitudes and two longitudinal phases.

4 Performance

Three different input sources are utilized to analyze the performance of the CORDIC hardware. Figure 3 depicts these sources. Three cases are considered here: (a) a linear phase and constant magnitude input source, (b) a signal with decaying X component and (c) a signal with decaying Y component.

Figure 4 maps the errors (the difference between the expected values and the computed version) of magnitude and phase result of each input source. Here it can be seen that the hardware computation with 16-bit resolution produces the same results with some minor errors compared to the expected results. Tables 1 and 2 summarize the statistical parameters of the error for both magnitude and phase components.

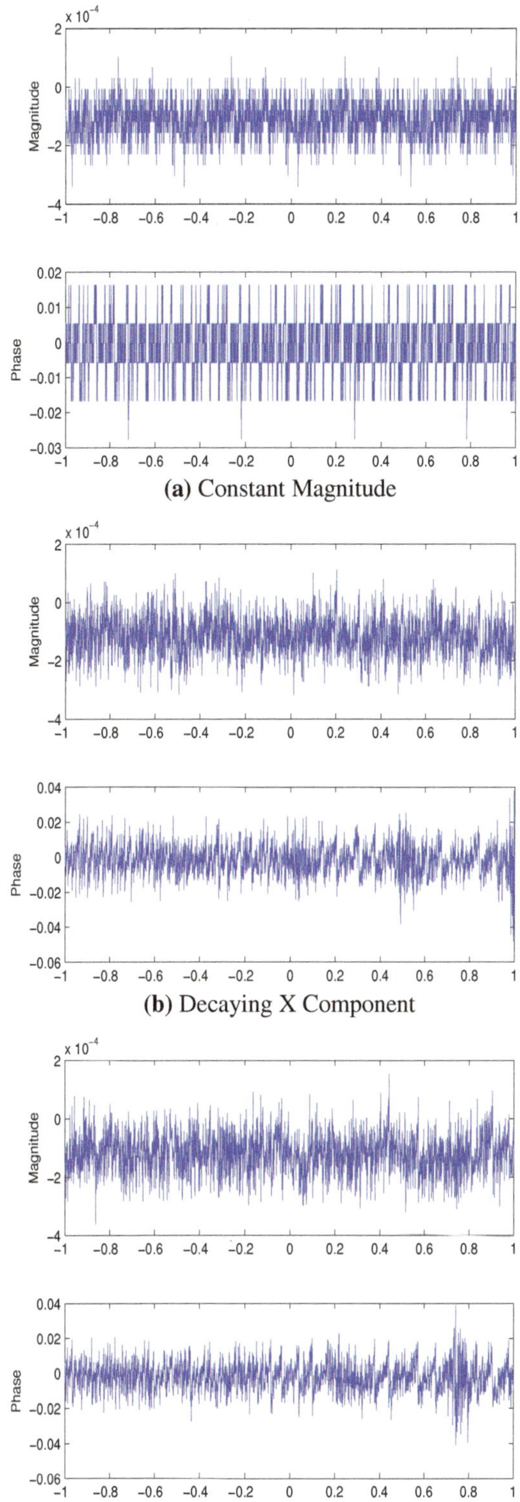

(a) Constant Magnitude

(b) Decaying X Component

(c) Decaying Y Component

Fig. 4. The difference between the expected and computed values of every given input source.

Table 2. Maximum error, absolute mean, and standard deviation (all in degree) of the resulting error for the phase component for each input source.

Source	Max Error	$\lvert\bar{x}\rvert$	σ
Const. Mag.	0.027466	0.007682	0.008817
Decaying X	0.047844	0.007344	0.009223
Decaying Y	0.040306	0.007084	0.008829

Table 3. Resource utilization on Xilinx XC2V2000-4.

	Usage/Cap.	Percentage
Slices	478/10752	4%
Flip-Flops	788/21504	3%
4 input LUTs	792/21504	3%
Max. Frequency	168.89 MHz	

Here we can see that by using 16-bit data representation to perform the computation, the introduced errors are tolerable. For the magnitude component, the maximum error is in order of 3×10^{-3}; and for the phase component, the maximum phase error does not exceed 0.05°.

5 Synthesis

The amplitude and phase detector and validation block are written in VHDL. The code has been synthesized using Xilinx ISE. As the target device, we use Xilinx XC2V2000-4 which is employed in our DSP chain. Table 3 details the resource usage of the implemented blocks on the target device. It is clear that the implemented block consumes only a small amount of resources and has high operation speed. Due to the pipeline structure, the amplitude and phase detector and the validation block can receive two input data streams as high as 337.78 MHz each (the core operated on 4 samples per clock cycle). Additionally, the computational latency determined by the number of pipeline stages (1+1+1+16+2+1=22) is very low (i.e. 130 ns).

6 Conclusions

Contrary to the DSP implementation which is only capable to process data streams as high as 1.5 MHz, the dedicated hardware block is 225 times faster and introduces only a very small amount of computational latency. Thus, the decimation by 19 which was taken on our original implementation is no longer required.

References

Andraka, R.: A survey of CORDIC algorithms for FPGA based computers, FPGA 98 Monterey CA USA, 1998.

Hind, M., Rajan, V., and Sweeney, P.: Phase shift detection: a problem classification, in: IBM Research Report, 2003.

Iverson, J.: Digital Control Technology Enhances Power System Reliability and Performance, in: Technical Information from Cummins Power Generation, Cummins Power Generation, 2004.

Klingbeil, H.: A Fast DSP-based Phase-detector for Closed-loop RF Control in Synchrotrons, IEEE Transaction on Instrumentation and Measurement, 2004.

Context-based user grouping for multi-casting in heterogeneous radio networks

C. Mannweiler, A. Klein, J. Schneider, and H. D. Schotten
Chair for Wireless Communications and Navigation, University of Kaiserslautern, Germany

Abstract. Along with the rise of sophisticated smartphones and smart spaces, the availability of both static and dynamic context information has steadily been increasing in recent years. Due to the popularity of social networks, these data are complemented by profile information about individual users. Making use of this information by classifying users in wireless networks enables targeted content and advertisement delivery as well as optimizing network resources, in particular bandwidth utilization, by facilitating group-based multi-casting. In this paper, we present the design and implementation of a web service for advanced user classification based on user, network, and environmental context information. The service employs simple and advanced clustering algorithms for forming classes of users. Available service functionalities include group formation, context-aware adaptation, and deletion as well as the exposure of group characteristics. Moreover, the results of a performance evaluation, where the service has been integrated in a simulator modeling user behavior in heterogeneous wireless systems, are presented.

1 Introduction

Group communication is of particular interest in wireless (access) networks, one example being the domain of multi-casting, i.e. simultaneously delivering the same content to multiple recipients. Chalmers and Almeroth (2001), Janic and Van Mieghem (2009), as well as Baumung and Zitterbart (2009), among others, have shown that multi-casting technologies can considerably contribute to more efficiently exploiting available network resources when applied under appropriate circumstances. A typical use case for exploiting multi-casting is a large-scale sport event with tens of thousands of spectators that can be grouped according to their user profile and context. Based on that classification, different groups can receive adapted content. However, a user classification service, as required in this use case, can also be exploited for selecting people with particular characteristics. Consumers with an affinity to a certain product category could be identified for specific marketing purposes. Both use cases would require a reliable means of data privacy and security as well as the user's approval.

The remainder of this paper is organized as follows: Sect. 2 presents the algorithms used for our user classification service and Sect. 3 briefly discusses implementation aspects. Section 4 evaluates the algorithms according to a comprehensive set of criteria. Section 5 concludes the paper with a short summary and outlook.

Correspondence to: C. Mannweiler
(mannweiler@eit.uni-kl.de)

2 Classification methods

In this section, we will briefly introduce the clustering methods used for a context-aware user classification service. This includes the algorithm of the respective method as well as relevant characteristics such as hard vs. soft mapping or variable vs. fixed number of clusters. The notation employed in this paper is defined below:

- $\boldsymbol{C}$: Set of all clusters, $\boldsymbol{C} = \{\boldsymbol{C}_1, \boldsymbol{C}_2, ..., \boldsymbol{C}_k\}$
- $\boldsymbol{N}$: Set of all cluster centers (nodes), $\boldsymbol{N} = \{N_1, N_2, ..., N_k\}$
- $\boldsymbol{K}$: Context space, $\boldsymbol{K} \subseteq \mathbb{R}^l$
- $\boldsymbol{X}$: Set of observations (user's context) to classify, $\boldsymbol{X} = \{\boldsymbol{x}_1, \boldsymbol{x}_2, ..., \boldsymbol{x}_n\}, \boldsymbol{x}_j \in \boldsymbol{K}$
- n: Number of observations (users) to classify
- k: Number of clusters (nodes)
- l: Dimension of context space
- $\boldsymbol{\Omega}$: Tuple of a cluster $\boldsymbol{C}_i$ and an observation $\boldsymbol{x}_j$, $\boldsymbol{\Omega} = (\boldsymbol{C}_i, \boldsymbol{x}_j)$
- $\boldsymbol{\mu}_i$: Center of cluster $\boldsymbol{C}_i$, $\boldsymbol{\mu}_i \in \mathbb{R}^l$
- $\boldsymbol{w}_i$: Weighing vector of node N_i (cluster center), $\boldsymbol{w}_i \in \mathbb{R}^l$
- t: Iteration counter

2.1 Methods with fixed number of cluster centers

Common to the methods presented in this subsection is the requirement to set the number of clusters before the start of the classification algorithm. During runtime, the number of clusters does not change.

2.1.1 K-Means algorithm

K-Means (MacQueen, 1967) is an algorithm that establishes a hard mapping between $\boldsymbol{C}_i$ and an observation $\boldsymbol{x}_j$, i.e. an observation is unambiguously associated with one cluster. Across multiple iterations, the following error function is minimized:

$$E = \sum_{i=1}^{k} \sum_{\boldsymbol{x}_j \in \boldsymbol{C}_i} ||\boldsymbol{x}_j - \boldsymbol{\mu}_i||^2 \tag{1}$$

The iteration steps of the algorithm are:

1. *(Random) initialization of k cluster centers (nodes)* – cluster centers are randomly initialized in the l-dimensional context space.

2. *Mapping of observations to centers* – each observation is mapped to the closest node based on the selected distance metric such as Euclidean distance.

3. *Update of cluster centers* – position of nodes is recomputed based on the observations that are assigned to the node. Go back to step 2.

The algorithm terminates as soon as a specified termination criterion is reached, e.g. the error function falls below a threshold. For further details on K-Means, the reader is referred to Steinhaus (1957).

2.1.2 Neural Gas

The Neural Gas algorithms establishes a graph of k nodes that, independently of each other, move through the context space during the iterations. The core idea is to present available observations to the graph and to accordingly adjust the cluster centers. In brief, the steps of the algorithms are as follows:

1. *(Random) initialization of k cluster centers (nodes)* – Cluster centers are randomly initialized in the l-dimensional context space.

2. *Presentation of an observation* – An observation $\boldsymbol{x}_c \in \boldsymbol{X}$ is randomly chosen and presented to the graph. (An observation can be drawn multiple times.)

3. *Sorting of nodes* – Nodes are sorted according to their Euclidean distance to $\boldsymbol{x}_c$.

4. *Adjustment of node positions* – The position $\boldsymbol{w}_i$ of the graph's nodes within the context space is adjusted according to the following equation:

$$\boldsymbol{w}_i(t+1) = \boldsymbol{w}_i(t) + l(t) \cdot h_d(i) \cdot (\boldsymbol{x}_c - \boldsymbol{w}_i(t)) \tag{2}$$

where $l(t)$ is the learning rate of the current iteration ($l(t) > l(t-1)$ for all t) and $h_{\mathrm{d}}(i)$ the adjustment for an individual node depending on its ranking in step 3. If the termination criterion (e.g. number of iterations) has not been reached yet, the next iteration can begin, starting with step 2. Otherwise, observations are assigned to the closest node.

For further details on the algorithm, the reader is referred to Martinez and Schulten (1991) and Fritzke (1997).

2.1.3 K-Fixed-Means

None of the presented algorithms considers fixed cluster centers. However, for some of the use cases described in Sect. 1, this is an important alternative for user classification. Therefore, we have developed the so-called K-Fixed-Means algorithm, where the position of a cluster center is fix except for a defined tolerance interval (for each dimension of the context space) within which the cluster center can move. Moreover, the maximum distance to a center is limited, i.e. some observations may not be assigned to any node at all. Figure 1 depicts the idea of tolerance space (purple rectangle) and maximum distance (blue circle). The basic steps of the algorithm are:

1. *Initialization* – k cluster centers (nodes) are initialized at the determined (and fix) positions.

2. *Mapping of observations to centers* – each observation is mapped to the closest node based on the selected distance metric such as Euclidean distance. Non-assigned observations (i.e. those whose distance is above the defined thresholds) are collected in a set $\boldsymbol{U}$.

3. *Adjustment of cluster centers* – for each observation in $\boldsymbol{U}$, it is checked whether there exists a node that, if moved within its tolerance room, can accommodate the given observation. If their exists such a node, it is moved according the following equation:

$$\boldsymbol{w}_i(t+1) = \boldsymbol{w}_i(t) + l(t) \cdot d(\boldsymbol{x}_c, \boldsymbol{w}_i) \cdot (\boldsymbol{x}_c - \boldsymbol{w}_i(t)) \tag{3}$$

where $l(t)$ is the learning rate and $d(\boldsymbol{x}_c, \boldsymbol{w}_i)$ a factor that takes into account the different tolerance intervals. After all observations in $\boldsymbol{U}$ have been checked, the next iteration starts with step 2, unless the criterion for termination is met.

As for the Neural Gas algorithm, the learning rate decreases in later iterations.

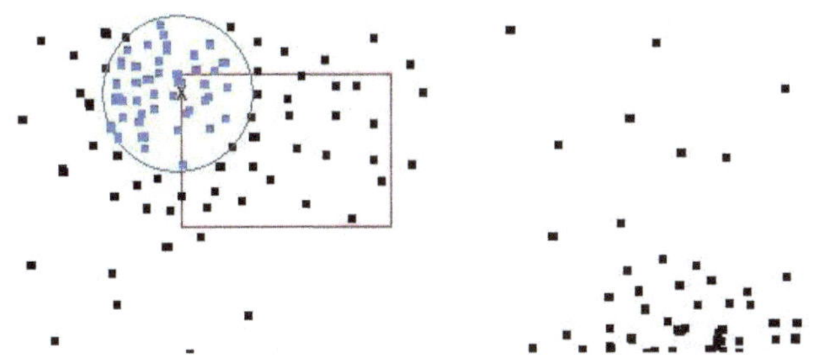

Fig. 1. Concept of K-Fixed-Means algorithm.

2.1.4 Fuzzy-C-Means

In contrast to the algorithms presented so far, Fuzzy-C-Means can assign an observation to several clusters. The degree u_{ij} of membership of an observation $\boldsymbol{x}_j$ to a single cluster $\boldsymbol{C}_i$ has to lie within the interval $]0,1]$ and the sum of all memberships of an observation must add up to 1, i.e. $\sum_{i=1}^{k} u_{ij} = 1$ for any j. The algorithm minimizes the following error function:

$$E = \sum_{i=1}^{k} \sum_{j=1}^{n} u_{ij}^{m} ||\boldsymbol{x}_j - \boldsymbol{\mu}_i||^2, 1 \leq m \leq \infty \tag{4}$$

The exponent m is the so-called "fuzzifier". The higher its value, the fuzzier the mappings of observations to nodes. In practice, values between 1 and 2.5 have proven to generate good clustering results (Bezdek, 1981). The basic steps of the algorithm are:

1. *Initialization* – (random) initialization of values u_{ij}
2. *(Re)calculation of cluster centers* – for the current iteration step, the cluster centers are calculated according to

$$\boldsymbol{\mu}_i = \frac{\sum_{j=1}^{n} u_{ij}^{m} \boldsymbol{x}_j}{\sum_{j=1}^{n} u_{ij}^{m}} \tag{5}$$

3. *Recalculation of degrees u_{ij} of membership* – the degree of membership of an observation $\boldsymbol{x}_j$ to a cluster $\boldsymbol{C}_i$ is calculated according to the following equation:

$$u_{ij} = \frac{1}{\sum_{l=1}^{k} \frac{||\boldsymbol{x}_j - \boldsymbol{\mu}_i||}{||\boldsymbol{x}_j - \boldsymbol{\mu}_l||}^{\frac{2}{m-1}}} \tag{6}$$

4. *Test of termination criterion* – if the termination criterion, e.g. sum of changes of u_{ij} for all combinations i, j is below a defined threshold, is not satisfied yet, another iteration is performed starting with step 2.

The presented algorithm converges to a local minimum which is not necessarily the optimal solution. Moreover, the results depend on the initialization of u_{ij}.

2.2 Methods with variable number of cluster centers

In contrast to the algorithms presented so far, this class of algorithms is capable of adjusting the number of cluster centers during runtime. Hence, not only the assignment of observations to clusters but also the number of clusters is optimized. Using the algorithms from the previous section, this could only be achieved by several runs with different amounts of clusters.

2.2.1 Growing Neural Gas

The Growing Neural Gas algorithm is an extended version of the Neural Gas algorithm as presented in the previous section. By the insertion and aging of edges between cluster centers (nodes) according to a set of rules, the topology of the underlying data shall be reflected more precisely. For a detailed description of the algorithm, the reader is referred to Fritzke (1995). Here, only a short overview of the algorithm steps is presented to sketch the basic idea:

1. *Initialization* – two nodes are randomly put in the context space (without an edge between them).
2. *Selection of observation and calculation of closest nodes* – from the set of observations, one is picked (randomly) and the closest node (N_1) as well as the second closest (N_2) are determined based on the Euclidean distance
3. *Insertion of edges* – in case there is no edge between N_1 and N_2, it is inserted. In any case, the age of the edge is set to 0.
4. *Calculation of a node's statistical error value* – for every node, the statistical error E_{N_i} is stored. It represents the total error of all observations assigned to that node. For N_1 (as determined in step 2), the value is updated by adding its Euclidean distance to the current observation $\boldsymbol{x}_c$:

$$E_{N_1}(t) = E_{N_1}(t-1) + ||\boldsymbol{w}_{N_1} - \boldsymbol{x}_c||. \tag{7}$$

5. *Adaptation of node positions* – the position of N_1 as well as its direct topological neighbors is adapted as follows:

$$\boldsymbol{w}_i(t+1) = \boldsymbol{w}_i(t) + l \cdot (\boldsymbol{x}_c - \boldsymbol{w}_i(t)). \tag{8}$$

The learning rate l, though being constant during all iterations, is different for N_1 and its neighbors, i.e. for the neighbors, it is approximately two orders of magnitude lower than for N_1.

6. *Aging of edges* – for all edges originating from N_1, the age is incremented by 1. If the age has passed a defined threshold, the edge is removed. If this results in a node without any edges, it is removed as well.

	Allows for Variable Number of Clusters	Allows for Fuzzy Clustering	Allows for Clustering of Nominal Data	Allows for Pre-defined Cluster Centers
K-Means	No	No	Yes	No
Neural Gas	No	No	No	No
Growing Neural Gas	Yes	No	No	No
Fuzzy-C-Means	No	Yes	Yes	No
K-Fixed-Means	No	No	No	Yes

Fig. 2. Overview of algorithm characteristics.

7. *Insertion of a new node* – after a defined number of iterations has passed, the node $N_{E_{\max}}$ with the highest accumulated error $E_{\max}$ is determined. Among its direct topological neighbors, the one with the highest accumulated error $E_{n_{\max}}$ is chosen and a new node is placed midway the selected nodes. The edge between $N_{E_{\max}}$ and $N_{E_{n,\max}}$ is removed and edges between the new node and $N_{E_{\max}}$ and $N_{E_{n,\max}}$, respectively, are inserted. The accumulated error of the old nodes is decreased by a factor α. The accumulated error of the new node is set to arithmetic mean of these two (updated) values.

8. *Reduction of accumulated error* – for all nodes, the accumulated error is decreased by a factor $\beta \ll \alpha$.

9. *Test of termination criterion* – if the termination criterion (e.g. current number of nodes) is not fulfilled, the next iteration starts at step 2.

A summarizing overview of the employed algorithms and their characteristics is given in Fig. 2.

3 Implementation

Implementation of the user classification service has been guided by several objectives, most importantly:

- high availability of the service
- high scalability
- simple extensibility
- usage of widely adopted protocols and data representation formats
- adequate latency behavior

Taking these aspects into account, we designed a service creation and delivery environment utilizing the JavaEE platform in conjunction with the JBOSS Application Server (version 5.1.0.GA). Service functionality can be reached via http requests and, optionally, additional transmission of XML data. The service hence implements a so-called RESTful (Tyagi, 2006) interface and can be made available to basically any hardware/software platform and (almost) independently of the kind of access network. Available functionality currently includes group formation, context-aware group adaptation, and deletion as well as the provisioning of characteristics of active groups.

4 Results and performance evaluation

For evaluating the performance of the presented algorithms, the following set of criteria has been defined:

- Quality of clustering results – quality is given by the total classification error of a given result.
- Temporal performance – this criteria measures the total time necessary for a classification.
- Stability – evaluates similarity of outcomes for multiple runs with identical data set.
- Flexibility – are algorithms capable of handling other than metric data?
- Implementation – evaluates the amount of effort necessary for algorithm implementation

4.1 Quality and temporal performance of clustering methods

For a comprehensive evaluation within the given multicasting scenario, every algorithm has been executed 100 times for any given input parameter combination. Input parameters included number of entities (840 and 8400), number of groups to be formed (4, 6, 9, 13, 19, 28), and dimension of context space (2 and 5). We analyzed total execution time of an algorithm, its variance as well as total accumulated error. In summary, most important observations and results are (a more detailed performance analysis of the individual algorithms can be found in Appendix A):

4.1.1 K-Means

K-Means is among the best performing algorithms. However, computation times significantly increase with higher

	Stability	Flexibility	Implementation Effort
K-Means	High	Very high	Low
Neural Gas	High	Medium	Medium
Growing Neural Gas	High	Medium	High
Fuzzy-C-Means	Very High	High	Low
K-Fixed-Means	Deterministic	Low	Medium

Fig. 3. Summary evaluation of classification algorithms.

number of users and groups. In terms of the absolute classification error, it achieves the best result of all analyzed algorithms with an average of 133.52 across 100 runs of the standard scenario.

4.1.2 Neural Gas

Similar to K-Means, Neural Gas is a relatively fast algorithm. Moreover, runtime decreases significantly for larger amount of groups because assignment of observations, in contrast to K-Means, is done only once and not in every iteration. Moreover, with higher numbers of nodes, their final position is reached much faster. The average classification error was 133.54.

4.1.3 Growing Neural Gas

In terms of both temporal behavior and dependency on entity as well as group count, the algorithm behaves very similar to Neural Gas. Moreover, its average classification error was only slightly higher, averaging at 137.74.

4.1.4 Fuzzy-C-Means

With regard to performance, the algorithm performs worse with increasing numbers of observations, variables per observation and number of groups. Due to its high computational effort, the algorithm cannot handle more than ten groups in a timely manner. For the calculation of the average classification error (134.69), observations have been assigned to the node they have had the highest affiliation to.

4.1.5 K-Fixed-Means

Overall, algorithm runtime is on a satisfying level and comparable to that of K-Means. For five variables (i.e. a five-dimensional context space), the algorithm arrives in a stable state earlier since (with the given set of test observations) the cluster centers reach their final position faster. A comparison of the accumulated classification error does not make sense since some observations were not classified at all.

4.2 Overall evaluation

A brief summary of the remaining evaluation criteria is depicted in Fig. 3.

Overall, the K-Means algorithm disposes of the most appropriate characteristics in the given multi-casting use case. Not only good results in terms of quality and performance but also its capability to handle nominal data makes it the default choice for classification requests. Neural Gas is the preferred algorithm for large entity and group counts; however, it should not be used for clustering observations with nominal data. In case that the number of groups is not known yet, Growing Neural Gas, despite its difficult parametrization, is the best choice. Fuzzy-C-Means is especially recommended if the number of groups remains small and the structure of the observation set is rather complex. Finally, the K-Fixed-Means algorithm should be chosen if cluster centers, i.e. group characteristics, are pre-determined and should only be marginally changed during execution.

5 Conclusions

This paper has presented an evaluation of different user classification algorithms for enabling group communication and multi-casting in wireless networks. Classification tests have been performed based on simulated context information about users (such as location, music taste, and age) with different numbers of both users and expected groups. K-Means and (Growing) Neural Gas as well as the newly developed K-Fixed-Means have consistently recognized the basic structure within the set of users and produced fast and stable classification results with low total errors.

A major aspect of future work is the development of classification methods that can handle nominal (and ordinal) data since most of the interesting user data enabling multi-casting (such as a user's profile) fall into these categories. Moreover, the implemented service will be verified in a testbed environment where efficiency gains based on the realization of multi-casting will be analyzed quantitatively.

Appendix A

Algorithms performance results

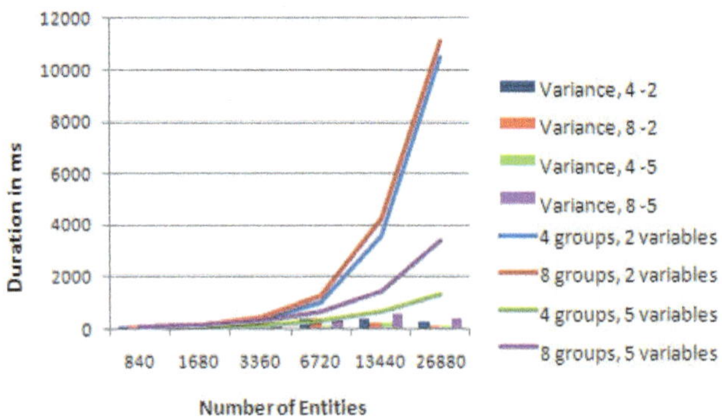

Fig. A1. Performance of K-Fixed-Means algorithm.

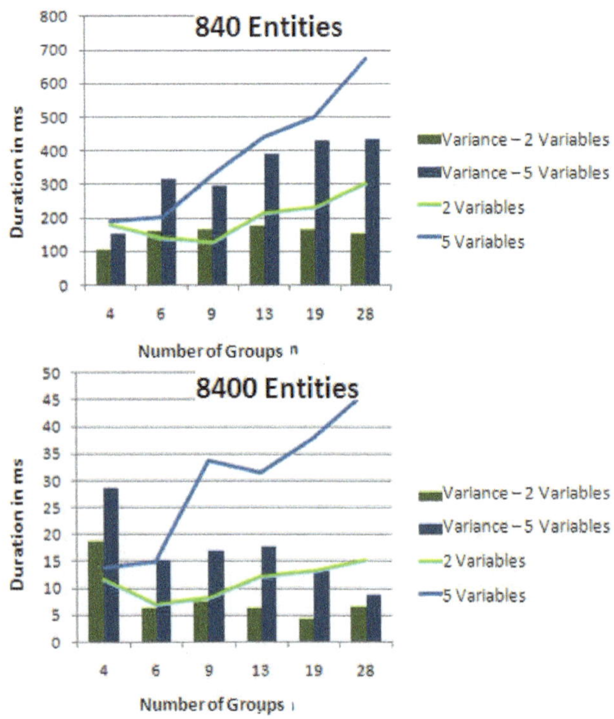

Fig. A2. Performance of K-Means algorithm.

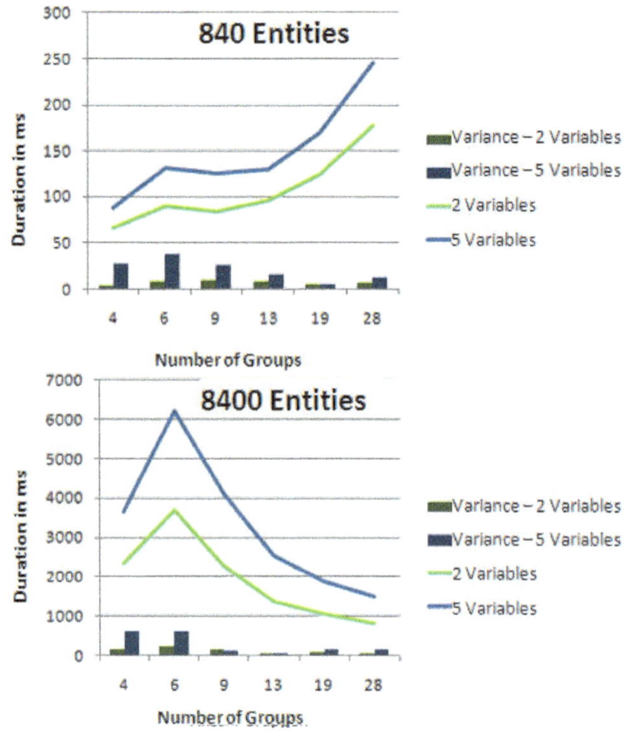

Fig. A3. Performance of Neural Gas algorithm.

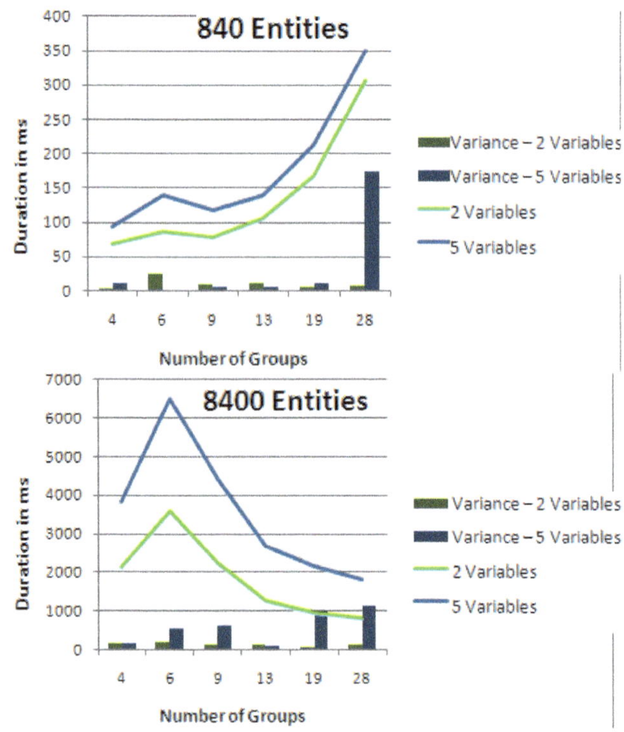

Fig. A4. Performance of Growing Neural Gas algorithm.

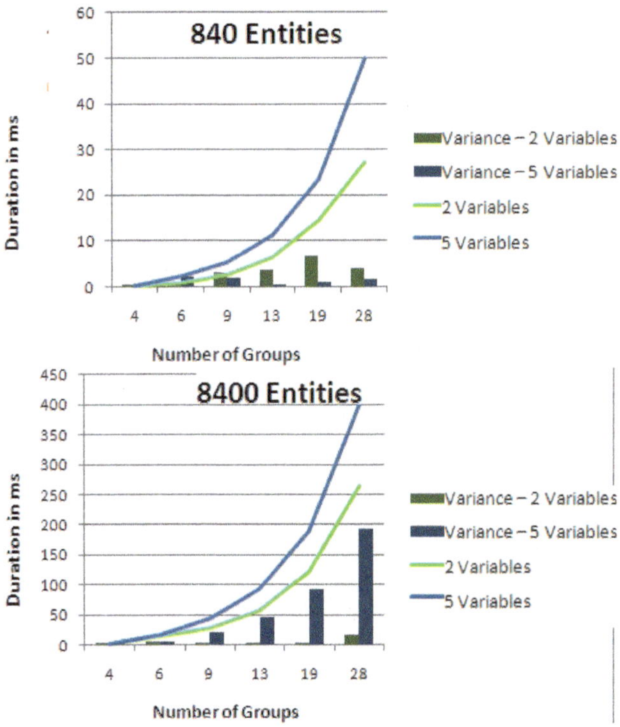

Fig. A5. Performance of Fuzzy-C-Means algorithm.

Acknowledgements. Parts of this work were funded by the Federal Ministry of Education and Research of the Federal Republic of Germany (Foerderkennzeichen 01 BK 0808, GLab). The authors alone are responsible for the content of the paper.

References

Baumung, P. and Zitterbart, M.: MAMAS – Mobility-aware Multicast for Ad-hoc Groups in Self-organizing Networks, in: Basissoftware für drahtlose Ad-hoc- und Sensornetze, edited by: Zitterbart, M. and Baumung, P., pp. 33–48, Universitaetsverlag Karlsruhe, March 2009.

Bezdek, J. C.: Pattern Recognition with Fuzzy Objective Function Algorithms. Plenum Press, New York, 1981.

Chalmers, R. and Almeroth, K.: Modeling the Branching Characteristics and Efficiency Gains of Global Multicast Trees, Proceedings of IEEE Infocom, Anchorage (AK), 2001.

Fritzke, B.: A growing neural gas network learns topologies, in: Advances in Neural Information Processing Systems 7, edited by: Tesauro, G., Tesauro, G., Touretzky, D. S., and Leenet, T. K., MIT Press, Cambridge (MA), pp. 625–632, 1995.

Fritzke, B.: Some Competitive Learning Methods. Java Paper, 1997, http://www.neuroinformatik.ruhr-uni-bochum.de/VDM/research/gsn/JavaPaper/, accessed 11 November 2010.

Janic, M. and Van Mieghem, P.: The Gain and Cost of Multicast Routing Trees. International Conference on Systems, Man and Cybernetics (IEEE SMC 2004), The Hague, 2004.

MacQueen, J.: Some methods for Classification and Analysis of Multivariate Observations: Proceedings of the Fifth Berkeley Symposium on Mathematical Statistics and Probability, pp. 281–297, 1967.

Martinez, T. M. and Schulten, K. J.: A neural gas network learns topologies, in: Artificial Neural Networks, edited by: Kohonen, T., Mäkisara, K., Simula, O., Kangas, J. Elsevier, North-Holland, pp. 397–402, 1991.

Steinhaus, H.: Sur la division des corps matiriels en parties, Bull. Acad. Polon. Sci., 4(12), 801–804, 1956.

Tyagi, S.: RESTful Web Services. Oracle Technology Network, 2006, http://www.java.sun.com/developer/technicalArticles/WebServices/restful/, accessed 28 November 2010.

Physical layer simulation results for IEEE 802.15.3c with different channel models

M. Liso Nicolás, M. Jacob, and T. Kürner

Institut für Nachrichtentechnik, Technische Universität Braunschweig, Braunschweig, Germany

Abstract. This paper investigates the performance of the 60 GHz IEEE 802.15.3c physical layer (PHY) specification in terms of bit error rate (BER) against signal to noise ratio. Two PHY modes of the standard have been implemented and simulated, i.e., Single Carrier and High Speed Interface. The first mode uses single carrier (SC) block transmission and the second mode uses orthogonal frequency division multiplexing (OFDM). One of the main issues in the new 60 GHz standards is multipath propagation, which plays an important role in the link quality. Thus, we have tested the PHY with the IEEE standard channel model, ray tracing simulations and real 60 GHz measurements.

1 Introduction

During the last years, wireless communication systems using the 60 GHz frequency band have attracted much attention. This unlicensed band, ranging from 57 to 66 GHz, is usually divided into four 2-GHz channels, and aims to satisfy the need for increasingly higher data rates required by new wireless multimedia applications. Given the high free space losses and attenuation through building materials at the millimetre-wave 60 GHz spectrum, only short-range indoor applications are considered.

Four different specifications for WLAN/WPAN applications using the 60 GHz band have already been issued: IEEE 802.15.3c by IEEE 802.15 Task Group 3c (TG3c) (TG3c Standard, 2009), ECMA-387 (ECMA-387, 2008), Wireless HD (Wireless HD, 2010) and WiGig (WiGig, 2010). Moreover, the IEEE 802.11 Task Group ad (TGad) has recently released a first draft (TGad Draft, 2009).

In order to evaluate the system performance of different 60 GHz standards, we have developed a simulation environment in MATLAB/Simulink. In a first step, we have implemented the PHY of the IEEE 802.15.3c (TG3c) specification. This standard includes three different physical layer designs, with both SC and OFDM modes.

An extensive work has been done on channel modelling in TG3c, since the propagation of the signal through 60 GHz wireless channels was not accurately described when the standardisation started. Hence, we have included this channel model in our simulation environment. Moreover, we have also evaluated different PHY modes with channel impulse responses (CIR) coming from 60 GHz channel measurements and ray tracing simulations.

This paper is organised as follows: in Sect. 2 we present a general view of the TG3c standard. In Sect. 3 we give an explanation of the 60 GHz channel modelling. In the next section we explain the simulation environment and the parameters we have used. In Sect. 5 the results from the simulation are presented and evaluated. Finally, Sect. 6 concludes this paper.

2 802.15.3c overview

In 2009, the TG3c working group finished the standardisation process, releasing the IEEE 802.15.3c PHY specification for millimetre wave communications. It includes three different PHY modes: Single Carrier (SC PHY), High Speed Interface PHY (HSI PHY) and Audio/Visual PHY (AV PHY). In our research, we have focused on HSI and SC PHY.

SC PHY is thought for line-of-sight (LOS) and non-line-of-sight (NLOS) and is able to work with and without equalisation. This PHY mode consists of 14 different modulation and coding schemes (MCS), with data rates of up to 5.3 Gbps. It utilises Reed-Solomon (RS) and low-density parity-check (LDPC) codes, whereas in the paper only MCSs with LDPC are considered.

Correspondence to: M. Liso Nicolás
(liso@ifn.ing.tu-bs.de)

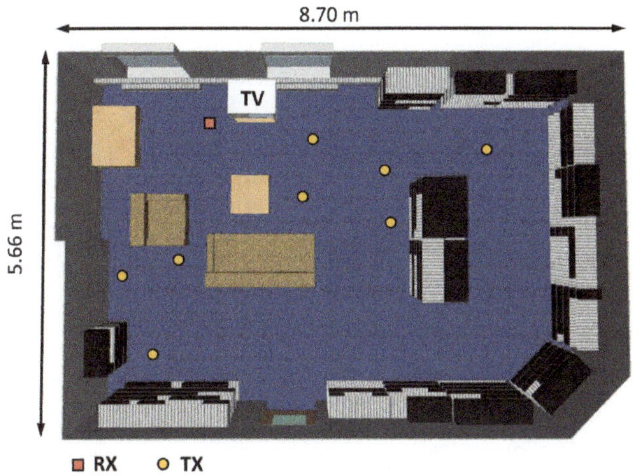

Fig. 1. Living room scenario with Rx/Tx positions.

On the other hand, HSI PHY is conceived for NLOS and utilises OFDM. It includes 12 MCSs, with a maximum data rate equal to 5.8 Gbps, and uses only LDPC.

The fact that a single carrier physical layer could outperform a multicarrier scheme in presence of multipath is a new topic that has been addressed at the 60 GHz standardisation. The result of the comparison depends on several factors like RF impairments, channel coding and multipath characteristics. A good overview of SC in mm-wave wireless networks can be found in Kato et al. (2009) and Funada et al. (2009), and a comparison between SC and OFDM in Hoffmann et al. (2009).

3 Channel modelling

Multipath propagation in millimetre-wave indoor communications poses a considerable problem, since it can dramatically degrade the link quality. Therefore, special attention has to be paid to this issue.

The TG3c developed a 60 GHz channel model that covers typical environments for the planned use cases like residential, office, library, desktop and kiosk scenarios. In order to implement a model that fits the reality as much as possible, the TG3c utilised real 60 GHz channel measurements as base for its work. Thus, CIRs are built with a combination of a two-ray channel model and an extension of the Saleh-Valenzuela (SV) model (Saleh and Valenzuela, 1987) to the angular domain (TG3c Channel Modelling, 2007). Hence, TG3c proposed a statistical channel model dependent on the temporal and spatial domain, where the signals arrive at the receiver first in a LOS component (calculated with a two-ray model), and then in groups called clusters (modified SV model). This 60 GHz channel model was included in our simulation environment, so that we can evaluate the system performance in different scenarios with different modulation and coding schemes.

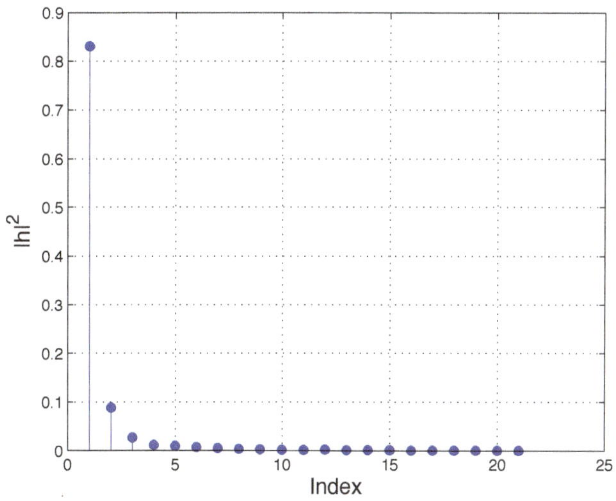

Fig. 2. CIR obtained from 60 GHz channel measurement.

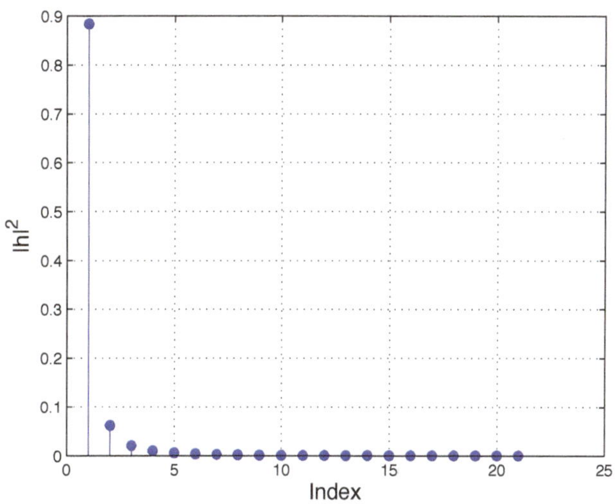

Fig. 3. CIR obtained from ray tracing simulation.

In addition, wideband channel measurements have been performed in a living room (Jacob and Kürner, 2009a, b). All measurements were conducted in frequency domain. The channel impulse responses were then calculated via inverse Fourier transform. The measurement setup consists of an Agilent E8361A vector network analyzer with external transmitting and receiving test heads. For the measurements we used circular horn antennas with a half-power beam width (HPBW) of approximately 20° at receiver and transmitter. Figure 1 shows a schematic of the living room where the measurements were performed. During the measurements the receiver position was kept fixed near a television set. The transfer functions were then measured at different positions in the room. As it can be seen, there is always line of sight between both antennas.

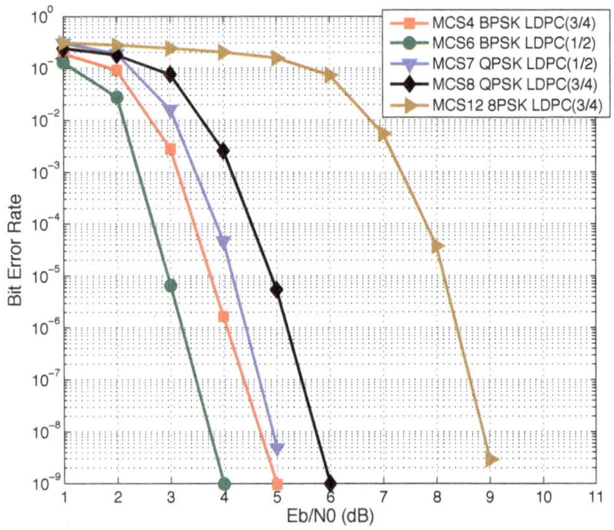

Fig. 4. BER against Eb/N0 for AWGN channel and SC PHY.

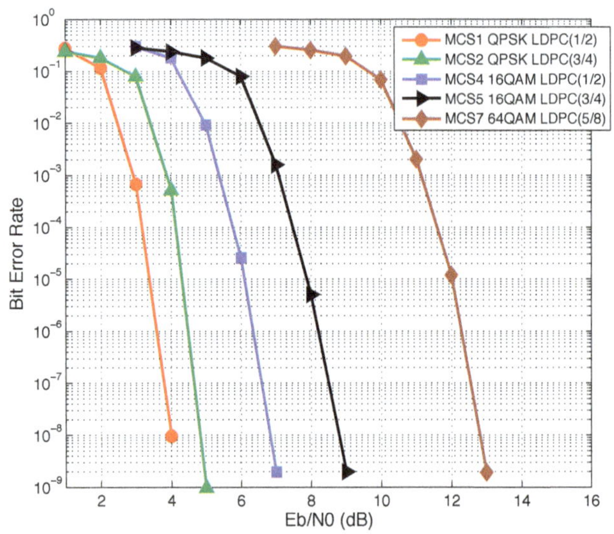

Fig. 5. BER against Eb/N0 for AWGN channel and HSI PHY.

Given this setup, a ray tracing tool (Kürner and Jacob, 2009) was used to model the wave propagation deterministically by using a 3-D model of the living room (see Fig. 1). For the modelling of the dielectric properties of fitment, walls etc. material parameters from literature have been used (BROADWAY, 2002). The 3-D antenna pattern of the horn antennas has been modelled by analytical expressions (Green, 2006).

Hence, real measurements, deterministic and statistical channel models are included in our work.

4 Simulation parameters

For our simulations we have considered SC and HSI PHYs. Tables 1 and 2 summarise the MCSs included in our work.

Table 1. Single Carrier PHY MCSs parameters.

MCS	Bit rate (Mbps)	Channel coding
4	1320	LDPC(672,504), $R = 3/4$
6	880	LDPC(672,336), $R = 1/2$
7	1760	LDPC(672,336), $R = 1/2$
8	2640	LDPC(672,504), $R = 3/4$
12	3960	LDPC(672,504), $R = 3/4$

Table 2. High Speed Interface PHY MCSs parameters.

MCS	Bit rate (Mbps)	Channel coding
1	1540	LDPC(672,336), $R = 1/2$
2	2310	LDPC(672,504), $R = 3/4$
4	3080	LDPC(672,336), $R = 1/2$
5	4620	LDPC(672,504), $R = 3/4$
7	5775	LDPC(672,504), $R = 3/4$

First, we evaluate the performance of SC and HSI using an AWGN channel. Then, we assess the goodness of SC PHY in a LOS environment. For this case, we have implemented single carrier frequency domain equalisation (SC-FDE) with zero forzing (ZF) as explained in Lei and Huang (2009) and Lei et al. (2008) using a guard interval of 64 samples. This guard length should be enough to cope with the multipath propagation present in the selected LOS scenario. For the channel estimation, Golay sequences (Golay, 1961) are used in time domain to estimate the channel.

For the TG3c channel model, we concentrate our work on the TG3c 1.3 channel model, where LOS in a residential environment is considered, since it is the model that most resembles the measurement scenario. The transmitting antenna HPBW is fixed to 30° and the receiving antenna to 20°. Eight different channel realisations where simulated for different values of signal to noise ratio and five modulation and coding schemes.

In case of the channel data from the channel measurements and the ray tracing simulations, the channel impulse responses according to the positions in Fig. 1 have been used. As for the TG3c simulation, eight CIRs obtained in the measurement campaign were utilised, as well as eight CIRs from the ray tracing simulation of the corresponding measured points. As an example, Figs. 2 and 3 show two CIRs obtained by measurement and ray tracing for one point in the room. The shape of the CIRs is due to the hight directivity of the antennas, the existence of line-of-sight and the short distance between transmitter and receiver.

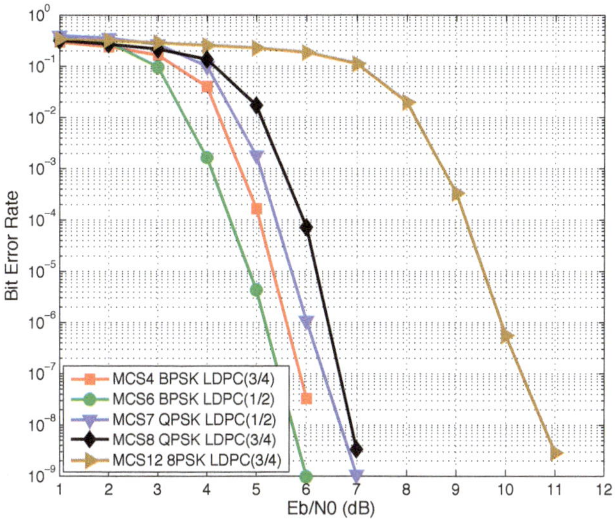

Fig. 6. BER against Eb/N0 for 60 GHz channel measurements and SC PHY.

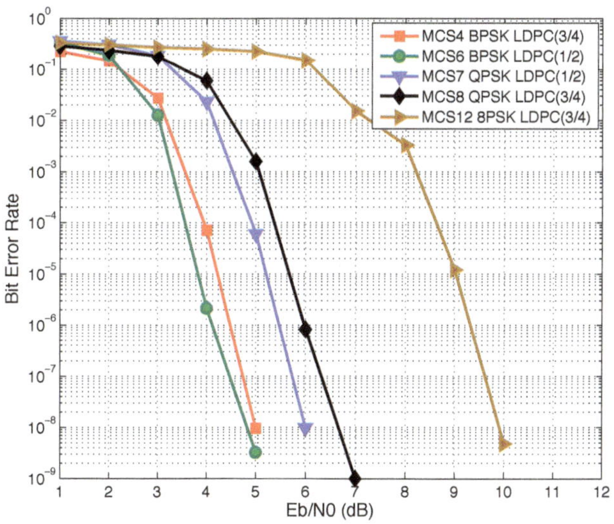

Fig. 7. BER against Eb/N0 for ray tracing simulation with SC PHY.

5 Simulation results

Figures 4 and 5 show the bit error rate obtained for different values of the relation energy per bit to spectral noise density (Eb/N0) for SC and HSI in case of AWGN channel. In case of perfect channel estimation and equalisation, the results of the simulation with 60 GHz channel realisations should resemble these figures. At both figures can be seen that, for the same modulationorder, by decreasing the code rate from 3/4 to 1/2 a gain of up to 1 dB is achieved.

After the simulation with the multipath channels, the BER values due to the best six of the eight CIRs where averaged. The results of the averaging are presented in Figs. 6, 7 and 8, which illustrate the BER obtained for the 60 GHz measurements, ray tracing and TG3c model. It is shown that a degradation of up to 2 dB is obtained when compared with the AWGN channel. This is partly caused by the ZF equalisation method, which on the one hand removes inter symbol interference in the received signal due to multipath, and on the other hand elevates the Gaussian noise after the equaliser. Besides, the time domain channel estimation performs quite well, so that its contribution to the degradation is very small. Moreover, the shift to the right in the curves of Figs. 6–8 when compared with Fig. 4 is also caused by the reduction of the spectrum efficiency introduced by the guard interval necessary for the channel equalisation.

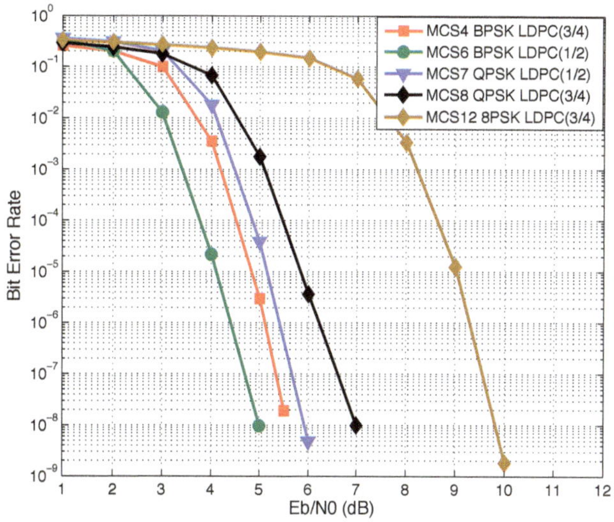

Fig. 8. BER against Eb/N0 for TG3c CM1.3 with SC PHY.

Figure 9 shows the values of Eb/N0 required to reach a BER below 10^{-7}. This value was selected because it is a good compromise between the different requirements in terms of quality of service in the planned use cases. Comparing the channel models (TG3c, Ray Tracing, Measurements), slight differences can be found. For the difference between TG3c and the other two data sets this can be explained by the fact that the scenarios and the used antennas are similar but of course not identical. The best results in terms of BER are obtained for the TG3c channel model, and the worst for the 60 GHz channel measurements. Nevertheless, the values obtained by the simulations with the multipath channels are very similar to each other, with differences below 1.1 dB. For the selected threshold (BER below 10^{-7}) a maximum data rate of up to 3.47 Gbps (MCS 12) can be achieved for Eb/N0 values better than 11.3 dB. Besides this, the figures illustrate that the gain obtained by reducing the code rate for a fixed modulation order is smaller than in the case of an AWGN channel.

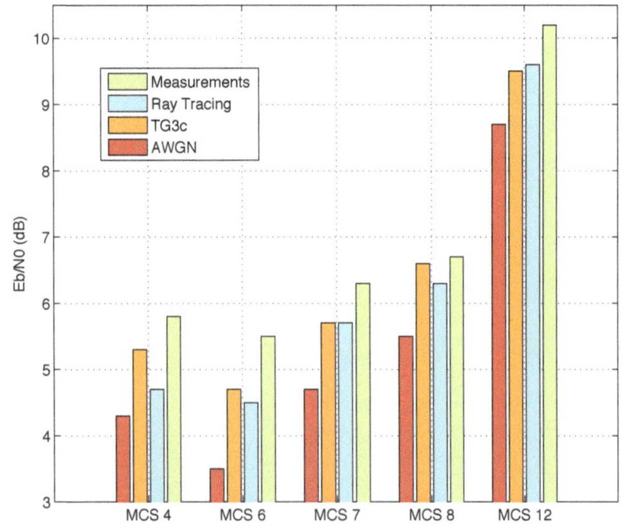

Fig. 9. Eb/N0 values (dB) for BER below 10^{-7}.

6 Conclusions

We have presented and evaluated two different physical layer modes of the IEEE 802.15.3c standard. This has been done first for an AWGN channel, and second with different kinds of channel impulse responses obtained from measurements and statistical and deterministic models.

Our simulations show that when an AWGN channel is considered, which would be the case with perfect channel estimation and equalisation, HSI and SC PHY performance is similar. For the other three channels included, SC PHY with SC-FDE presents a good overall performance, and allows multigigabit wireless communication with reasonable values of signal to noise ratio.

In future work, we will improve the channel equalisation method and we will include NLOS propagation. For this case, we will implement a beam forming algorithm, and will evaluate its impact on the system performance.

Acknowledgements. This work has been carried out within the European Medea+project QStream "Ultra-high data-rate wireless communication" and financed by the German Ministry for Economics within the ProInno funding scheme.

References

BROADWAY, WP1 Study: Functional System Parameters description, Tech. Rep., iST-2001-32686, 2002.

ECMA-378 Standard, High Rate 60 GHz PHY, MAC and HDMI PAL, http://www.ecma-international.org/publications/standards/Ecma-387, 2008.

Funada, R., Harada, H., Shoji, Y., Kimura, R., Nishiguchi, Y., Lei, M., Choi, C.-S., Kojima F., Pyo, C.-W., Lan, Z., Lakkis, I., Umehira, M., and Kato, S.: A design of single carrier based PHY for IEEE 802.15.3c standard, IEEE PIMRC, 1–5, 2007.

Golay, M.: Complementary series, IRE Transactions on Information Theory, 7, 1961.

Green, H. E.: The radiation pattern of a conical horn, J. Electromagnetic Waves Applications, 20(9), 1149–1160, 2006.

Hoffmann, O., Kays, R., and Reinhold, R.: Coded Performance of OFDM and SC PHY of IEEE 802.15.3c for Different FEC Types, IEEE GLOBECOM Workshops, 1–3, 2009.

IEEE 802.15 TG3c Working Group: TG3c Channel Modeling Subcommittee Final Report, http://www.ieee802.org/15/pub/TG3.html, 2007.

IEEE 802.15 TG3c Working Group: Part 15.3: Wireless Medium Access Control (MAC) and Physical Layer (PHY) Specifications for High Data Rate Wireless Personal Area Networks, (WPANs), http://www.ieee802.org/15/pub/TG3.html, 2009.

IEEE 802.11 TGad Working Group: Draft Standard for Information Technology, Telecommunications and Information Exchange Between Systems, Local and Metropolitan Area Networks, Specific Requirements, Part 11: Wireless LAN Medium Access Control (MAC) and Physical Layer (PHY) Specifications, Amendment 6: Enhancements for Very High Throughput in the 60 GHz Band, http://www.ieee802.org/11/pub/TGad.html, 2009.

Jacob, M. and Kürner, T.: IEEE 802.11 TGad Working Group: Measurement campaign at 60 GHz in the living room environment at TUBS, https://mentor.ieee.org/802.11/documents, 2009a.

Jacob, M. and Kürner, T.: Radio channel characteristics for broadband WLAN/WPAN applications between 67 and 110 GHz, 3rd European Conference onAntennas and Propagation, 2663–2667, 2009b.

Kato, S., Harada, H., Funada, R., Baykas, T., Sum, C. S., Wang, J., and Rahman, M. A.: Single Carrier Transmission for Multi-Gigabit 60-GHz WPAN Systems, IEEE Journal on Selected Areas in Communications, 8, 1466–1478, 2009.

Kürner, T. and Jacob, M.: Application of ray tracing to derive channel models for future multi-gigabit systems, International Conference onElectromagnetics in Advanced Applications, 517–520, 2009.

Lei, M. and Huang, Y.: CFR and SNR Estimation Based on Complementary Golay Sequences for Single-Carrier Block Transmission in 60-GHz WPAN, IEEE WCNC, 1–5, 2009.

Lei, M., Lakkis, I., Harada, H., and Kato, S.: MMSE-FDE based on Estimated SNR for Single-Carrier Block Transmission (SCBT) in Multi-Gbps WPAN (IEEE 802.15.3c), IEEE International Conference on Communications Workshop, 52–56, 2008.

Saleh, A. and Valenzuela, R.: A Statistical Model for Indoor Multipath Propagation, IEEE Journal on Selected Areas in Communications, 5, 128–137, 1987.

The Wireless HD specification: http://www.wireslesshd.org, 2010.

Design and realization of a broadband single-side-band mixer with a very short settling time

S. Haßler[1,*], **T. Reichthalhammer**[1], **and E. Biebl**[1]

[1]Fachgebiet Höchstfrequenztechnik, Technische Universität München, Munich, Germany
[*]now at: Siemens AG, Erlangen, Germany

Abstract. To achieve high range resolution in synthetic aperture radar imaging a frequency synthesizer with high bandwidth is a possible solution. To operate in the required frequency band an LF-signal usually has to be upconverted. In this paper we describe the design and realization of a broadband Single-Side-Band Mixer with a very short settling time between frequency steps of arbitrary length inside a high bandwidth. Compared to already existing SSB-Mixers, our novel concept has three major advantages: At first, the mixer could be used in combination with an arbitrary signal source. Due to a modular circuit concept it is possible to use the system for different input frequency ranges. Moreover, just by changing single modules, the output frequency-range can be adapted to individual requirements. Thirdly, as a main advantage, the system is able to generate a high frequency output span with a very fast settling time between frequency steps. Even with applied steps up to 400 MHz, the settling time remains below 3 μs, which is more than 5 times faster than the settling time of similar synthesizers.

1 Introduction

A few requirements have been claimed for developing the single side band mixer: A fast frequency sweep, a high output spectrum and a high sideband supression. For synthetic radar imaging with a stepped frequency synthesizer, for example, frequency ramps have to be generated and therefore the system has to be able to switch very fast between discrete frequency steps of up to 400 MHz. Within this time the sideband suppression has to rise as much as possible. Due to a modular concept, the system could be adapted to different signal sources and various RF-frequency-ranges (Haßler, 2009).

Correspondence to:
T. Reichthalhammer
(reichthalhammer@tum.de)

2 System design

There are different concepts how to convert frequency with a high sideband suppression. In Fig. 1 there is shown a fundamental circuit concept of our mixer. The signal is devided into two channels with a relative phase difference of 90°.

The signals amplitude is equalized by an AGC. Due to the requirement of a very fast settling time, the time-constant of the AGC is a fundamental criterion of choice. A broadband AGC unit is necessary to allow a high input spectrum. The output level of the AGC is compared to the setpoint and the difference converted into a current which charges an external capacitor. The voltage at the capacitance adjusts the amplification factor. The smaller the capacitance the lower the settling time but also the higher the probabilty that the system oscillates. With our system we achieve a settling time of 1 μs. The influence of the amplitude differences between the two signal paths on the sideband suppression has been evaluated using a simulation. To show the influence of amplitude differences between the two signal paths on the sideband suppression, the single sideband mixer was simulated. For this simulation, a phase difference of 90° between the I- and Q-signal was assumed. Figure 2 shows the output spectrum of the single sideband mixer during the settling time after a 400 MHz step. Figure 3 shows the output spectrum after the settling time. The different sideband suppression is obvious.

Each signal is mixed with the VCO signal, that is splitted in an inphase and a quadrature component as well. Assuming an equivalent signal level and a phase difference is 90° Fig. 4 shows the spectra of the input signals AI1 and AQ1 together with the output signal.

3 Amplitude and phase error

Referring to Fig. 1, the input signals AI1 and AQ1 can be written as

$$A_{I1}(t) = A_{I1}\cos(w_{\text{in}}t + \Delta\Phi_I) \quad (1)$$

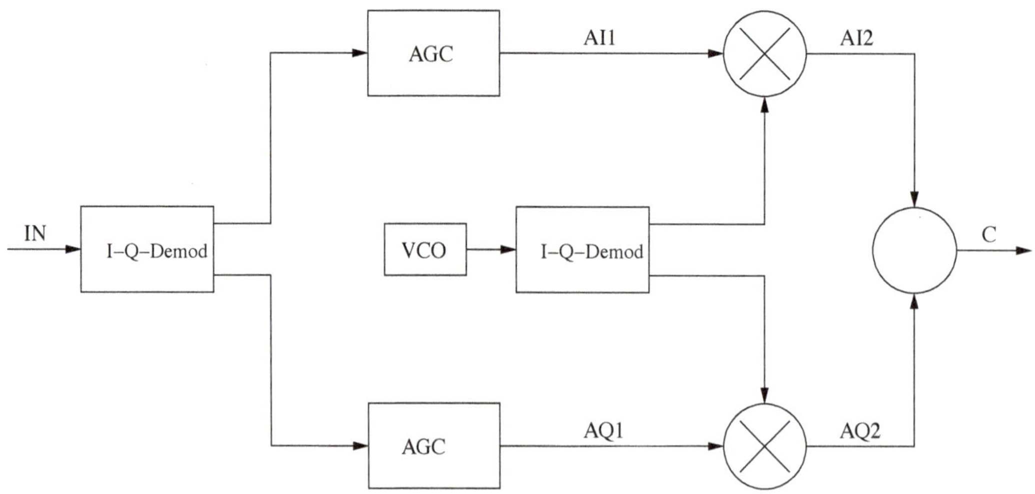

Fig. 1. Systemconcept: single sideband mixer.

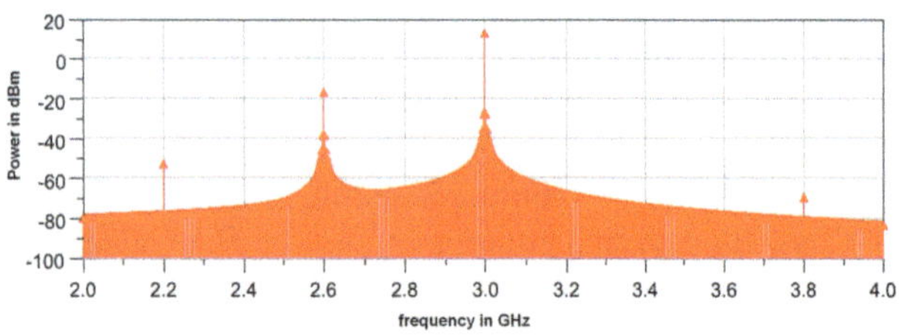

Fig. 2. Output spectrum of FFT during settling time.

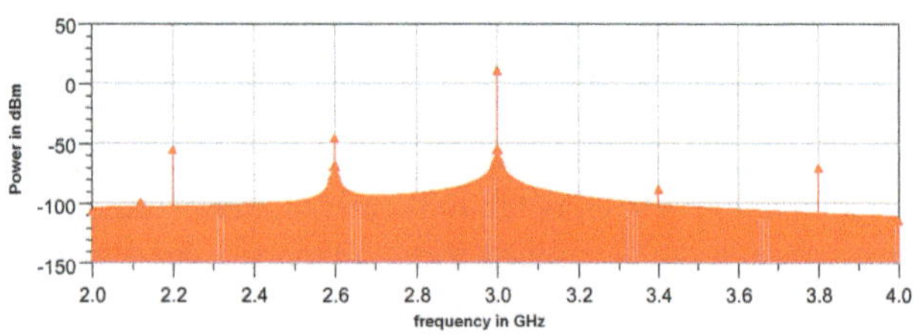

Fig. 3. Output spectrum of FFT after settling time.

and

$$A_{Q1}(t) = A_{Q1}\cos(w_{\text{in}}t + \Delta\Phi_Q). \tag{2}$$

After a few steps, the output signal

$$C(t) = A_{I1}(t) + A_{Q1}(t)$$

can be calculated for the upper and lower sideband

$$|C_{\text{USB}}| = \sqrt{a^2 - b^2} \tag{3}$$

$$|C_{\text{LSB}}| = \sqrt{a^2 + b^2} \tag{4}$$

where

$$a = \frac{A_{Q1}B_Q\sin(\Delta\Phi_Q)}{2}$$

and

$$b = \frac{A_{I1}B_I}{2} + \frac{A_{Q1}B_Q\cos(\Delta\Phi_Q)}{2}.$$

In Fig. 5 the sideband suppression is shown in dependence on the relative phase error and the amplitude deviation in both channels.

Obviously both, an amplitude and a phase deviation have a strong influence on the sideband suppression. A phase error of 3° approximately deteriotrates the signal as much as an amplitude error of about 2 dB.

4 Generation of a broad band I/Q signal

There are a variety of possibilities to generate a broadband I/Q signal. Under consideration of the requirements of a wideband input spectrum and a very fast settling time and regarding previous results of the analysis of phase- and amplitude errors, three concepts have been proven as applicable.

4.1 RC-CR phase shifter

The first possibility to generate a 90° phase difference, is an RC-CR combination. Identical resistors and capacitors assumed, this construction always provides a relative phase difference of 90°. With the transfer functions of the low- and highpass (Eqs. 5, 6) it can be shown, that the relative phase difference between this two filters remains 90° over the whole frequency range.

$$H_{\text{TP}} = \frac{1}{1 + j\omega RC} \tag{5}$$

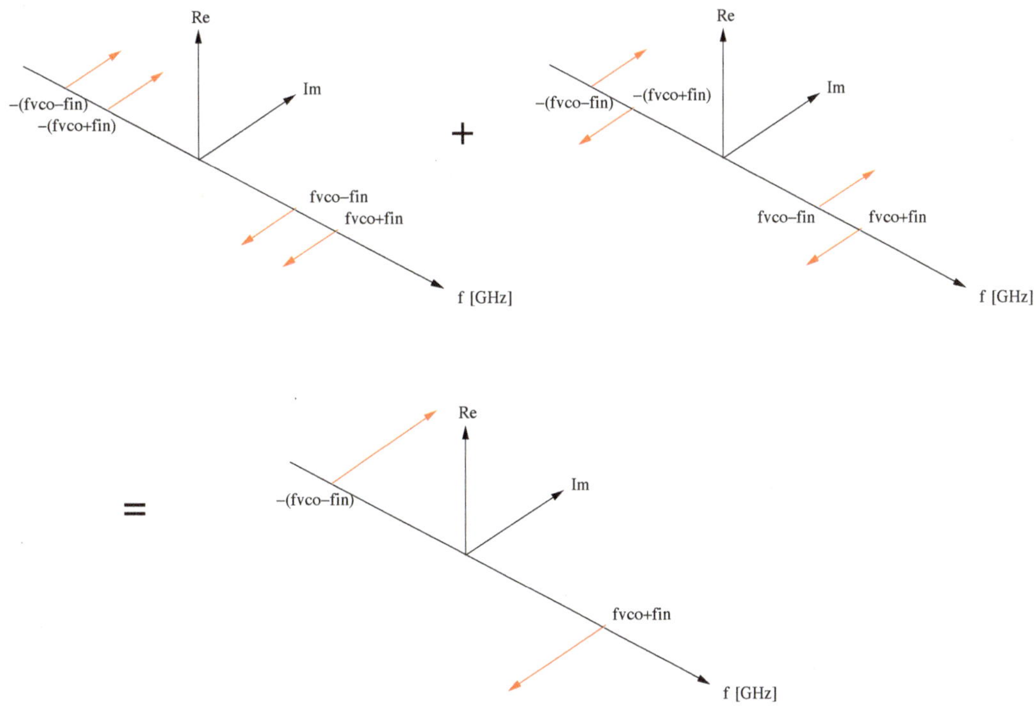

Fig. 4. Spectrum of the inphase and quadrature signals after frequency conversion.

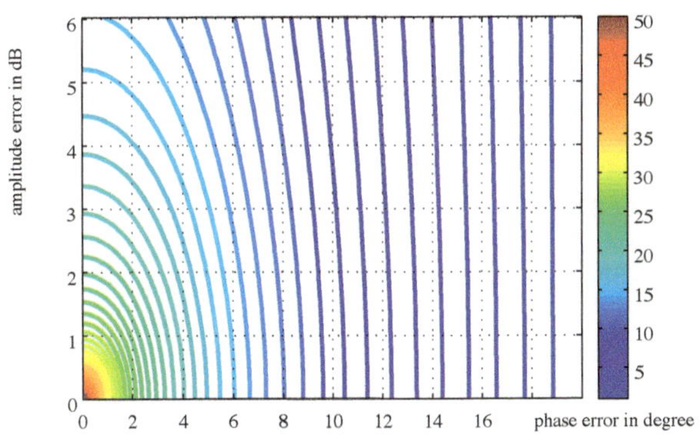

Fig. 5. Influence of phase- and amplitude errors on the side band suppression.

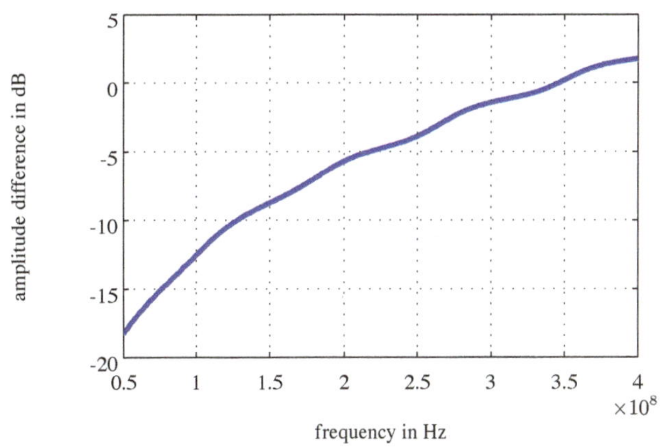

Fig. 6. Amplitude difference of I and Q channel of RC-CR element.

$$H_{\mathrm{HP}} = \frac{j\omega RC}{1 + j\omega RC} \tag{6}$$

$$\begin{aligned} &\text{phase difference} = \angle H_{\mathrm{TP}} - \angle H_{\mathrm{HP}} = \\ &-\arctan(\mathrm{wRC}) - \frac{\pi}{2} + \arctan(\mathrm{wRC}) = -90^\circ \end{aligned} \tag{7}$$

An issue of this type of phase shifter is the antipodal change of the amplitude levels in both signal channels. To achieve good results in sideband suppression the amplitudes of both signals have to be equal. This is done by an automatic-gain-control unit (AGC). In Fig. 6 the difference between the in-phase and quadrature channel is depicted. For a frequency step from 400 MHz to 50 MHz the maximum difference is approximately 18 dB.

An increasing difference of the amplitudes causes a higher time constant of the AGC. To minimize the deterioration of the fast settling time, the amplitude influence of the phase shifter should be kept as small as possible. The phase response of this RC-CR component, which is designed for an input frequency range from 50 MHz to 400 MHz, is shown in Fig. 7.

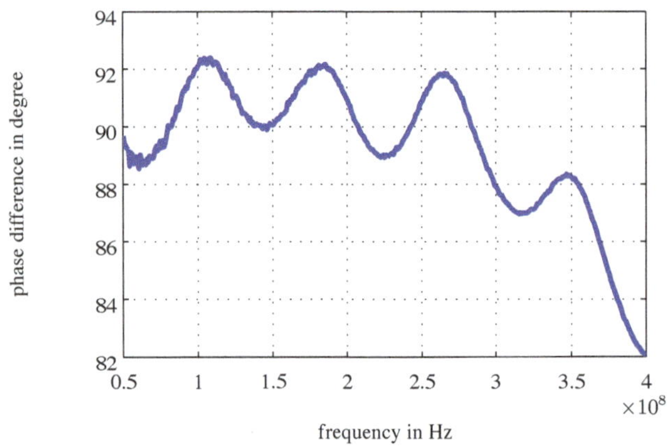

Fig. 7. Phase difference of I and Q channel of RC-CR element.

Up to a frequency of 360 MHz the phase error remains smaller than 2.5°.

4.2 Delay line phase shifter

Another concept for phase shifting is shown in Fig. 8. The generation of a 90° phase difference is realized exclusively with transducers and delay lines. Every transducer is used to split a single-ended signal into a differential signal. For mathematical description, the input signal is written vectorial.

$$\mathbf{IN}(t) = e^{j\omega t}\begin{pmatrix}\cos\varphi\\ \sin\varphi\end{pmatrix} = e^{j\omega t}\begin{pmatrix}1\\ 0\end{pmatrix} \tag{8}$$

The input phase offset can be set to zero without further restrictions and the time dependence can be ignored. Assuming that the upper output of every balun is delayed by 180° B1 and B2 can be written as

$$\mathbf{B1} = \begin{pmatrix}\cos(\varphi_{B1} = 180°)\\ \sin(\varphi_{B1} = 180°)\end{pmatrix} = \begin{pmatrix}-1\\ 0\end{pmatrix} \tag{9}$$

$$\boldsymbol{B2} = \begin{pmatrix}\cos(\varphi_{B2} = 0°)\\ \sin 0(\varphi_{B2} = 0°)\end{pmatrix} = \begin{pmatrix}1\\ 0\end{pmatrix}. \tag{10}$$

Each signal is divided by a transducer. The signals C2 and C4 are delayed and compared to C1 and C3. To get a relative phase difference of 90° independent of frequency between the output signals Port2 and Port3, the length of the delay line can be choosen arbitrarily. The amplitude difference of the output signals increases, if the electrical length of the delay line or phase difference between C2 and C1, and as well between C4 and C3, becomes smaller. With $\Delta\Phi_{C2}$ and $\Delta\Phi_{C4}$ being the phase differences caused by the delay lines C2 and C4, the output signals at Port2 and Port3 can be written as:

$$|\mathbf{Port2}| = \left|\begin{pmatrix}\cos\Delta\varphi_{C2} - 1\\ \sin\Delta\varphi_{C2}\end{pmatrix}\right| \tag{11}$$

$$|\mathbf{Port3}| = \left|\begin{pmatrix}\cos\Delta\varphi_{C4} + 1\\ \sin\Delta\varphi_{C4}\end{pmatrix}\right| \tag{12}$$

Using the dot product (13), it can be shown that the signal vectors at the output ports are perpendicular to each other and therefore the phase difference is 90°, assumed that both delay lines cause the same phase difference.

$$\mathbf{Port2} \bullet \mathbf{Port3} = \cos^2(\Delta\varphi) - 1 + \sin^2(\Delta\varphi) \equiv 0 \tag{13}$$

The attenuation of the delay lines can be neglected and the phase difference of all balun-output-signals accounts for 180°. Due to the frequency dependence of the electrical length of the delay line (Pozar, 1998) the phase difference is

$$\Delta\varphi = \frac{360° \cdot l_{\text{delay}}}{\lambda} = \frac{360° \cdot l_{\text{delay}}\sqrt{\epsilon_r}\, f}{c} \tag{14}$$

Figure 9 shows the variation of the absolute amplitudes at Port2 and Port3, as well as the amplitude difference between the two output signals versus the phase delay caused by the delay lines.

To minimize the amplitude deviation of the output signals at a high frequency step it is necessary to scale the delay line. The phase delay should remain within 50° to 150° inside a frequency range of 50 MHz to 400 MHz. Considering the attenuation of the delay lines Fig.10 visualizes the result.

Measurements of the transfer coefficients S21 are shown in Figs. 11 and 12. Figure 11 shows the absolute value of S21. A frequency step of 350 MHz causes an amplitude deviation of 25 dB.

Regarding Fig. 12 the measured phase deviation is up to 15°. Compared to theory and to the RC-CR module, this component shows a poorer behavior, since a lot of small degradations are accumulated.

4.3 Active allpass as phase shifter

The third possibilty generating an I/Q signal is an active allpass filter. The absolute value of the transfer function is a constant one what in turn makes an AGC dispensable. To maintain equivalent amplitude levels in both signal paths an active allpass filter is used to get a phase difference of 90° at the output. But an allpass with a fixed capacitor generates a phase delay of 90° at merely one frequency point. So the capacitor values of the allpass have to be controlled. In our circuit this is done by a varicap. In addition the allpass in the upper signal path only adds a phase offset to the signal, while the relative phase difference between the two signal paths is generated through the allpass in the lower signal path. The phase response of an allpass is

$$\angle \mathrm{H}(jw) = -2\arctan(wRC) \tag{15}$$

The required capacitor value of the varicap C_{var} can be calculated in order to get a phase delay of 90° independent of the frequency.

$$C_{\text{var}} = \frac{-\left(\frac{\tan(\frac{\pi}{4})}{2\pi f R}\right) C_{\text{fix}}}{\left(\frac{\tan(\frac{\pi}{4})}{2\pi f R}\right) - C_{\text{fix}}} \tag{16}$$

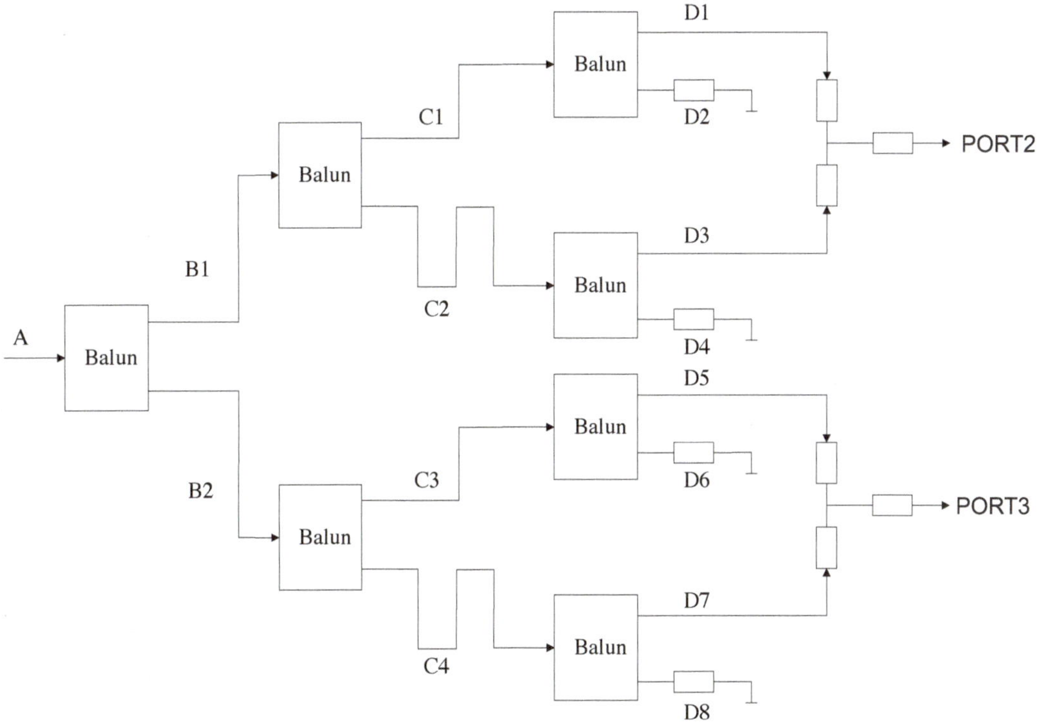

Fig. 8. Phase shifter consisting of transducers and delay lines.

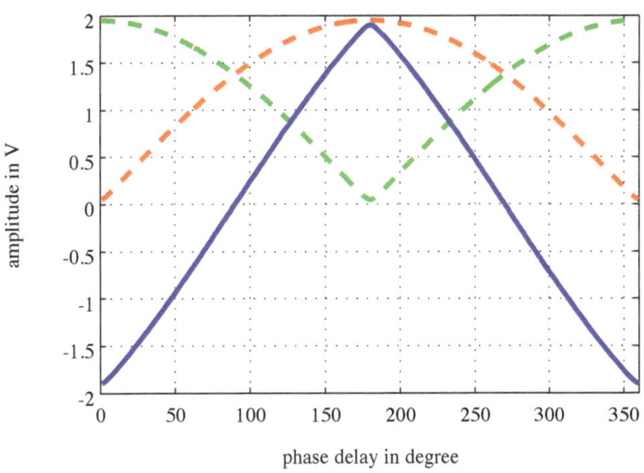

Fig. 9. Simulation of amplitude behaviour at the output of the phase shifter with delay lines, green dashed: PORT2, red dashed: PORT3, blue cont: difference.

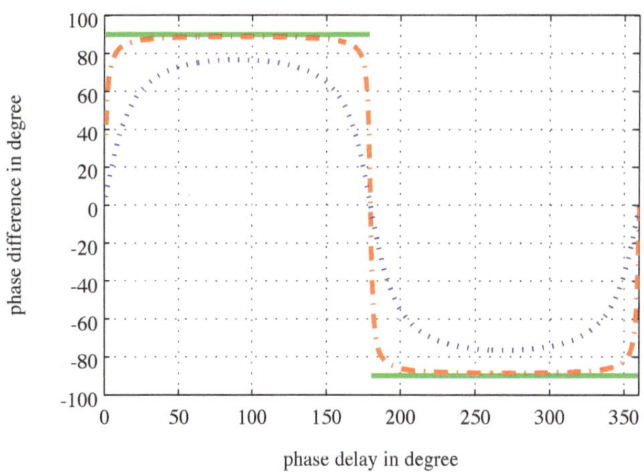

Fig. 10. Influence of attenuation of delay lines on phase difference between output signals, solid line: 0dB, dashed dotted: 0.1 dB, dotted: 1 dB.

To avoid a phase drift a closed loop control is used. The following nonlinear closed loop Eq. (17) adjusts the varicap to get a constant phase shift of 90°.

$$\Delta\varphi(C(\mathrm{U_{var}})) = \Delta\varphi(C(\mathrm{PI}(\mathrm{U_{90^\circ}} - U_{\Delta\varphi}))) \quad (17)$$

where $\mathrm{PI}(\mathrm{U_{90^\circ}} - U_{\Delta\phi})$ and $(C(U_{\mathrm{var}}))$ are the transfer functions of the PI-controller and the active allpass, respectively. The nonlinearity is a reason not to design a stand-alone solution with an allpass. But the transfer function of the allpass could be linearized if the control range of the varicap was smaller. This can be realized by a combination of an RC-CR element and an allpass. In this case the allpass only has to compensate a residual phase shift of about 5°.

Figure 13 shows the large signal schematic of the control loop. The control variable and the output variable is the same one φ_{out1}.

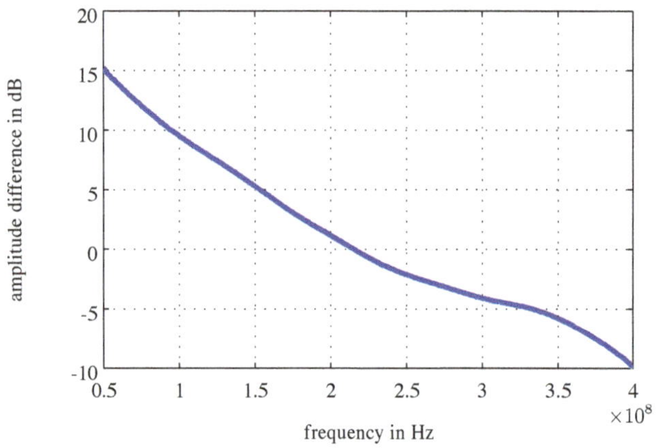

Fig. 11. Measured amplitude difference between the two output signals of the phase shifter with delay lines.

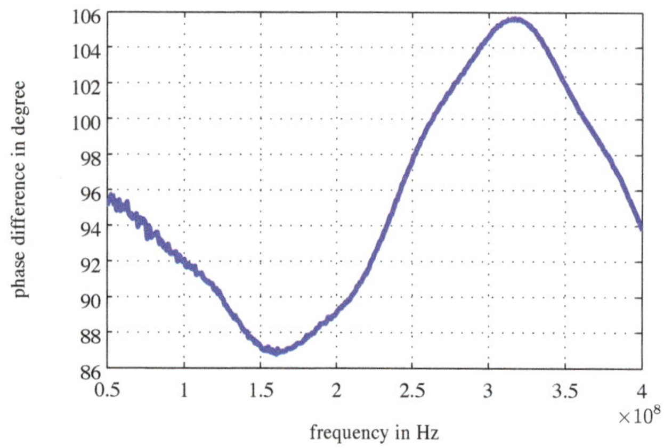

Fig. 12. Measured phase difference between the two output signals of the phase shifter with delay lines.

With an offset voltage U_{off} the operation point is adjusted. The phase delay can be written as

$$\varphi_{out1}(U_{var}, f) = \varphi_{out1}(U_{off} + PI(U_{90^\circ} - U_{\Delta\varphi}, f)). \tag{18}$$

The transfer functions of the allpass and the phase detector have to be linearized. Linearizing the allpass the input signal U_{var} can be written as

$$U_{var} = U_{var_0} + \Delta U_{var}. \tag{19}$$

The allpass transfer function differentiated with respect to ΔU_{var} yields

$$\varphi_{out1}(U_{var}, f) = \varphi_{out1}(U_{var}, f) + \varphi'_{out1}(U_{var}, f) \cdot \Delta U_{var} \tag{20}$$

in the operating point. The phase difference detected by the phase detector is

$$\begin{aligned}\Delta\varphi(U_{var}, f) &= \varphi_{out1}(U_{var}, f) - \varphi_{out2}(U_{var}, f) = \\ &\varphi_{out1}(U_{offset}, f) + \Delta\varphi_{out1}(\Delta U_{var}, f) \\ &-\varphi_{out2}(U_{offset}, f) = \varphi'_{out1}(U_{var}, f) \cdot \Delta U_{var}\end{aligned} \tag{21}$$

Now the input signal of the PI-controller is

$$U_{err}(f) = U_{90} - U_{\Delta\varphi_{off}} - m \cdot \Delta\varphi(f), \tag{22}$$

where $m \cdot \Delta\varphi(f)$ ist the characteristic line of the phase detector. Taking into account the PI-controller the control voltage U_{var} yields

$$U_{var}(f) = \left(K_P + \frac{K_I}{s}\right) \cdot U_{err}(f) + U_{off} \tag{23}$$

where K_I and K_P are the intergrated and the proportional part of the controller, respectively. Finally the transfer function of the control loop.

$$H(s) = \frac{\Delta\varphi(f)}{W(s)} = \frac{-(K_P + \frac{K_I}{s})\varphi'_{out}(U_{off}, f)m}{1 + (K_P + \frac{K_I}{s})\varphi'_{out}(U_{off}, f)m} \tag{24}$$

where $W(s)$ is an interference signal concerning the phase difference. It is used to simulate the change of phase difference between the two channels.

5 Frequency conversion and signal superposition

For upconverting the frequency in both signal paths a mixer (Fig. 1) with a wide output spectrum and a high LO suppression was chosen. The LO consists of a VCO with a PLL. The phase difference of 90° is done by a discrete power splitter and the superposition of the signals at the output by a resistive combiner.

6 Simulation and measurements

6.1 Phase stability

The control loop was simulated in Matlab (Angermann and Beuschel, 2005). A settling time lower than $3\,\mu s$ can be achieved. Figure 14 shows the control behaviour after an applied phase deviation, that is similar to a frequency step of 350 MHz at the input. The time delay to adjust the manipulated phase to 90° can be seen clearly.

6.2 Measurement of side band suppression

The setup for the measurement is as follows: The input signal is splitted by a resistive divider. For the inphase and quadrature signal generation an RC-CR filter is used followed by an AGC with a settling time below $3\,\mu s$. An allpass as described in Sect. 4.3 is resigned but could improve the results. The achievable sideband suppression is shown in Fig. 15.

7 Conclusions

We designed a broadband single side band mixer with a very short settling time. The modular setup allows simple changing. For generating an I/Q signal with high bandwidth three methods have been analyzed. An RC-CR element, a phase shifter with transducers and delay lines and finally an active allpass filter. Each individual method has its advantages but

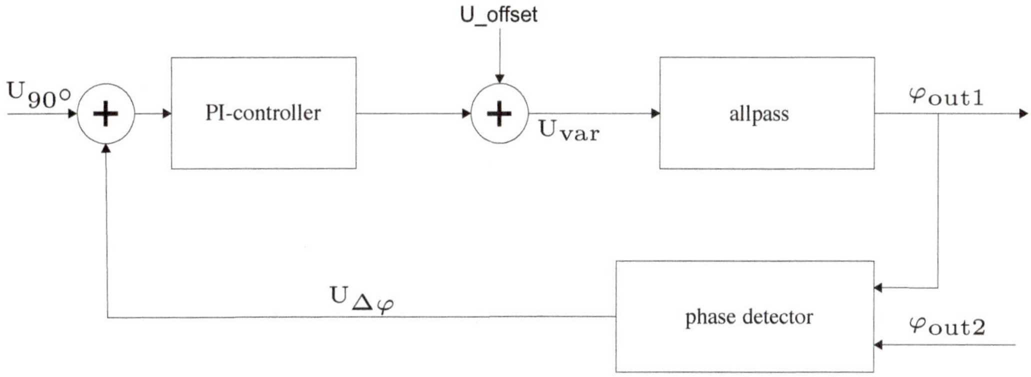

Fig. 13. Control loop of the active phase shifter.

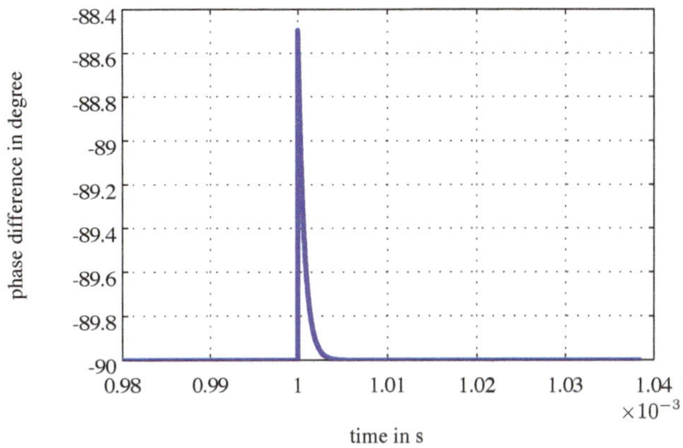

Fig. 14. Matlab/Simulink simulation of the small signal control loop.

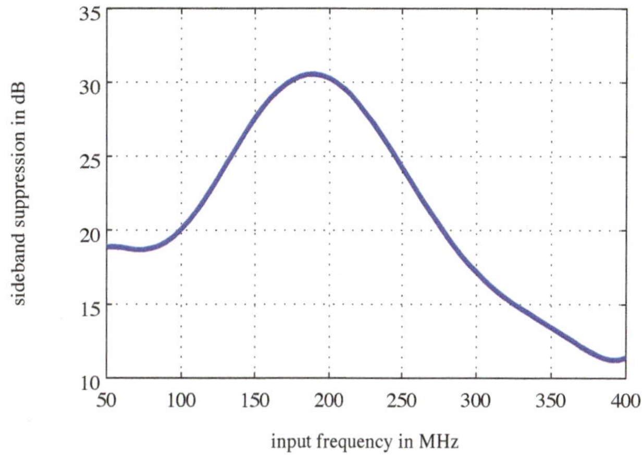

Fig. 15. Sideband suppression with VCO=2.8 GHz.

the RC-CR filter was chosen, since the phase error is best inside the required bandwidth and equalizing the amplitudes up to a difference of 18 dB can be obtained by a fast AGC. In combination with an active allpass filter the phase error could be even improved. The described combination was build up and evaluated. It is shown, that a settling time of less than 3 μs is possible within the bandwidth of 400 MHz. Hence the signal source can be used as synthsizer for stepped frequency radar and could even be used as a hopped frequency signal source.

References

Angermann, A. and Beuschel, M. R. U. W.: Matlab – Simulink-Stateflow, 4th edn., 2005.

Haßler, S.: Entwicklung eines breitbandigen Einseitenbandmischers mit sehr kurzer Einschwingzeit, 2009.

Pozar, D. M.: Microwave Engineering, 2nd edn., 1998.

Non-linear behaviour of charge-pump phase-locked loops

C. Wiegand[1,2]**, C. Hedayat**[1]**, and U. Hilleringmann**[1,2]

[1]Fraunhofer ENAS, Department Advanced System Engineering (ASE), Warburger Str. 100, 33098 Paderborn, Germany
[2]University of Paderborn, Department of Sensor Technology Warburger Str. 100, 33098 Paderborn, Germany

Abstract. The analysis of the mixed analogue and digital structure of charge-pump phase-locked loops (CP-PLL) is a challenge in modelling and simulation. In most cases the system is designed and characterized using its continuous linear model or its discrete linear model neglecting its non-linear switching behaviour. I.e., the time-varying model is approximated by a time-invariant representation using its average dynamics. Depending on what kind of phase detector is used, the scopes of validity of these approximations are different. Here, a preeminent characterization and simulation technique based on the systems event-driven feature is presented, merging the logical and analogue inherent characteristics of the system. In particular, the high-grade non-linear locking process and the dead-zone are analyzed.

1 Introduction

Phase-locked loops (PLL) are widely used and can be utilized to synchronize an oscillator in phase and frequency. If the PLL is absolutely synchronized, the phase error between the input signal and the oscillator signal is minimal. As soon as a phase error occurs, the oscillator will be readjusted until the phase error is again minimal (Gardner, 1980, 1966; Best, 1993; Shu and Sanchez-Sinencio, 2005). Depending on what kind of phase detector (PD) is used, the lock margin sensitively varies (Best, 1993; Wolaver, 1991). If the PLL is stable and the loop bandwidth is wide enough, the locked system has a dynamic, which can be described by an averaged/linearized model. By using a tri state phase and frequency detector (PFD), the PLL is locked if the absolute phase error remains smaller than 2π (Best, 1993; Wolaver, 1991; Den Dulk, 1988). Generally, there are two operating ranges in CP-PLL systems: the linear operating range and the non-linear operating range. The linear operating range is defined by the linear model of the PLL and can be assigned as that range where the PLL is locked. Even if the PLL is locked and all non-idealities (dead-zone, non-linear characteristic of the VCO etc.) are neglected, the PLL possesses a weak non-linear behaviour caused by the logical switching of the PFD. Furthermore, while the PLL is out-of-lock, the system shows a high-grade non-linear behaviour. The non-linear operating range is also called the pull-in range. For this purpose, the PLL is characterized by considering the switching of the PFD. In addition a new lock detection circuitry delivering reliable information about the status of the output clock signal is presented. This is a direct consequence of the characterization of the switching behaviour. The proposed method monitors and detects when the phase error between the reference and the synthesized output clock leaves (out-of lock) or enters (locked) the phase and frequency detector's linear range ($\varphi_e \in [-2\pi, 2\pi]$). For this, the digital information of the PFD monitoring the closed loop is scanned along the transient progress of the system. We will focus here on the case of a tri state PFD. A methodology is presented, which allows to identify and to characterize a CP-PLL by means of its inherent non-linear switching behaviour.

Beside the switching of the PLL, another non-linearity is considered and characterized. The dead-zone due to the PFD and the charge-pump, mostly considered as a static non-linearity embedded into the linear continuous approximation, represents a dynamic non-ideality caused by the delays of the PFD's logic and the charge-pump's slew-rate (Hedayat et al., 1999). An advanced PFD model for accurate simulations of CP-PLL systems is presented.

In Sect. 2.1 the general structure of the CP-PLL is envisaged. Section 2.2 presents the continuous linear model of a second order high-gain phase-locked loop. The non-linear behaviour (without non-idealities) and the linear model are compared and characterized in Sect. 2.3. For this, special

Correspondence to: C. Wiegand
(christopher.wiegand@enas-pb.fraunhofer.de)

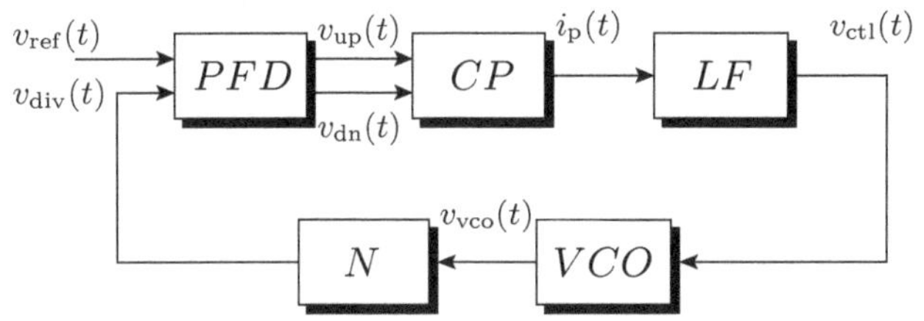

Fig. 1. Schematic representation of a charge-pump phase-locked loop.

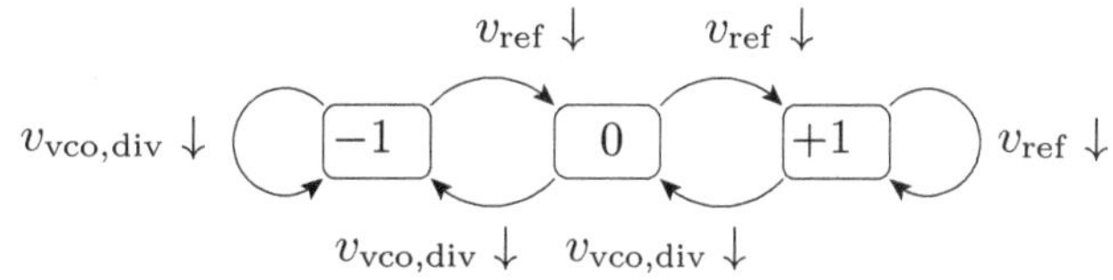

Fig. 2. Finite state machine of the phase and frequency detector driven by the falling edges of the reference signal $v_{ref}(t)$ and the VCO or the divider output signal $v_{vco}(t)$ respectively $v_{div}(t)$.

PFD state sequences are considered. The last subsection of Sect. 2 is dedicated to a particular non-linearity. A behavioural non-linear model of the dead-zone is introduced. In Sect. 3 a direct implication (application) of Sect. 2.3 is presented.

2 Modelling

The PLL offers various models and various non-linearities. In the following, the main focus is pointed upon the non-linear behaviour of the PFD. For this, the switching properties and the linear model are compared. Additionally, an advanced PFD model is introduced.

2.1 Architecture

For simplicity, an integer-N PLL frequency synthesizer architecture is considered (see Fig. 1). The PLL consists of five functional blocks.

- **PFD:** The phase and frequency detector is a purely digital device having two output signals ($v_{up}(t)$ and $v_{dn}(t)$) and is driven by the falling (or rising) edges of its input signals (reference signal $v_{ref}(t)$ and VCO signal $v_{vco}(t)$ [if no divider is used] respectively the divider output signal $v_{div}(t)$) estimating the phase error between $v_{ref}(t)$ and $v_{vco,div}(t)$.
- **CP:** The charge-pump delivers a current $i_p(t)$, driven by the PFD outputs $v_{up}(t)$ and $v_{dn}(t)$.
- **LF:** The loop filter is a low-pass filter, which converts the current i_p into a voltage $v_{ctl}(t)$ filtering the alternating current component.
- **VCO:** The oscillating output $v_{vco}(t)$ of the VCO is controlled by the loop filter's output voltage $v_{ctl}(t)$.
- **N:** The divider is a periodic or a modulo N counter, allowing frequency multiplication $f_{vco} = f_{div} \cdot N$.

2.2 Linearized continuous model

It is assumed, that the system is stable. This is mainly assured when the denominator of the following phase transfer function (second order PLL) is hurwitzian and the reference frequency f_{ref} is notably higher than the loop natural frequency $f_n = \omega_n/(2\pi)$ (Gardner, 1980). In general, it can be said, that if a second order CP-PLL is locked (the absolute phase error $|\varphi_e|$ remains smaller than 2π), the linear continuous approximation can be used to describe the average dynamic behaviour. It holds

$$H(s) = \frac{2\zeta\omega_n s + \omega_n^2}{s^2 + 2\zeta\omega_n s + \omega_n^2}, \tag{1}$$

where

$$\zeta = \frac{R_1 C_1}{2}\omega_n \tag{2}$$

denotes the damping factor and $\omega_n = \sqrt{K/C_1}$ represents the natural frequency, where $K = K_v I_p$ is called the loop gain. Obviously, the linear model can not describe the non-linear pull-in process.

2.3 Non-linear model and characterization

In the following, the non-linear signature of the switched differential equation system caused by the finite state machine is characterized. For this, the VCO is considered as an ideal integrator. Non-idealities like dead-zone and non-linear characteristic of the VCO are neglected. With this assumption, the only non-linearity of the loop is related to the switching behaviour of the digital PFD. The traditional PFD (tri state PFD) is a sequential circuit, driven by the falling or rising edges of its input signals (reference signal v_{ref} and VCO respectively divider signal $v_{vco,div}$, where the subscripts "vco" or "div" denote a PLL without divider respectively with divider). In a more abstract view, the PFD can be represented by a finite state machine consisting of three states. Figure 2 shows the state graph of the PFD, driven by a given type of edges (falling edges or rising edges of v_{ref} and $v_{vco,div}$). The states of the PFD are represented by the logical output signals v_{up} and v_{dn} and can be defined with $\{v_{up} = 0; v_{dn} = 1\} \hat{=} \mathcal{S}_{-1}$, $\{v_{up} = 0; v_{dn} = 0\} \hat{=} \mathcal{S}_0$ and $\{v_{up} = 1; v_{dn} = 0\} \hat{=} \mathcal{S}_{+1}$. It is obvious, that the charge-pump is controlled by these states, i.e. by the PFD output signals v_{up} and v_{dn}. The interrelation between the states of the PFD and the solution of the differential equation of the switching system is presented in Eqs. (3) and (5).

$$\boldsymbol{x}_{n+1} = \boldsymbol{\Phi}(t_n - t_{n+1})\boldsymbol{x}_n +$$

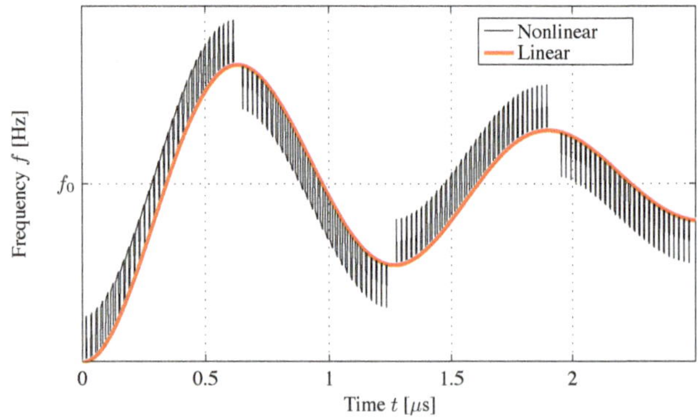

Fig. 3. While the second order PLL delivers a phase error between $-2\pi < \varphi_e < 2\pi$ the PLL is called locked and the linear approximation provides a good average conformity.

$$\int_{t_n}^{t_{n+1}} \boldsymbol{\Phi}(t_n-\tau)\boldsymbol{b}i_p(\mathcal{S}(\tau))\mathrm{d}\tau \tag{3}$$

$\boldsymbol{\Phi}(\cdot)$ represents the matrix exponential of the state matrix, $\mathbf{b}$ is the control vector, the vector $\boldsymbol{x}$ is denoted as the vector of state variables and $i_p(\mathcal{S}(\tau))$ is the charge-pump current controlled by the states $\mathcal{S}(\cdot)$ of the PFD, where $\mathcal{S}(\cdot) \in \{\mathcal{S}_{+1}, \mathcal{S}_0, \mathcal{S}_{-1}\}$ and is controlled by the reference signal $v_{ref}(t)$ and the output signal $v_{vco,div}$. Therefore, $\mathcal{S}(\tau)$ is a function of $v_{ref}(\tau)$ and $v_{vco,div}(\tau)$:

$$\mathcal{S}(\tau) = \mathcal{S}(v_{ref}(\tau), v_{vco,div}(\tau)) \tag{4}$$

Obviously, there are three states modulating the charge-pump current into three constant values. It holds $i_p(\mathcal{S}(\tau)) \in \{+I_p, 0, -I_p\}$. Thus, Eq. (5) results:

$$\boldsymbol{x}_{n+1} = \boldsymbol{\Phi}(t_n - t_{n+1})\boldsymbol{x}_n + \begin{cases} +I_p \int_{t_n}^{t_{n+1}} \boldsymbol{\Phi}(t_n-\tau)\boldsymbol{b}\mathrm{d}\tau & , \mathcal{S}_{+1} \\ 0 & , \mathcal{S}_0 \\ -I_p \int_{t_n}^{t_{n+1}} \boldsymbol{\Phi}(t_n-\tau)\boldsymbol{b}\mathrm{d}\tau & , \mathcal{S}_{-1} \end{cases} \tag{5}$$

By using a first order passive $R-C$ loop filter (Gardner, 1980; Hedayat et al., 1997; Daniels and Farrell, 2008), a second order loop results.

As it can be observed on Fig. 3, when the loop is locked (i.e. the absolute phase error $|\varphi_e|$ remains smaller than 2π) and stable with a reference frequency f_{ref} notably higher than the loop natural frequency, the linearized continuous-time model represents a good approximation of the average behaviour. In addition to the damped sinusoidal dynamics, the non-linear model possesses a comp-shaped behaviour (illustrated in Fig. 3). This comb-shaped structure indicates the state $\mathcal{S}(t)$ of the PFD. The heights of the peaks is correlated to the product of the loop filters resistor R_1 and the charge-pump current $i_p(\mathcal{S}(\tau))$. All the peaks of the comb-shaped structure directed upwards are associated to the logical state $\mathcal{S}_{+1}$. The peaks directed downwards are related to the PFD state $\mathcal{S}_{-1}$. When the linear model matches with the non-linear one, the PFD is at the state $\mathcal{S}_0$. Considering Fig. 3 two state sequences can be identified. The PFD sequence corresponding to a digital overshoot is defined as

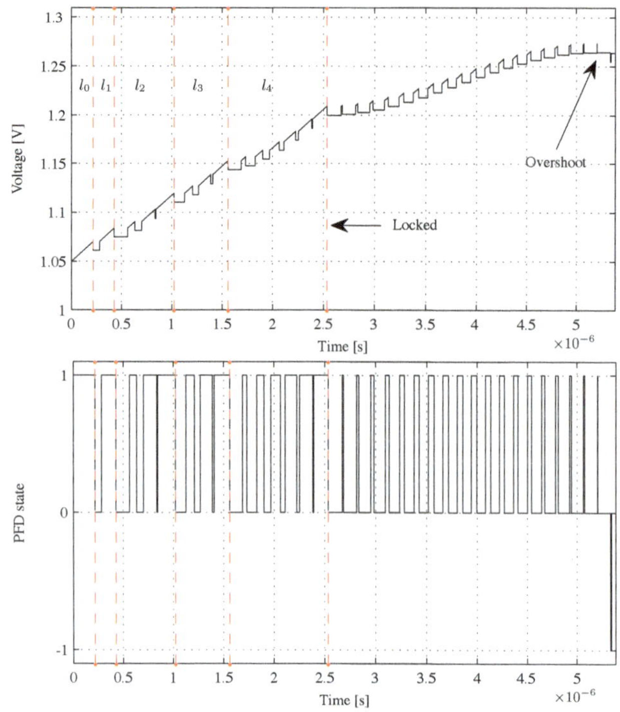

Fig. 4. The pull-in process is divided into five ranges l_i ($i \in \{0,1,2,3,4\}$) and the course of events of the PFD's states $\mathcal{S}$.

$$\mathcal{S}_\uparrow = \{\mathcal{S}_{+1}, \mathcal{S}_0, \mathcal{S}_{-1}\}, \tag{6}$$

where the subscripted down arrow emphasizes the direction and the state sequence of a digital undershoot is defined as

$$\mathcal{S}_\downarrow = \{\mathcal{S}_{-1}, \mathcal{S}_0, \mathcal{S}_{+1}\}. \tag{7}$$

Generally an overshoot or an undershoot occurs if the damping ζ is less than 1.2. In the case of a third order or a high order PLL system (the second order filter consists of the first order passive loop filter with a parallel smoothing capacitor, see Shu and Sanchez-Sinencio, 2005; Hedayat et al., 1999), the state sequences $\mathcal{S}_\downarrow$ and $\mathcal{S}_\uparrow$ do not identify the real frequency f_{vco} (voltage v_{ctl}) overshoot or undershoot. Only zero crossings of the phase error φ_e are detected by these state sequences, because the phase and frequency detector operates on the phase of the reference (v_{ref}) and divider (resp. VCO) output signal ($v_{vco,div}$). For this reason, the logical signature can be used for characterizing the dynamic behaviour of the PLL.

While the PLL is out-of-lock (phase error $|\varphi_e| > 2\pi$), the linear model is no more valid. Therefore the pull-in process has to be analyzed by means of the non-linear pulse width modulated (PWM) model (see Eq. 5). It is obvious, that the

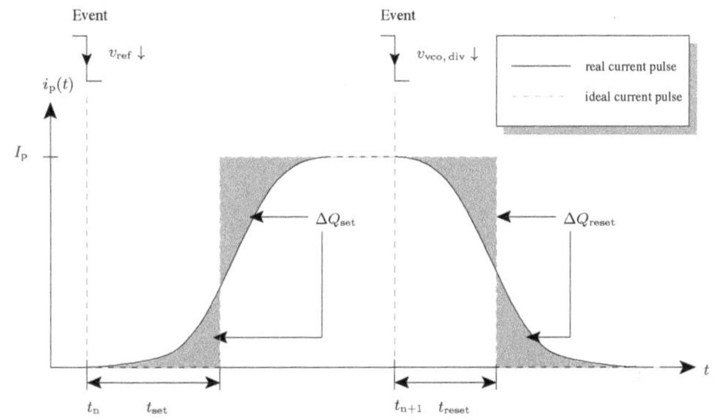

Fig. 5. Charge-pump current with slew-rate and delay propagation from the inputs of the PFD to the output of the charge-pump.

state signal $\mathcal{S}(t)$ is a PWM signal (see Fig. 4). The wider the pulse width is, the bigger is the absolute phase error $|\varphi_e|$ (detected by the PFD). The detectable phase error is limited by the PFD, even if any frequency can be theoretically reached. Thus, a phase error of $|\varphi_e| = 2\pi$ is equivalent to the occurrence of two successive falling edges of the reference signal v_{ref}, respectively the output or divider signal $v_{vco,div}$. Hence, two important state sequences can be identified, where

$$\mathcal{S}_{\not\downarrow} = \{\mathcal{S}_-, \mathcal{S}_-\} \tag{8}$$

and

$$\mathcal{S}_{\not\uparrow} = \{\mathcal{S}_+, \mathcal{S}_+\} \tag{9}$$

are denoted as cycle slips of the down branch ($\mathcal{S}_{\not\downarrow}$) respectively of the up branch ($\mathcal{S}_{\not\uparrow}$). The pull-in process is shown in Fig. 4, where five ranges are emphasized and labeled with l_i ($i \in \{0,1,2,3,4\}$). Evidently, if the last cycle slip occurs, the PLL is locked and the absolute phase error remains less than 2π. By considering the logical outputs of the PLL, the transient behaviour can be monitored, without observing the analogue outputs. Assuming that a cycle slip is identified and an overshoot (undershoot) appears subsequently, the PLL is locked. However, the occurrence of an overshoot or an undershoot does not imply, that the PLL is locked. If there is no knowledge in terms of a detected cycle slip, an overshoot and an undershoot must be identified. To ensure that the PLL is locked, one of the following state sequences must be detected:

$$\mathcal{S}_\circ = \{\mathcal{S}_\downarrow, \mathcal{S}_\uparrow\} \vee \{\mathcal{S}_\uparrow, \mathcal{S}_\downarrow\} \vee \{\mathcal{S}_{\not\downarrow}, \mathcal{S}_\downarrow\} \vee \{\mathcal{S}_{\not\uparrow}, \mathcal{S}_\uparrow\}. \tag{10}$$

2.4 Dead-zone

Ideally, the PFD can be modelled as a finite state machine (see Fig. 2), consisting of three states and operating on the triggering edges (falling or rising edges) of the input signal v_{ref} and the divider output signal v_{div} (or the VCO output signal v_{vco} when the feedback comprises no divider). Every state machine can be realized by using flip-flop circuits.

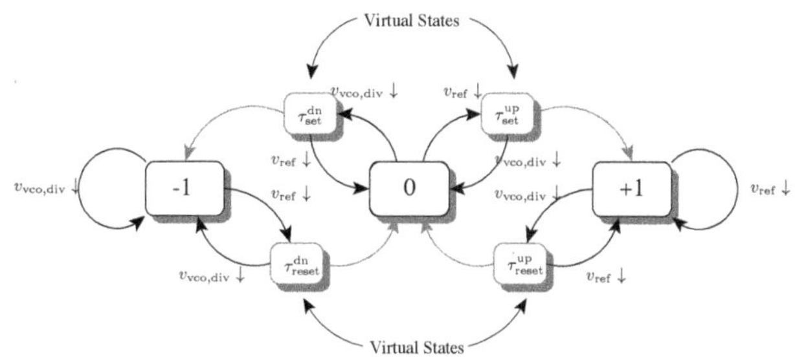

Fig. 6. Modified PFD model describing delay propagation from the inputs to the output.

Thus, the PFD operates by setting and resetting the flip-flops with the incoming signals v_{ref} and v_{div} (v_{vco}). While a triggering edge is identified, the PFD is not able to change the state (the current) instantaneously. Thus, the flip-flop is set by the triggering edge. But, when the phase error is small, the second triggering edge will reset the flip-flop before the signal has been propagated to the output. Correspondingly, the phase error between the incoming signals will not arise at the PFD output (respectively the charge-pump output). Beside the presented delay propagation from the inputs to outputs of the PFD, the waveform cannot offer edges with an infinite slope. Thus, a different charge is injected to the loop filters capacitance compared to the ideal charge injection. A schematic waveform is viewed in Fig. 5, describing the delay propagation and the charge injection. The aspect of the delay propagation can be modelled introducing virtual states within the state machine. If the charges ΔQ_{set} and ΔQ_{reset} are equal to zero, the new state machine (see Fig. 6) can be used for accurate simulations of the non-linear behaviour. If $\Delta Q_{set} \neq 0$ and $\Delta Q_{reset} \neq 0$, the discrepancy of the charge injection to the loop filter's capacitance can be interpreted as the decrease or the increase of the current pulse integration and therefore this can be modelled as a time delay.

The modified PFD model described in Fig. 6 represents a simple possibility to include the non-linear characteristic of the dead-zone into the simulation model of Eq. (3) or into the event-driven modelling techniques (see Hedayat et al., 1997, 1999; Van Paemel, 1994).

3 Application and simulation

Conventional digital lock detection techniques are based on counting methods. These methods are heuristic procedures based on the maximum lock time and the maximum settling time. Because of this, for every application a new counter has to be designed. Additionally, the locking point and settling point in time is increasing if the frequency step applied to the system is increasing. Using an identification algorithm of the presented state sequences (see Eq. 10), a simple purely digital lock and settling detection circuitry can be developed. The circuit is divided into three subcircuits, identifying the

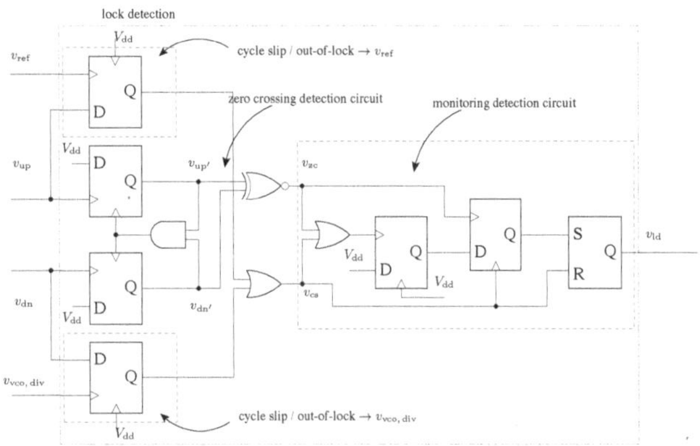

Fig. 7. Schematic representation of the monitoring circuit based on the state sequence identification.

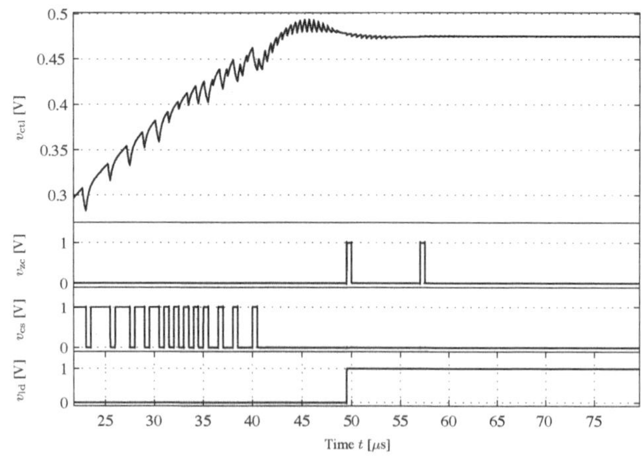

Fig. 8. Course of events of the important lock detection signals and the VCO control voltage v_{ctl} of a third order CP-PLL.

state sequences of Eq. (3). The cycle slip (out-of-lock) detection circuit consists of two D-flip-flops and an OR gate (see Fig. 7). Assuming that the UP signal v_{up} (resp. the DN signal v_{dn}) is true, the occurence of a reference (resp. output) triggering edge makes the $v_{up'}$ (resp. $v_{dn'}$) to be set to one. This situation $\{v_{up} = 1, v_{up'} = 1\}$ (resp. $\{v_{dn} = 1, v_{dn'} = 1\}$) corresponds to a phase error larger than 2π and the PLL is considered as out-of-lock. The zero crossing detection circuit is depicted in Fig. 7. This circuit is like a conventional tri state PFD, where its outputs are combined through an exclusive OR gate. It appears that $v_{up'}$ (resp. $v_{dn'}$) can be understood as the envelope of v_{up} (resp. v_{dn}). More precisely, these envelopes are used as inputs of the exclusive OR gate and thereby make it possible to detect zero crossing of the phase error (digital overshoot or undershoot).

The lock detection monitoring circuit (see Fig. 7) is driven by the outputs of the cycle slip (out-of-lock) detection circuit and the zero crossing detection circuit. This monitoring circuit consists of two D flip-flops, an RS flip-flop and an OR gate and verifies Eq. (3). In Fig. 8 a Cadence/Spectre Simulation with 90nm technology of ST-Microelectronics of a third order CP-PLL is presented. By means of the proposed lock detection circuit, the digital information delivers a simple possibility to track the transient behaviour of the VCO control voltage v_{ctl} (resp. the output frequency f_{vco}) and the dynamics of the system. The control voltage and the important logical outputs of the detection circuitry are depicted in Fig. 8. The lock detection circuit delivers three important signals: phase zero crossing detection v_{zc}, cycle slip detection v_{cs} and lock detection v_{ld}.

4 Conclusions

In this work the non-linear behaviour of a charge-pump phase-locked loop was characterized and modelled. The focus was pointed upon the digital phase and frequency detector, i.e. the switching signature, the dead-zone of the PFD and the charge-pump. An accurate model of the non-linear PFD was presented for efficient simulations of mixed signal systems within event-driven modelling techniques. In addition, a simple and robust lock detection circuit based on monitoring zero crossing (overshoot/undershoot) events and cycle slip identification was presented, by using only the digital information of the system (i.e. the input and output signals of the phase and frequency detector).

Furthermore, the detection circuit can be used to implement fast locking phase-locked loops with course and fine tuning elements. Further work will focus in the optimization and an the further developement of the event-driven and non-linear modelling and simulation techniques for accurate and efficient design methodologies.

References

Best, R.: Theorie und Anwendungen des Phase-locked Loops, vde-Verl.; Aarau/Schweiz: ATVerl., 1993.

Daniels, B. and Farrell, R.: Nonlinear Analysis of the 2nd Order Digital Phase Locked loop, ISSC, Galway, 2008.

Den Dulk, R.: Digital PLL Lock-Detection Circuit, Electronics Letters, 24, 880–882, 1988.

Gardner, F. M.: Phaselock Techniques, John Wiley & Sins, Inc., 1966.

Gardner, F. M.: Charge-Pump Phase-Lock Loops, Communications, IEEE Transactions on, 28, 1849–1858, 1980.

Hedayat, C. D., Hachem, A., Leduc, Y., and Benbassat, G.: High-Level Modeling Applied to the Second-Order Charge-Pump PLL Circuit, Texas Instruments Technical Journal, 14, 1997.

Hedayat, C. D., Hachem, A., Leduc, Y., and Benbassat, G.: Modeling and Characterization of the Third Order Charge-Pump PLL: A Fully Event-Driven Approach, Analog Integrated Circuits and Signal Processing, 19, 24–45, 1999.

Shu, K. and Sanchez-Sinencio, E.: CMOS PLL Synthesizers: Analysis and Design, Springer, 2005.

Van Paemel, M.: Analysis of a Charge-Pump PLL: A New Model, Communications, IEEE Transactions on, 42, 2490–2498, doi:10.1109/26.297861, 1994.

Wolaver, D. H.: Phase-Locked Loop Circuit Design, Advanced Reference Series & Biophysics and Bioengineering Series, Englewood Cliffs, New Jersey 07632: Prentice Hall, 1991.

Performance of combined constellation shaping and bit interleaved coded modulation with iterative decoding (BICM-ID)

T. Arafa, W. Sauer-Greff, and R. Urbansky

University of Kaiserslautern, Communications Engineering, Gottlieb-Daimler-Str., 67653 Kaiserslautern, Germany

Abstract. The increasing demand of achieving high data rates in modern communication systems requires highly efficient bandwidth utilization. For this purpose, multilevel modulation schemes are used in association with forward error correction (FEC) codes in order to approach the channel capacity. However, there is a gap between the capacity of a uniform signal constellation and the Shannon unconstrained capacity. This gap can be reduced by applying constellation shaping. The task of shaping is to modify a uniform distributed signal constellation towards a Gaussian like distribution. In this paper, we investigate different approaches to combine the constellation shaping with a bit interleaved coded modulation with iterative decoding (BICM-ID) system. Simulation results show that this combination can offer a shaping gain up to 0.6 dB.

1 Introduction

Since the capacity of the additive white Gaussian noise (AWGN) channel was found by Shannon, most research in information theory is interested in designing a communication system that operates close to Shannon's limit. The channel capacity is the maximum mutual information between channel input and output. The capacity of the AWGN channel for transmitted signals with Gaussian distribution is called as Shannon limit and it can be expressed with Shannon's formula (Shannon, 1948): $C_{2\text{–dim}} = \log_2(1 + \frac{E_s}{N_o})$ bit/s/Hz. Figure 1 shows the AWGN channel capacities with equiprobable M-QAM, for $M = 4, 16, ..., 1024$, compared with the Shannon limit. We observe from Fig. 1 that in the low SNR regime (<5 dB), the reduction in capacity due to using uniform signal constellation is negligible. Thus, in this case the binary coding techniques can approach the capacity limit. On the other hand, high order modulation schemes like M-QAM ($M > 4$) have to be used for good channel conditions in order to increase the spectral efficiency. However, the use of equiprobable signal constellations leads to a loss of $\pi e/6$ (1.53 dB) due to using a uniform distribution rather than a Gaussian distribution over the signal set (Forney and Ungerboeck, 1998). Therefore, the use of powerful channel coding with uniform constellation signal is not enough to achieve the capacity limit. The constellation shaping technique that produces a Gaussian-like distribution over constellation points has also to be used to close the gap of 1.53 dB.

Higher level modulation schemes such as M-QAM can be used to achieve spectral efficient high data rate transmission over wireless, wireline, and optical links without increasing bandwidth. However, an increased density of the symbol constellation decreases its robustness against noise, which results in a high BER of the demapper. Therefore, the use of high order modulation schemes strongly requires channel coding to guarantee low BER at high bandwidth efficiency. Since an erroneous detected symbol may result in a burst of bit errors and channel codes are often sensitive to burst error, a permutation of the bit sequence has to be introduced between coding and mapping, so-called interleaving. The concatenation of channel coding, interleaving and modulation is denoted as a bit interleaved coded modulation (BICM). Iteratively demapping and decoding BICM (BICM-ID) can improve the performance by exchanging the soft information between soft-demapper and channel decoder.

Constellation shaping is a method to shape an M-QAM signal constellation to follow the Gaussian distribution. Actually, two techniques for constellation shaping have been mainly proposed in the literature. The first approach is to transmit the constellation points with unequal probability (Fischer, 2002; Khoo et al., 2006). The other is based on optimizing of the constellation points, i.e. using equiprobable signal constellations with unequal spacing (Sommer and Fettweis, 2000; Fischer, 2002; Djordjevic et al., 2010).

Correspondence to: T. Arafa
(arafa@eit.uni-kl.de)

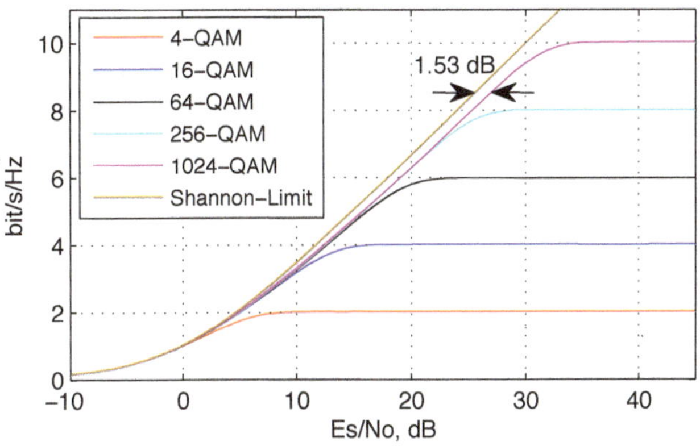

Fig. 1. The ideal AWGN channel capacities of equiprobable M-QAM signals compared with that of Gaussian signal.

The remainder of this paper is organized as follows. The concept of BICM-ID is briefly introduced in Sect. 2. In Sect. 3, the constellation shaping with different approaches including unequal probability and unequal spacing constellation points are presented. Since all BICM-ID systems suffer from an error floor, the inner precoding that can remove the error floor without adding extra redundancy bits is applied to the system in Sect. 4. Then, the chain of FEC coding, inner precoding and mapping will be optimized by means of extrinsic information transfer (EXIT) chart analysis. Simulation results and discussions are provided in Sect. 5. Finally, conclusions are drawn in Sect. 6.

2 BICM-ID concept

It is well accepted that the concatenation of code systems with iterative decoding can approach the channel capacity. The BICM-ID scheme is a special case of serial concatenated code systems where the turbo principle is applied between the inner decoder, i.e. the soft-demapper and the outer channel decoder (ten Brink et al., 1998).

The block diagram of the BICM-ID scheme is depicted in Fig. 2. At the transmitter, the sequence of k data bits is firstly encoded into a codeword of n bits by a recursive systematic convolutional (RSC) encoder where the code rate is $R=\frac{k}{n}$, which can be obtained, e.g., by puncturing. The encoded bits are randomly interleaved by a bit-wise interleaver. After interleaving, a group of m coded bits is mapped onto one complex symbol of $M=2^m$ according to a mapping scheme. Therefore, the spectral efficiency of this system is given by

$$\eta = R \cdot m \qquad \text{bit/s/Hz}. \tag{1}$$

The originated complex symbols are corrupted by noise through the AWGN channel, so the received symbols are $\underline{r}=\underline{y}+\underline{n}$ where $\underline{n}$ is a complex circular symmetric Gaussian noise with zero mean and variance $N_o/2$ in each dimension. At the receiver, and based on channel observations $\underline{r}$

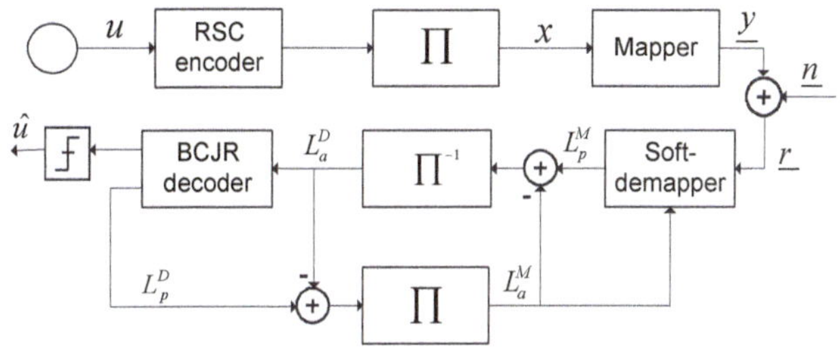

Fig. 2. BICM-ID system model.

and a-priori information of unmapped bits L_{a}^{M} (at first iteration $L_{\mathrm{a}}^{M}=0$), the soft-demapper calculates the a-posteriori log-likelihood ratio (LLR) for each of the m coded bits per symbol by using (ten Brink et al., 1998)

$$L_p^M(x_k|\underline{r}) = L_a^M(x_k) + \ln \frac{\sum\limits_{\underline{y}\in\chi_k^1} \exp[-\frac{|\underline{r}-\underline{y}|^2}{N_o} + \sum\limits_{j=1,j\neq k}^{m} \mu_j^{-1}(\underline{y}) L_{\mathrm{a}}^M(x_j)]}{\sum\limits_{\underline{y}\in\chi_k^0} \exp[-\frac{|\underline{r}-\underline{y}|^2}{N_o} + \sum\limits_{j=1,j\neq k}^{m} \mu_j^{-1}(\underline{y}) L_{\mathrm{a}}^M(x_j)]}, \tag{2}$$

where $\mu_j^{-1}(\underline{y})$ gives the j-th bit of the symbol $\underline{y}$ while χ_k^1, χ_k^0 is the set of symbols having the k-th bit set to 1 and 0, respectively. Then, the extrinsic information of the soft-demapper $L_{\mathrm{e}}^M = L_{\mathrm{p}}^M - L_{\mathrm{a}}^M$ is deinterleaved to become the a-priori input L_{a}^D to the outer soft-in/soft-out decoder, which gives the a-posteriori value L_{p}^D based on the BCJR algorithm (Hagenauer, 1997). The extrinsic information of the outer decoder $L_{\mathrm{e}}^D = L_{\mathrm{p}}^D - L_{\mathrm{a}}^D$ is passed through the interleaver and fed back as a-priori information L_{a}^M to the soft-demapper for the next iteration.

As we have seen, only extrinsic information is exchanged between inner soft-demapper and outer decoder at the iteratively decoded BICM system. Therefore, the performance of this system can be analyzed and predicted by using the EXIT chart (ten Brink, 1999). It is an efficient tool to visualize the convergence behaviour and the BER floor of a concatenated coding system. Moreover, it graphically shows the relationship between the input mutual information against the output mutual information from the soft-demapper and decoder which help us to select the coding and mapping schemes that match each other.

In this context, different EXIT characteristics of mapping can be simply constructed by an irregular modulation (modulation doping) that applies Gray as well as anti-Gray mapping according to the doping ratio

$$\alpha = \frac{N_{\text{anti-Gray}}}{N_{\text{anti-Gray}} + N_{\text{Gray}}} \tag{3}$$

within the same transmitted block (Szczecinski et al., 2005).

3 Constellation shaping

Constellation shaping is used to obtain a shaping gain over uniform signaling. The ultimate shaping gain for high data rates is 1.53 dB. Essentially, two approaches have been proposed for signal shaping. The first approach is to shape the signal constellation to have a non-uniform probability distribution. With the other approach, the signal constellation is non-uniformly spaced.

3.1 Shaping with non-uniform probability distribution over constellation points

The basic idea of this approach is to transmit the symbols with unequal probability. In this case, the points with low energy, i.e. the points near the centre of the constellation are more frequently transmitted than the points with high energy. For example, Khoo et al. (2006) proposed to divide the 16-QAM constellation into three sub-constellations with different power levels. The first sub-constellation contains four points with lowest energy, the eight points with medium energy forms the second sub-constellation and the third has four points with highest energy. Although every four bits are mapped onto one symbol, the first two bits can also be used to select one of these sub-constellations as follows: if we have 00 the first sub-constellation is chosen, 01 or 10 leads to the second and the third is selected for incoming bits 11. As we can see, we can obtain unequal probability over symbols if the first two bits fulfill this requirement $P_r(0) > P_r(1)$. This can be achieved by modifying the basic BICM-ID model as depicted in Fig. 3. The encoded sequence by a rate-R_c RSC encoder is divided into four sequences by a serial-to-parallel (S/P) converter. Each sequence of the first and second composed K bits is interleaved to $m_{1,2}$ and then broken into L successive K_s-bit blocks, where $K = L \cdot K_s$. These L successive blocks are serially fed to a rate-R_s shaping encoder (S-ENC) to generates L successive N_s-bit blocks, which result in N-bit sequences $s_{1,2}$, where $R_s = \frac{K_s}{N_s}$. The shaping encoder is designed so that the probability of a zero at the sequences $s_{1,2}$ is maximized. Then, these two N-bit sequences are interleaved to $x_{1,2}$ and passed to the mapper. Each of the third and fourth sequence already includes N bits and they are also interleaved to $x_{3,4}$ and passed to mapper. Thus, there are four N-bit sequences at the input of the mapper to generate N symbols, which fulfill the unequal probability distribution. This transmission system is defined with spectral efficiency

$$\eta = 2R_c(R_s + 1) \qquad \text{bit/s/Hz.} \tag{4}$$

At the receiver, the third and fourth sequences are directly processed between soft-demapper and channel decoder. A shaping decoder has to be used for the first and second sequences. It accepts L blocks of a-priori information for info bits $L_a^S(m_1(k), m_2(k))$, $k = 1, ..., K_s$ and for code bits $L_a^S(s_1(n), s_2(n))$, $n = 1, ..., N_s$ which are generated from the

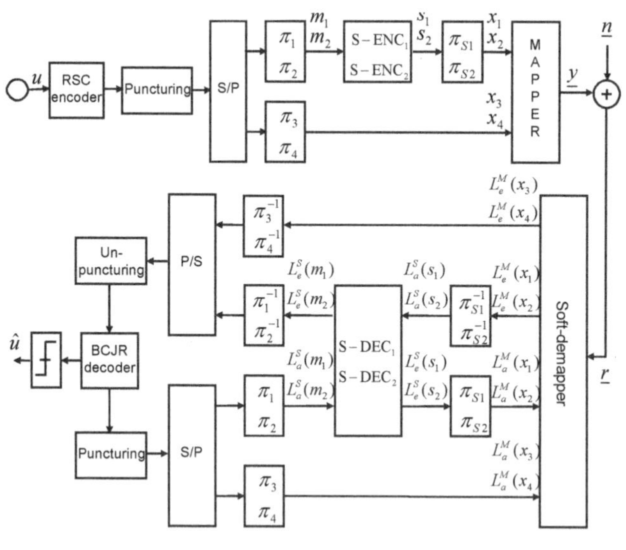

Fig. 3. BICM-ID system to shape 16-QAM with a shaping coding.

RSC decoder and soft-demapper, respectively, and calculates using these values the L blocks of extrinsic information for info bits $L_e^S(m_1(k), m_2(k))$ and for code bits $L_e^S(s_1(n), s_2(n))$ which are passed to the RSC decoder and soft-demapper, respectively. Thus, the shaping decoder calculates the extrinsic information of both info bits and code bits based on the maximum a-posteriori algorithm (Khoo et al., 2006)

$$L_e^S(m(k)) = \ln \frac{\sum\limits_{M \in \Omega_k^1} \exp(A)}{\sum\limits_{M \in \Omega_k^0} \exp(A)}, \tag{5}$$

where

$$A = \sum_{l=1, l \neq k}^{K_s} t_l \cdot L_a^S(m(l)) + \sum_{n=1}^{N_s} t_n \cdot (L_a^S(s(n)) - L_n), \tag{6}$$

Ω_k^1, Ω_k^0 is the set of all possible blocks M of info having the k-th bit set to 1 and 0, respectively, t_l is the value of the l-th bit in the block M under consideration, t_n is the value of the n-th bit in the codeword S associated to this particular block, $L_n = \ln \frac{P_r\{s(n)=0\}}{P_r\{s(n)=1\}}$ is the a-priori knowledge of bit $s(n)$ before the first iteration starts. Analogously

$$L_e^S(s(n)) = \ln \frac{\sum\limits_{S \in \Delta_n^1} \exp(B)}{\sum\limits_{S \in \Delta_n^0} \exp(B)}, \tag{7}$$

where

$$B = \sum_{k=1}^{K_s} t_k \cdot L_a^S(m(k)) + \sum_{l=1, l \neq n}^{N_s} t_l \cdot (L_a^S(s(l)) - L_l), \tag{8}$$

Δ_n^1, Δ_n^0 is the set of all possible codewords S having the n-th bit set to 1 and 0, respectively, t_l is the value of the l-th bit

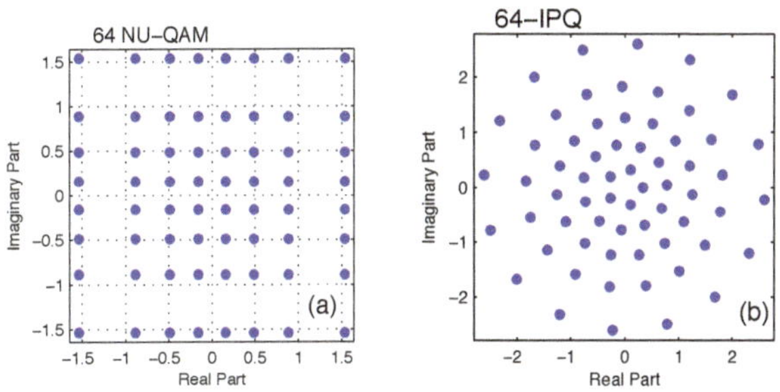

Fig. 4. Signal constellations of **(a)** 64-NU-QAM and **(b)** 64-IPQ.

in the codeword S under consideration, and t_k is the value of the k-th bit in the block M associated to this particular codeword.

3.2 Shaping with non-uniform spacing of constellation points

It was proved in Sommer and Fettweis (2000), Fischer (2002), and Djordjevic et al. (2010) that by using equally likely signal points with unequal spacing the capacity limit can also be approached. In this method, the shaping gain can be obtained without any additional computational effort.

For optimizing the rectangular M-QAM constellation points, each dimension (i.e. N-ASK with $N = \sqrt{M}$) is optimized as follows. The probability of each N-ASK point is $1/N$, but the coordinates x_i, $i = 1,2,...,N$, of the points are chosen appropriately. Therefore, the Gaussian cumulative distribution function (CDF), which has value between zero and one, is partitioned into N equal parts with $N+1$ boundaries $\frac{i}{N}$, $i = 0,1,...,N$. The coordinates x_i of the signal points are then given as the centroids of these parts $\frac{\frac{i}{N}+\frac{i-1}{N}}{2} = \frac{2i-1}{2N}$, namely,

$$x_i = \mathrm{CDF}^{-1}\left(\frac{2i-1}{2N}\right) = \sqrt{2}\,\mathrm{erf}^{-1}\left(\frac{2i-1-N}{N}\right), \tag{9}$$

where $i = 1,2,...,N$. For example, Fig. 4a shows the non-uniform QAM signal constellation with 64 points (64-NU-QAM) while the channel capacity for this constellation is depicted in Fig. 5. We can see that the 64-NU-QAM constellation is close to the optimum and outperforms the classical 64-QAM if the capacity is less than 5 bit/s/Hz.

On the other hand, the two dimensions of the circular QAM constellation can be jointly optimized. The optimized complex symbols have to follow a complex Gaussian distribution. The amplitude and phase of a complex Gaussian random variable are independent and conform a Rayleigh and a uniform distribution, respectively. By using this idea, the optimal circular signal constellation is generated based on an iterative polar quantization IPQ procedure (Djordjevic et al., 2010). In IPQ, the constellation points are distributed over the circles of radius determined by the Rayleigh distribution. Therefore, the circular M-QAM constellation is optimized

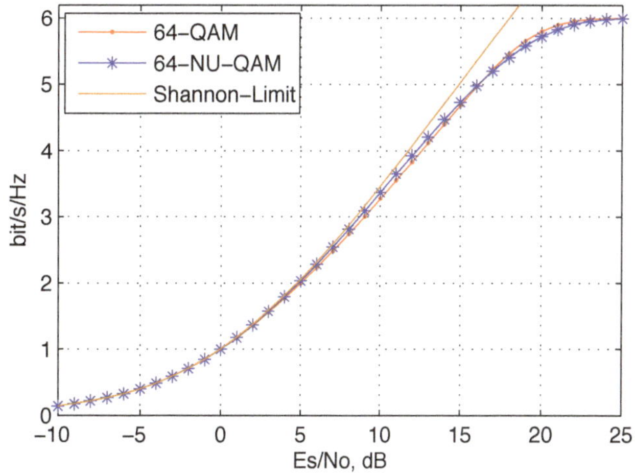

Fig. 5. Channel capacity of 64-NU-QAM versus 64-QAM.

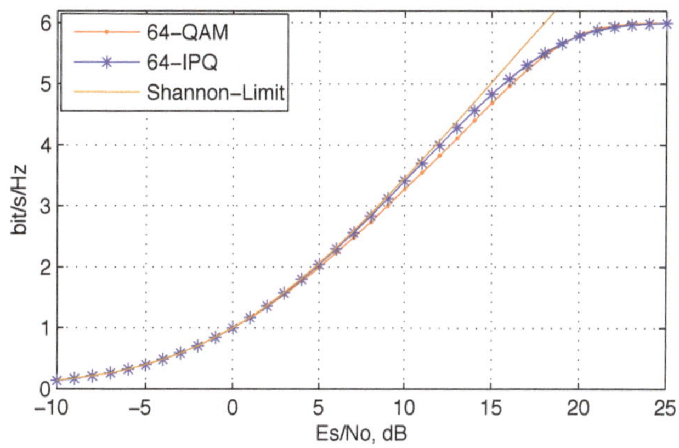

Fig. 6. Channel capacity of 64-IPQ versus 64-QAM.

to get an M-IPQ constellation based on the minimum mean square quantization error (Djordjevic et al., 2010). The optimum number of points L_i per i-th circle with radius m_i is determined by (Djordjevic et al., 2010)

$$L_i = \frac{\sqrt[3]{m_i^2 \int_{r_i}^{r_{i+1}} p(r)\mathrm{d}r}}{\sum_{i=2}^{L_r} \frac{1}{M}\sqrt[3]{m_i^2 \int_{r_i}^{r_{i+1}} p(r)\mathrm{d}r}}, \qquad i = 1,2,...,L_r, \tag{10}$$

where L_r is the number of circles in the constellation and $p(r)$ is the Rayleigh distribution function. The radius of the i-th circle is determined by

$$m_i = \frac{2\sin(\Delta\theta_i/2)\int_{r_i}^{r_{i+1}} r p(r)\mathrm{d}r}{\Delta\theta_i \int_{r_i}^{r_{i+1}} p(r)\mathrm{d}r}, \qquad \Delta\theta_i = \frac{2\pi}{L_i}. \tag{11}$$

The limits of integration in Eqs. (10) and (11) are determined by

$$r_i = \frac{\pi(m_i^2 - m_{i-1}^2)}{2[m_i L_i \sin(\Delta\theta_i/2) - m_{i-1} L_{i-1} \sin(\Delta\theta_{i-1}/2)]}. \tag{12}$$

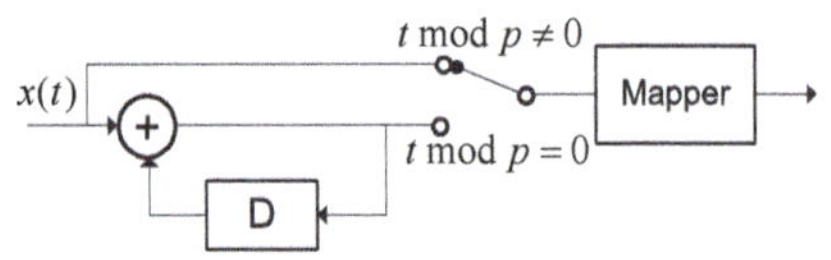

Fig. 7. Additional inner rate-one RSC code with doping.

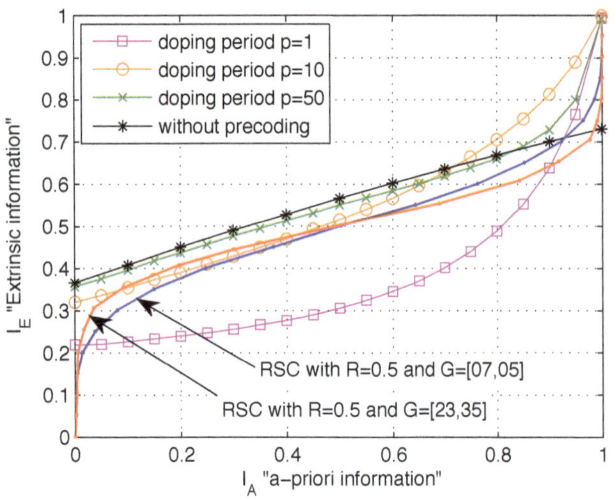

Fig. 8. EXIT chart of 64-IPQ mapping with different doping periods at $E_b/N_o = 5.25$ dB.

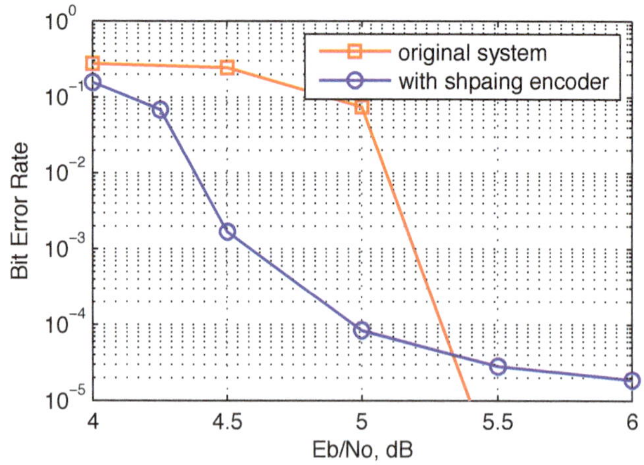

Fig. 9. Performance of BICM-ID with and without shaping encoder after 10 iterations.

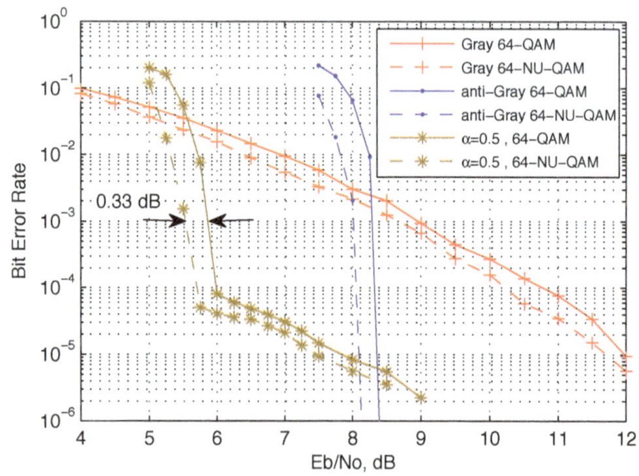

Fig. 10. BER of BICM-ID with different 64-QAM mapping schemes after 14 iterations.

In Fig. 4b we depict the 64-IPQ constellation. The corresponding channel capacity is shown in Fig. 6. As we can observe from this figure, the 64-IPQ constellation is very close to the Shannon limit and outperforms significantly the classical 64-QAM for low and medium SNRs.

4 Error floor problem at BICM-ID systems

In general, all BICM-ID systems suffer highly from an error floor. The reason of this phenomenon can be explained using an EXIT chart. The EXIT-characteristic of soft-demappers mostly does not reach to the point (1,1), i.e. the intersection between the EXIT curves of soft-demapper and decoder is at $I_e^M < 1$ for $I_a^M \approx 1$. This means that even for ideal a-priori information, the obtained extrinsic information is not perfect and some bit errors remain.

This problem can be solved if the tail of the soft-demapper curve (for $I_a^M > 0.5$) is bent up to reach point (1,1) without lowering it for $I_a^M < 0.5$. This can be achieved by using a doping precoder (Pfletschinger and Sanzi, 2004). By this approach, a rate-1 RSC is inserted between interleaver and mapper as depicted in Fig. 7. The output of the inner encoder is doped where every p-th info bit is replaced by a coded bit and thus the rate is maintained. This scheme is defined with a doping period p, which has to be chosen rather high to achieve the mentioned requirement as shown in Fig. 8.

5 Simulation results

Figure 9 shows the BER performance of the BICM-ID systems applying an anti-Gray 16-QAM mapping scheme without and with a shaping encoder at spectral efficiency $\eta = 2$ bit/s/Hz. To ensure that both schemes operate at the same spectral efficiency, we used a rate 1/2 RSC code with generator polynomials in octal form $G = [23, 35]$ for the original system and a rate 2/3 punctured RSC code with a rate 1/2 shaping encoder for the system with shaping coding. The shaping encoder receives the info bits blockwise with $K_s = 7$ and generates the corresponding codeword with $N_s = 14$ which results in moderate complexity. As can be seen from Fig. 9, the system with a shaping encoder is 0.61 dB better than the original system at BER $= 10^{-3}$. However, this scheme shows an error floor occurring at BER $\approx 10^{-5}$ due to the relatively poor error capabilities of a punctured rate 2/3 RSC code.

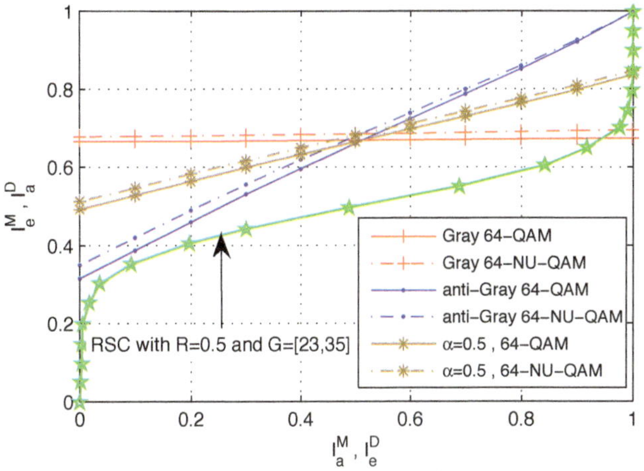

Fig. 11. EXIT chart of different 64-QAM mapping schemes at $E_b/N_o = 8\,\text{dB}$.

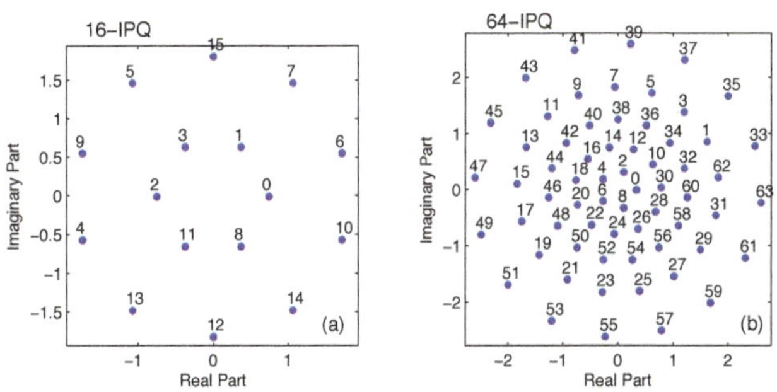

Fig. 12. Proposed mapping schemes for constellations of **(a)** 16-IPQ; **(b)** 64-IPQ.

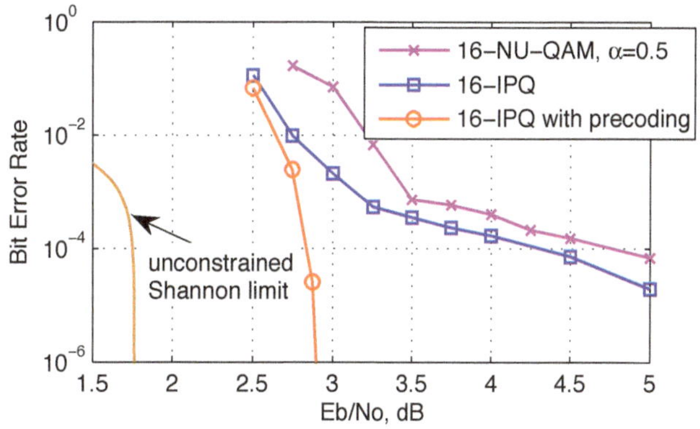

Fig. 13. BER of BICM-ID without and with precoding for 16-IPQ mapping after 25 iterations.

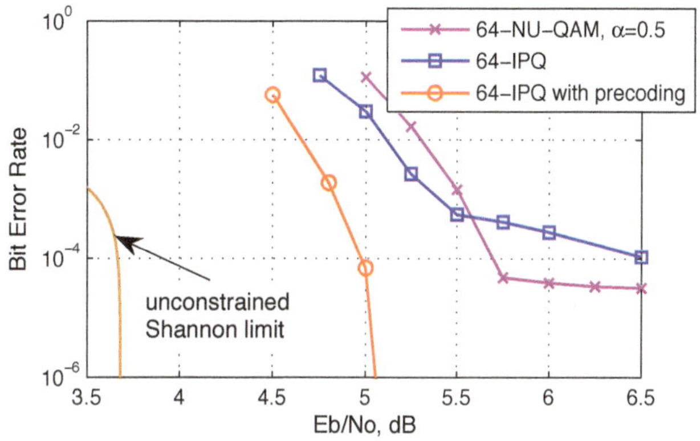

Fig. 14. BER of BICM-ID without and with precoding for 64-IPQ mapping after 25 iterations.

The performance of a BICM-ID system for both uniform (64-QAM) and non-uniform (64-UN-QAM) spacing of constellation points with different mapping schemes, Gray, anti-Gray and modulation doping ($\alpha = 0.5$) is depicted in Fig. 10. The corresponding EXIT chart is shown in Fig. 11. The simulation results show that the optimized rectangular constellation 64-NU-QAM offers an improvement of 0.33 dB for all mapping schemes. On the other hand, the modulation doping ($\alpha = 0.5$) combines the benefits of both Gray and anti-Gray mapping and outperforms the both, where it reaches to the turbo cliff at low SNR but suffer then from error floor as explained in Sect. 4. These results match to the EXIT chart where the soft-demapper curve is slightly shifted towards the top by using 64-NU-QAM for Gray, anti-Gray and modulation doping which means an increase in the capacity.

Finally, the performance of BICM-ID systems with 16-IPQ and 64-IPQ mapping are evaluated. Since constructing Gray and anti-Gray mapping schemes for arbitrary M-IPQ constellations is not easy, we try to construct the mapping schemes for 16-IPQ and 64-IPQ so that the proposed schemes match the outer RSC coding with a memory size of 4. The proposed mapping schemes for 16-IPQ and 64-IPQ are depicted in Fig. 12. The use of 16-IPQ and 64-IPQ offers an improvement of 0.32 dB and 0.12 dB at BER=10^{-3} compared with 16-NU-QAM and 64-NU-QAM with doping ratio $\alpha = 0.5$, respectively.

On the other hand, 16-IPQ and 64-IPQ mapping schemes, as expected, suffer from an error floor. It can be removed by using precoding with a doping period $p = 50$. Moreover, by using precoding with selecting a memory-2 RSC encoder for matching the coding transfer curve to the new mapper curve with precoding as shown in Fig. 8, the performance can be improved as depicted in Fig. 13 and Fig. 14. The performance is 1.09 dB and 1.27 dB away from the unconstrained Shannon limit at BER $= 10^{-3}$ for 2 and 3 bit/s/Hz, respectively, with moderate complexity.

6 Conclusions

In this contribution we presented the combined constellation shaping and BICM-ID system. A shaping gain of 0.6 dB can be obtained with unequal probability over constellation points by adding a shaping encoder which results in high complexity compared to the original system. On the other hand, the optimized unequal spacing of points for both rectangular and circular constellations achieves approximately the capacity and can be applied to a BICM-ID system without any increase in complexity.

References

Djordjevic, I. B., Batshon, H. G., Xu, L., and Wang, T.: Coded Polarization-Multiplexed Iterative Polar Modulation (PM-IPM) for Beyond 400 Gb/s Serial Optical Transmission, Optical Fiber Communication Conference (OFC), San Diego, California, USA, 21–25 March 2010, OMK2, 2010.

Fischer, R. F. H.: Precoding and Signal Shaping for Digital Transmission, John Wiley & Sons Inc., New York, USA, 2002.

Forney Jr., G. D. and Ungerboeck, G.: Modulation and Coding for Linear Gaussian Channels, IEEE Transactions Information Theory, 44, 2384–2415, 1998.

Hagenauer, J.: The Turbo Principle: Tutorial Introduction and State of the Arts, Symposium on Turbo Codes, Brest, France, September 1997, 1–11, 1997.

Khoo, B. K., Le Goff, S. Y., Sharif, B. S., and Tsimenidis, C. C.: Bit-Interleaved Coded Modulation with Iterative Decoding Using Constellation Shaping, IEEE Transactions on Communications, 54, 1517–1520, 2006.

Pfletschinger, S. and Sanzi, F.: Iterative Demapping for OFDM with Zero-Padding or Cyclic Prefix, IEEE International Conference on Communications, Paris, France, 20–24 June 2004, 842–846, 2004.

Shannon, C. E.: A Mathematical Theory of Communication, Bell System Technical Journal, 27, 379–423 and 623–656, 1948.

Sommer, D. and Fettweis, G. P.: Signal Shaping by Non-Uniform QAM for AWGN Channels and Applications Using Turbo Coding, International ITG Conference on Source and Channel Coding, Munich, Germany, 17–19 January 2000, 81–86, 2000.

Szczecinski, L., Chafnaji, H., and Hermosilla, C.: Modulation Doping for Iterative Demapping of Bit-Interleaved Coded Modulation, IEEE Communications Letters, 9, 1031–1033, 2005.

ten Brink, S.: Convergence of Iterative Decoding, IEEE Electronics Letters, 35, 1117–1119, 1999.

ten Brink, S., Speidel, J., and Yan, R.-H.: Iterative Demapping and Decoding for Multilevel Modulation, IEEE International Conference on Global Communications (GLOBECOM), Sydney, Australia, 8–12 November 1998, 579–584, 1998.

Channel equalization using watermark as a training sequence

M. K. Samee and J. Götze

Information Processing Lab, Dortmund University of Technology, Otto-Hahn-Str. 4, 44227 Dortmund, Germany

Abstract. In this paper blind equalization algorithms based on digital watermarking are presented. Training sequence is sent along with the data in the form of watermark. Robust CDMA based watermarking algorithm is used to watermark the data. Watermark, which is spread over the data, does not consume extra bandwidth. So with the help of extracted watermark channel can be equalized just by using simple mean square algorithm. Two algorithms for equalization are proposed in this paper. In algorithm I the receiver does not require to know training sequence in advance. Using algorithm II, the sender does not require to send training sequence either, and data can still be equalized using simple LMS algorithm at low computational complexity.

1 Introduction

Intersymbol Interference (ISI) is a very common type of distortion which a signal undergoes during transmission (Proakis, 2000). Trained equalization is a technique widely used to cope with ISI. A training sequence is sent over the channel and with the help of received and already known training sequence channel is equalized. Traditional trained equalization scheme works with simple algorithms like Least Mean Squares (LMS) (Haykin, 1996). Most crucial drawback of traditional trained equalization is that it consumes extra bandwidth. To save the bandwidth, some blind equalization techniques, which do not require training sequences, are used to equalize channels. These blind equalization schemes are usually based on computationally complex algorithms like Constant Modulus Algorithm (CMA) (Litwin, 1999). So the cost of saving bandwidth is increased computational complexity.

Correspondence to: M. K. Samee
(kashif.samee@tu-dortmund.de)

Digital Watermarking is a well known technique used for authentication, copyrights protection and applications like avoiding illicit copying etc. Watermark is simply a sequence of some bits which is hidden in host data (e.g. videos, images and someother multimedia data). If some meaningful watermark is embedded in host data watermarking can be used for some other purposes as well. Tracing watermarks are used for blind quality assessment for multimedia communication in (Campisi, 2003). A multipurpose public-key cryptosystem based on image watermarking is developed in (Ding, 2008). End-to-end QoS provision and control in wireless communication by means of digital watermarking is presented in (Benedetto, 2006). Digital watermarking is used for dynamic and adaptive equalization in (Maity, 2007; Neto, 2006) respectively. However in (Maity, 2007; Neto, 2006) receiver should know the watermark (training sequence) in advance, which is not required in proposed scheme.

In comparison to blind equalization, traditional trained equalization is known to be computationally efficient. But the most crucial drawback of trained equalization is that it consumes extra bandwidth. Another disadvantage is that the receiver must have prior information of the original training sequence. There are some blind equalization algorithms, where training sequences are not used and data can still be equalized (Litwin, 1999; Zarzoso, 2005). To design a blind equalization algorithm convergence of the algorithm and computational complexity are two important issues. Constant Modulus Algorithm (CMA) (Litwin, 1999) is one of well known algorithm used for blind equalization. Disadvantage for CMA is its slow convergence. Therefore, researchers tried to lessen the computational complexity of CMA and tried to make convergence faster (Sun, 2007; Dabeer, 2003).

In this work, a completely different approach to blind equalization is taken by using digital watermarking. The proposed method does not require training sequences to be known in advance and works with standard trained equalization algorithms (e.g. LMS) at low computational complexity.

Furthermore, proposed scheme can simultaneously be used for blind equalization and usual watermarking applications (Podilchuk, 2001).

Proposed schemes can be looked from two different viewpoints. First, how watermarking can be used for blind equalization (the receiver is not required to know the training sequence). Second, how digital watermarking can be used to correct errors in transmitted images. Additionally the proposed algorithms can simultaneously be used for blind equalization and usual watermarking applications. In this paper, two watermarking based equalization algorithms are presented, for two different scenarios:

1. when channel specific training sequence or specific watermark is required,
2. when watermark/training sequence can be any sequence (can be a part of data).

In algorithm I, training sequence is embedded in the data in the form of watermark. This training sequence is also sent over the channel. On receiving side, the watermark is extracted and with the help of received training sequence and extracted watermark channel can be equalized. In this algorithm receiver does not required to know the training sequence in advance. Consider the following scenario; An image (data) distributor distributes watermarked images (data) containing different watermarks to several clients. Distributor also sends the watermark along with the watermarked image (data) through the channel. On the receiving sides, receivers only have a watermark extractor (all the receivers use the same extracting software). Therefore, the receivers can extract the watermark and use it as a reference training sequence. The watermark which distributor has sent through the channel can be treated as received training sequence. Now, the channel can be estimated and hence received, erroneous, watermarked images can be corrected.

In algorithm II, a chunk of data, which is a part of the data actually to be transmitted through the channel, is used as a watermark. This chunk is embedded in the data as a watermark. With the help of received chunk and extracted chunk (extracted watermark) the channel can be equalized blindly. Watermarking algorithm used for these blind equalization schemes, must have the following two qualities:

- Robust (to withstand distortions due to transmission).
- Carry enough payload (which can be used as training sequence).

CDMA based spread spectrum watermarking scheme (Samee, 2007) fulfills the above requirements. The main advantage of spread spectrum watermarking is that each watermark bit is embedded in a number of pixels. Because of that it is proved to be robust in transmission. By using multiple orthogonal spreading codes, CDMA based watermarking scheme can carry more payload (large watermark). Normalized Least Mean Square (NLMS) algorithm (Haykin, 1996) is used in this scheme for equalization. Other equalization algorithms e.g. Least Mean Square (LMS) and Recursive Least Square (RLS) can also be used.

This paper is organized as follows: Sect. 2 presents an overview of previously described CDMA based watermarking algorithm. In Sect. 3 blind equalization scheme using digital watermarking is described. Section 4 presents the experimental results and Sect. 5 concludes this paper.

2 Overview of the watermarking scheme

An oblivious watermarking scheme is presented in (Samee, 2007), which is further improved by using by-part interleaving in (Samee, 2008). In this watermarking scheme every bit of the watermark is hidden in mutually similar frequency coefficients in the DCT domain. These specially selected frequency coefficients are preferably in middle frequency range. Every vector of the modified frequency coefficients $\boldsymbol{i}'_j$ is formed according to:

$$\boldsymbol{i}'_j = \boldsymbol{i}_j + \alpha[b_1\boldsymbol{s}_1 + b_2\boldsymbol{s}_2 + \ldots + b_k\boldsymbol{s}_k] \quad (1)$$

where $\boldsymbol{i}_j$ is a vector of similar frequency coefficients, b_i is a watermark bit, α is the gain factor and $\boldsymbol{s}_i$ are the spreading codes. Watermark bits can be extracted by using:

$$b_i = \text{sign} < \boldsymbol{i}'_j, \boldsymbol{s}_i > \quad if\ |\boldsymbol{i}_j \cdot \boldsymbol{s}_i^T| < |\alpha \boldsymbol{s}_i \cdot \boldsymbol{s}_i^T| \quad (2)$$

For complete description of the watermarking scheme see (Samee, 2007). This scheme is proved to be robust against many intentional and unintentional attacks. Transmission of watermarked image through a channel can be seen as an unintentional attack. The watermark easily survives the transmission attack by using this watermarking scheme. The use of multiple spreading codes makes this scheme able to carry large watermark.

3 Watermarking-based blind equalization

Figure 1 shows how trained equalization works. In order to equalize the data receiver must know the reference training sequence in advance. Error $e(n)$ is calculated with the help of received training sequence and reference training sequence. Now by minimizing $e(n)$, the inverse of the channel is estimated and weights (taps) $\boldsymbol{w}(n)$ are updated. The received data can be equalized by using $\boldsymbol{w}(n)$.

Figure 2 shows algorithm I for blind equalization, where a training sequence is hidden (or superimposed) in the data by using digital watermarking. After transmission, if this hidden training sequence is extracted from the received data without much errors, the extracted training sequence can be used as reference training sequence (which is supposed to be provided by the sender).

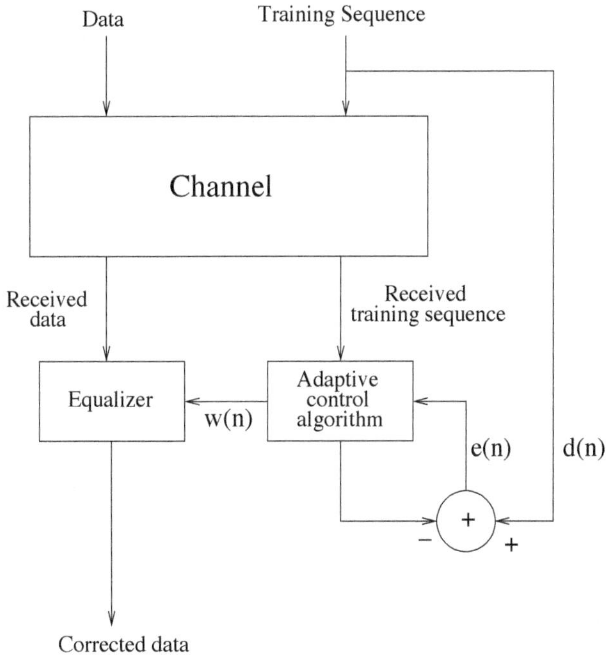

Fig. 1. Traditional trained equalization technique.

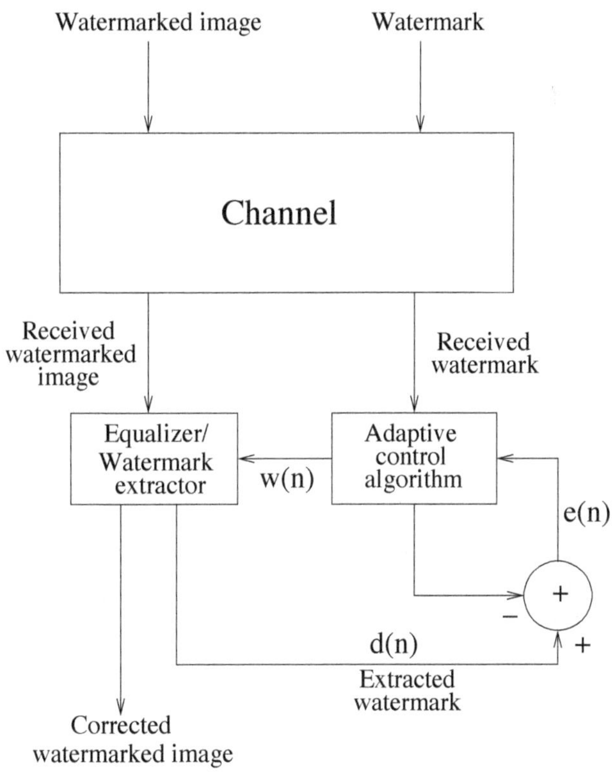

Fig. 2. Algorithm I for blind equalization.

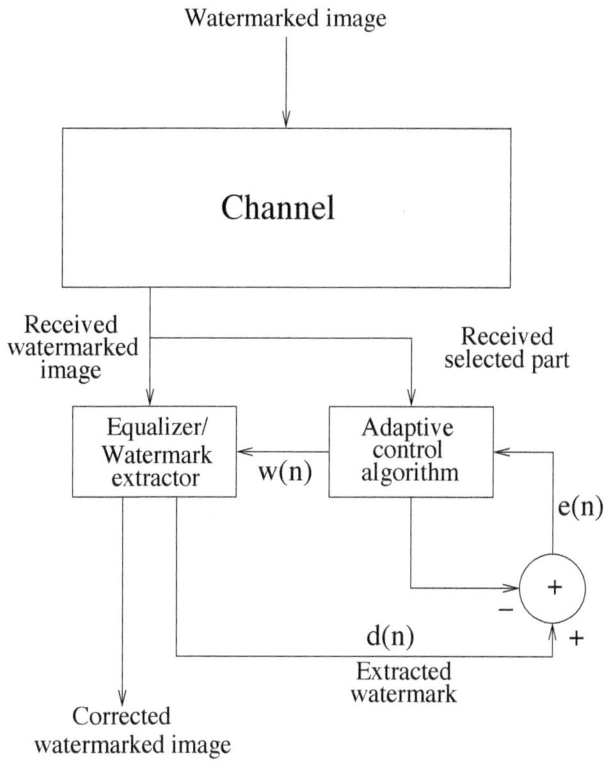

Fig. 3. Algorithm II for blind equalization.

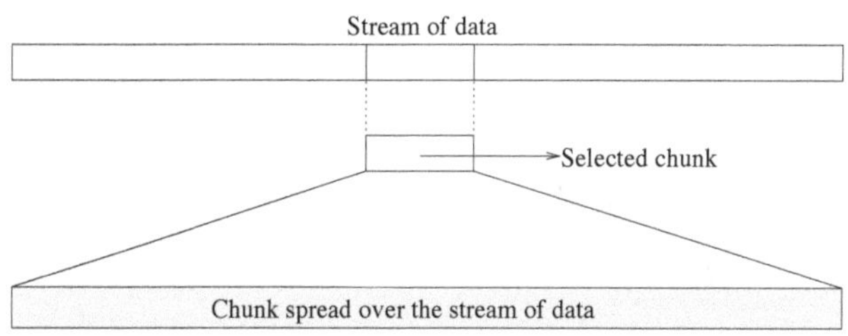

Fig. 4. Selecting a chunk of data and hidding it in the stream of data.

Figure 3 shows algorithm II for blind equalization, where a selected chunk of data is hidden (or superimposed) in the entire stream of data (Fig. 4). After transmission, this hidden chunk of data is extracted from the received data. This extracted chunk can be used as a reference training sequence and the received chunk can be used as a received training sequence. Thus $e(n)$ can be calculated and NLMS algorithm can be used to update $\boldsymbol{w}(n)$. Hence received data can be equalized blindly.

For equalization Normalized LMS algorithm is used. This algorithm works as follows Haykin (1996):

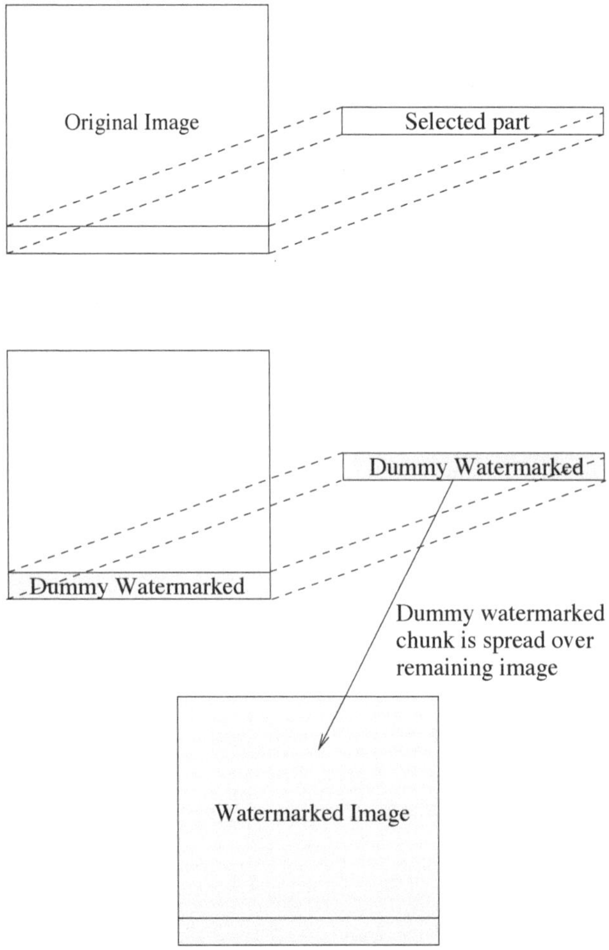

Fig. 5. Forming a watermarked image, considering implementation problem.

Table 1. BER in extracted watermarks using algorithm I.

	Received images		Equalized images	
channels	Lena %	Milk drop %	Lena %	Milk drop %
1	8.2	8.59	0	0
2	16.8	16.8	0	0
3	0.39	10.16	0	0
4	2.54	10.55	0	0

$$Parameter: \; M = \text{number of taps}$$
$$a = \text{positive constant}$$
$$\tilde{\mu} = \text{adaption constant (step-size parameter)}$$
$$0 < \tilde{\mu} < 2$$

Initialization: if prior knowledge on the tap-weight vector $\hat{\boldsymbol{w}}(n)$ is available, use it to select an appropriate value for $\hat{\boldsymbol{w}}(0)$. Otherwise, set $\hat{\boldsymbol{w}}(0) = \mathbf{0}$.

Data:

(a) Given: $\boldsymbol{u}(n)$: M-by-1 tap-input vector at time n
$d(n)$: desired response at time n

(b) To be computed: $\hat{\boldsymbol{w}}(n+1)$ = estimate of tap-weight vector at time n+1

$$Computation: n = 0, 1, 2, \ldots$$

$$e(n) = d(n) - \hat{\boldsymbol{w}}^H(n)\boldsymbol{u}(n)$$
$$\hat{\boldsymbol{w}}(n+1) = \hat{\boldsymbol{w}}(n) + \frac{\tilde{\mu}}{a + \| \boldsymbol{u}(n) \|^2}\boldsymbol{u}(n)e^*(n) \tag{3}$$

Normalized LMS is a modified version of LMS algorithm. LMS algorithm also works fine in proposed blind equalization method. However simulations have shown that Normalized LMS algorithm works slightly better than LMS algorithm. LMS algorithm experiences a gradient noise amplification problem when $\boldsymbol{u}(n)$ is large, because the correction applied to the tap-weight vector $\hat{\boldsymbol{w}}(n)$ at iteration $n+1$ is directly proportional to the tap-input vector $\boldsymbol{u}(n)$. In normalized LMS algorithm correction applied to tap-weight vector $\hat{\boldsymbol{w}}(n)$ at iteration $n+1$ is normalized with respect to squared Euclidean norm of tap-input vector $\boldsymbol{u}(n)$ at iteration n. That is why it is called "normalized" LMS.

3.1 Implementation problem

A very crucial problem arose while implementing proposed scheme using digital watermarking. A part of an image is selected, and later it is spread over the whole image forming watermarked image. Obviously, watermarked image is slightly different from the original image. When this watermarked image is transmitted, channel can not be equalized. Because, the extracted watermark is a part of the original image and received part is from watermarked image. Hence both of them are different and the channel can not be equalized using extracted watermark and received part of the watermarked image.

One of the simplest solutions to this problem is to hide the selected part in the remaining image (other than selected part). However, this simple solution causes problems for usual watermarking applications, like copyrights protection etc. From general watermarking applications point of view, it is not good to leave a reference to the original image. Watermark should be spread over the whole image. If reference to the original image is present in the watermarked image, some denoising algorithm can detect watermark as noise. So watermark can be removed easily. Therefore, this scheme can not be used for usual watermarking applications.

Table 2. Lena received and equalized using algorithm I.

channels	PSNR Received dB	PSNR Equalized dB	Erroneous bits Received %	Erroneous bits Equalized %
1	29.0929	77.5595	8.94	0.01
2	28.9833	69.6869	12.2	0.09
3	22.0178	Infinite	33.58	0.00
4	13.8834	42.9292	9.29	0.00

Table 3. Milk drop received and equalized using algorithm I.

channels	PSNR Received dB	PSNR Equalized dB	Erroneous bits Received %	Erroneous bits Equalized %
1	29.2870	74.1472	8.51	0.03
2	29.1856	92.3162	11.65	0.00
3	23.0038	Infinite	33.58	0.00
4	13.8618	21.8373	9.29	1.27

Table 4. BER in extracted watermarks using algorithm II.

channels	Received images Lena %	Received images Milk drop %	Equalized images Lena %	Equalized images Milk drop %
1	4.76	8.93	0	0
2	6.75	10.52	0	0
3	0.2	4.37	0	0
4	24.40	17.26	0	0

Table 5. Lena received and equalized using algorithm II.

channels	PSNR Received dB	PSNR Equalized dB	Erroneous bits Received %	Erroneous bits Equalized %
1	26.8461	87.1311	9.67	0.00
2	26.8677	86.5184	13.04	0.00
3	22.5406	102.3162	33.60	0.00
4	12.3285	88.8920	8.06	0.00

This problem is solved by watermarking selected portion with some dummy bits first and then hide this dummy watermarked selected portion into remaining image (Fig. 5). Now this watermarked image has same level of noise over whole image and has no reference to the original image. Therefore, this watermarked image can simultaneously be used for blind equalization as well as for usual watermarking applications.

3.2 Possible security issue

As watermark is a part of the watermarked image, it can be considered as a possible security issue. By using proposed scheme, watermark can be any part of watermarked image, which is very difficult to locate for an evedropper. Furthermore, because of proposed implementation scheme noise level (watermark intensity) is same over the whole image. So it is difficult to distinguish between selected part, watermark, and remaining watermarked image.

4 Experimental results

In all the experiments 512×512 (8 bits/pixel, gray scale) images (Lena Fig. 6a and Milk drop Fig. 6b) are used. 512 bits long watermark is hidden in 3rd, 4th, 5th and 6th rows of 8×8 blocks of DCT coefficients (Samee, 2007). Two spreading codes, each of length 512, are used to spread the watermark bits. All the images are watermarked at 40 dB (here $\alpha = 0.01$). The quality of an image is measured in terms of Peak Signal to Noise Ratio (PSNR).

$$\text{PSNR} = 10\log\frac{255^2}{\text{MSE}}\ \text{dB} \tag{4}$$

Using MATLAB, these images are transmitted over four different channels:

- channel 1: {0.986, 0.845, 0.237, 0.123+0.310i}
- channel 2: {0.986, 0.845, 0.537, 0.323, 0.123}
- channel 3: Frequency-flat (single path) Rayleigh fading channel, sample time 1×10^{-5} and maximum Doppler shift 0.09 Hz.
- channel 4: Frequency-flat (single path) Rician fading channel, sample time 1×10^{-5}, maximum Doppler shift 0.99 Hz and Rician factor equal to 1.

4.1 Algorithm I

First of all, training sequence is embedded in an image as watermark. 512 bits long training sequence (watermark) is sent over the channel followed by the watermarked image. On the receiving side, watermark is extracted from the watermarked image. The extracted watermark has a few erroneous bits (table 1) but the BER is good enough to use it as a reference training sequence. Normalized LMS with 8 weights, $\tilde{\mu} = 0.01$ and $\hat{\boldsymbol{w}}(0) = \mathbf{0}$ are used. The calculated weights $\hat{\boldsymbol{w}}(\boldsymbol{n})$ are used to equalize the received image. Received and corresponding equalized Lena images are shown in Fig. 7. Errors and PSNR values of received and equalized

Table 6. Milk drop received and equalized using algorithm II.

channels	PSNR Received dB	PSNR Equalized dB	Erroneous bits Received %	Erroneous bits Equalized %
1	27.2704	83.6830	4.56	0.00
2	27.3037	74.7575	4.76	0.03
3	22.6232	102.3162	10.32	0.00
4	13.2914	102.3162	14.88	0.00

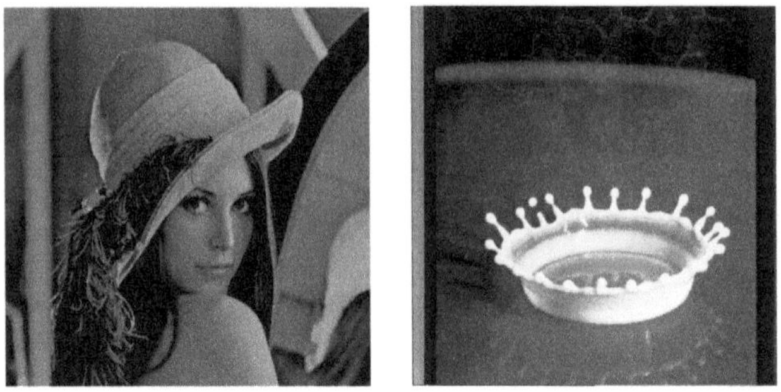

(a) Lena image. **(b)** Milk drop.

Fig. 6. Original Lena and Milk drop images.

images for both Lena and Milkdrop are shown in Tables 2 and 3, respectively. Here, second and third columns show PSNR value of received and equalized images with respect to watermarked images. Fourth and fifth columns show percentage of erroneous bits in received and equalized images. Column 4 and 5 of Table 1 show the BER in the watermarks extracted from equalized images. These watermarks are error free and easily be used for usual watermarking applications.

4.2 Algorithm II

First eight rows of an image are watermarked with eight dummy bits. First 504 bits are selected from the dummy watermarked part and hidden in remaining rows of the image. These watermarked images (watermarked at 40 dB) are transmitted over the same four channels, mentioned above.

On the receiving side, watermark is extracted from the watermarked image. The watermark extracted from the received image, which forms the training sequence, has a few erroneous bits (column 2 and 3 of Table 4). The received selected part is treated as the received training sequence. $\hat{\boldsymbol{w}}(n)$ are calculated using Normalized LMS with $M = 8$ weights, step size $\tilde{\mu} = 0.01$ and $\hat{\boldsymbol{w}}(0) = \mathbf{0}$. The calculated weights $\hat{\boldsymbol{w}}(n)$ are used to equalize the received image. Errors and PSNR values of received and corresponding equalized images are shown in Tables 5 and 6. Here, second and third columns show PSNR value of received and equalized images with respect to watermarked images. Fourth and fifth columns show percentage of erroneous bits in received and equalized images. Column 4 and 5 of Table 4 show the BER in the watermarks extracted from equalized images.

(a) Lena image. **(b)** Watermarked Lena.

(c) Received through channel 1. **(d)** Equalized image for (c).

(e) Received through channel 2. **(f)** Equalized image for (e).

(g) Received through channel 3. **(h)** Equalized image for (g).

(i) Received through channel 4. **(j)** Equalized image for (i).

Fig. 7. Lena received and equalized.

5 Conclusions

How watermarking can be used to equalize transmission channels only using the received data is discussed in this paper. Simulations have shown that proposed schemes can correct almost all the errors from received watermarked images. An important advantage of this scheme is that it works at the complexity of traditional trained equalization methods unlike other very complex blind equalization methods. Proposed schemes can simultaneously be used for blind equalization as well as for usual watermarking applications. Further research will be done to develop a similar scheme for radio frequency communication signals. Furthermore, the suitability of different watermarking schemes for the proposed blind equalization algorithm can be studied.

References

John G. Proakis: Digital Communications, fourth edition, McGraw Hill, 2000.

Simon Haykin: Adaptive Filter Theory, third edition, Prentice Hall, 1996.

Litwin Jr., L. R.: Blind Channel Equalization, IEEE Potentials, 18(4), 9–12, 1999.

Campisi, P., Carli, M., Giunta, G., and Neri, A.: Blind Quality Assessment System for Multimedia Communications Using Tracing Watermarking, IEEE Transactions on Signal Processing [see also IEEE Transactions on Acoustics, Speech, and Signal Processing], 51(4), 996–1002, 2003.

Ding, Y. W., Lin, Z., and Wang, L.: A Multipurpose Public-Key Cryptosystem Based Image Watermarking, 4th International Conference on Wireless Communications, Networking and Mobile Computing, WiCOM '08., 2008.

Benedetto, F., Giunta, G., and Neri, A.: End-to-End QoS Provision and Control in Wireless Communication Systems by Means of Digital Watermarking Signal Processing, 3rd International Symposium on Wireless Communication Systems, ISWCS '06. 655–659, 2006.

Maity, S. P., Kundu, M. K., and Maity, S.: An Efficient Digital Watermarking Scheme for Dynamic Estimation of Wireless Channel Condition, International Conference on Computing: Theory and Applications. ICCTA '07, 671–675, 2007.

Neto, M. U., Gomes, L. de C. T., Romano, J. M. T., and Bonnet, M.: Adaptive Equalization based on Watermarking, 2006 International Telecommunications Symposium, 743–748, 2006.

Zarzoso, V., and Comon, P.: Blind and Semi-Blind Equalization Based on the Constant Power Criterion, IEEE Transactions on Signal Processing, 53(11), 4363–4375, 2005.

Sun, L., and Zhao, C.: Optimizing Blind Equalization Intelligent Algorithm for Wireless Communication Systems, 3rd International Conference on Natural Computation, ICNC 2007, 3, 146–149, 2007.

Dabeer, O., and Masry, E.: Convergence Analysis of the Constant Modulus Algorithm, IEEE Transactions on Information Theory, 49(6), 1447–1464, 2003.

Podilchuk, C.I., and Delp, E.J.: Digital Watermarking: Algorithms and Applications, IEEE Signal Processing Magazine, 18(4), 33-46, 2001.

Samee, M. K., and Götze, J.: Increased Robustness and Security of Digital Watermarking Using DS-CDMA, Proceedings of the 7th IEEE International Symposium on Signal Processing and Information Technology, 189–193, 2007.

Samee, M. K., Götze, J., Ruan, S. J., and Pai, Y. T.: Digital Watermarking: Spreading Code versus Channel Coding, Proceedings of the 3rd IEEE International Symposium on Communications, Control and Signal Processing, Malta, 1409–1412, 2008.

Low complexity Turbo synchronization without initial carrier synchronization

U. Wasenmüller, C. Gimmler, and N. Wehn

Microelectronic Systems Design Research Group, University of Kaiserslautern, Erwin-Schrödinger Str., 67663 Kaiserslautern, Germany

Abstract. Wireless data transmission results in frequency and phase offsets of the signal in the receiver. In addition the received symbols are corrupted by noise. Therefore synchronization and channel coding are vital parts of each receiver in digital communication systems. By combining the phase and frequency synchronization with an advanced iterative channel decoder (inner loop) like turbo codes in an iterative way (outer loop), the communications performance can be increased. This principal is referred to as turbo synchronization. For turbo synchronization an initial estimate of phase and frequency offset is required. In this paper we study the case, where the initial carrier synchronization is omitted and an approach with trial frequencies is chosen. We present novel techniques to minimize the number of trial frequencies to be processed. The communications performance and effort of our method is demonstrated. Furthermore the implementation complexity of the whole system is shown on a Xilinx FPGA.

1 Introduction

Synchronization and channel decoding are vital parts of every digital receiver for wireless communication. The transmission over a wireless channel results in timing, frequency and phase offsets. In addition, the received symbols are corrupted by noise. Task of the synchronization is to present data bits to the channel decoder, where the negative influences of timing, frequency and phase offset are eliminated. A well known advanced scheme for channel coding is the use of turbo codes. The turbo encoder delivers a stream of systematic bits and two parity bit streams by a recursive systematic convolutional encoding of the user data bits and an interleaved version of the user data bits. Turbo code decoding is done in an iterative algorithm based on the maximum a posteriori principle. Communication systems with Turbo codes can operate at very low signal-to-noise ratios (SNR).

Frequency offset synchronization and phase offset synchronization (carrier synchronization) is typically performed only once before channel decoding. The variance of the phase and frequency estimation depends on the SNR as well as on the number of symbols available for estimation of the phase and frequency offset. The mentioned variances influence the decoder performance heavily; i.e. large variances lead to an unacceptable performance degradation of the used decoder.

In our paper we focus on carrier synchronization in conjunction with turbo code decoding. The method of joined iterative turbo code decoding and synchronization is called turbo synchronization. Turbo synchronization allows correct decoding also for larger variances of frequency and phase offset. This can be used to decrease the number of known symbols for synchronization purposes and thus increases the user data rate for a given bandwidth. However, also the advanced turbo synchronization technique needs an initial coarse carrier synchronization and can tolerate frequency offset estimation errors up to a small limit.

In the situation of unacceptable estimation errors of the initial carrier synchronization tentative decoding with several trial frequency offsets must be performed. The effort of the receiver depends now on the number of investigated trial frequency offsets. We demonstrate novel techniques for minimizing the number of trial frequency offsets to be processed. To prove the feasibility of the system approach an implementation of the whole system is presented. The key facts of the hardware architecture for achieving low complexity for the principally expensive usage of trial frequencies are illustrated. Furthermore we analyze the additional effort for the turbo decoder caused by turbo synchronization and processing of trial frequency offsets and demonstrate the

Correspondence to: U. Wasenmüller
(wasenmueller@eit.uni-kl.de)

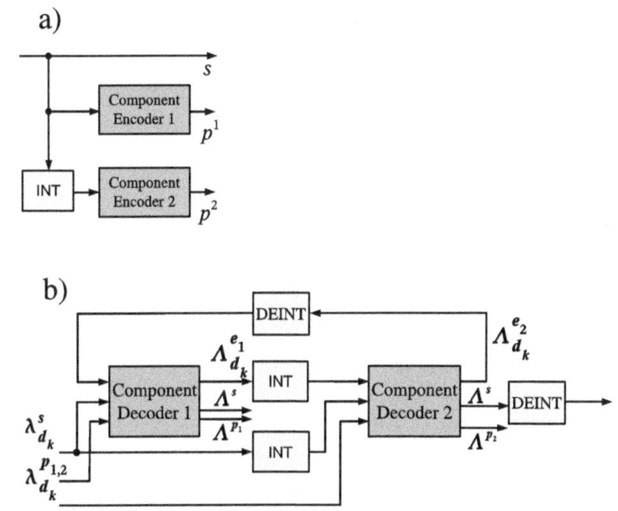

Fig. 1. **(a)** Turbo encoder, **(b)** Turbo decoder.

communications performance of the targeted communication system.

In Sect. 2 we summarize turbo encoding as well as the the principle of turbo code decoding. In the following Sect. 3 we explain the iterative synchronization and the interrelation with the decoding algorithm. In Sect. 4 we give an short overview about the target communication system and explain the use of trial frequencies for frequency offset estimation. Our proposed method for minimizing the number of trials to be performed is presented. In Sect. 5 we analyze the communications performance of our proposed method. The hardware architecture is explained and hardware complexity of turbo synchronization without initial carrier synchronization is analyzed. Communications performance is determined by bit true models. Implementation complexities and results are given based on Xilinx devices. The paper is concluded in Sect. 6.

2 Turbo codes

With the introduction of binary turbo codes by Berrou in 1993 (Berrou et al., 1993) near optimum error correction became possible. Due to these error correction capabilities, binary and duo-binary turbo codes allow for low frame error rates (FER) at a low signal-to-noise ratio (SNR), outperforming the widely used convolutional codes. Because of this advantage turbo codes are now part of a large number of communication standards.

Turbo codes generally consist of a serial or parallel concatenation of two codes, so called component codes, and an interleaver. While the first component code encodes the information in the original order, the second one gets the information in a permuted order, see Fig. 1a). In all standards convolutional codes are used as component codes.

Decoding of turbo codes is an iterative process where probabilistic information is exchanged between component decoders (Berrou, 2003). Iterative decoding implies a big challenge with respect to low latency and high throughput requirements.

A possible realization of a decoder for turbo codes is given in Fig. 1b). The two component decoders that decode the two component codes are connected via interleaver and deinterleaver. They use *log likelihood ratios* (LLR) $\lambda^s_{d_k}, \lambda^{p^1}_{d_k}$ and $\lambda^{p^2}_{d_k}$ of the systematic and parity information to compute the extrinsic information $\Lambda^{e1}_{d_k}$ and $\Lambda^{e2}_{d_k}$ on the information bits. The iterative exchange of $\Lambda^{e1}_{d_k}$ and $\Lambda^{e2}_{d_k}$ between these component decoders is referred to as turbo principle. One *(full) iteration* is done if Decoder 1 and Decoder 2 have run once. If only one decoder has calculated new information, we call this one *half iteration*.

Both component decoders optimize the *maximum a posteriori probability* (MAP) criterion. However, in hardware implementations the suboptimal Max-Log MAP algorithm with *extrinsic scaling factor* (ESF) is more suitable. In comparison to the optimal algorithm the Max-Log MAP results in a performance loss of less than 0.2 dB (Robertson et al., 1995, 1997). Furthermore, the Max-Log MAP algorithm does not require knowledge of the SNR in contrast to the optimal Log MAP algorithm (Worm et al., 2000).

The Max-Log MAP algorithm consists of a forward and a backward recursion. It computes for each possible information or parity bit d_k an *a posteriori probability* (APP) LLR $\Lambda^s, \Lambda^{p1}, \Lambda^{p2}$.

3 Turbo synchronization

The synchronization consists of the estimation of the unknown parameters of timing, frequency and phase offset, and the elimination of all possible negative influences introduced by these parameters. We focus on the frequency and phase synchronization of bursts with linear modulation (e.g. QPSK,16-QAM) in conjunction with turbo decoding. We assume, that the steps of gain control, timing synchronization and burst detection are properly carried out before. The received sample sequence r is given in the complex baseband according to Eq. (1):

$$r(l) = s(l) \cdot e^{j(2\pi f_o l + \Phi)} + n(l) \qquad l = 0, 1, \ldots, L-1 \tag{1}$$

The sample sequence r with L elements is based on modulation symbols $s(l)$ with one sample per symbol and symbol duration T, and is disturbed by a noise sequence n. The unknown parameters of frequency offset f_o and phase offset Φ have to be estimated for every received sample sequence. These parameters are considered fixed during an estimation interval; the sample sequence has to be corrected accordingly to the estimated frequency offset $\tilde{f}_o$ and estimated phase offset $\tilde{\Phi}$.

The synchronization with turbo synchronization is done in two main steps. Initially, a (coarse) carrier synchronization is performed. In case of larger variances of the estimation parameters a degradation in the decoding performance will occur. The principle of turbo synchronization is to improve the estimation parameters for synchronization (fine synchronization) with the additional use of tentative decoder decisions. The improved carrier synchronization is used to provide better input data for the decoding process. This process is done iteratively after each decoder iteration. We concentrate first on the step of fine synchronization, which will be used for turbo synchronization.

Frequency and phase offset can be optimally estimated on an unmodulated carrier. With the assumption, that the transmitted symbol sequence s of the burst is known the effect of the modulation by each transmitted symbol $s(l)$ can be removed by:

$$\tilde{r}(l) := r(l)\cdot s^*(l) \qquad l=0,1,\dots,L-1 \tag{2}$$

However, it must be considered, that usually the symbols of the burst are unknown or only some symbols, used for supporting the burst detection or supporting the coarse synchronization are known. Thus we replace the transmitted symbol sequence s by an estimated symbol sequence s_e. The estimation of the transmitted symbol sequence is provided by the turbo decoder.

The fine estimation of frequency and phase offset is based on the average phase $\tilde{\phi}_0$ of the front part and on the average phase $\tilde{\phi}_1$ of the rear part of the burst with a modulation removal by the estimated symbol sequence s_e. This is formally given by

$$\tilde{\phi}_k := \sum_{l=0+k\cdot L/2}^{L/2-1+k\cdot L/2} r(l)\cdot s_e^*(l) \qquad k=0,1 \tag{3}$$

With the two phase values of Eq. (3) the estimate of the frequency offset can be calculated with

$$\tilde{f}_0 = \frac{\arg(\tilde{\phi}_0\cdot\tilde{\phi}_1^{\,*})}{2\pi\cdot L} \tag{4}$$

The estimate of the phase offset is calculated with the help of Eq. (4)

$$\tilde{\phi} = \arg(\tilde{\phi}_0+\tilde{\phi}_1) - L\cdot\tilde{f}_0\cdot\pi \tag{5}$$

The first decoder iteration is based on the LLR values $\lambda^s_{d_k}$, $\lambda^{p^1}_{d_k}$ and $\lambda^{p^2}_{d_k}$ calculated with the symbols of the coarse synchronized received sequence. For the iterative fine synchronization an estimate of the transmitted symbols is used, which is produced by the turbo code decoder after each iteration. The estimate of the transmitted symbols is gathered by the APP LLR of the decoder. A turbo code decoder computes APP LLR values Λ^s of the systematic bits by default.

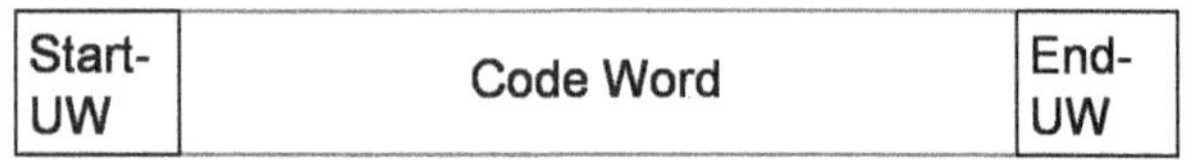

Fig. 2. Considered burst structure.

In turbo synchronization applications the decoder must additionally calculate the APP LLR values $\Lambda^{p1,2}$ for the parity bits.

To reduce the effect of using some erroneous reference symbols in the fine synchronization, soft values are used for the reference symbols. The values for the quadrature components of the estimated symbols are calculated by a tanh operation on the APP LLR. Provided that the k-th transmitted QPSK symbol $s(k)$ contains the bits $d_k^{(I)}$ and $d_k^{(Q)}$ of the code word the estimated symbol $s_e(k)$ is calculated as:

$$s_e(k) = \tanh\frac{\Lambda(d_k^{(I)})}{2} + j\tanh\frac{\Lambda(d_k^{(Q)})}{2} \tag{6}$$

With the tentative soft values of systematic bits and parity bits the sequence s_e is generated and the described fine synchronization process can be carried out.

The received sequence r is corrected with the new estimates of frequency and phase offset. A new synchronized received sequence r is calculated after each full decoder iteration. Turbo decoding and fine synchronization run in parallel in our architecture to avoid throughput degradation by turbo synchronization. For the n-th iteration of the decoder the LLR values $\bar{\Lambda}^s$, $\bar{\Lambda}^{p1}$, and $\bar{\Lambda}^{p2}$ of the transmitted bits are calculated on base of the fine synchronized sequence, which used the APP LLR values of the $(n-2)$-th iteration. A discussion of the effects of the schedule regarding the update of LLR values can be found in Alles et al. (2007).

4 Evaluation of a grid of trial frequency offsets

For using the principle of turbo synchronization an initial coarse carrier synchronization with a sufficiently small variance of the estimation parameters is needed. The methods for the carrier synchronization depend on the existing communications system. The burst structure of the targeted communication system is depicted in Fig. 2. The burst consists at the start and the end of a sequence of known symbols – start unique word and end unique word respectively. The unknown symbols in the mid of the burst correspond to the code word of the turbo code encoded information sequence. A lot of algorithms exist to estimate frequency and phase offset with known symbols (data aided estimation) or with the unknown data symbols (blind estimation) as described in Meyr et al. (1998); Mengali and D'Andrea (1997). The variance of the estimation parameters depends on the SNR, the number of symbols and the use of known or unknown symbols.

In our targeted communication system the variance of frequency offset estimation with data aided estimation methods as well as with blind estimation methods is not sufficiently small to perform afterwards a turbo synchronization with satisfying communications performance. This is caused by the targeted SNR range below zero dB and the low number of symbols (known and unknown).

Thus a grid of trial frequency offsets is tentatively used. These trial frequency offsets are evaluated by the decoding to decide on the best estimation. The effort is now determined by the number of trial frequencies to be processed.

To perform synchronization and decoding a set of trial frequency offsets $\tilde{f}_i$ is used. For using this method it is required to estimate a corresponding phase offset $\tilde{\phi}$ for each trial frequency offset. Applying the trial frequency offset to to the received sequence r is given by:

$$r_{\tilde{f}_i} = r(l) \cdot e^{-j(\tilde{f}_i \cdot l)} \qquad l = 0, 1, \cdots L-1 \tag{7}$$

For maximum likelihood estimation of the phase offset the well known V&V algorithm (Viterbi and Viterbi, 1983) is used. A correlation of the modified received sequence $r_{\tilde{f}_i}$ with the unique word at the start and end of the burst is performed, which is given by:

$$k_{\tilde{f}_i} = \sum_{l \in UW} r_{\tilde{f}_i}(l) \cdot u^*_{UW}(l) \tag{8}$$

The estimation of the phase offset $\tilde{\phi}$ for the trial frequency offset is given by the argument of the phasor $k_{\tilde{f}_i}$ of Eq. (8):

$$\tilde{\phi}_{f_i} = \arg(k_{\tilde{f}_i}) \tag{9}$$

As mentioned a criterion is required to decide on the best frequency offset. The received sequence r will be corrected with the trial frequency $\tilde{f}_i$ and the belonging phase offset $\tilde{\phi}$. Based on this new sequence the LLR values for the turbo decoder can be calculated. The decoder performs the iterations as described in Sect. 2. For a decision on the best trial frequency offset an estimation s_e of the transmitted symbol sequence is used. The estimate of the transmitted symbol sequence is achieved as described in Sect. 3. For every trial frequency the correlation of received symbol sequence r and the estimated transmitted symbol sequence s_e is used. Correlation is a measurement of similarity and thus the trial frequency producing the greatest similarity between the two mentioned sequences is chosen. The correlation operation is given by

$$c_{\tilde{f}_i} = \sum_{l=0}^{L-1} r_{\tilde{f}_i}(l) \cdot s^*_\mathrm{e}(l) \tag{10}$$

The selection process is done with

$$\tilde{f} = \tilde{f}_k \quad \text{with} \quad c_{\tilde{f}_k} \geq c_{\tilde{f}_i} \quad \text{for all i} \tag{11}$$

For the selection process of $\tilde{f}$ it is sufficient to use the sequence s_e in Eq. (11), which is achieved after one or two turbo decoder iterations.

The computational effort for the evaluation of trial frequencies is dominated by the total number of turbo decoder iterations which must be performed. Thus it is desirable to minimize the number of trial frequency offsets to be evaluated by the decoder. The calculated phasor $k_{\tilde{f}_i}$ for phase estimation in Eq. (8) can be used to exclude trial frequency offsets from consideration before the decoding step. The magnitude of the correlation is a measure for similarity and thus only trial frequency offsets f_i with

$$|c_{\tilde{f}_i}| \geq C_{Tr} \tag{12}$$

will be processed. It must be emphasized, that the best estimation of the frequency offset is not given by the maximum correlation value of Eq. (8).

5 Results

Our approach for synchronization and decoding without initial carrier synchronization is validated per software simulations with bit true models in sense of communications performance. In addition the architecture of the system is presented. The complexity of the components for implementation on Xilinx FPGA is shown and briefly analyzed.

5.1 Communications performance

The communications performance of our system is demonstrated for a burst with QPSK modulation. The start unique word and end unique word contains 40 and 24 symbols, respectively. The code word uses 1248 symbols; it is based on turbo encoding with rate 1/3 and 16 states. A grid of 61 trial frequencies is used for this burst type. The trial frequencies are chosen to cover the maximum relative frequency offset of 6×10^{-3} and to allow a fine synchronization by the principle of turbo synchronization.

The communications performance expressed as a frame error rate (FER) is shown in Fig. 3 for different scenarios in an additive white gaussian noise channel. The graph with label "Blind" shows the FER for a blind carrier synchronization. Use of turbo synchronization produces no visible improvement in the figure. It is evident, that the carrier estimation parameters cause a unacceptable performance. The graph with label "no threshold" and the graph with label "2800" coincide. In the targeted communication system a FER of 10^{-3} is specified at SNR of -0.7 dB for ideal synchronization and optimal decoding. Hence the achieved communications performance is comparable to a system with perfect knowledge of the synchronization parameters. The effort advantage of the threshold criterion can be checked by Fig. 4. Using this threshold only approximately 60% of the trial frequencies have to be processed.

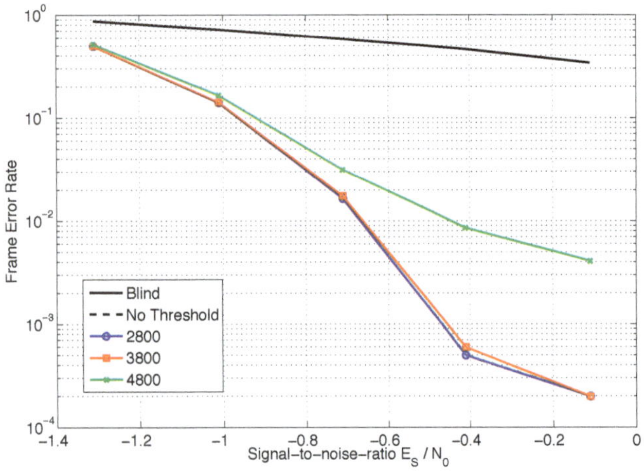

Fig. 3. Communications performance.

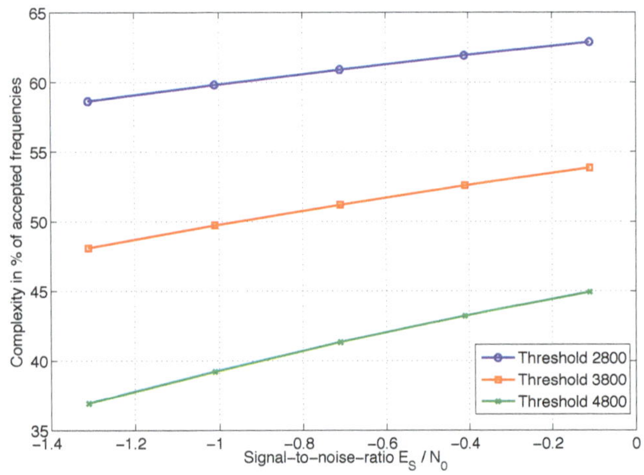

Fig. 4. Trial frequencies evaluated by decoder.

With the threshold parameter a trade off between communications performance and implementation performance can be regulated. The graphs with label "3800" and "4800" respectively show the FER performance for different values of the threshold. The computational effort for the different threshold values is shown in Fig. 4 as a percentage of the number of considered frequency offset values. For the threshold label "3800" approximately 50% of the trial frequencies have to be evaluated. It is possible to choose the threshold value for a defined SNR operation point. This is demonstrated with the graphs with label "3800". For SNR values above −0.2 dB the performance is identical to the method without threshold. Depending on the exact value of the threshold about 40% to 60% of the trial frequencies can be excluded before the decoding step.

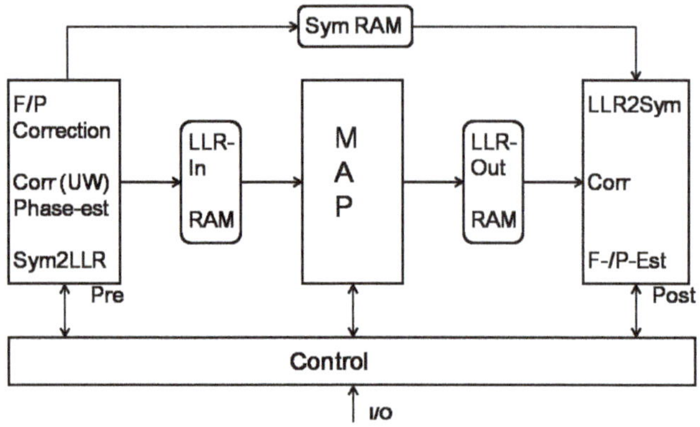

Fig. 5. Architecture.

Table 1. Key parameter of the turbo decoder.

	Binary Turbo decoder
Architecture	Serial MAP, 3 ACS units
Trellis states	16
Infoword size	128–5124 bits
Codeword size	128–15 372 bits
Code rates	$^1/_3 \ldots {}^9/_{10}$
Parallelism	1
Input quantization	6 bit
Algorithm	MaxLog-MAP
ESF	0.75
Max. iterations	8
Technology	FPGA
Clock frequency	223 MHz
Payload Thrpt. [Mbit/s]	4.67–14.02
Latency [μs]	10.36–414.7
Payload Bit/Cycle	0.02–0.06

5.2 Architecture and implementation

The architecture of the system is depicted in Fig. 5. Objective of the architecture is to allow a hardware sharing of functionalities of turbo synchronization and trial frequency processing. Central part is the MAP component, which performs a half iteration of turbo code decoding. Component Pre includes the frequency and phase correction of the received sequence r, which is used for the step of fine synchronization in turbo synchronization as well as for processing of trial frequencies. The phase estimation according to Eq. (8) for the processing of trial frequencies is also included in component Pre as well as the determination, whether the actual trial frequency f_i will be excluded. The calculation of LLR values including in the demapping and depuncturing is used for providing the input data for the MAP component.

Unit Post carries out the central correlation operation according to Eqs. (3) and (10), respectively, which is used for fine synchronization as well as trial frequency processing. Both functionalities must use the transformation of LLR-Out

Table 2. Details of implementation complexity.

	MAP	Pre	Post	Ctrl	RAM	Sum
LUT	4372	2478	1974	909	1506	11 329
Xilinx Virtex-5 XC5VSX95T @ 223 MHz						
LUT	4372	2478	1974	909	1506	11 329
FF	2280	1243	1340	608	1064	6625
MAC	0	10	20	0	0	30
BRAMs	4	3	4	1	68	80

values to symbols as described for QPSK symbols in Eq. (6). The small part for frequency and phase estimation according to Eqs. (4) and (5) in component Post is used only for fine synchronization. All shown RAM blocks are double buffered to allow a parallel processing of the main components. Therefore different trial frequencies are processed in the components Pre, MAP and Post. The step of fine synchronization is carried out in component Pre and Post on the results of iteration n, while the MAP decoder executes the iteration $n+2$.

The architecture of the turbo decoder is a state-of-the-art SMAP architecture with three recursion units which run in parallel. The key parameter are summarized in Table 1. For more details the reader is referred to May et al. (2007).

The resources for the components are presented in Table 2 for an implementation in a Xilinx Virtex-5 FPGA. Component Post is needed only for turbo synchronization and trial frequency evaluation respectively. Approximately half of the resources of component Pre are used for turbo synchronization and trial frequency evaluation, respectively.

6 Conclusions

In this paper we presented a novel method to reduce the complexity for turbo synchronization without initial carrier synchronization. To perform the decoding and synchronization steps without initial carrier synchronization a grid of trial frequencies is needed. The computational effort for processing of trial frequencies can be reduced by elimination of trial frequencies with a simple threshold comparison in the phase estimation step. Furthermore by applying turbo synchronization the grid of trial frequencies can be kept more coarse than with traditional decoding. The proposed hardware architecture allows a sharing of the components for evaluation of trial frequencies and for synchronization and decoding.

The components for trial frequency evaluation as well as for synchronization and turbo decoding can work concurrently. The mentioned features result in a low complexity system. The achieved communications performance is comparable to a system with perfect knowledge of the synchronization parameters.

References

Alles, M., Lehnigk-Emden, T., Wasenmüller, U., and Wehn, N.: Implementation Issues of Turbo Synchronization with Duo-Binary Turbo Decoding, in: Proc. 19th Annual IEEE International Symposium on Personal, Indoor and Mobile Radio Communications (PIMRC) 2007, Athens, Greece, 2007.

Berrou, C.: The Ten-Year-Old Turbo Codes are Entering into Service, IEEE Communications Magazine, 41, 110–116, 2003.

Berrou, C., Glavieux, A., and Thitimajshima, P.: Near Shannon Limit Error-Correcting Coding and Decoding: Turbo-Codes, in: Proc. 1993 International Conference on Communications (ICC '93), pp. 1064–1070, Geneva, Switzerland, 1993.

May, M., Neeb, C., and Wehn, N.: Evaluation of High Throughput Turbo-Decoder Architectures, in: Proc. IEEE International Symposium on Circuits and Systems (ISCAS 2007), New Orleans, USA, 2007.

Mengali, U. and D'Andrea, A.: Synchronization Techniques for Digital Receivers, Plenum Publishing Corporation, New York, 1997.

Meyr, H., Moeneclaey, M., and Fechtel, S. A.: Digital Communication Receivers, John Wiley & Sons Inc., 1998.

Robertson, P., Villebrun, E., and Hoeher, P.: A Comparison of Optimal and Sub-Optimal MAP decoding Algorithms Operating in the Log-Domain, in: Proc. 1995 International Conference on Communications (ICC '95), pp. 1009–1013, Seattle, Washington, USA, 1995.

Robertson, P., Hoeher, P., and Villebrun, E.: Optimal and Sub-Optimal Maximum a Posteriori Algorithms Suitable for Turbo Decoding, European Transactions on Telecommunications (ETT), 8, 119–125, 1997.

Viterbi, A. J. and Viterbi, A. M.: Nonlinear Estimation of PSK Modulated Carrier Phase with Application to Burst Digital Transmission, IEEE Transactions on Information Theory, 32, 543–551, 1983.

Worm, A., Hoeher, P., and Wehn, N.: Turbo-Decoding without SNR Estimation, IEEE Communications Letters, 4, 193–195, 2000.

44

Automated parametrical antenna modelling for ambient assisted living applications

R. Kazemzadeh[1], **W. John**[1,2], **and W. Mathis**[1]

[1]Institute for Theoretical Electrical Engineering, Hannover, Germany
[2]SIL System Integration Laboratory GmbH (R+D Consulting), Paderborn, Germany

Correspondence to: R. Kazemzadeh (kazemzadeh@tet.uni-hannover.de)

Abstract. In this paper a parametric modeling technique for a fast polynomial extraction of the physically relevant parameters of inductively coupled RFID/NFC (radio frequency identification/near field communication) antennas is presented. The polynomial model equations are obtained by means of a three-step procedure: first, full Partial Element Equivalent Circuit (PEEC) antenna models are determined by means of a number of parametric simulations within the input parameter range of a certain antenna class. Based on these models, the RLC antenna parameters are extracted in a subsequent model reduction step. Employing these parameters, polynomial equations describing the antenna parameter with respect to (w.r.t.) the overall antenna input parameter range are extracted by means of polynomial interpolation and approximation of the change of the polynomials' coefficients. The described approach is compared to the results of a reference PEEC solver with regard to accuracy and computation effort.

1 Introduction

For an energy-efficient and secure wireless medical data transmission in future Mobile Ambient Assisted Living (MAS) devices, novel antenna concepts in the near field region need to be investigated. This requires the examination of a wide variety of antenna systems during the design process to find the optimum configuration for a specific application. Thus, an adequate method that is able to accurately characterize the antennas on the one hand, and to allow for fast and efficient parameter sweeps on the other, is needed. While several approaches exist for the efficient determination of the coupling between spiral coils, the modeling of the actual antennas leads to more complex models, since phenomenon like eddy current losses need to be considered. This means a limiting factor for the applicability of accurate, but computationally expensive full wave 3-D solvers. Other approaches that can be used for a more efficient extraction of antenna parameters, such as analytical expressions or complexity reduction techniques, have the common benefit of reducing the computational effort to a certain degree compared to the full-wave solution, but show individual drawbacks: rules of thumb do not provide sufficient accuracy and analytical expressions exist only for a limited number of geometries usable for antenna modeling. By using Model Order Reduction (MOR) algorithms for the complexity reduction of RLC network models, such as PEEC models, the physical properties of the original model can usually not be preserved. In the reluctance-based method (Devgan et al., 2000) as well as the Fast Multipole Method (FMM) (Antonini, 2003) a simulation speed-up is being achieved by neglecting the weakly coupled elements. Due to the high aspect ratios as well as the close proximity of the current cells of inductively coupled coil models in the high frequency regime, the applicability of these two methods is limited (Scholz, 2010). In Scholz et al. (2010), an iterative procedure, described in more detail in Sect. 2, is presented where the modeling of the antenna impedances and the coupling between the antennas is separated. While this approach allows for fast spatial parameter sweeps as soon as the parameters for the equivalent antenna circuit models have been extracted, the necessity for the user to perform preceding reduction steps means a drawback.

Here, a parametric modeling technique for a fast polynomial approximation of the macro-model parameters of a single coil, depicted in Fig. 1, is being presented (see Sect. 3). The method is based on the PEEC method as well as the reduced self-impedance broadband models of Scholz et al. (2010). A major advantage is the wide range of validity

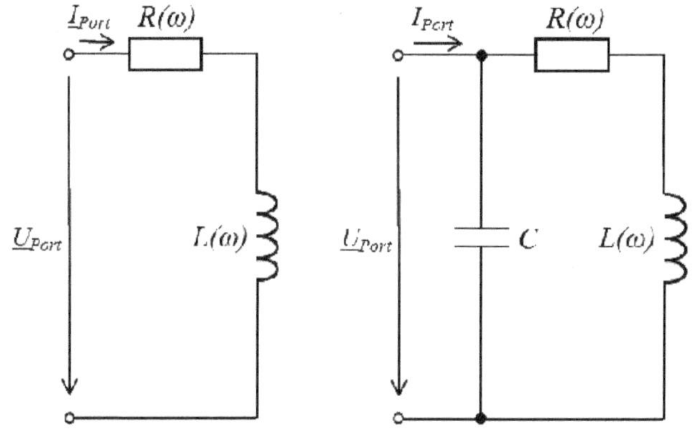

Fig. 1. MQS (left) and QS (right) equivalent circuit model of a coil.

of the established model equations, allowing for the determination of antenna RLC parameters of an entire antenna class, thereby freeing the designer from time consuming model reduction steps. In a case study, the DC antenna inductance for the class of rectangular coils with rectangular cross section is modeled and compared to the results of a reference PEEC solver concerning accuracy and simulation time (see Sect. 4).

2 PEEC and reduced broadband self-impedance models

As the parametric modeling technique presented here is based on PEEC as well as reduced broadband self-impedance models, the two methods are outlined in short in the following.

2.1 PEEC models

The PEEC method is an integral equation based numerical method for the electromagnetic characterization of arbitrary three-dimensional interconnection structures, developed by Ruehli (1974). By means of the Mixed Potential Integral Equations (MPIE), the electric field is expressed as a function of the electric potential $\phi(\underline{r},t)$ and vector potential $\underline{A}(\underline{r},t)$:

$$\underline{E} = -\mathrm{grad}\varphi - j\omega\underline{A}. \quad (1)$$

The electric field in Eq. (1) is substituted with ohm's law $\underline{J}/\kappa$ and the potentials $\phi(\underline{r},t)$ and $\underline{A}(\underline{r},t)$ are substituted with the Green's functions containing the source terms ρ and $\underline{J}$. The interconnection structure is then partitioned into inductive/resistive volume cells containing the currents and capacitive surface cells containing the surface charges. After developing the source terms by constant basis functions and applying the Galerkin method, the terms in Eq. (1) can be interpreted as partial resistances, capacitances and inductances, where the capacitive and inductive mutual couplings are being considered. Thus, a transition to the network domain is obtained where the partial elements form an equivalent circuit that can be analyzed by circuit solvers, e.g. SPICE. A system of equations can then be set up, e.g. by means of the Modified Nodal Analysis (MNA) or a mesh-based approach, where the system matrices are set up by means of the partial elements and the source terms form the states of the system. The found solutions for the source terms may then be used to determine the EM fields of the considered problem.

As the inductively coupled coil antennas considered here are designed for the 13.56 MHz frequency range, retardation effects can be neglected, making it sufficient to use the non-retarded PEEC method based on the quasi static assumption of Maxwell's equations.

2.2 Reduced broadband self-impedance models

In Scholz et al. (2010), an antenna parameter extraction technique is presented where at first the antenna impedance of each individual antenna is determined in an iterative procedure, followed by the calculation of the mutual inductive coupling between the antennas. Finally, the extracted parameters are assembled to a *reduced broadband self-impedance model*.

In the first step, a magneto quasi static (MQS) and a quasi static (QS) high complexity antenna model is extracted by means of a number of PEEC simulations, whereas in the first case the capacitive elements are being neglected (see Fig. 1, left) and incorporated in the second case (see Fig. 1, right). By means of the obtained magneto quasi static and quasi static antenna impedances Z_{MQS} and Z_{QS}, the lumped inductance L and resistance R, as well as the parasitic capacitance C of the coil (see Fig. 1) can be extracted from the PEEC models, using network based extraction-techniques (Scholz et al., 2010). In the second step, the mutual coupling M between the antennas is efficiently calculated by means of a filamentary approximation of the coil's segments (Grover, 2004) as well as a modified version of the well-known Greenhouse method (Greenhouse, 1974) adapted for the PEEC method (Reinhold et al., 2007).

Based on this procedure, the required input parameters for broadband models valid from DC to the first resonance can be obtained. The frequency dependent behavior of the impedance is then approximated by means of a ladder model obtained from e.g. a continued fraction expansion of the expression describing $Z(\omega)$ (see Engin et al., 2005; Scholz et al., 2010).

While the calculation of the mutual inductance M can be performed very efficient with the aid of the before mentioned measures, the extraction of the RLC parameters for the actual antennas demands several costly PEEC simulations. In Sect. 3, an approach for the automated extraction of polynomials allowing for the fast approximation of the RLC coil parameters is presented that releases the designer from having to perform preceding reduction steps.

Table 1. Input parameters and range.

Parameter	Range
Outer length l_x	20–100 mm
Outer length l_y	20–100 mm
Trace width w	1.0–2.6 mm
Track spacing s	1.0–2.6 mm
Trace height h	35 μm
Number of turns N_t	1–3

3 Automated parametrical modelling of rectangular spiral inductors

The automated parametrical antenna modelling technique presented in this contribution is illustrated by means of the approximation of the coil's DC inductance L_{DC} in the following, but can be applied for the approximation of any other RLC antenna parameter as well. This is due to the fact that the procedure for the extraction of the polynomial equations used for the approximation of a certain antenna parameter is the same for any parameter, making it a fully automated approach. A main advantage is the wide range of validity of the polynomial equations, making it possible to approximate a certain antenna parameter for any input parameter combination with one set of polynomial equations only. Thus, by means of the extracted equations, the antenna parameters for a whole class of antennas are being calculated. Here, the class of rectangular coils with rectangular cross section is focused on.

As mentioned earlier, the presented technique is based on the PEEC method as well as the *reduced broadband self-impedance models*. The input parameters needed for the extraction of the polynomial equations are obtained by a number of PEEC simulations performed within the antennas' geometrical input parameter range (see Table 1), followed by the parameter extraction methodology of the reduced broadband self-impedance models outlined in Sect. 2.2. The extracted parameters, in this example the DC inductance L_{DC}, serve as the input values for the program source code used for the extraction of the polynomial equations.

To be able to determine the DC inductance L_{DC} of the coil over the whole geometrical input parameter range, it is necessary to define a minimum step size Δ_{Step} for each parameter of Table 1. The step size needs to be chosen so that the change of the sought antenna parameter L_{DC} w.r.t. a respective input parameter smaller than the chosen step size does not exceed a value of interest for the user. Here it is chosen $\Delta_{Step,l} = 1$ mm for the outer lengths l_x, l_y and $\Delta_{Step,w} = \Delta_{Step,s} = 0.1$ mm for the track width w and the track spacing s. Thus, the input parameter range is subdivided into $N_{l_x} = N_{l_y} = 81$ steps for l_x and l_y and $N_w = N_s = 17$ steps for w and s, leading to a number of input parameter combinations of

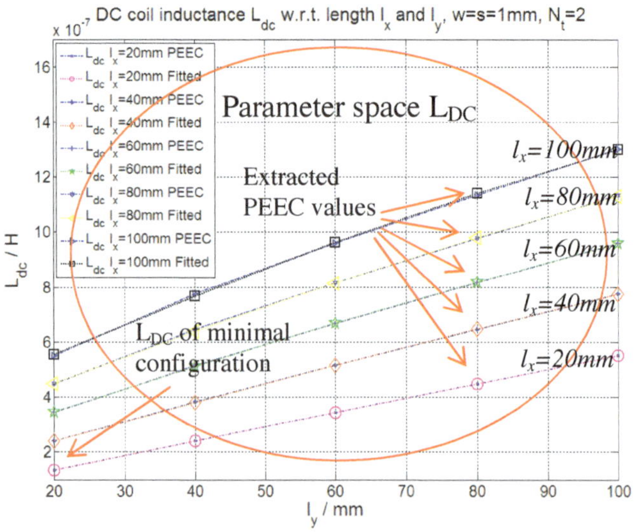

Fig. 2. Parameter space of DC inductance L_{DC} w.r.t. input parameters l_x, l_y.

$$N_{Komb} = N_{l_y\text{Steps}} \cdot N_{l_x\text{Steps}} \cdot N_{w\text{Steps}} \cdot N_{s\text{Steps}} \approx 1.89 \times 10^6. \quad (2)$$

In Fig. 2, the parameter space of the DC inductance $L_{DC}(l_x, l_y)$ w.r.t. two of geometrical input parameters is shown, namely the two outer lengths l_x and l_y.

The array of curves describes the DC inductance w.r.t. the outer length l_y, and every curve of the array represents a discrete value of the outer length l_x (see legend of Fig. 2). The dots mark the extracted inductance values from the PEEC simulations. The curves are approximated by polynomial interpolation using polynomials of order two. While Fig. 2 illustrates the modelling of the DC inductance w.r.t. only two input parameters, the main challenge is to model the inductance w.r.t. all five geometrical input parameters listed in Table 1 (trace height not counted). Obviously, every geometrical input parameter added to the modelling procedure needs two be incorporated w.r.t. all input parameters that were regarded in the preceding modelling steps. Starting point of the modelling procedure is to choose a minimal antenna configuration, describing the smallest antenna that can be modelled. The values of the minimal configuration correspond to the smallest values of the parameter range in Table 1 for each input parameter (see also Fig. 2).

In the first modelling step, the DC inductance is modelled w.r.t. the outer length l_y. For all other geometrical input parameters, the values of the minimal antenna configuration are chosen. Using polynomial interpolation, a polynomial $L_{DC}(l_y)$ is obtained:

$$L_{DC}(l_y) = \sum_{n=0}^{2} K_{n,l_y} \cdot l_y^n \quad (3)$$

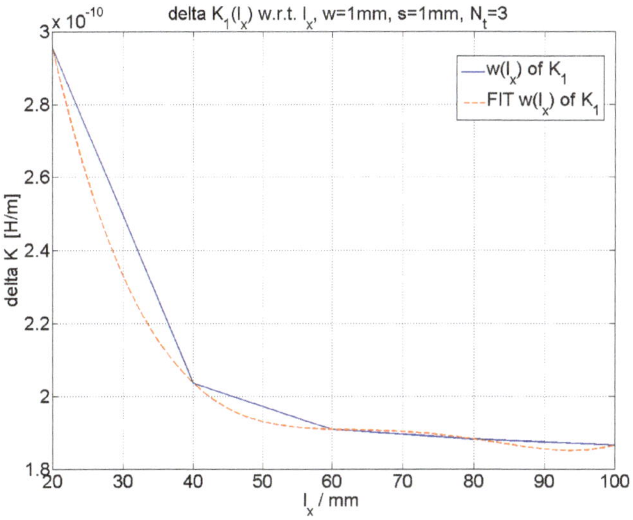

Fig. 3. Interpolation (dashed) of the change of coefficient $\Delta K_1(l_x)$ (continuous).

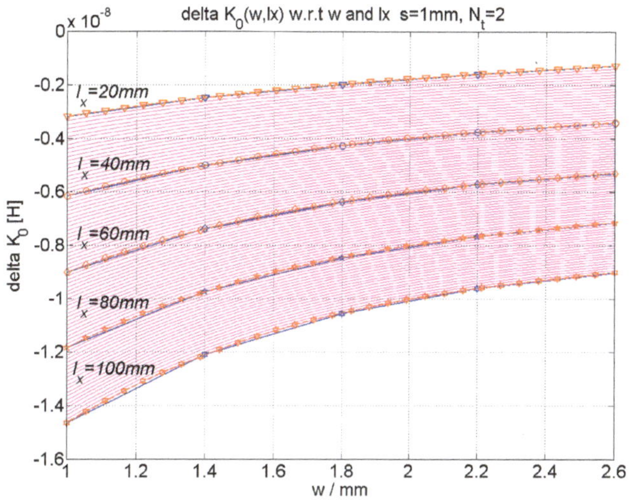

Fig. 4. Interpolation (dashed marked) of the change of coefficient $\Delta K_0(w, l_x)$ (marked and unmarked continuous).

Equation (3) describes the bottommost curve in Fig. 2. Next, the second length parameter, l_x, is accounted for by investigating the change of the coefficients $K_{2,l_y} - K_{0,l_y}$ in Eq. (3) w.r.t. l_x. First, the rest of the curves of the array in Fig. 2 are modeled by means of polynomial interpolation accordingly to the extraction of Eq. (3). Then, the PEEC simulations and L_{DC} parameter extraction is performed again for l_x, this time with a displacement of l_x of the step size $\Delta_{\text{Step},l} = 1$ mm compared to the original simulations. Finally, the change of the coefficients $K_{2,l_y} - K_{0,l_y}$ is determined by comparison of the original coefficients with the coefficients obtained for the displaced curves at five discrete equidistant points over the input range of the parameter l_x.

In Fig. 3, the piece-wise linear curve represents the determined change of coefficient $\Delta K_{1,m}(l_x)$, where the indices m describes the subdivision of the range of $\Delta K_1(l_x)$ into $N_{l_x\text{Steps}}$ (see Eq. 2) discrete values. This curve is fitted by means of a polynomial of third order, illustrated by the dashed curve. The change of the original coefficients $K_{2,l_y} - K_{0,l_y}$ w.r.t. l_x is expressed by means of a new set of coefficients $K_{2,l_x} - K_{0,l_x}$, see Eq. (4). For a certain l_x, these coefficients are obtained by summing the first M coefficient changes $\Delta K_{1,m}(l_x)$, where M is defined according to Eq. (4) and $l_{x,\text{Min}}$ describes the lower bound of the input parameter range of l_x.

$$K_{n,l_x} = \sum_{m=1}^{M} \Delta K_{n,m}(l_x) \text{ with } n=0,1,2 \text{ and } M = \frac{l_x - l_{x,\text{Min}}}{\Delta_{\text{step},l}} \quad (4)$$

To obtain the DC inductance $L_{\text{DC}}(l_x, l_y)$ w.r.t. l_y and l_x, the new coefficients $K_{2,l_x} - K_{0,l_x}$ are added to the respective coefficients $K_{2,l_y} - K_{0,l_y}$ according to Eq. (5):

$$L_{\text{DC}}(l_y, l_x) = \sum_{n=0}^{2} \left(K_{n,l_y} + K_{n,l_x}\right) \cdot l_y^n \quad (5)$$

To suggest for the influence of the track width w, this parameter needs to be incorporated in the modeling process dependent on the previously considered input parameter l_x. Other than this, a similar procedure as for the incorporation of the parameter l_x is pursued. The change of the coefficients $K_{2,l_y} - K_{0,l_y}$ and $K_{2,l_x} - K_{0,l_x}$ in Eq. (4) w.r.t. w and l_x is determined by repeating the original PEEC simulations with w displaced by the step size $\Delta_{\text{Step},w} = 0.1$ mm and calculation of the change of the coefficients by comparing the displaced with the original results. In Fig. 4, the continuous marked curves represent the change of coefficient $\Delta K_{0,m}(w, l_x)$, where each individual curve describes the change of coefficient w.r.t. the track width w for a certain discrete value of l_x over the input range of l_x.

The dashed marked curves describe the polynomial approximation of the original curves by means of third order polynomials. An array of thin lined curves between two dashed marked curves illustrates the approximation of the value range between two adjacent dashed marked curves. As two adjacent dashed marked curves correspond to a length difference of $\Delta l_x = 20$ mm in this example, the value range is subdivided by means of the step size $\Delta_{\text{Step},l} = 1$ mm, resulting in 19 thin lined curves between every pair of dashed marked curves. This is achieved by a linear subdivision of the value range between two respective polynomial coefficients of the same degree, belonging to two adjacent dashed marked curves. Following the procedure of the incorporation of the input parameter l_x, a new set of coefficients $K_{2,w} - K_{0,w}$ is extracted for a certain track width to be modeled. $K_{2,w} - K_{0,w}$ are again obtained by summation of the first M coefficient changes $\Delta K_{2,m}(w, l_x) - \Delta K_{0,m}(w, l_x)$, see Eq. (6).

$$K_{n,w} = \sum_{m=1}^{M} \Delta K_{n,m}(w, l_x) \text{ with } n=0,1,2 \text{ and } M = \frac{w - w_{\text{Min}}}{\Delta\text{step}} \quad (6)$$

To obtain the DC inductance $L_{\mathrm{DC}}(w, l_x, l_y)$ w.r.t. w, l_y and l_x, the new coefficients $K_{2,w} - K_{0,w}$ are added to the respective coefficients $K_{2,l_y} - K_{0,l_y}$ and $K_{2,l_x} - K_{0,l_x}$ according to Eq. (7):

$$L_{\mathrm{DC}}(l_y, l_x, w) = \sum_{n=0}^{2} \left(K_{n,l_y} + K_{n,l_x} + K_{n,w}\right) \cdot l_y^n \qquad (7)$$

The inclusion of the track space s in the modeling of the DC inductance $L_{\mathrm{DC}}(s, w, l_x, l_y)$ follows a similar procedure as described for the input parameters l_x and w. Again, the incorporation of s needs to be accomplished w.r.t. the input parameters that were regarded for in the preceding modeling steps, namely l_x and w. This is realized in two steps by means of the extraction of the coefficient change $\Delta K_{2,m}(s, w, l_x) - \Delta K_{0,m}(s, w, l_x)$ through comparison of the original with displaced PEEC simulations (displacement $\Delta_{\mathrm{Step},s} = 0.1$ mm):

first, a coarse subdivision of the parameter range of l_x and w in e.g. 5 subsections is carried out. Then, for every subdivision point of w the changes of coefficients $\Delta K_{2,m}(s, w, l_x) - \Delta K_{0,m}(s, w, l_x)$ are determined at every subdivision point of l_x. Next, after approximating the three functions $\Delta K_{2,m}(s, w, l_x) - \Delta K_{0,m}(s, w, l_x)$ by means of third order polynomials, for every adjacent pair of coefficients of the same degree of these polynomials, a linear subdivision is undertaken corresponding to the subdivision of the range between two subdivision points of l_x. By this, an approximation of the change of coefficients $\Delta K_{2,m}(s, w, l_x) - \Delta K_{0,m}(s, w, l_x)$ over the whole value range of the input parameter l_x is obtained at each of the five discrete subdivision points of the input parameter w.

In the second step, approximations over the whole input range of w, i.e. between the subdivision points of w, need to be found. To that end, the family of polynomials describing the change of coefficients $\Delta K_{2,m}(s, w, l_x) - \Delta K_{0,m}(s, w, l_x)$ over the whole value range of l_x at one subdivision point w is compared to the corresponding family of polynomials of the adjacent subdivision point w. This is carried out for all pairs of subdivision points w, again by means of linear subdivision of adjacent pairs of coefficients of the same degree.

By means of this two-step procedure, a new set of coefficients $K_{2,s} - K_{0,s}$ in Eq. (8) is obtained from the summation of the change of coefficients $\Delta K_{2,m}(s, w, l_x) - \Delta K_{0,m}(s, w, l_x)$.

$$K_{n,s} = \sum_{m=1}^{M} \Delta K_{n,m}(s, w, l_x) \text{ with } n = 0, 1, 2 \text{ and } M = \frac{s - s_{\mathrm{Min}}}{\Delta \mathrm{step}} \qquad (8)$$

To determine the DC inductance $L_{\mathrm{DC}}(s, w, l_x, l_y)$, the a new set of coefficients $K_{2,s} - K_{0,s}$ is added to the coefficients extracted in the preceding steps.

$$L_{\mathrm{DC}}(l_y, l_x, w) = \sum_{n=0}^{2} \left(K_{n,l_y} + K_{n,l_x} + K_{n,w} + K_{n,s}\right) \cdot l_y^n \qquad (9)$$

Table 2. Comparison of simulation time.

Method	Turns N_t	Simulation Time
Parametric modeling	arbitrary	0.25 s
PEEC	1	7.52 s
PEEC	3	67.04 s
PEEC	5	183.40 s

The remaining input parameter to be included in the modeling process is the number of turns N_t. Contrary to the input parameters considered so far, this parameter attains only integers. No intermediate values have to be considered. Thus, it seems natural to incorporate this parameter differently. While it might be possible to incorporate the last input parameter N_t in the same manner as the other parameters, it was decided here for reasons of simplification to repeat the modeling procedure described in the preceding N_t number of times. This means that N_t equations of type (9) describe a class of antennas to be modeled.

The polynomial equation extraction procedure outlined here was illustrated using the example of the DC inductance L_{DC}, but is applicable for any other antenna parameter without modification of the program code. Thus, the input parameter extraction for the broadband ladder models described in Sect. 2.2 is straightforward.

4 Validation of the method

In the following, the validation of the presented modelling approach is made by means of comparison with the PEEC method and the *reduced self-impedance broadband models*.

Whereas in the latter, the designer needs to perform a reduction of the full PEEC models to obtain the reduced model for every antenna configuration, this is not necessary for the automated parametrical modelling technique: Here, the designer obtains a set of polynomial equations valid for a class of antennas to extract the physically relevant parameters. All time consuming reduction steps were carried out in the course of the geometrical input parameter determination needed for the equation extraction, as described at the beginning of Sect. 3. The actual equation extraction is fully automated for all antenna parameters and demands only a few seconds. Regarding the computational effort of the parameter extraction with the new method, the simulation time for three different spiral coils is shown in Table 2 in comparison with full PEEC simulations. For all three coils holds: $w = s = 2$ mm, $l_x = l_y = 50$ mm and $h = 35\,\mu$m. The PEEC simulations were performed by means of a simulator designed in the course of this work. All simulations were performed on an Intel Core i3 Processor with 4 GByte RAM. The simulations were carried out at a single frequency point

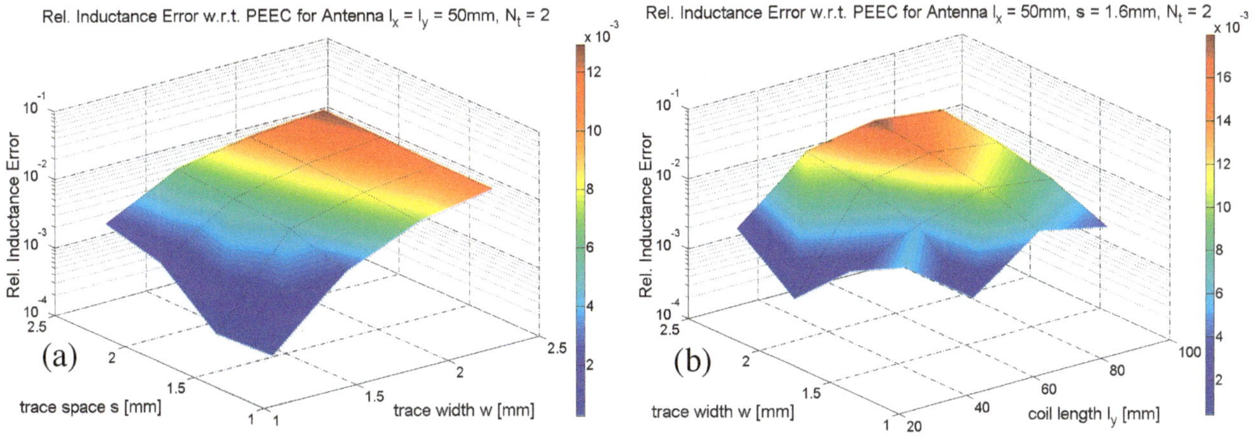

Fig. 5. Relative inductance error w.r.t. trace space s and trace width w **(a)** and w.r.t. trace width w and length l_y **(b)**.

at $f_{DC} = 1$ Hz. While the simulation time raises exponentially with the number of turns for the PEEC simulations, it remains constant for the parametrical modeling technique, as it does not depend on the configuration of the antenna. The total simulation time is less than a second for the new approach, while the PEEC simulation time exceeds one minute for a three turn antenna (Table 2). Concerning the accuracy of the new approach, 256 simulations were performed with the PEEC reference solver over the input parameter range. The geometrical input parameters were chosen as far away as possible from the input parameters of the original PEEC simulations, on which the equation extraction was based on. It is assumed that the modeling error is largest for these values. Figure 5a shows the relative inductance error w.r.t. the trace width w and the trace space s for a two-turn antenna with the outer lengths $l_x = l_y = 50$ mm. The rel. error lies below ~ 1.3 % for all values. It rises with increasing trace width and space and is more sensitive to the trace width than to the trace space. Figure 5b shows the relative inductance error w.r.t. the trace width w and the outer length l_x for a two-turn antenna with $l_x = 50$ mm and $s = 1.6$ mm. Here, the rel. error lies below ~ 1.7 % for all values. The maximum relative error w.r.t. all 256 reference values is ~ 2.2 % and the mean relative error over all values is ~ 0.6 %. Thus, a very good agreement between the extracted parameters of the parametrical modeling technique and the PEEC reference values is found.

5 Conclusions

In this contribution a parametrical modelling approach for the fast characterization of inductively coupled coils was presented. Based on the PEEC method and the reduced broadband self-impedance models, polynomial equations for the determination of the coil's DC inductance were extracted. The approach is fully automated in the sense that it is applicable for the extraction of all physically relevant RLC coil parameters without any modification. Further, its usage is not limited to the PEEC method, but can be applied universally to any numerical method that is able to characterize antennas by means of lumped element models. Numerical results show a good agreement with the PEEC method within the input parameter range. The mean relative error is ~ 0.6 % and the maximum relative error is ~ 2.2 % for a parameter sweep of 256 antenna configurations. With a computation time of below one second for the parameter extraction, the presented technique is very efficient compared to full-wave solvers. In contrast to other model reduction methods, the user is freed from having to perform reduction steps to obtain a reduced model. Another advantage is the easy handling for the user, as the geometrical input parameters simply need to be entered and no antenna geometries or script files need to be established. The extension to broadband models by means of physically motivated ladder-models is straightforward.

Acknowledgements. This work was performed in the frame of the BMBF/ENIAC project MAS Mobile Ambient Assisted Living Systems. This particular research was supported by BMBF (Germany) and ENIAC (EU) under grant 16 SV 4059. The responsibility for this publication is held by the authors only.

References

Antonini, G.: Fast Multipole Method for Time Domain PEEC Analysis; IEEE Trans. Mobile Comput., 2, 275–287, 2003.

Devgan, A., Ji, H., and Dai, W.: How to Efficiently Capture On-chip Inductance Effects – Introducing a New Circuit Element K; Proc. IEEE/ACM Int. Conf. on Computer Aided Design (ICCAD'00), 150–155, 2000.

Engin, A. E., Mathis, W., John, W., Sommer, G., and Reichl, H.: Closed-form Network Representations of Frequency-dependent RLGC Parameters, Int. J. Circuit Theory Appl., 33, 463–485, 2005.

Greenhouse, H. M.: Design of Planar Rectangular Microelectronic Inductors, IEEE Trans. Parts, Hybrids, Packag., PHP-10, 101–109, 1974.

Grover, F. W.: Inductance Calculations; D. Van Nostrand Co., New York, 1946; reprinted by Dover Publications, New York, 2004.

Reinhold, C., Scholz, P., John, W., and Hilleringmann, U.: Efficient Antenna Design of Inductive Coupled RFID-systems with High Power Demand, Journal of Communications, 2, 14–23, available at: http://www.academypublisher.com/jcm/vol02/no06/jcm02061423.pdf, 2007.

Ruehli, A. E.: Equivalent Circuit Models for Three-dimensional Multiconductor Systems, IEEE Trans. Microw. Theory Tech., MTT-22, 216–221, 1974

Scholz, P.: Analysis and Numerical Modeling of Inductively Coupled Antenna Systems, Ph.D. thesis, http://tuprints.ulb.tu-darmstadt.de/2354/, 2010.

Scholz, P., Ackermann, W., Weiland, T., Reinhold, C.: Antenna Modeling for Inductive RFID Applications Using the Partial Element Equivalent Circuit Method; IEEE Trans. Magn., 46, 2967–2970, 2010.

A context management system for a cost-efficient smart home platform

J. Schneider, A. Klein, C. Mannweiler, and H. D. Schotten

University of Kaiserslautern, Paul-Ehrlich-Street, Building 11, Germany

Institute for Wireless Communications and Navigation

Correspondence to: J. Schneider (schneider@eit.uni-kl.de)

Abstract. This paper presents an overview of state-of-the-art architectures for integrating wireless sensor and actuators networks into the Future Internet. Furthermore, we will address advantages and disadvantages of the different architectures. With respect to these criteria, we develop a new architecture overcoming these weaknesses. Our system, called Smart Home Context Management System, will be used for intelligent home utilities, appliances, and electronics and includes physical, logical as well as network context sources within one concept. It considers important aspects and requirements of modern context management systems for smart X applications: plug and play as well as plug and trust capabilities, scalability, extensibility, security, and adaptability. As such, it is able to control roller blinds, heating systems as well as learn, for example, the user's taste w.r.t. to home entertainment (music, videos, etc.). Moreover, Smart Grid applications and Ambient Assisted Living (AAL) functions are applicable. With respect to AAL, we included an Emergency Handling function. It assures that emergency calls (police, ambulance or fire department) are processed appropriately. Our concept is based on a centralized Context Broker architecture, enhanced by a distributed Context Broker system. The goal of this concept is to develop a simple, low-priced, multi-functional, and save architecture affordable for everybody. Individual components of the architecture are well tested. Implementation and testing of the architecture as a whole is in progress.

1 Introduction

Today, the importance of context information has increased and a lot of systems are already based on context data. Especially automated system, like Home Automation or Home Assisted Living are based on such data. Therefore it is necessary to develop appropriate context management architectures as well as concepts and mathematical methods to interpret the data and adapting the system. In this paper, we introduce a cost efficient Context Management Platform which is based on distributed Context Broker (CxB) architecture for enabling context aware system adaptation. But first it is important to define the notion "Context". According to Dey and Abowd (Dey, 1999), Context is "any information that can be used to characterize the situation of an entity. An entity is a person, place or object that is considered relevant to the interaction between a user and an application, including the user and application themselves."

For the so called Smart Home Platform (SHP), context consists of sensor data and data which is gathered from the internet (e.g. weather). Consequently our architecture must be able to handle all these kind of context data. Different kinds of architectures have been studied in the last years. Most of the architectures have been developed with a centralized broker architecture, e.g. CoBrA (Chen, 2003a) (Chen, 2003a) or (Schneider, 2010a).

Easy Meeting (Chen, 2004a,b) and MyCampusEasy (Gandon, 2003) (Gandon, 2004) are typical projects were researchers deal with reasoning. Easy Meeting is based on the Context Broker Architecture (CoBrA) and is a prototype of an intelligent meeting room. The Carnegie Mellon University has developed the MyCampus to provide context aware services for the community in the university. The e-Wallet is the key element, that supports access control and obfuscation rules for user's privacy.

The introduced approaches are based on a centralized Context Broker. A resulting disadvantage of such a system is the limited quantity of administrable clients. However, the basis for our system will be a centralized Broker system (compare CCast project, CCast, 2009). It breaks the system down into three main entities, a Context Provider (CxP), a Context Consumer (CxC) and a Context Broker. The CxB acts a broker for all kind of data and is aware of all CxPs and CxCs. Figure 1 shows the principal structure of the centralized CxB approach. This approach is not entirely new and

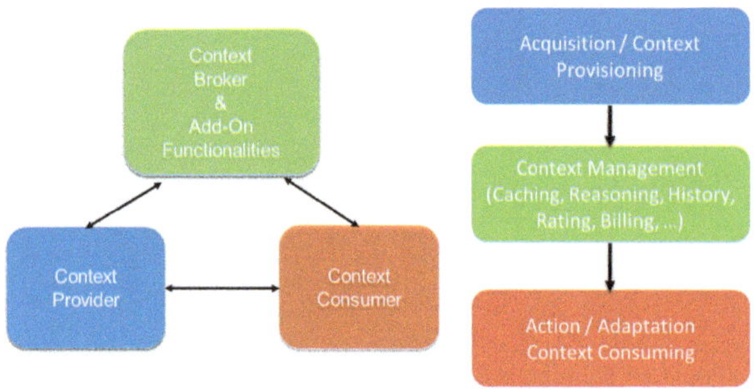

Fig. 1. Centralized Context Management architecture.

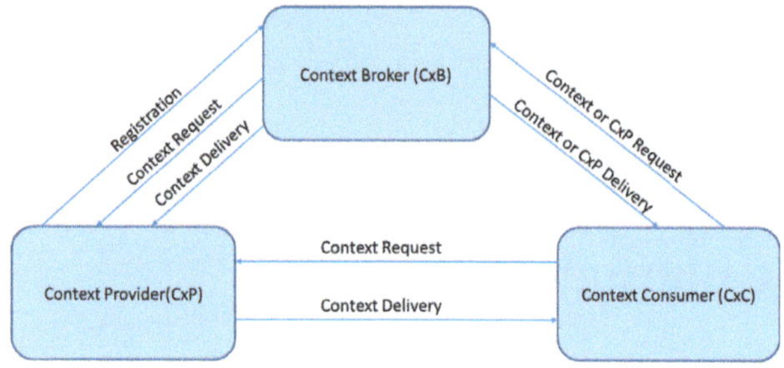

Fig. 2. Interfaces Centralized Context Management architecture.

has already been studied by several researchers and used in different projects.

In the case of large architectures, more precisely systems with hundreds of clients, an efficient management of the data is needed. A disadvantage of the centralized approach is the decreasing performance if the number of entities increases. Especially in the use case of a Smart Home Platform, sensor data and other context data (e.g. from the internet) will be used to trigger events. Additionally, the system intelligence is located into so called Services. Each Service implements a Context Consumer and a Context Provider Interface, furthermore is has an interface to subscribe for context data. This interface is needed to enable value-triggered events.

Hence a centralized architecture will not be able to handle all this amount of data. To overcome this weakness, we developed a distributed architecture which is based on the federation of sub context managements systems. The organization of these systems will be comparable to a Domain Naming Service (DNS) and was introduced in (Schneider, 2010b).

2 Decentralized context management architecture

In this section, we introduce our decentralised system approach in detail. But first it es necessary to define some requirements for the system:

- *Plug and Play:* will be fulfilled with respect to Schneider, 2010b and Schneider, 2009
- *Security and Privacy*

2.1 Interfaces

As basic concept we used the centralized broker architecture from (CCast, 2009). Figure 2 shows all necessary interfaces to deploy the system in a centralized manner.

As already explained, for the smart home platform use case a decentralized architecture is necessary. Therefore, some of the available entities need to be enhanced with additional interfaces and will be described in the following subsections.

Table 1. Interfaces of Centralized Context Broker.

Interfaces	Description	Data
Registration	Registration of available CxPs and verifying of the CxP list.	Bidirectional via TCP/IP
Context Transmission with CxP	Request and transmission of Context Data between CxP and CxB	Bidirectional via TCP/IP
Context Transmission with CxC	Request and transmission of Context Data between CxC and CxB	Bidirectional via TCP/IP

2.1.1 Context broker CxB

The centralized CxP implementation comprises of the following interfaces:

To use Context Broker in a distributed architecture, additional interfaces are necessary. One interface is needed to enable communication between two CxBs. But before this communication can be established, a CxB detection needs to be performed. Therefore, a Broker detection interface has been included. It is used to detect adjacent CxBs and to request their address to establish a communication. Additionally, each CxB has to provide an Provider Detection interface and a Consumer Detection Interface.

2.1.2 Context provider CxP

A Context Provider delivers various kind of data. It is not restricted to sensor data from physical sensors (such as temperature or light), but also allows to provide pre-processed data, for example system load or available bandwidth. For the centralized system, a CxP is equipped with the Interfaces listed in Table 2.

Table 2. Interfaces of Centralized Context Provider.

Interfaces	Description	Data
Registration	Registration to a CxB.	Bidirectional via TCP/IP
Context Transmission to CxB	Request and transmission of Context Data between CxP and CxB	Bidirectional via TCP/IP
Context Transmission with CxC	Request and transmission of Context Data between CxP and CxC	Bidirectional via TCP/IP

Table 3. Interfaces of Centralized Context Consumer.

Interfaces	Description	Data
Context Transmission to CxB	Request and transmission of Context Data between CxP and CxB	Bidirectional via TCP/IP
Context Transmission with CxP	Request and transmission of Context Data between CxP and CxC	Bidirectional via TCP/IP

With respect to the plug and play requirement, an additional Broker Detection Interface is necessary. This Interface is the counterpart to the Provider Detection Interface in a CxB. The detailed function of a CxB detection is described in Schneider, 2010a.

Additionally each Broker need an interface to communicate with other CxBs. But before a CxB is able to communicate with another CxB, a detection process is to be carried out. Therefore, a CxB sends a broadcast with its own address and waits for response.

Another issue is the implementation of a context subscription interface for enabling services. This interface must be able to allow CxCs to subscribe to context for periodic as well as event-triggered data transfer.

2.1.3 Context consumer CxC

A Context Consumer is an entity which uses context data for its actual purpose. This can be, for example, an actuator for steering a roller blind or the heater. But for developing of our system architecture, we didn't make any restrictions. This offers the possibility to include any kind of services into the platform. For the centralized approach, a CxC is equipped with the Interfaces, listed in Table 3.

This entity also needs an additional Broker detection interface, which is counterpart of the Consumer Detection Interface, located on a CxB.

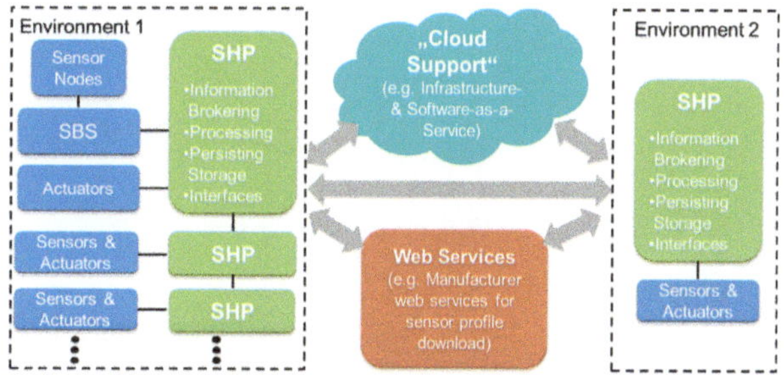

Fig. 3. Cloud-/Web-based Service Provisioning.

2.2 Services

An important issue is the integration and provision of services. Services are software components to implement any kind of functionality. Figure 3 shows an example the principal structure of providing cloud-/web-based services to distributed environment.

With this system, a user is able to add or remove functionalities to the system without any additional hardware. Further more functionalities can be moved to the "Cloud", in other words, external servers provide most of the necessary services and the user needs only the basic system. Further functionalities can be bought or rented if required.

A user also has the possibility to provide own sensor data or offer own services for public use. With this concept the privacy of a user is ensured and he can decide, which data/information should be private or public.

As already mentioned, the centralized broker architecture is not new and has been studied in several projects. Services offer the possibility to maintain various kinds of tools. A user can subscribe to a service and adjust the service behaviour according to his demand. Therefore the service has to request or subscribe for the necessary context information. With this data, a service is able to trigger an event based on the users adjustments. Each event will request the address of the appropriate CxC and query for the requested data. All kind of data transmission is performed via TCP/IP in a XML coded scheme.

This approach provides the possibility to move software components, especially computationally intensive processes to the "Cloud". For our system, it doesn't matter how the cloud looks like, or is implemented, respectively. We handle it as a black box but in most cases it will be implemented as a software-service on an external server located in the internet i.e. in the "Cloud".

2.3 Data transmission

An example for data transmission is given in Fig. 4. It shows the automatic detection of a Smart Home Platform (SHP).

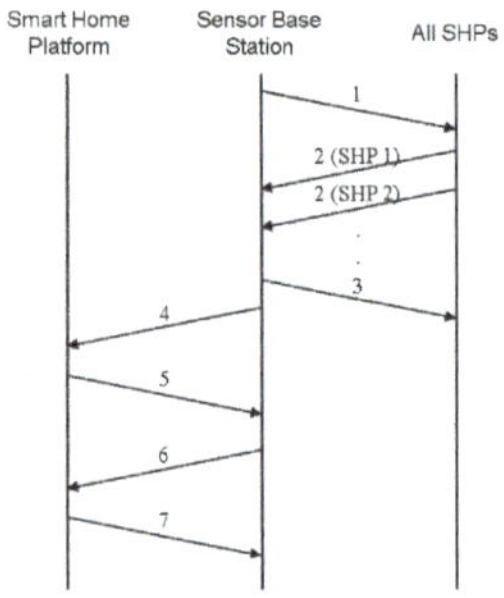

Number	Transmission Type	Port	Payload	Description
1	UDP: Broadcast	5000	getSHP	"get SHP" request
2	TCP: Acknowledge	5000	SHP IP Port	Acknowledgement from SHPs with appropriate meta data
3	UDP: Broadcast	5000	StopACK request	Tells other SMPs to stop ACK
4	TCP: Advertising	5001	SBS metadata sensor information	
5	TCP: Acknowledge	5001	Acknowledge expiry date	ACK of advertisement
6	TCP: Advertising	5001	SBS metadata sensor information	
7	TCP: Acknowledge	5001	Acknowledge expiry date	ACK of advertisement

Fig. 4. Automatic SHP detection.

As already explained, all the communication between the entities will be done by a XML derivate Language called ContextML (CML). In this paper we will not provide details on CML, however, refer interested to (Knappmeyer, 2010).

2.4 Context prediction and situation detection

To make a Smart Home Platform as attractive as possible for a consumer, most of the functions should be automatically performed and installation effort should be as low as possible. To deploy such a SHP in a fully automated manner, it is necessary to detect situation or to predict them. Consequently it is necessary to predict context information and to combine the data for gathering new findings. To achieve this goal, different concepts are available to recognize a situation. But first, we define the term "situation": "A situation is an abstraction for a pattern of observations made by a distributed system such as a sensor network" (Cardell, 2010). Therefore different recognition algorithms are available:

- *Finite State Machine*
- *Tree-based evaluation*
- *Distributed Commitment Machines*

The goal of such a complex system is the Combined observations of the individual sensor nodes yield spatial-temporal map of the monitored environment. With this data a service is able to adopt the system behaviour to the users preferences.

To detect a dynamic situation, it is essential to select an appropriate modelling method. Two approaches are adequate for this usage:

- *Construction of appropriate Bayesian networks*
- *Consideration of approximate inference methods (e.g. Monte Carlo algorithm)*

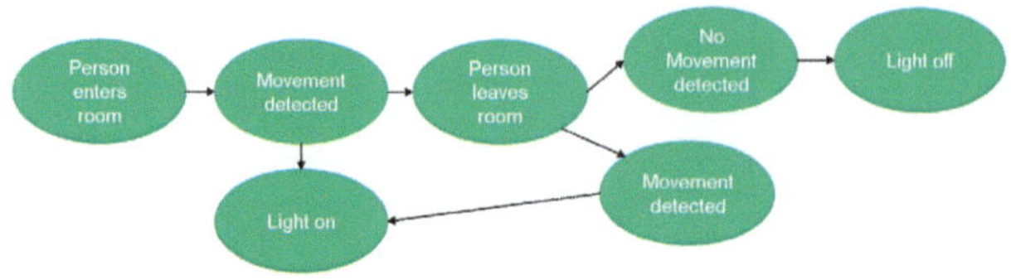

Fig. 5. Bayesian Network.

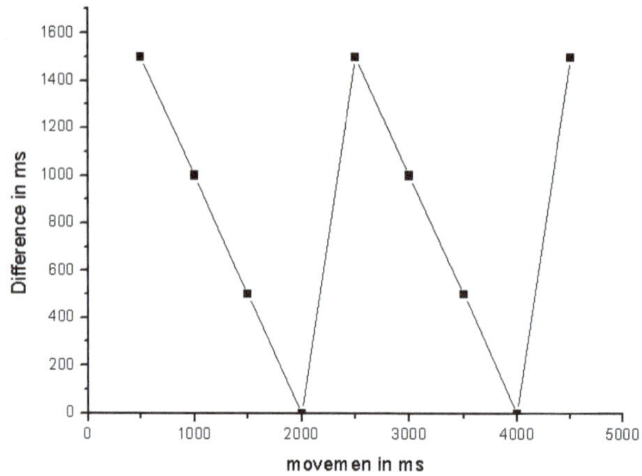

Fig. 6. Room light simulation.

Figure 5 shows an example for a Bayesian network and models the light automation. The room is equipped with different kind of sensors (e.g. movement, temperature, ...), consequently the system is able to detect if somebody is in the room. If movement detection is true, the light is switched on. A weakness respectively a problem is the time how long the room is in use. This results in the problem, how long should we turn on the light after a movement? If there is no movement, has the person left the room or is she/he still sitting inside the room? If we describe the system with a fixed on-time, we get the graph in Fig. 6.

To get an energy efficient system, the time difference between the time of movement and the light-on time should be minimized. Figure 6 shows that systems with fixed times are not able to provide satisfying results. Furthermore general concepts to provide adaptive system behaviour need to be developed. In case of a smart home platform, most of the system parameters depend on the users behaviour and are function of time. To ensure such a behaviour, we use a Finite State Machine and combine this with a Bayesian Network. The following subsections will describe our model in detail.

2.4.1 Finite state model

A Finite State Model is a mathematical model that is used to design computer programs and digital logic circuits. As already explained above, the system behaviour is simplified into a finite amount of states. A transition from one state to another can only initiated by a triggering event or a changing condition. As result, we sustain a model, similar to Fig. 5.

2.4.2 Bayesian network

In our case, Bayesian Networks can be seen as an extension of the finite state model. The additional part or the extensions are the conditional probability tables. To make such a system adaptive respectively intelligent, these table are not fixed and must depend on external parameters. In our movement example, time seems to be a very important parameter because most of the human behaviour depends on the daytime. Therefore appropriate model for the userB4s behaviour have to be found.

3 Conclusions

In this paper we introduced the so called Smart Home Platform and listed the requirements to the Context Management Platform. Furthermore, we have shown, that it is possible to extend available centralized broker architectures to distributed architectures with adding interfaces to the available components. A benefit of this approach is, that available interfaces can be used without any change. The new introduced interfaces can use the available communication protocol which is based on ContextML. Additionally the architecture has been extended with services that offer the user to add functionalities to the system without any change of the hardware.

Furthermore, we have shown approaches to predict Context and to detect situations. This enables an intelligent behaviour of the Smart Home Platform

Acknowledgements. This work has been supported by the Federal Ministry of Education and Research of the Federal Republic of Germany (Foerderkennzeichen 01 BK 0808, GLab). The authors alone are responsible for the content of the paper.

References

Dey, A. K. and Abowd, G. D.: Towards a Better Understanding of Context and Context-Awareness, College of Computing, Georgia Institute of Technology, Tech. Rep. GIT-GVU-99-22, 1999.

Context Casting C-CAST, available at: http://www.ICT-CCAST.eu, Last access: October 2011.

Chen, H., Finin, T., Anupam, J.: Semantic Web in a Pervasive Context-Aware Architecture, Artificial Intelligence in Mobile System, 2003a.

Chen, H. and Finin, T.: An ontology for context-aware pervasive computing environments, The Knowledge Engineering Review page 197–207, 2003b.

Chen, H., Finin, T., Joshi, A., Kagal, L., Perich, F., and Chakraborty, D.: Intelligent agents meet the semantic web in smart spaces, IEEE Internet Computing page 69–79, November–December 2004a.

Chen, H., Perich, F., Chakraborty, D., Finin, T., and Joshi, A.: Intelligent agents meet semantic web in a smart meeting room, AAMAS 04: Proceedings of the Third International Joint Conference on Autonomous Agents and Multiagent Systems, 854–861, IEEE Computer Society, Los Alamitos CA USA, 2004b.

Gandon, F. and Sadeh, N. M.: A semantic e-wallet to reconcile privacy and context awareness, Lecture Notes in Computer Science: The SemanticWeb – ISWC 2003, 385–401, 20–23 October 2003.

Gandon, F. and Sadeh, N. M.: Semantic web technologies to reconcile privacy and context awareness, J. Web Semant., 241–260, 2004.

Schneider, J., Klein, A., Mannweiler, C., and Schotten, H. D.: Environmental Context Detection for Context-Aware Systems, ICST's Global Community Magazine, 2010a.

Schneider, J., Mannweiler, C., Klein, A., and Schotten, H. D.: Einbindung von Sensoren und Sensornetzwerken in das Future Internet, in ITG-Fachbericht der 15. ITG Fachtagung Mobilkommunikation, Osnabrück, Germany, May 2010b.

Schneider, J., Mannweiler, C., Klein, A., and Schotten, H. D.: Erfassung von Umgebungskontext und Kontextmanagement, ITG-Fachbericht Band 215 der 14. ITG Fachtagung Mobilkommunikation, Osnabrueck, Germany, May 2009.

Knappmeyer, M., Liaquat, S., Kiani, Fra, C., Moltchanov, B., and Baker, N.: ContextML: A Light-Weight Context Representation and Context Management Schema, IEEE International Symposium on Wireless Pervasive Computing (ISWPC), Modena, Italy, May 2010.

Cardell-Oliver, R. and Liu, W.: Representation and Recognition of Situations in Sensor Networks, Communications Magazine, IEEE The University of Western Australia, Mach 2010.

Permissions

All chapters in this book were first published in ARS, by Copernicus Publications; hereby published with permission under the Creative Commons Attribution License or equivalent. Every chapter published in this book has been scrutinized by our experts. Their significance has been extensively debated. The topics covered herein carry significant findings which will fuel the growth of the discipline. They may even be implemented as practical applications or may be referred to as a beginning point for another development.

The contributors of this book come from diverse backgrounds, making this book a truly international effort. This book will bring forth new frontiers with its revolutionizing research information and detailed analysis of the nascent developments around the world.

We would like to thank all the contributing authors for lending their expertise to make the book truly unique. They have played a crucial role in the development of this book. Without their invaluable contributions this book wouldn't have been possible. They have made vital efforts to compile up to date information on the varied aspects of this subject to make this book a valuable addition to the collection of many professionals and students.

This book was conceptualized with the vision of imparting up-to-date information and advanced data in this field. To ensure the same, a matchless editorial board was set up. Every individual on the board went through rigorous rounds of assessment to prove their worth. After which they invested a large part of their time researching and compiling the most relevant data for our readers.

The editorial board has been involved in producing this book since its inception. They have spent rigorous hours researching and exploring the diverse topics which have resulted in the successful publishing of this book. They have passed on their knowledge of decades through this book. To expedite this challenging task, the publisher supported the team at every step. A small team of assistant editors was also appointed to further simplify the editing procedure and attain best results for the readers.

Apart from the editorial board, the designing team has also invested a significant amount of their time in understanding the subject and creating the most relevant covers. They scrutinized every image to scout for the most suitable representation of the subject and create an appropriate cover for the book.

The publishing team has been an ardent support to the editorial, designing and production team. Their endless efforts to recruit the best for this project, has resulted in the accomplishment of this book. They are a veteran in the field of academics and their pool of knowledge is as vast as their experience in printing. Their expertise and guidance has proved useful at every step. Their uncompromising quality standards have made this book an exceptional effort. Their encouragement from time to time has been an inspiration for everyone.

The publisher and the editorial board hope that this book will prove to be a valuable piece of knowledge for researchers, students, practitioners and scholars across the globe.

List of Contributors

F. Gössel
Dresden University of Technology, Fakultät für Verkehrswissenschaften „Friedrich List", Lehrstuhl Informationstechnik für Verkehrssysteme, D-01062 Dresden, Germany

E. Michler
Dresden University of Technology, Fakultät für Verkehrswissenschaften „Friedrich List", Lehrstuhl Informationstechnik für Verkehrssysteme, D-01062 Dresden, Germany

B. Wrase
Dresden University of Technology, Fakultät für Verkehrswissenschaften „Friedrich List", Lehrstuhl Informationstechnik für Verkehrssysteme, D-01062 Dresden, Germany

T. Thiessen
Leibniz Universität Hannover, Institut für Theoretische Elektrotechnik, Appelstr. 9A, 30167 Hannover, Germany

M. Prochaska
NXP Semiconductors Germany GmbH Automotive Innovation Center Stresemannallee 101, 22529 Hamburg, Germany

A. Miskiewicz
.Infineon technologies AG, Neubiberg, Germany
University of Erlangen-Nuremberg, Erlangen, Germany

A. Holm
Infineon technologies AG, Neubiberg, Germany

R. Weigel
University of Erlangen-Nuremberg, Erlangen, Germany

P. Beinschob
Department of Signal Processing and Communications, Helmut-Schmidt-Universität/University of the Federal Armed Forces Hamburg, Germany

M. Lieberei
Department of Signal Processing and Communications, Helmut-Schmidt-Universität/University of the Federal Armed Forces Hamburg, Germany

U. Zölzer
Department of Signal Processing and Communications, Helmut-Schmidt-Universität/University of the Federal Armed Forces Hamburg, Germany

D. Rinas
Dortmund University of Technology, Dortmund, Germany

S. Frei
Dortmund University of Technology, Dortmund, Germany

D. Würfel
Fraunhofer Institute for Microelectronic Circuits and Systems, Duisburg, Germany

M. Ruß
Fraunhofer Institute for Microelectronic Circuits and Systems, Duisburg, Germany

R. Lerch
Fraunhofer Institute for Microelectronic Circuits and Systems, Duisburg, Germany

D. Weiler
Fraunhofer Institute for Microelectronic Circuits and Systems, Duisburg, Germany

P. Yang
Fraunhofer Institute for Microelectronic Circuits and Systems, Duisburg, Germany

H. Vogt
Fraunhofer Institute for Microelectronic Circuits and Systems, Duisburg, Germany

O. Lange
Robert Bosch GmbH, Robert-Bosch-Strasse 200, 31139 Hildesheim, Germany

B. Yang
Chair of System Theory and Signal Processing, Universität Stuttgart, Pfaffenwaldring 47, 70569 Stuttgart, Germany

P. Beinschob
Department of Signal Processing and Communications, Helmut-Schmidt-Universität/University of the Federal Armed Forces Hamburg, Holstenhofweg 85, 22043 Hamburg, Germany

U. Zölzer
Department of Signal Processing and Communications, Helmut-Schmidt-Universität/University of the Federal Armed Forces Hamburg, Holstenhofweg 85, 22043 Hamburg, Germany

P. Beinschob
Department of Signal Processing and Communications, Helmut-Schmidt-Universität/University of the Federal Armed Forces, Hamburg, Holstenhofweg 85, 22043 Hamburg, Germany

U. Zölzer
Department of Signal Processing and Communications, Helmut-Schmidt-Universität/University of the Federal Armed Forces, Hamburg, Holstenhofweg 85, 22043 Hamburg, Germany

H. Günther
Technische Universität Dortmund, Dortmund, Germany

U. Hilger
Technische Universität Dortmund, Dortmund, Germany

S. Frei
Technische Universität Dortmund, Dortmund, Germany

M. Isikhan
IMMS gGmbH, Erfurt, Germany

A. Richter
IMMS gGmbH, Erfurt, Germany

P. Teichmann
Institute for Technical Electronics, Technische Universität München, Arcisstr. 21, 80333 München, Germany

C. Friederich
Institute for Technical Electronics, Technische Universität München, Arcisstr. 21, 80333 München, Germany

D. Schmitt-Landsiedel
Institute for Technical Electronics, Technische Universität München, Arcisstr. 21, 80333 München, Germany

T. Nick
Information Processing Lab, TU Dortmund University, Otto-Hahn-Str. 4, 44227 Dortmund, Germany

J. Götze
Information Processing Lab, TU Dortmund University, Otto-Hahn-Str. 4, 44227 Dortmund, Germany

Y. He
Institute for Integrated Systems, Ruhr University Bochum, 44780 Bochum, Germany

A. Bilgic
KROHNE Messtechnik GmbH, Ludwig-Krohne-Str. 5, 47058 Duisburg, Germany

J. Schneider
University of Kaiserslautern, Institute for Wireless Communication and Navigation, Paul-Ehrlich-Straße - Building 11, 67663 Kaiserslautern, Germany

A. Klein
University of Kaiserslautern, Institute for Wireless Communication and Navigation, Paul-Ehrlich-Straße - Building 11, 67663 Kaiserslautern, Germany

C. Mannweiler
University of Kaiserslautern, Institute for Wireless Communication and Navigation, Paul-Ehrlich-Straße - Building 11, 67663 Kaiserslautern, Germany

H. D. Schotten
University of Kaiserslautern, Institute for Wireless Communication and Navigation, Paul-Ehrlich-Straße - Building 11, 67663 Kaiserslautern, Germany

G. S. Sangha
Department of Microwave Techniques, University of Ulm, Germany

M. H. W. Hoffmann
Department of Microwave Techniques, University of Ulm, Germany

C. C. Sun
Dortmund University of Technology, Information Processing Lab, Otto-Hahn-Str. 4, 44227 Dortmund, Germany

J. Götze
Dortmund University of Technology, Information Processing Lab, Otto-Hahn-Str. 4, 44227 Dortmund, Germany

Lj. Radić-Weissenfeld
Leibniz University Hannover, Institute of Electromagnetic Theory, Hannover, Germany
Fraunhofer Research Institution for Electronic Nanosystems, Advanced System Engineering, Paderborn, Germany

S. Ludwig
Leibniz University Hannover, Institute of Electromagnetic Theory, Hannover, Germany
Fraunhofer Research Institution for Electronic Nanosystems, Advanced System Engineering, Paderborn, Germany

W. Mathis
Leibniz University Hannover, Institute of Electromagnetic Theory, Hannover, Germany

T. Corbach
Helmut-Schmidt-University, University of the Federal Armed Forces Hamburg, Germany

M. Holters
Helmut-Schmidt-University, University of the Federal Armed Forces Hamburg, Germany

U. Zölzer
Helmut-Schmidt-University, University of the Federal Armed Forces Hamburg, Germany

S. Ludwig
Leibniz University Hannover, Institute of Electromagnetic Theory, Hannover, Germany
Fraunhofer Research Institution for Electronic Nano Systems, Advanced System Engineering, Paderborn, Germany

W. Mathis
Leibniz University Hannover, Institute of Electromagnetic Theory, Hannover, Germany

J. O. Hinz
Department of Signal Processing and Communications, Helmut-Schmidt-Universität/University of the Federal Armed Forces Hamburg, Holstenhofweg 85, 22043 Hamburg, Germany

U. Zölzer
Department of Signal Processing and Communications, Helmut-Schmidt-Universit¨at/University of the Federal Armed Forces Hamburg, Holstenhofweg 85, 22043 Hamburg, Germany

P. H¨acker
Chair of System Theory and Signal Processing, Universit¨at Stuttgart, Pfaffenwaldring 47, 70550 Stuttgart, Germany

S. Uhlich
Chair of System Theory and Signal Processing, Universit¨at Stuttgart, Pfaffenwaldring 47, 70550 Stuttgart, Germany

B. Yang
Chair of System Theory and Signal Processing, Universit¨at Stuttgart, Pfaffenwaldring 47, 70550 Stuttgart, Germany

F. X. Nsabimana
Department of Signal Processing and Communications, Helmut Schmidt University, Hamburg, Germany

V. Subbaraman
Department of Signal Processing and Communications, Helmut Schmidt University, Hamburg, Germany

U. Z¨olzer
Department of Signal Processing and Communications, Helmut Schmidt University, Hamburg, Germany

L. Nicolosi
TU Dresden, Faculty of Electrical Engineering and Information Technology, Institute of Principles of Electrical and Electronic Engineering, Germany

R. Tetzlaff
TU Dresden, Faculty of Electrical Engineering and Information Technology, Institute of Principles of Electrical and Electronic Engineering, Germany

J. Nuckelt
Institut f¨ur Nachrichtentechnik, Technische Universit¨at Braunschweig, Germany

M. Schack
Institut f¨ur Nachrichtentechnik, Technische Universit¨at Braunschweig, Germany

T. K¨urner
Institut f¨ur Nachrichtentechnik, Technische Universit¨at Braunschweig, Germany

F. Hageb¨olling
Department of Signal Processing and Communications, Helmut Schmidt University - University of the Federal Armed Forces Hamburg, Holstenhofweg 85, 22043 Hamburg, Germany

U. Z¨olzer
Department of Signal Processing and Communications, Helmut Schmidt University - University of the Federal Armed Forces Hamburg, Holstenhofweg 85, 22043 Hamburg, Germany

M. Muxfeldt
Department of Signal Processing and Communications, Helmut-Schmidt-Universit¨at/University of the Federal Armed Forces Hamburg, Holstenhofweg 85, 22043 Hamburg, Germany

P. Beinschob
Department of Signal Processing and Communications, Helmut-Schmidt-Universit¨at/University of the Federal Armed Forces Hamburg, Holstenhofweg 85, 22043 Hamburg, Germany

U. Z¨olzer
Department of Signal Processing and Communications, Helmut-Schmidt-Universit¨at/University of the Federal Armed Forces Hamburg, Holstenhofweg 85, 22043 Hamburg, Germany

C. Mannweiler
Chair for Wireless Communications and Navigation, University of Kaiserslautern, Germany

C. Lottermann
Chair for Wireless Communications and Navigation, University of Kaiserslautern, Germany

A. Klein
Chair for Wireless Communications and Navigation, University of Kaiserslautern, Germany

J. Schneider
Chair for Wireless Communications and Navigation, University of Kaiserslautern, Germany

H. D. Schotten
Chair for Wireless Communications and Navigation, University of Kaiserslautern, Germany

S. More
Lehrstuhl f¨ur Technische Elektronik,Technische Universit¨at M¨unchen, Germany

M. Fulde
Infineon Technologies AG, Villach, Austria

F. Chouard
Lehrstuhl f¨ur Technische Elektronik,Technische Universit¨at M¨unchen, Germany

D. Schmitt-Landsiedel
Lehrstuhl f¨ur Technische Elektronik,Technische Universit¨at M¨unchen, Germany

S. Lange
Institut f¨ur Mikroelektronik- und Mechatronik- Systeme gemeinn¨utzige GmbH, Erfurt, Germany

T. Reich
Institut f¨ur Mikroelektronik- und Mechatronik- Systeme gemeinn¨utzige GmbH, Erfurt, Germany

J. Nowak
Institut f¨ur Mikroelektronik- und Mechatronik- Systeme gemeinn¨utzige GmbH, Erfurt, Germany

B. Dimov
Institut f¨ur Mikroelektronik- und Mechatronik- Systeme gemeinn¨utzige GmbH, Erfurt, Germany

M. Meister
Institut f¨ur Mikroelektronik- und Mechatronik- Systeme gemeinn¨utzige GmbH, Erfurt, Germany

E. Hennig
Institut f¨ur Mikroelektronik- und Mechatronik- Systeme gemeinn¨utzige GmbH, Erfurt, Germany

D. Huo
Bell Laboratories, 67 Whippany Road, Whippany, NJ 07981, USA

M. C. Neuenhahn
Chair for Electrical Engineering and Computer Systems, RWTH Aachen Univ., Schinkelstraße 2, 52062 Aachen, Germany

J. Schleifer
Chair for Electrical Engineering and Computer Systems, RWTH Aachen Univ., Schinkelstraße 2, 52062 Aachen, Germany

H. Blume
Chair for Electrical Engineering and Computer Systems, RWTH Aachen Univ., Schinkelstraße 2, 52062 Aachen, Germany

T. G. Noll
Chair for Electrical Engineering and Computer Systems, RWTH Aachen Univ., Schinkelstraße 2, 52062 Aachen, Germany

R. Kazemzadeh
Leibniz University of Hannover Institute of Electromagnetic Theory, Hannover, Germany
Fraunhofer Research Institution for Electronic Nano Systems, Advanced System Engineering, Paderborn, Germany

S. Ludwig
Leibniz University of Hannover Institute of Electromagnetic Theory, Hannover, Germany
Fraunhofer Research Institution for Electronic Nano Systems, Advanced System Engineering, Paderborn, Germany

Lj. Radi´c-Weissenfeld
Leibniz University of Hannover Institute of Electromagnetic Theory, Hannover, Germany
Fraunhofer Research Institution for Electronic Nano Systems, Advanced System Engineering, Paderborn, Germany

W. Mathis
Leibniz University of Hannover Institute of Electromagnetic Theory, Hannover, Germany

T. Lehnigk-Emden
Microelectronic Systems Design Research Group, University of Kaiserslautern, Erwin-Schroedinger Str., 67663 Kaiserslautern, Germany

U. Wasenm¨ uller
Microelectronic Systems Design Research Group, University of Kaiserslautern, Erwin-Schroedinger Str., 67663 Kaiserslautern, Germany

C. Gimmler
Microelectronic Systems Design Research Group, University of Kaiserslautern, Erwin-Schroedinger Str., 67663 Kaiserslautern, Germany

N. Wehn
Microelectronic Systems Design Research Group, University of Kaiserslautern, Erwin-Schroedinger Str., 67663 Kaiserslautern, Germany

D. W¨ urfel, D. Weiler
Fraunhofer Institute for Microelectronic Circuits and Systems, Duisburg, Germany

B. J. Hosticka
Fraunhofer Institute for Microelectronic Circuits and Systems, Duisburg, Germany

H. Vogt
Fraunhofer Institute for Microelectronic Circuits and Systems, Duisburg, Germany

A. Guntoro
Institute of Microelectronic Systems, Department of Electrical Engineering and Information Technology, Technische Universit¨at Darmstadt, Germany

M. Glesner
Institute of Microelectronic Systems, Department of Electrical Engineering and Information Technology, Technische Universit¨at Darmstadt, Germany

C. Mannweiler
Chair for Wireless Communications and Navigation, University of Kaiserslautern, Germany

A. Klein
Chair for Wireless Communications and Navigation, University of Kaiserslautern, Germany

J. Schneider
Chair for Wireless Communications and Navigation, University of Kaiserslautern, Germany

H. D. Schotten
Chair for Wireless Communications and Navigation, University of Kaiserslautern, Germany

M. Liso Nicol´as
Institut f¨ur Nachrichtentechnik, Technische Universit¨at Braunschweig, Braunschweig, Germany

M. Jacob
Institut f¨ur Nachrichtentechnik, Technische Universit¨at Braunschweig, Braunschweig, Germany

T. K¨urner
Institut f¨ur Nachrichtentechnik, Technische Universit¨at Braunschweig, Braunschweig, Germany

S. Haßler
1Fachgebiet H¨ochstfrequenztechnik, Technische Universit¨at M¨unchen, Munich, Germany

T. Reichthalhammer
Fachgebiet H¨ochstfrequenztechnik, Technische Universit¨at M¨unchen, Munich, Germany

E. Biebl
Fachgebiet H¨ochstfrequenztechnik, Technische Universit¨at M¨unchen, Munich, Germany

C.Wiegand
Fraunhofer ENAS, Department Advanced System Engineering (ASE), Warburger Str. 100, 33098 Paderborn, Germany
University of Paderborn, Department of Sensor Technology Warburger Str. 100, 33098 Paderborn, Germany

C. Hedayat
Fraunhofer ENAS, Department Advanced System Engineering (ASE), Warburger Str. 100, 33098 Paderborn, Germany

U. Hilleringmann
Fraunhofer ENAS, Department Advanced System Engineering (ASE), Warburger Str. 100, 33098 Paderborn, Germany
University of Paderborn, Department of Sensor Technology Warburger Str. 100, 33098 Paderborn, Germany

T. Arafa
University of Kaiserslautern, Communications Engineering, Gottlieb-Daimler-Str., 67653 Kaiserslautern, Germany

W. Sauer-Greff
University of Kaiserslautern, Communications Engineering, Gottlieb-Daimler-Str., 67653 Kaiserslautern, Germany

R. Urbansky
University of Kaiserslautern, Communications Engineering, Gottlieb-Daimler-Str., 67653 Kaiserslautern, Germany

M. K. Samee
Information Processing Lab, Dortmund University of Technology, Otto-Hahn-Str. 4, 44227 Dortmund, Germany

J. G¨otze
Information Processing Lab, Dortmund University of Technology, Otto-Hahn-Str. 4, 44227 Dortmund, Germany

U. Wasenm¨ uller
Microelectronic Systems Design Research Group, University of Kaiserslautern, Erwin-Schr¨odinger Str., 67663 Kaiserslautern, Germany

C. Gimmler
Microelectronic Systems Design Research Group, University of Kaiserslautern, Erwin-Schr¨odinger Str., 67663 Kaiserslautern, Germany

N.Wehn
Microelectronic Systems Design Research Group, University of Kaiserslautern, Erwin-Schr¨odinger Str., 67663 Kaiserslautern, Germany

R. Kazemzadeh
Institute for Theoretical Electrical Engineering, Hannover, Germany

W. John
Institute for Theoretical Electrical Engineering, Hannover, Germany
SIL System Integration Laboratory GmbH (R+D Consulting), Paderborn, Germany

W. Mathis
Institute for Theoretical Electrical Engineering, Hannover, Germany

J. Schneider
University of Kaiserslautern, Paul-Ehrlich-Street, Building 11, Germany
Institute for Wireless Communications and Navigation

A. Klein
University of Kaiserslautern, Paul-Ehrlich-Street, Building 11, Germany Institute for Wireless Communications and Navigation

C. Mannweiler
University of Kaiserslautern, Paul-Ehrlich-Street, Building 11, Germany Institute for Wireless Communications and Navigation

H. D. Schotten
University of Kaiserslautern, Paul-Ehrlich-Street, Building 11, Germany Institute for Wireless Communications and Navigation